# Digital Systems Design Using VHDL®

Second Edition

## Charles H. Roth, Jr
*The University of Texas at Austin*

## Lizy Kurian John
*The University of Texas at Austin*

CENGAGE
Learning™

Australia • Brazil • Japan • Korea • Mexico • Singapore • Spain
United Kingdom • United States

CENGAGE
Learning™

**Digital Systems Design Using VHDL®**
**(International Student Edition)**
**Charles H. Roth, Jr. and**
**Lizy Kurian John**

Director, Global Engineering
Program: Chris Carson

Senior Developmental Editor:
Hilda Gowans

Marketing Services Coordinator:
Lauren Bestos

Director, Content and Media
Production: Renate McCloy

Production Service: RPK Editorial
Services, Inc.

Copyeditor: Patricia Daly

Proofreader: Erin Wagner

Indexer: Shelly Gerger-Knechtl

Compositor: Integra

Senior Art Director: Michelle Kunkler

Internal Designer: Carmela Pereira

Cover Designer: Andrew Adams

Cover Image: © 2007 JupiterImages and
its Licensors. All Rights Reserved

Text Permissions Researcher:
Vicki Gould

Senior First Print Buyer: Doug Wilke

For product information and technology assistance, contact us at **Cengage Learning Customer & Sales Support, 1-800-354-9706.**

For permission to use material from this text or product, submit all requests online at **www.cengage.com/permissions.** Further permissions questions can be emailed to **permissionrequest@cengage.com.**

International Student Edition:
ISBN-13: 978-0-495-24470-7
ISBN-10: 0-495-24470-8

Système Internationale:
ISBN-13: 978-0-495-24470-7
ISBN-10: 0-495-24470-8

**Cengage Learning**
200 First Stamford Place, Suite 400
Stamford, CT 06902
USA

Cengage Learning is a leading provider of customized learning solutions with office locations around the globe, including Singapore, the United Kingdom, Australia, Mexico, Brazil, and Japan. Locate your local office at: **www.cengage.com/country.**

Cengage Learning products are represented in Canada by Nelson Education Ltd.

For your course and learning solutions, visit **www.cengage.com/engineering.**

Purchase any of our products at your local college store or at our preferred online store **www.cengagebrain.com.**

Printed in the United States of America
3 4 5 6 15 14 13 12

# Contents

● ● ● ● ● ● ● ● ● ● ● ●

**Preface**   vii

## Chapter 1 Review of Logic Design Fundamentals   1

1.1   Combinational Logic   1
1.2   Boolean Algebra and Algebraic Simplification   3
1.3   Karnaugh Maps   7
1.4   Designing with NAND and NOR Gates   11
1.5   Hazards in Combinational Circuits   12
1.6   Flip-Flops and Latches   14
1.7   Mealy Sequential Circuit Design   17
1.8   Moore Sequential Circuit Design   25
1.9   Equivalent States and Reduction of State Tables   28
1.10  Sequential Circuit Timing   30
1.11  Tristate Logic and Busses   41

## Chapter 2 Introduction to VHDL   51

2.1   Computer-Aided Design   51
2.2   Hardware Description Languages   54
2.3   VHDL Description of Combinational Circuits   57
2.4   VHDL Modules   61
2.5   Sequential Statements and VHDL Processes   67
2.6   Modeling Flip-Flops Using VHDL Processes   69
2.7   Processes Using Wait Statements   73
2.8   Two Types of VHDL Delays: Transport and Inertial Delays   75
2.9   Compilation, Simulation, and Synthesis of VHDL Code   77
2.10  VHDL Data Types and Operators   82
2.11  Simple Synthesis Examples   84
2.12  VHDL Models for Multiplexers   87
2.13  VHDL Libraries   90
2.14  Modeling Registers and Counters Using VHDL Processes   95
2.15  Behavioral and Structural VHDL   101

2.16  Variables, Signals, and Constants     111
2.17  Arrays     114
2.18  Loops in VHDL     117
2.19  Assert and Report Statements     119


## Chapter 3 Introduction to Programmable Logic Devices     137

3.1  Brief Overview of Programmable Logic Devices     137
3.2  Simple Programmable Logic Devices (SPLDs)     140
3.3  Complex Programmable Logic Devices (CPLDs)     156
3.4  Field-Programmable Gate Arrays (FPGAs)     160


## Chapter 4 Design Examples     190

4.1  BCD to 7-Segment Display Decoder     191
4.2  A BCD Adder     192
4.3  32-Bit Adders     194
4.4  Traffic Light Controller     201
4.5  State Graphs for Control Circuits     204
4.6  Scoreboard and Controller     205
4.7  Synchronization and Debouncing     208
4.8  A Shift-and-Add Multiplier     210
4.9  Array Multiplier     216
4.10 A Signed Integer/Fraction Muliplier     219
4.11 Keypad Scanner     231
4.12 Binary Dividers     239


## Chapter 5 SM Charts and Microprogramming     260

5.1  State Machine Charts     260
5.2  Derivation of SM Charts     265
5.3  Realization of SM Charts     275
5.4  Implementation of the Dice Game     279
5.5  Microprogramming     283
5.6  Linked State Machines     297


## Chapter 6 Designing with Field Programmable Gate Arrays     310

6.1  Implementing Functions in FPGAs     310
6.2  Implementing Functions Using Shannon's Decomposition     316
6.3  Carry Chains in FPGAs     321
6.4  Cascade Chains in FPGAs     323
6.5  Examples of Logic Blocks in Commercial FPGAs     324
6.6  Dedicated Memory in FPGAs     326

6.7    Dedicated Multipliers in FPGAs    332
6.8    Cost of Programmability    333
6.9    FPGAs and One-Hot State Assignment    335
6.10   FPGA Capacity: Maximum Gates Versus Usable Gates    337
6.11   Design Translation (Synthesis)    338
6.12   Mapping, Placement, and Routing    349

# Chapter 7 Floating-Point Arithmetic    361

7.1    Representation of Floating-Point Numbers    361
7.2    Floating-Point Multiplication    367
7.3    Floating-Point Addition    377
7.4    Other Floating-Point Operations    383

# Chapter 8 Additional Topics in VHDL    389

8.1    VHDL Functions    389
8.2    VHDL Procedures    393
8.3    Attributes    395
8.4    Creating Overloaded Operators    399
8.5    Multi-Valued Logic and Signal Resolution    400
8.6    The IEEE 9-Valued Logic System    405
8.7    SRAM Model Using IEEE 1164    408
8.8    Model for SRAM Read/Write System    410
8.9    Generics    413
8.10   Named Association    414
8.11   Generate Statements    415
8.12   Files and TEXTIO    417

# Chapter 9 Design of a RISC Microprocessor    429

9.1    The RISC Philosophy    429
9.2    The MIPS ISA    432
9.3    MIPS Instruction Encoding    438
9.4    Implementation of a MIPS Subset    441
9.5    VHDL Model    449

# Chapter 10 Hardware Testing and Design for Testability    468

10.1   Testing Combinational Logic    468
10.2   Testing Sequential Logic    473
10.3   Scan Testing    476
10.4   Boundary Scan    479
10.5   Built-In Self-Test    490

## Chapter 11 Additional Design Examples    507

11.1  Design of a Wristwatch    507
11.2  Memory Timing Models    518
11.3  A Universal Asynchronous Receiver Transmitter (UART)    526

## Appendix A    545
## VHDL Language Summary

## Appendix B    553
## IEEE Standard Libraries

## Appendix C    555
## TEXTIO PACKAGE

## Appendix D    557
## Projects

## References    568

## Index    571

# Preface

● ● ● ● ● ● ● ● ● ● ● ● ●

This textbook is intended for a senior-level course in digital systems design. The book covers both basic principles of digital system design and the use of a hardware description language, VHDL, in the design process. After basic principles have been covered, design is best taught by using examples. For this reason, many digital system design examples, ranging in complexity from a simple binary adder to a microprocessor, are included in the text.

Students using this textbook should have completed a course in the fundamentals of logic design, including both combinational and sequential circuits. Although no previous knowledge of VHDL is assumed, students should have programming experience using a modern high-level language such as C. A course in assembly language programming and basic computer organization is also very helpful, especially for Chapter 9.

Because students typically take their first course in logic design two years before this course, most students need a review of the basics. For this reason, Chapter 1 includes a review of logic design fundamentals. Most students can review this material on their own, so it is unnecessary to devote much lecture time to this chapter. However, a good understanding of timing in sequential circuits and the principles of synchronous design is essential to the digital system design process.

Chapter 2 starts with an overview of modern design flow. It also summarizes various technologies for implementation of digital designs. Then, it introduces the basics of VHDL, and this hardware description language is used throughout the rest of the book. Additional features of VHDL are introduced on an as-needed basis, and more advanced features are covered in Chapter 8. From the start, we relate the constructs of VHDL to the corresponding hardware. Some textbooks teach VHDL as a programming language and devote many pages to teaching the language syntax. Instead, our emphasis is on how to use VHDL in the digital design process. The language is very complex, so we do not attempt to cover all its features. We emphasize the basic features that are necessary for digital design and omit some of the less-used features. Use of standard IEEE VHDL libraries is introduced in this chapter and only IEEE standard libraries are used throughout the text.

VHDL is very useful in teaching top-down design. We can design a system at a high level and express the algorithms in VHDL. We can then simulate and debug the designs at this level before proceeding with the detailed logic design. However, no design is complete until it has actually been implemented in hardware and the hardware has been tested. For this reason, we recommend that the course include some lab exercises in which designs are implemented in hardware. We introduce simple programmable logic devices (PLDs) in Chapter 3 so that real hardware can be used early in the course if desired. Chapter 3 starts with an overview of programmable logic devices and presents simple programmable logic devices first, followed by an introduction to complex programmable logic devices (CPLDs) and Field Programmable Gate Arrays (FPGAs). There are many products in the market, and it is good for students to learn about commercial products. However, it is more important for them to understand the basic principles in the construction of these programmable devices. Hence we present the material in a generalized fashion, with references to specific products as examples. The material in this chapter also serves as an introduction to the more detailed treatment of FPGAs in Chapter 6.

Chapter 4 presents a variety of design examples, including both arithmetic and non-arithmetic examples. Simple examples such as a BCD to 7-segment display decoder to more complex examples such as game scoreboards, keypad scanners and binary dividers are presented. The chapter presents common techniques used for computer arithmetic, including carry look-ahead addition, and binary multiplication and division. Use of a state machine for sequencing the operations in a digital system is an important concept presented in this chapter. Synthesizable VHDL code is presented for the various designs. A variety of examples are presented so that instructors can select their favorite designs for teaching.

Use of sequential machine charts (SM charts) as an alternative to state graphs is presented in Chapter 5. We show how to write VHDL code based on SM charts and how to realize hardware to implement the SM charts. Then, the technique of microprogramming is presented. Transformation of SM charts for different types of microprogramming is discussed. Then, we show how the use of linked state machines facilitates the decomposition of complex systems into simpler ones. The design of a dice-game simulator is used to illustrate these techniques.

Chapter 6 presents issues related to implementing digital systems in Field Programmable Gate Arrays. A few simple designs are first hand-mapped into FPGA building blocks to illustrate the mapping process. Shannon's expansion for decomposition of functions with several variables into smaller functions is presented. Features of modern FPGAs like carry chains, cascade chains, dedicated memory, and dedicated multipliers are then presented. Instead of describing all features in a selected commercial product, the features are described in a general fashion. Once students understand the general principles, they will be able to understand and use any commercial product they have to work with. This chapter also presents an introduction to the processes and algorithms in the software design flow. Synthesis, mapping, placement, and routing processes are briefly described. Optimizations during synthesis are illustrated.

Basic techniques for floating-point arithmetic are described in Chapter 7. A simple floating-point format with 2's complement numbers is presented and then the IEEE standard floating-point formats are presented. A floating-point multiplier example is presented starting with development of the basic algorithm, then simulating the system using VHDL, and finally synthesizing and implementing the system using an FPGA. Some instructors may prefer to cover Chapter 8 and 9 before teaching Chapter 7. Chapter 7 can be omitted without loss of any continuity.

By the time students reach Chapter 8, they should be thoroughly familiar with the basics of VHDL. At this point we introduce some of the more advanced features of VHDL and illustrate their use. The use of multi-valued logic, including the IEEE-1164 standard logic, is one of the important topics covered. A memory model with tri-state output busses is presented to illustrate the use of the multi-valued logic.

Chapter 9 presents the design of a microprocessor, starting from the description of the instruction set architecture (ISA). The processor is an early RISC processor, the MIPS R2000. The important instructions in the MIPS ISA are described and a subset is then implemented. The design of the various components of the processor, such as the instruction memory module, data memory module and register file are illustrated module by module. These components are then integrated together and a complete processor design is presented. The model can be tested with a test bench, or can be synthesized and implemented on an FPGA. In order to test the design on an FPGA, one will need to write input-output modules for the design. This example requires understanding of the basics of assembly language programming and computer organization.

The important topics of hardware testing and design for testability are covered in Chapter 10. This chapter introduces the basic techniques for testing combinational and sequential logic. Then scan design and boundary-scan techniques, which facilitate the testing of digital systems, are described. The chapter concludes with a discussion of built-in self-test (BIST). VHDL code for a boundary-scan example and for a BIST example is included. The topics in this chapter play an important role in digital system design, and we recommend that they be included in any course on this subject. Chapter 10 can be covered any time after the completion of Chapter 8.

Chapter 11 presents three complete design examples that illustrate the use of VHDL synthesis tools. First, a wristwatch design is presented. It shows the progress of a design from a textual description to a state diagram and then a VHDL model. This example illustrates modular design. The test bench for the wristwatch illustrates the use of multiple procedure calls to facilitate the testing. The second example describes the use of VHDL to model RAM memories. The third example, a serial communications receiver-transmitter, should easily be understood by any student who has completed the material through Chapter 8.

For instructors who used the first edition of this text, here is a mapping to help them understand the changes in the second edition. The homegrown library BITLIB is not used in this edition of the book. The IEEE numeric-bit library is used first until multi-valued logic is introduced in Chapter 8. The multi-valued IEEE numeric-std library is used thereafter. All code has been converted to use IEEE standard libraries instead of the BITLIB library.

| Chapter 1 | Simpler Mealy and Moore designs added. More detailed descriptions added to sequential circuit timing section. |
|---|---|
| Chapter 2 | Overview of design flow and design technologies added. Functions and procedures from old Chapter 2 moved to Chapter 8. Inertial delays and transport delays moved from Chapter 8 to Chapter 2. Synthesis is introduced in Chapter 2 and all code presented is generally synthesizeable. |
| Chapter 3 | Contains first part of old Chapter 3. New material on CPLDs and FPGAs added. The design examples from old Chapter 3 (traffic light, keypad scanner) are moved to Chapter 4. |
| Chapter 4 | Several new examples are added. Old Chapter 4 examples are largely retained, but converted to synthesizeable code. Two examples from old Chapter 3 are now here. |
| Chapter 5 | Added more detailed treatment of microprogramming. |
| Chapter 6 | New material on FPGAs in a generalized fashion, without making it specific to any commercial product, but drawing examples from several commercial devices. A brief treatment of software design flow including principles of mapping, placement, routing added. |
| Chapter 7 | IEEE floating point standards and floating point adder design added. |
| Chapter 8 | Functions and procedures from old Chapter 2 moved to here. Many sections from old Chapter 8 are still here. A memory model previously in old Chapter 9 presented as example of multi-valued logic design in new Chapter 8 |
| Chapter 9 | This chapter is new. MIPS instruction set and design of a MIPS processor presented. Memory models from old Chapter 9 are moved to Chapter 8 or 11. Bus model from old Chapter 9 omitted. |
| Chapter 10 | Added details on boundary scan and STUMPS architecture. |
| Chapter 11 | A new design (wristwatch) added. Memory timing models from old Chapter 9 appear here now. UART design from old Chapter 11 retained. Microcontroller design is omitted. |

This book is the result of many years of teaching a senior course in digital systems design at the University of Texas at Austin. Throughout the years, the technology for hardware implementation of digital systems has kept changing, but many of the same design principles are still applicable. In the early years of the course, we handwired modules consisting of discrete transistors to implement our designs. Then integrated circuits were introduced, and we were able to implement our designs using breadboards and TTL logic. Now we are able to use FPGAs and CPLDs to realize very complex designs. We originally used our own hardware description language together with a simulator running on a mainframe computer. When PCs came along, we wrote an improved hardware description language and implemented a simulator that ran on PCs. When VHDL was adopted as an IEEE standard and became widely used in industry, we switched to VHDL. The widespread availability of high-quality commercial CAD tools now enables us to synthesize complex designs directly from the VHDL code.

All of the VHDL code in this textbook has been tested using the ModelSim simulator. The ModelSim software is available in a student edition, and we recommend its use in conjunction with this text. The CD that accompanies this text provides a link for downloading the ModelSim student edition and an introductory tutorial to help students get started using the software. All of the VHDL code in this textbook is available on the CD. The CD also contains two software packages, LogicAid and SimUaid, which are useful in teaching digital system design. Instruction manuals and examples of using this software are on the CD.

● ● ● ● ● ● ● ● ● ● ●
# Acknowledgments

We would like to thank the many individuals who have contributed their time and effort to the development of this textbook. Over many years we have received valuable feedback from the students in our digital systems design courses. We would especially like to thank the faculty members who reviewed the previous edition and offered many suggestions for its improvement. These faculty include:

Gang Feng, University of Wisconsin, Platteville
Elmer. A. Grubbs, University of Arizona
Marius Z. Jankowski, University of Southern Maine
Chun-Shin Lin, University of Missouri – Columbia
Peter N. Marinos, Duke University
Maryam Moussavi, California State University, Long Beach
Aaron Striegel, University of Notre Dame
Peixin Zhong, Michigan State University

Special thanks go to Ian Burgess at Mentor Graphics for arranging the ModelSim student version. We also wish to acknowledge the help from Chris Carson, Hilda Gowans, Rose Kernan and Kamilah Reid Burrell during various steps of the publication process. It was a pleasure to work with all. We also take this opportunity to express our gratitude to the student assistants who helped with the word processing, VHDL code testing, CD, and illustrations: Ciji Isen, Roger Chen, William Earle, Manish Kapadia, Matt Morgan, Elizabeth Norris, and Raman Suri.

*Charles. H. Roth, Jr*

*Lizy K. John.*

# CHAPTER 1

# Review of Logic Design Fundamentals

This chapter reviews many of the logic design topics normally taught in a first course in logic design. Some of the review examples that follow are referenced in later chapters of this text. For more details on any of the topics discussed in this chapter, the reader should refer to a standard logic design textbook such as Roth, *Fundamentals of Logic Design*, 5th Edition (Thomson Brooks/Cole, 2004). First, we review combinational logic and then sequential logic. Combinational logic has no memory, so the present output depends only on the present input. Sequential logic has memory, so the present output depends not only on the present input but also on the past sequence of inputs. The sections on sequential circuit timing and synchronous design are particularly important, since a good understanding of timing issues is essential to the successful design of digital systems.

## 1.1 Combinational Logic

Some of the basic gates used in logic circuits are shown in Figure 1-1. Unless otherwise specified, all the variables that we use to represent logic signals will be two-valued, and the two values will be designated 0 and 1. We will normally use positive logic, for which a low voltage corresponds to a logic 0 and a high voltage corresponds to a logic 1. When negative logic is used, a low voltage corresponds to a logic 1 and a high voltage corresponds to a logic 0.

For the AND gate of Figure 1-1, the output $C = 1$ if and only if the input $A = 1$ *and* the input $B = 1$. We will use a raised dot or simply write the variables side by side to indicate the AND operation; thus $C = A$ AND $B = A \cdot B = AB$. For the OR gate, the output $C = 1$ if and only if the input $A = 1$ *or* the input $B = 1$ (inclusive OR). We will use + to indicate the OR operation; thus $C = A$ OR $B = A + B$. The NOT gate, or inverter, forms the complement of the input; that is, if $A = 1$, $C = 0$, and if $A = 0$, $C = 1$. We will use a prime (') to indicate the complement (NOT) operation, so $C =$ NOT $A = A'$. The exclusive-OR (XOR) gate has an output $C = 1$ if $A = 1$ and $B = 0$ or if $A = 0$ and $B = 1$. The symbol $\oplus$ represents exclusive OR, so we write

$$C = A \text{ XOR } B = AB' + A'B = A \oplus B \qquad (1\text{-}1)$$

The behavior of a combinational logic circuit can be specified by a truth table that gives the circuit outputs for each combination of input values. As an example,

FIGURE 1-1: **Basic Gates**

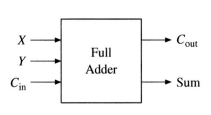

AND: $C = A\,B$

OR: $C = A + B$

NOT: $C = A'$

Exclusive OR: $C = A \oplus B$

consider the full adder of Figure 1-2, which adds two binary digits ($X$ and $Y$) and a carry ($C_{in}$) to give a sum (*Sum*) and a carry out ($C_{out}$). The truth table specifies the adder outputs as a function of the adder inputs. For example, when the inputs are $X = 0$, $Y = 0$ and $C_{in} = 1$, adding the three inputs gives $0 + 0 + 1 = 01$, so the sum is 1 and the carry out is 0. When the inputs are 011, $0 + 1 + 1 = 10$, so *Sum* $= 0$ and $C_{out} = 1$. When the inputs are $X = Y = C_{in} = 1$, $1 + 1 + 1 = 11$, so *Sum* $= 1$ and $C_{out} = 1$.

FIGURE 1-2: **Full Adder**

| $X$ | $Y$ | $C_{in}$ | $C_{out}$ | *Sum* |
|---|---|---|---|---|
| 0 | 0 | 0 | 0 | 0 |
| 0 | 0 | 1 | 0 | 1 |
| 0 | 1 | 0 | 0 | 1 |
| 0 | 1 | 1 | 1 | 0 |
| 1 | 0 | 0 | 0 | 1 |
| 1 | 0 | 1 | 1 | 0 |
| 1 | 1 | 0 | 1 | 0 |
| 1 | 1 | 1 | 1 | 1 |

(a) Full adder module

(b) Truth table

We will derive algebraic expressions for *Sum* and $C_{out}$ from the truth table. From the table, *Sum* $= 1$ when $X = 0$, $Y = 0$, and $C_{in} = 1$. The term $X'Y'C_{in}$ equals 1 only for this combination of inputs. The term $X'YC_{in}' = 1$ only when $X = 0$, $Y = 1$, and $C_{in} = 0$. The term $XY'C_{in}'$ is 1 only for the input combination $X = 1$, $Y = 0$, and $C_{in} = 0$. The term $XYC_{in}$ is 1 only when $X = Y = C_{in} = 1$. Therefore, *Sum* is formed by ORing these four terms together:

$$Sum = X'Y'C_{in} + X'YC_{in}' + XY'C_{in}' + XYC_{in} \qquad (1\text{-}2)$$

Each of the terms in this sum of products (SOP) expression is 1 for exactly one combination of input values. In a similar manner, $C_{out}$ is formed by ORing four terms together:

$$C_{out} = X'YC_{in} + XY'C_{in} + XYC_{in}' + XYC_{in} \qquad (1\text{-}3)$$

Each term in Equations (1-2) and (1-3) is referred to as a *minterm*, and these equations are referred to as *minterm expansions*. These minterm expansions can also be written in *m*-notation or decimal notation as follows:

$$Sum = m_1 + m_2 + m_4 + m_7 = \Sigma m(1, 2, 4, 7)$$
$$C_{out} = m_3 + m_5 + m_6 + m_7 = \Sigma m(3, 5, 6, 7)$$

The decimal numbers designate the rows of the truth table for which the corresponding function is 1. Thus *Sum* $= 1$ in rows 001, 010, 100, and 111 (rows 1, 2, 4, 7).

A logic function can also be represented in terms of the inputs for which the function value is 0. Referring to the truth table for the full adder, $C_{out} = 0$ when $X = Y = C_{in} = 0$. The term $(X + Y + C_{in})$ is 0 only for this combination of inputs. The term $(X + Y + C_{in}')$ is 0 only when $X = Y = 0$ and $C_{in} = 1$. The term $(X + Y' + C_{in})$ is 0 only when $X = C_{in} = 0$ and $Y = 1$. The term $(X' + Y + C_{in})$ is 0 only when $X = 1$ and $Y = C_{in} = 0$. $C_{out}$ is formed by ANDing these four terms together:

$$C_{out} = (X + Y + C_{in})(X + Y + C_{in}')(X + Y' + C_{in})(X' + Y + C_{in}) \qquad (1\text{-}4)$$

$C_{out}$ is 0 only for the 000, 001, 010, and 100 rows of the truth table and, therefore, must be 1 for the remaining four rows. Each of the terms in the Product of Sums (POS) expression in Equation (1-4) is referred to as a *maxterm*, and (1-4) is called a *maxterm expansion*. This *maxterm expansion* can also be written in decimal notation as

$$C_{out} = M_0 \cdot M_1 \cdot M_2 \cdot M_4 = \Pi M(0, 1, 2, 4)$$

where the decimal numbers correspond to the truth table rows for which $C_{out} = 0$.

## 1.2 Boolean Algebra and Algebraic Simplification

The basic mathematics used for logic design is Boolean algebra. Table 1-1 summarizes the laws and theorems of Boolean algebra. They are listed in dual pairs; for example, Equation (1-10D) is the dual of (1-10). They can be verified easily for two-valued logic by using truth tables. These laws and theorems can be used to simplify logic functions so they can be realized with a reduced number of components.

A very important law in Boolean algebra is the *DeMorgan's law*. DeMorgan's laws stated in Equations (1-16, 1-16D) can be used to form the complement of an expression on a step-by-step basis. The generalized form of DeMorgan's law in Equation (1-17) can be used to form the complement of a complex expression in one step. Equation (1-17) can be interpreted as follows: To form the complement of a Boolean expression, replace each variable by its complement; also replace 1 with 0, 0 with 1, OR with AND, and AND with OR. Add parentheses as required to assure the proper hierarchy of operations. If AND is performed before OR in $F$, then parentheses may be required to assure that OR is performed before AND in $F'$.

*Example*     Find the complement of F if

$$F = X + E'K \left( C \left( AB + D' \right) \cdot 1 + WZ' \left( G'H + 0 \right) \right)$$
$$F' = X' \left( E + K' + \left( C' + \left( A' + B' \right) D + 0 \right) \left( W' + Z + \left( G + H' \right) \cdot 1 \right) \right)$$

Additional parentheses in $F'$ were added when an AND operation in $F$ was replaced with an OR. The dual of an expression is the same as its complement, except that the variables are not complemented.

**TABLE 1-1: Laws and Theorems of Boolean Algebra**

Operations with 0 and 1:

| | | | |
|---|---|---|---|
| $X + 0 = X$ | (1-5) | $X \cdot 1 = X$ | (1-5D) |
| $X + 1 = 1$ | (1-6) | $X \cdot 0 = 0$ | (1-6D) |

Idempotent laws:

| | | | |
|---|---|---|---|
| $X + X = X$ | (1-7) | $X \cdot X = X$ | (1-7D) |

Involution law:

$$(X')' = X \qquad \text{(1-8)}$$

Laws of complementarity:

| | | | |
|---|---|---|---|
| $X + X' = 1$ | (1-9) | $X \cdot X' = 0$ | (1-9D) |

Commutative laws:

| | | | |
|---|---|---|---|
| $X + Y = Y + X$ | (1-10) | $XY = YX$ | (1-10D) |

Associative laws:

| | | | |
|---|---|---|---|
| $(X + Y) + Z = X + (Y + Z)$ | (1-11) | $(XY)Z = X(YZ) = XYZ$ | (1-11D) |
| $\quad = X + Y + Z$ | | | |

Distributive laws:

| | | | |
|---|---|---|---|
| $X(Y + Z) = XY + XZ$ | (1-12) | $X + YZ = (X + Y)(X + Z)$ | (1-12D) |

Simplification theorems:

| | | | |
|---|---|---|---|
| $XY + XY' = X$ | (1-13) | $(X + Y)(X + Y') = X$ | (1-13D) |
| $X + XY = X$ | (1-14) | $X(X + Y) = X$ | (1-14D) |
| $(X + Y')Y = XY$ | (1-15) | $XY' + Y = X + Y$ | (1-15D) |

DeMorgan's laws:

| | | | |
|---|---|---|---|
| $(X + Y + Z + \cdots)' = X'Y'Z' \cdots$ | (1-16) | $(XYZ \ldots)' = X' + Y' + Z' + \cdots$ | (1-16D) |
| $[f(X_1, X_2, \ldots, X_n, 0, 1, +, \cdot)]' = f(X_1', X_2', \ldots, X_n', 1, 0, \cdot, +)$ | | | (1-17) |

Duality:

| | | | |
|---|---|---|---|
| $(X + Y + Z + \cdots)^D = XYZ \cdots$ | (1-18) | $(XYZ \cdots)^D = X + Y + Z + \cdots$ | (1-18D) |
| $[f(X_1, X_2, \ldots, X_n, 0, 1, +, \cdot)]^D = f(X_1, X_2, \ldots, X_n, 1, 0, \cdot, +)$ | | | (1-19) |

Theorem for multiplying out and factoring:

| | | | |
|---|---|---|---|
| $(X + Y)(X' + Z) = XZ + X'Y$ | (1-20) | $XY + X'Z = (X + Z)(X' + Y)$ | (1-20D) |

Consensus theorem:

| | | | |
|---|---|---|---|
| $XY + YZ + X'Z = XY + X'Z$ | (1-21) | $(X + Y)(Y + Z)(X' + Z)$ | (1-21D) |
| | | $\quad = (X + Y)(X' + Z)$ | |

Four ways of simplifying a logic expression using the theorems in Table 1-1 are as follows:

1. *Combining terms.* Use the theorem $XY + XY' = X$ to combine two terms. For example,

$$ABC'D' + ABCD' = ABD' \ [X = ABD', Y = C]$$

When combining terms by this theorem, the two terms to be combined should contain exactly the same variables, and exactly one of the variables should appear complemented in one term and not in the other. Since $X + X = X$, a given term may be duplicated and combined with two or more other terms. For example, the expression for $C_{out}$ in Equation (1-3) can be simplified by combining the first and fourth terms, the second and fourth terms, and the third and fourth terms:

$$
\begin{aligned}
C_{out} &= (X'YC_{in} + XYC_{in}) + (XY'C_{in} + XYC_{in}) + (XYC'_{in} + XYC_{in}) \\
&= YC_{in} + XC_{in} + XY
\end{aligned}
\tag{1-22}
$$

Note that the fourth term in Equation (1-3) was used three times.

The theorem can still be used, of course, when $X$ and $Y$ are replaced with more complicated expressions. For example,

$$(A + BC)(D + E') + A'(B' + C')(D + E') = D + E'$$
$$[X = D + E', Y = A + BC, Y' = A'(B' + C')]$$

2. *Eliminating terms.* Use the theorem $X + XY = X$ to eliminate redundant terms if possible; then try to apply the consensus theorem ($XY + X'Z + YZ = XY + X'Z$) to eliminate any consensus terms. For example,

$$A'B + A'BC = A'B \ [X = A'B]$$
$$A'BC' + BCD + A'BD = A'BC' + BCD \ [X = C, Y = BD, Z = A'B]$$

3. *Eliminating literals.* Use the theorem $X + X'Y = X + Y$ to eliminate redundant literals. Simple factoring may be necessary before the theorem is applied. For example,

$$
\begin{aligned}
A'B + A'B'C'D' + ABCD' &= A'(B + B'C'D') + ABCD' && \text{(by (1-12))} \\
&= A'(B + C'D') + ABCD' && \text{(by (1-15D))} \\
&= B(A' + ACD') + A'C'D' && \text{(by (1-10))} \\
&= B(A' + CD') + A'C'D' && \text{(by (1-15D))} \\
&= A'B + BCD' + A'C'D' && \text{(by (1-12))}
\end{aligned}
$$

The expression obtained after applying 1, 2, and 3 will not necessarily have a minimum number of terms or a minimum number of literals. If it does not and no further simplification can be made using 1, 2, and 3, deliberate introduction of redundant terms may be necessary before further simplification can be made.

4. *Adding redundant terms.* Redundant terms can be introduced in several ways, such as adding $XX'$, multiplying by $(X + X')$, adding $YZ$ to $XY + X'Z$

(consensus theorem), or adding $XY$ to $X$. When possible, the terms added should be chosen so that they will combine with or eliminate other terms. For example,

$$WX + XY + X'Z' + WY'Z' \qquad \text{(Add } WZ' \text{ by the consensus theorem.)}$$
$$= WX + XY + X'Z' + WY'Z' + WZ' \qquad \text{(Eliminate } WY'Z'.)$$
$$= WX + XY + X'Z' + WZ' \qquad \text{(Eliminate } WZ'.)$$
$$= WX + XY + X'Z'$$

When multiplying out or factoring an expression, in addition to using the ordinary distributive law (1-12), the second distributive law (1-12D) and theorem (1-20) are particularly useful. The following is an example of multiplying out to convert from a product of sums to a sum of products:

$$(A + B + D)(A + B' + C')(A' + B + D')(A' + B + C')$$
$$= (A + (B + D)(B' + C'))(A' + B + C'D') \qquad \text{(by (1-12D))}$$
$$= (A + BC' + B'D)(A' + B + C'D') \qquad \text{(by (1-20))}$$
$$= A(B + C'D') + A'(BC' + B'D) \qquad \text{(by (1-20))}$$
$$= AB + AC'D' + A'BC' + A'B'D \qquad \text{(by (1-12))}$$

Note that the second distributive law (1-12D) and theorem (1-20) were applied before the ordinary distributive law. Any Boolean expression can be factored by using the two distributive laws (1-12 and 1-12D) and theorem (1-20). As an example of factoring, read the steps in the preceding example in the reverse order.

The following theorems apply to exclusive-OR:

$$X \oplus 0 = X \tag{1-23}$$
$$X \oplus 1 = X' \tag{1-24}$$
$$X \oplus X = 0 \tag{1-25}$$
$$X \oplus X' = 1 \tag{1-26}$$
$$X \oplus Y = Y \oplus X \qquad \text{(commutative law)} \tag{1-27}$$
$$(X \oplus Y) \oplus Z = X \oplus (Y \oplus Z) = X \oplus Y \oplus Z \quad \text{(associative law)} \tag{1-28}$$
$$X(Y \oplus Z) = XY \oplus XZ \qquad \text{(distributive law)} \tag{1-29}$$
$$(X \oplus Y)' = X \oplus Y' = X' \oplus Y = XY + X'Y' \tag{1-30}$$

The expression for *Sum* in Equation (1-2) can be rewritten in terms of exclusive-OR by using Equations (1-1) and (1-30):

$$Sum = X'(Y'C_{in} + YC'_{in}) + X(Y'C'_{in} + YC_{in})$$
$$= X'(Y \oplus C_{in}) + X(Y \oplus C_{in})' = X \oplus Y \oplus C_{in} \tag{1-31}$$

The simplification rules that you studied in this section are important when a circuit has to be optimized to use a smaller number of gates. The existence of equivalent forms also helps when mapping circuits into particular target devices where only certain types of logic (e.g., NAND only or NOR only) are available.

• • • • • • • • • • • •

# 1.3 **Karnaugh Maps**

*Karnaugh maps* (K-maps) provide a convenient way to simplify logic functions of three to five variables. Figure 1-3 shows a four-variable Karnaugh map. Each square in the map represents one of the 16 possible minterms of four variables. A 1 in a square indicates that the minterm is present in the function, and a 0 (or blank) indicates that the minterm is absent. An X in a square indicates that we don't care whether the minterm is present or not. *Don't cares* arise under two conditions: (1) The input combination corresponding to the don't care can never occur, and (2) the input combination can occur, but the circuit output is not specified for this input condition.

The variable values along the edge of the map are ordered so that adjacent squares on the map differ in only one variable. The first and last columns and the

**FIGURE 1-3:**
**Four-Variable**
**Karnaugh Maps**

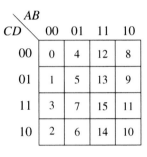

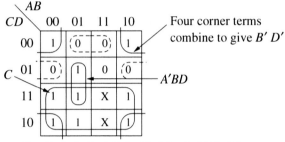

$F = \Sigma m\,(0, 2, 3, 5, 6, 7, 8, 10, 11) + \Sigma d\,(14, 15)$
$= C + B'\,D' + A'\,BD$

(a) Location of minterms          (b) Looping terms

top and bottom rows of the map are considered to be adjacent. Two 1's in adjacent squares can be combined by eliminating one variable using $xy + xy' = x$. Figure 1-3 shows a four-variable function with nine minterms and two don't cares. Minterms $A'BC'D$ and $A'BCD$ differ only in the variable $C$, so they can be combined to form $A'BD$, as indicated by a loop on the map. Four 1's in a symmetrical pattern can be combined to eliminate two variables. The 1's in the four corners of the map can be combined as follows:

$$(A'B'C'D' + AB'C'D') + (A'B'CD' + AB'CD') = B'C'D' + B'CD' = B'D'$$

as indicated by the loop. Similarly, the six 1's and two X's in the bottom half of the map combine to eliminate three variables and form the term $C$. The resulting simplified function is

$$F = A'BD + B'D' + C$$

The minimum sum-of-products representation of a function consists of a sum of prime implicants. A group of one, two, four, or eight adjacent 1's on a map represents a prime implicant if it cannot be combined with another group of 1's to eliminate a variable. A prime implicant is essential if it contains a 1 that is not contained in any other prime implicant. When finding a minimum sum of products from a map, essential prime implicants should be looped first, and then a minimum number of prime implicants to cover the remaining 1's should be looped. The Karnaugh map shown in Figure 1-4 has five prime implicants and three essential prime implicants. $A'C'$ is essential because minterm $m_1$ is not covered by any other prime implicant. Similarly, $ACD$ is essential because of $m_{11}$, and $A'B'D'$ is essential because of $m_2$. After looping the essential prime implicants, all 1's are covered except $m_7$. Since $m_7$ can be covered by either prime implicant $A'BD$ or $BCD$, $F$ has two minimum forms:

$$F = A'C' + A'B'D' + ACD + A'BD$$

and

$$F = A'C' + A'B'D' + ACD + BCD$$

When don't cares (X's) are present on the map, the don't cares are treated like 1's when forming prime implicants, but the X's are ignored when finding a minimum

FIGURE 1-4:
**Selection of Prime Implicants**

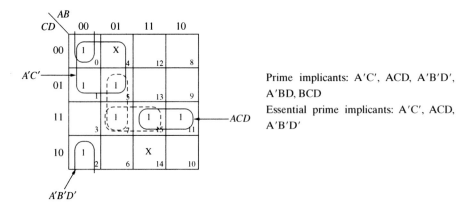

Prime implicants: $A'C'$, $ACD$, $A'B'D'$, $A'BD$, $BCD$

Essential prime implicants: $A'C'$, $ACD$, $A'B'D'$

set of prime implicants to cover all the 1's. The following procedure can be used to obtain a minimum sum of products from a Karnaugh map:

1. Choose a minterm (a 1) that has not yet been covered.
2. Find all 1's and X's adjacent to that minterm. (Check the $n$ adjacent squares on an $n$-variable map.)
3. If a single term covers the minterm and all the adjacent 1's and X's, then that term is an essential prime implicant, so select that term. (Note that don't cares are treated like 1's in steps 2 and 3 but not in step 1.)
4. Repeat steps 1, 2, and 3 until all essential prime implicants have been chosen.
5. Find a minimum set of prime implicants that cover the remaining 1's on the map. (If there is more than one such set, choose a set with a minimum number of literals.)

To find a minimum product of sums from a Karnaugh map, loop the 0's instead of the 1's. Since the 0's of $F$ are the 1's of $F'$, looping the 0's in the proper way gives the minimum sum of products for $F'$, and the complement is the minimum product of sums for $F$. For Figure 1-3, we can first loop the essential prime implicants of $F'$ ($BC'D'$ and $B'C'D$, indicated by dashed loops) and then cover the remaining 0 with $AB$. Thus the minimum sum for $F'$ is

$$F' = BC'D' + B'C'D + AB$$

from which the minimum product of sums for $F$ is

$$F = (B' + C + D)(B + C + D')(A' + B')$$

### 1.3.1 Simplification Using Map-Entered Variables

Two four-variable Karnaugh maps can be used to simplify functions with five variables. If functions have more than five variables, *map-entered variables* can be used. Consider a truth table as in Table 1-2. There are six input variables (A, B, C, D, E, F) and one output variable (G). Only certain rows of the truth table have been specified. To completely specify the truth table, 64 rows will be required. The input combinations not specified in the truth table result in an output of 0.

TABLE 1-2: **Partial Truth Table for a Six-Variable Function**

| A | B | C | D | E | F | G |
|---|---|---|---|---|---|---|
| 0 | 0 | 0 | 0 | X | X | 1 |
| 0 | 0 | 0 | 1 | X | X | X |
| 0 | 0 | 1 | 0 | X | X | 1 |
| 0 | 0 | 1 | 1 | X | X | 1 |
| 0 | 1 | 0 | 1 | 1 | X | 1 |
| 0 | 1 | 1 | 1 | 1 | X | 1 |
| 1 | 0 | 0 | 1 | X | 1 | 1 |
| 1 | 0 | 1 | 0 | X | X | X |
| 1 | 0 | 1 | 1 | X | X | 1 |
| 1 | 1 | 0 | 1 | X | X | X |
| 1 | 1 | 1 | 1 | X | X | 1 |

Karnaugh map techniques can be extended to simplify functions such as this using map-entered variables. Since $E$ and $F$ are the input variables with the most number of don't cares (X), a Karnaugh map can be formed with $A$, $B$, $C$, $D$ and the remaining two variables can be entered inside the map. Figure 1-5 shows a four-variable map with variables $E$ and $F$ entered in the squares in the map. When $E$ appears in a square, this means that if $E = 1$, the corresponding minterm is present in the function $G$, and if $E = 0$, the minterm is absent. The fifth and sixth rows in the truth table result in the $E$ in the box corresponding to minterm 5 and minterm 7. The seventh row results in the $F$ in the box corresponding to minterm 9. Thus, the map represents the six-variable function

$$G(A, B, C, D, E, F) = m_0 + m_2 + m_3 + Em_5 + Em_7 + Fm_9 + m_{11} + m_{15}$$
$$(+ \text{ don't care terms})$$

where the minterms are minterms of the variables $A$, $B$, $C$, $D$. Note that $m_9$ is present in $G$ only when $F = 1$.

Next we will discuss a general method of simplifying functions using map-entered variables. In general, if a variable $P_i$ is placed in square $m_j$ of a map of function $F$, this means that $F = 1$ when $P_i = 1$ and the variables are chosen so that $m_j = 1$. Given a

FIGURE 1-5:
**Simplification Using Map-Entered Variables**

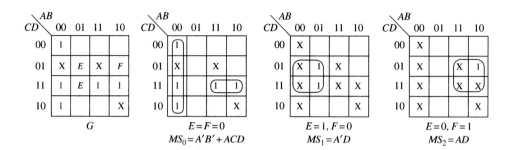

map with variables $P_1, P_2, \ldots$ entered into some of the squares, the minimum sum-of-products form of $F$ can be found as follows: Find a sum-of-products expression for $F$ of the form

$$F = MS_0 + P_1 MS_1 + P_2 MS_2 + \cdots \qquad (1\text{-}32)$$

where

- $MS_0$ is the minimum sum obtained by setting $P_1 = P_2 = \cdots = 0$.
- $MS_1$ is the minimum sum obtained by setting $P_1 = 1, P_j = 0$ $(j \neq 1)$, and replacing all 1's on the map with don't cares.
- $MS_2$ is the minimum sum obtained by setting $P_2 = 1, P_j = 0$ $(j \neq 2)$, and replacing all 1's on the map with don't cares.

Corresponding minimum sums can be found in a similar way for any remaining map-entered variables.

The resulting expression for $F$ will always be a correct representation of $F$. This expression will be a minimum sum provided that the values of the map-entered variables can be assigned independently. On the other hand, the expression will not generally be a minimum sum if the variables are not independent (for example, if $P_1 = P_2'$).

For the example of Figure 1-5, maps for finding $MS_0$, $MS_1$, and $MS_2$ are shown, where $E$ corresponds to $P_1$ and $F$ corresponds to $P_2$. Note that it is not required to draw a map for $E = 1, F = 1$, because $E = 1$ already covers cases with $E = 1, F = 0$ and $E = 1, F = 1$. The resulting expression is a minimum sum of products for $G$:

$$G = A'B' + ACD + EA'D + FAD$$

After some practice, it should be possible to write the minimum expression directly from the original map without first plotting individual maps for each of the minimum sums.

• • • • • • • • • • • •

## 1.4 Designing With NAND and NOR Gates

In many technologies, implementation of NAND gates or NOR gates is easier than that of AND and OR gates. Figure 1-6 shows the symbols used for NAND and NOR gates. The *bubble* at a gate input or output indicates a complement. Any logic function can be realized using only NAND gates or only NOR gates.

FIGURE 1-6: **NAND and NOR Gates**

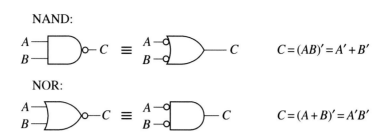

NAND:

$$C = (AB)' = A' + B'$$

NOR:

$$C = (A + B)' = A'B'$$

Conversion from circuits of OR and AND gates to circuits of all NOR gates or all NAND gates is straightforward. To design a circuit of NOR gates, start with a product-of-sums representation of the function (circle 0's on the Karnaugh map). Then find a circuit of OR and AND gates that has an AND gate at the output. If an AND gate output does not drive an AND gate input and an OR gate output does not connect to an OR gate input, then conversion is accomplished by replacing all gates with NOR gates and complementing inputs if necessary. Figure 1-7 illustrates the conversion procedure for

$$Z = G(E + F)(A + B' + D)(C + D) = G(E + F)[(A + B')C + D]$$

Conversion to a circuit of NAND gates is similar, except the starting point should be a sum-of-products form for the function (circle 1's on the map), and the output gate of the AND-OR circuit should be an OR gate.

FIGURE 1-7:
**Conversion to NOR Gates**

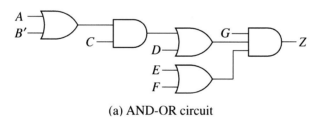

(a) AND-OR circuit

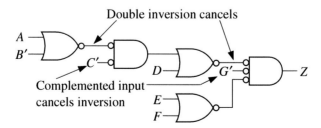

(b) Equivalent NOR-gate circuit

Even if AND and OR gates do not alternate, we can still convert a circuit of AND and OR gates to a NAND or NOR circuit, but it may be necessary to add extra inverters so that each added inversion is canceled by another inversion. The following procedure may be used to convert to a NAND (or NOR) circuit:

1. Convert all AND gates to NAND gates by adding an inversion bubble at the output. Convert OR gates to NAND gates by adding inversion bubbles at the inputs. (To convert to NOR, add inversion bubbles at all OR gate outputs and all AND gate inputs.)
2. Whenever an inverted output drives an inverted input, no further action is needed, since the two inversions cancel.
3. Whenever a noninverted gate output drives an inverted gate input or vice versa, insert an inverter so that the bubbles will cancel. (Choose an inverter with the bubble at the input or output, as required.)
4. Whenever a variable drives an inverted input, complement the variable (or add an inverter) so the complementation cancels the inversion at the input.

In other words, if we always add bubbles (or inversions) in pairs, the function realized by the circuit will be unchanged. To illustrate the procedure, we will convert Figure 1-8(a) to NANDs. First, we add bubbles to change all gates to NAND gates (Figure 1-8(b)). The highlighted lines indicate four places where we have added only a single inversion. This is corrected in Figure 1-8(c) by adding two inverters and complementing two variables.

**FIGURE 1-8:**
**Conversion of AND-OR Circuit to NAND Gates**

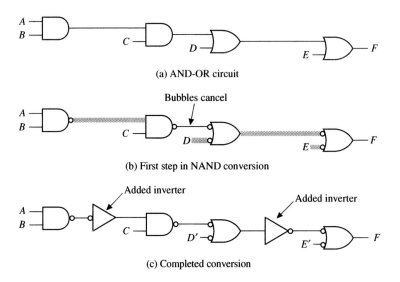

(a) AND-OR circuit

(b) First step in NAND conversion

(c) Completed conversion

● ● ● ● ● ● ● ● ● ● ● ●

## 1.5 Hazards in Combinational Circuits

When the input to a combinational circuit changes, unwanted switching transients may appear in the output. These transients occur when different paths from input to output have different propagation delays. If, in response to an input change and for some combination of propagation delays, a circuit output may momentarily go to 0 when it should remain a constant 1, we say that the circuit has a static 1-*hazard*. Similarly, if the output may momentarily go to 1 when it should remain a 0, we say that the circuit has a static 0-*hazard*. If, when the output is supposed to change from 0 to 1 (or 1 to 0), the output may change three or more times, we say that the circuit has a *dynamic hazard*.

Consider the two simple circuits in Figure 1-9. Figure 1-9(a) shows an inverter and an OR gate implementing the function $A + A'$. Logically, the output of this circuit is expected to be a 1 always; however, a delay in the inverter gate can cause static hazards in this circuit. Assume a nonzero delay for the inverter and that the value of $A$ just changed from 1 to 0. There is a short interval of time until the inverter delay has passed when both inputs of the OR gate are 0 and hence the output of the circuit may momentarily go to 0. Similarly, in the circuit in Figure 1-9(b), the expected output is always 0; however, when $A$ changes from 1 to 0, a momentary 1 appears at the output of the inverter because of the delay. This circuit hence has a static 0-hazard. The hazard occurs because both $A$ and $A'$ have the same value for a short duration after $A$ changes.

**FIGURE 1-9: Simple Circuits Containing Hazards**

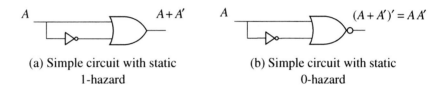

(a) Simple circuit with static 1-hazard

(b) Simple circuit with static 0-hazard

A static 1-hazard occurs in a sum-of-product implementation when two minterms differing by only one input variable are not covered by the same product term. Figure 1-10(a) illustrates another circuit with a static 1-hazard. If $A = C = 1$, the output should remain a constant 1 when $B$ changes from 1 to 0. However, as shown in Figure 1-10(b), if each gate has a propagation delay of 10 ns, $E$ will go to 0 before $D$ goes to 1, resulting in a momentary 0 (a 1-hazard appearing in the output $F$). As seen on the Karnaugh map, there is no loop that covers both minterm $ABC$ and $AB'C$. So if $A = C = 1$ and $B$ changes from 1 to 0, $BC$ immediately becomes 0, but until an inverter delay passes, $AB'$ does not become a 1. Both terms can momentarily go to 0, resulting in a glitch in $F$. If we add a loop corresponding to the term $AC$ to the map and add the corresponding gate to the circuit (Figure 1-10(c)), this eliminates the hazard. The term $AC$ remains 1 while $B$ is changing, so no glitch can appear in the output. In general, nonminimal expressions are required to eliminate static hazards.

To design a circuit that is free of static and dynamic hazards, the following procedure may be used:

1. Find a sum-of-products expression ($F^t$) for the output in which every pair of adjacent 1s is covered by a 1-term. (The sum of all prime implicants will always satisfy this condition.) A two-level AND-OR circuit based on this $F^t$ will be free of 1-, 0-, and dynamic hazards.
2. If a different form of circuit is desired, manipulate $F^t$ to the desired form by simple factoring, DeMorgan's laws, and so on. **Treat each $x_i$ and $x_i'$ as independent variables to prevent introduction of hazards.**

FIGURE 1-10:
**Elimination of
1-Hazard**

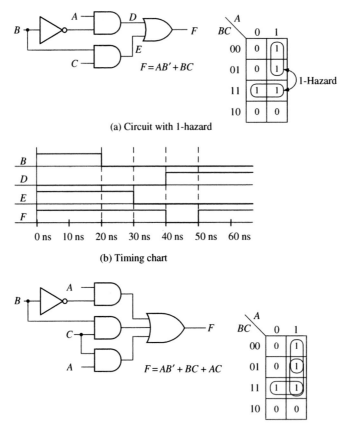

(a) Circuit with 1-hazard

(b) Timing chart

$F = AB' + BC + AC$

(c) Circuit with hazard removed

Alternatively, you can start with a product-of-sums expression in which every pair of adjacent 0s is covered by a 0-term.

Given a circuit, one can identify the static hazards in it by writing an expression for the output in terms of the inputs exactly as it is implemented in the circuit and manipulating it to a sum-of-products form, treating $x_i$ and $x_i'$ as independent variables. A Karnaugh map can be constructed and all implicants corresponding to each term circled. If any pair of adjacent 1's is not covered by a single term, a static 1-hazard can occur. Similarly, a static 0-hazard can be identified by writing a product-of-sums expression for the circuit.

● ● ● ● ● ● ● ● ● ● ● ●

## 1.6 Flip-Flops and Latches

Sequential circuits commonly use flip-flops as storage devices. There are several types of flip-flops, such as Delay (D) flip-flops, J-K flip-flops, Toggle (T) flip-flops, and so on. Figure 1-11 shows a clocked D flip-flop. This flip-flop can change state in response to the rising edge of the clock input. The next state of the flip-flop after the rising edge of the clock is equal to the $D$ input before the rising edge. The *characteristic equation* of the flip-flop is therefore $Q^+ = D$, where $Q^+$ represents the next state of the $Q$ output after the active edge of the clock and $D$ is the input before the active edge.

**FIGURE 1-11:**
**Clocked D Flip-Flop with Rising-Edge Trigger**

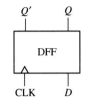

| D | Q | Q⁺ |
|---|---|---|
| 0 | 0 | 0 |
| 0 | 1 | 0 |
| 1 | 0 | 1 |
| 1 | 1 | 1 |

Figure 1-12 shows a clocked J-K flip-flop and its truth table. Since there is a bubble at the clock input, all state changes occur following the falling edge of the clock input. If $J = K = 0$, no state change occurs. If $J = 1$ and $K = 0$, the flip-flop is set to 1, independent of the present state. If $J = 0$ and $K = 1$, the flip-flop is always reset to 0. If $J = K = 1$, the flip-flop changes state. The characteristic equation, derived from the truth table in Figure 1-12, using a Karnaugh map is

$$Q^+ = JQ' + K'Q \qquad (1\text{-}33)$$

**FIGURE 1-12:**
**Clocked J-K Flip-Flop**

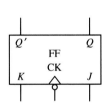

| J | K | Q | Q⁺ |
|---|---|---|---|
| 0 | 0 | 0 | 0 |
| 0 | 0 | 1 | 1 |
| 0 | 1 | 0 | 0 |
| 0 | 1 | 1 | 0 |
| 1 | 0 | 0 | 1 |
| 1 | 0 | 1 | 1 |
| 1 | 1 | 0 | 1 |
| 1 | 1 | 1 | 0 |

A clocked T flip-flop (Figure 1-13) changes state following the active edge of the clock if $T = 1$, and no state change occurs if $T = 0$. T flip-flops are particularly useful for designing counters. The characteristic equation for the T flip-flop is

$$Q^+ = QT' + Q'T = Q \oplus T \qquad (1\text{-}34)$$

A J-K flip-flop is easily converted to a T flip-flop by connecting $T$ to both $J$ and $K$. Substituting $T$ for $J$ and $K$ in Equation (1-33) yields Equation (1-34).

**FIGURE 1-13:**
**Clocked T Flip-Flop**

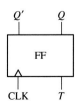

| T | Q | Q⁺ |
|---|---|---|
| 0 | 0 | 0 |
| 0 | 1 | 1 |
| 1 | 0 | 1 |
| 1 | 1 | 0 |

Two NOR gates can be connected to form an unclocked S-R (set-reset) flip-flop, as shown in Figure 1-14. An unclocked flip-flop of this type is often referred to as an S-R latch. If $S = 1$ and $R = 0$, the $Q$ output becomes 1 and $P = Q'$. If $S = 0$ and $R = 1$, $Q$ becomes 0 and $P = Q'$. If $S = R = 0$, no change of state occurs. If $R = S = 1$, $P = Q = 0$, which is not a proper flip-flop state, since the two outputs should always be complements. If $R = S = 1$ and these inputs are simultaneously changed to 0, oscillation may occur. For this reason, $S$ and $R$ are not allowed to be 1 at the same time. For purposes of deriving the characteristic equation, we assume that $S = R = 1$ never occurs, in which case $Q^+ = S + R'Q$. In this case, $Q^+$ represents the state after any input changes have propagated to the $Q$ output.

**FIGURE 1-14: S-R Latch**

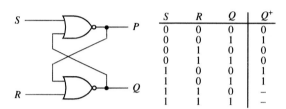

| S | R | Q | Q⁺ |
|---|---|---|---|
| 0 | 0 | 0 | 0 |
| 0 | 0 | 1 | 1 |
| 0 | 1 | 0 | 0 |
| 0 | 1 | 1 | 0 |
| 1 | 0 | 0 | 1 |
| 1 | 0 | 1 | 1 |
| 1 | 1 | 0 | – |
| 1 | 1 | 1 | – |

A gated D latch (Figure 1-15), also called a transparent D latch, behaves as follows: If the gate signal $G = 1$, then the $Q$ output follows the $D$ input ($Q^+ = D$). If $G = 0$, then the latch holds the previous value of $Q$ ($Q^+ = Q$). Essentially, the device will not respond to input changes unless $G = 1$; it simples "latches" the previous input right before $G$ became 0. Some refer to the D latch as a level-sensitive D flip-flop. Essentially, if the gate input $G$ is viewed as a clock, the latch can be considered as a device that operates when the clock level is high and does not respond to the inputs when the clock level is low. The characteristic equation for the D latch is $Q^+ = GD + G'Q$. Figure 1-16 shows an implementation of the D latch using gates. Since the $Q^+$ equation has a 1-hazard, an extra AND gate has been added to eliminate the hazard.

**FIGURE 1-15: Transparent D Latch**

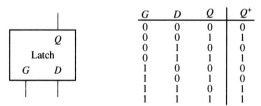

| G | D | Q | Q⁺ |
|---|---|---|---|
| 0 | 0 | 0 | 0 |
| 0 | 0 | 1 | 1 |
| 0 | 1 | 0 | 0 |
| 0 | 1 | 1 | 1 |
| 1 | 0 | 0 | 0 |
| 1 | 0 | 1 | 0 |
| 1 | 1 | 0 | 1 |
| 1 | 1 | 1 | 1 |

**FIGURE 1-16: Implementation of D Latch**

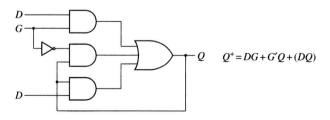

$Q^+ = DG + G'Q + (DQ)$

● ● ● ● ● ● ● ● ● ● ● ●
# 1.7 Mealy Sequential Circuit Design

There are two basic types of sequential circuits: Mealy and Moore. In a Mealy circuit, the outputs depend on both the present state and the present inputs. In a Moore circuit, the outputs depend only on the present state. A general model of a Mealy sequential circuit consists of a combinational circuit, which generates the outputs and the next state, and a state register, which holds the present state (see Figure 1-17). The state register normally consists of D flip-flops. The normal sequence of events is (1) the $X$ inputs change to a new value; (2) after a delay, the corresponding $Z$ outputs and next state appears at the output of the combinational circuit; and (3) the next state is clocked into the state register and the state changes. The new state feeds back into the combinational circuit and the process is repeated.

**FIGURE 1-17:**
**General Model of Mealy Sequential Machine**

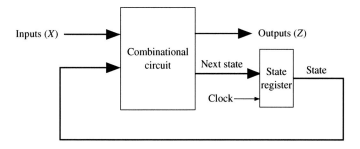

### 1.7.1 Mealy Machine Design Example 1: Sequence Detector

To illustrate the design of a clocked Mealy sequential circuit, let us design a sequence detector. The circuit has the form indicated in the block diagram in Figure 1-18.

**FIGURE 1-18: Block Diagram of a Sequence Detector**

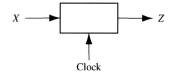

The circuit will examine a string of 0's and 1's applied to the $X$ input and generate an output $Z = 1$ only when the input sequence ends in 1 0 1. The input $X$ can change only between clock pulses. The output $Z = 1$ coincides with the last 1 in 1 0 1. The circuit does not reset when a 1 output occurs. A typical input sequence and the corresponding output sequence are

$$X = 0 \ 0 \ 1 \ 1 \ 0 \ 1 \ 1 \ 0 \ 0 \ 1 \ 0 \ 1 \ 0 \ 1 \ 0 \ 0$$
$$Z = 0 \ 0 \ 0 \ 0 \ 0 \ 1 \ 0 \ 0 \ 0 \ 0 \ 0 \ 1 \ 0 \ 1 \ 0 \ 0$$

Let us construct a *state graph* for this sequence detector. We will start in a reset state designated $S_0$. If a 0 input is received, we can stay in state $S_0$ as the input sequence we are looking for does not start with 0. However, if a 1 is received, the circuit should go to a new state. Let us denote that state as $S_1$. When in $S_1$, if we receive a 0, the circuit must change to a new state ($S_2$) to remember that the first two inputs of the desired sequence (1 0) have been received. If a 1 is received in state $S_2$, the desired input sequence is complete and the output should be a 1. The output will be produced as a Mealy output and will coincide with the last 1 in the detected sequence. Since we are designing a Mealy circuit, we are not going to go to a new state that indicates the sequence 101 has been received. When we receive a 1 in $S_2$, we cannot go to the start state since the circuit is not supposed to reset with every detected sequence. But the last 1 in a sequence can be the first 1 in another sequence; hence, we can go to state $S_1$. The partial state graph at this point is indicated in Figure 1-19.

**FIGURE 1-19:**
**Partial State Graph of the Sequence Detector**

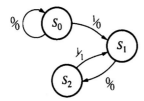

When a 0 is received in state $S_2$, we have received two 0's in a row and must reset the circuit to state $S_0$. If a 1 is received when we are in $S_1$, we can stay in $S_1$ because the most recent 1 can be the first 1 of a new sequence to be detected. The final state graph is shown in Figure 1-20. State $S_0$ is the starting state, state $S_1$ indicates that a sequence ending in 1 has been received, and state $S_2$ indicates that a sequence ending in 10 has been received. Converting the state graph to a state table yields Table 1-3. In row $S_2$ of the table, an output of 1 is indicated for input 1.

**FIGURE 1-20:**
**Mealy State Graph for Sequence Detector**

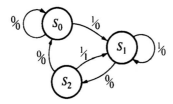

**TABLE 1-3: State Table for Sequence Detector**

| Present State | Next State | | Present Output | |
|---|---|---|---|---|
| | $X = 0$ | $X = 1$ | $X = 0$ | $X = 1$ |
| $S_0$ | $S_0$ | $S_1$ | 0 | 0 |
| $S_1$ | $S_2$ | $S_1$ | 0 | 0 |
| $S_2$ | $S_0$ | $S_1$ | 0 | 1 |

Next, *state assignment* is performed, whereby specific flip-flop values are associated with specific states. There are two techniques to perform state assignment (1) one-hot state assignment and (2) encoded state assignment. In one-hot state assignment, one flip-flop is used for each state. Hence three flip-flops will be required if this circuit is to be implemented using the one-hot approach. In encoded state assignment, just enough flip-flops to have a unique combination for each state are sufficient. Since we have three states, we need at least two flip-flops to represent all states. We will use encoded state assignment in this design. Let us designate the two flip-flops as $A$ and $B$. Let the flip-flop states $A = 0$ and $B = 0$ correspond to state $S_0$; $A = 0$ and $B = 1$ correspond to state $S_1$; and $A = 1$ and $B = 0$ correspond to state $S_2$. Now, the transition table of the circuit can be written as in Table 1-4.

**TABLE 1-4:**
**Transition Table for**
**Sequence Detector**

| | $A^+B^+$ | | $Z$ | |
|---|---|---|---|---|
| $AB$ | $X = 0$ | $X = 1$ | $X = 0$ | $X = 1$ |
| 00 | 00 | 01 | 0 | 0 |
| 01 | 10 | 01 | 0 | 0 |
| 10 | 00 | 01 | 0 | 1 |

From this table, we can plot the K-maps for the next states and the output Z. The next states are typically represented by $A^+$ and $B^+$. The three K-maps are shown in Figure 1-21.

**FIGURE 1-21:**
**K-Maps for Next**
**States and Output**
**of Sequence**
**Detector**

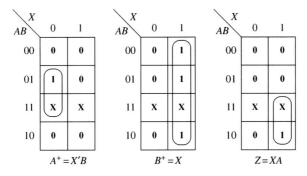

The next step is deriving the flip-flop inputs to obtain the desired next states. If D flip-flops are used, one simply needs to give the expected next state of the flip-flop to the flip-flop input. So, for flip-flops A and B, $D_A = A^+$ and $D_B = B^+$. The resulting circuit is shown in Figure 1-22.

### 1.7.2 Mealy Machine Design Example 2: BCD to Excess-3 Code Converter

As an example of a more complex Mealy sequential circuit, we will design a serial code converter that converts an 8-4-2-1 binary-coded-decimal (BCD) digit to an excess-3-coded decimal digit. The input ($X$) will arrive serially with the least significant bit

FIGURE 1-22:
**Circuit for Mealy
Sequence Detector**

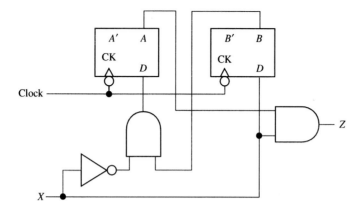

(LSB) first. The outputs will be generated serially as well. Table 1-5 lists the desired inputs and outputs at times $t_0, t_1, t_2$, and $t_3$. After receiving four inputs, the circuit should reset to its initial state, ready to receive another BCD digit.

TABLE 1-5: **Code
Converter**

| X Input (BCD) | | | | Z Output (excess-3) | | | |
|---|---|---|---|---|---|---|---|
| $t_3$ | $t_2$ | $t_1$ | $t_0$ | $t_3$ | $t_2$ | $t_1$ | $t_0$ |
| 0 | 0 | 0 | 0 | 0 | 0 | 1 | 1 |
| 0 | 0 | 0 | 1 | 0 | 1 | 0 | 0 |
| 0 | 0 | 1 | 0 | 0 | 1 | 0 | 1 |
| 0 | 0 | 1 | 1 | 0 | 1 | 1 | 0 |
| 0 | 1 | 0 | 0 | 0 | 1 | 1 | 1 |
| 0 | 1 | 0 | 1 | 1 | 0 | 0 | 0 |
| 0 | 1 | 1 | 0 | 1 | 0 | 0 | 1 |
| 0 | 1 | 1 | 1 | 1 | 0 | 1 | 0 |
| 1 | 0 | 0 | 0 | 1 | 0 | 1 | 1 |
| 1 | 0 | 0 | 1 | 1 | 1 | 0 | 0 |

The excess-3 code is formed by adding 0011 to the BCD digit. For example,

$$\begin{array}{r} 0\ \ 1\ \ 0\ \ 0 \\ +\ 0\ \ 0\ \ 1\ \ 1 \\ \hline 0\ \ 1\ \ 1\ \ 1 \end{array} \qquad \begin{array}{r} 0\ \ 1\ \ 0\ \ 1 \\ +\ 0\ \ 0\ \ 1\ \ 1 \\ \hline 1\ \ 0\ \ 0\ \ 0 \end{array}$$

If all of the BCD bits are available simultaneously, this code converter can be implemented as a combinational circuit with four inputs and four outputs. However, here the bits arrive sequentially, one bit at a time. Hence we must implement this code converter sequentially.

Let us now construct a state graph for the code converter (Figure 1-23(a)). Let us designate the start state as $S_0$. The first bit arrives and we need to add 1 to this bit, as it is the LSB of 0011, the number to be added to the BCD digit to obtain the

**FIGURE 1-23: State Graph and Table for Code Converter**

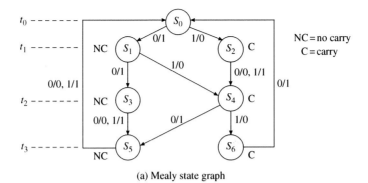

(a) Mealy state graph

|      | NS     |        | Z     |       |
|------|--------|--------|-------|-------|
| PS   | X = 0  | X = 1  | X = 0 | X = 1 |
| S0   | S1     | S2     | 1     | 0     |
| S1   | S3     | S4     | 1     | 0     |
| S2   | S4     | S4     | 0     | 1     |
| S3   | S5     | S5     | 0     | 1     |
| S4   | S5     | S6     | 1     | 0     |
| S5   | S0     | S0     | 0     | 1     |
| S6   | S0     | –      | 1     | –     |

(b) State table

excess-3 code. At $t_0$, we add 1 to the least significant bit, so if $X = 0, Z = 1$ (no carry), and if $X = 1, Z = 0$ (carry = 1). Let us use $S_1$ to indicate no carry after the first addition, and $S_2$ to indicate a carry of 1 after the addition to the LSB.

At $t_1$, we add 1 to the next bit, so if there is no carry from the first addition (state $S_1$), $X = 0$ gives $Z = 0 + 1 + 0 = 1$ and no carry (state $S_3$), and $X = 1$ gives $Z = 1 + 1 + 0 = 0$ and a carry (state $S_4$). If there is a carry from the first addition (state $S_2$), then $X = 0$ gives $Z = 0 + 1 + 1 = 0$ and a carry ($S_4$), and $X = 1$ gives $Z = 1 + 1 + 1 = 1$ and a carry ($S_4$).

At $t_2$, 0 is added to $X$, and transitions to $S_5$ (no carry) and $S_6$ are determined in a similar manner. At $t_3$, 0 is again added to $X$, and the circuit resets to $S_0$.

Figure 1-23(b) gives the corresponding state table. At this point, we should verify that the table has a minimum number of states before proceeding (see Section 1–9). Then state assignment must be performed. Since this state table has seven states, three flip-flops will be required to realize the table in encoded state assignment. In the one-hot approach, one flip-flop is used for each state. Hence seven flip-flops will be required if this circuit is to be implemented using the one-hot approach. The next step is to make a state assignment that relates the flip-flop states to the states in the table. In the sequence detector example, we simply did a straight binary state assignment. Here we are going to look for an optimal assignment. The best state assignment to use depends on a number of factors. In many cases, we should try to find an assignment that will reduce the amount of required logic. For some types of programmable logic, a straight binary state assignment will work just as well as any other. For programmable gate arrays, a one-hot assignment may be preferred. In recent years, with the abundance of transistors on silicon chips, the emphasis on optimal state assignment has been reduced.

In order to reduce the amount of logic required, we will make a state assignment using the following guidelines (see Roth, *Fundamentals of Logic Design*, 5th Ed. [Thomson Brooks/Cole, 2004] for details):

**I.** States that have the same next state (NS) for a given input should be given adjacent assignments (look at the columns of the state table).
**II.** States that are the next states of the same state should be given adjacent assignments (look at the rows).
**III.** States that have the same output for a given input should be given adjacent assignments.

Using these guidelines tends to clump 1's together on the Karnaugh maps for the next state and output functions. The guidelines indicate that the following states should be given adjacent assignments:

**I.** $(1, 2), (3, 4), (5, 6)$      (in the $X = 1$ column, $S_1$ and $S_2$ both have NS $S_4$; in the $X = 0$ column, $S_3$ and $S_4$ have NS $S_5$, and $S_5$ and $S_6$ have NS $S_0$)

**II.** $(1, 2), (3, 4), (5, 6)$      ($S_1$ and $S_2$ are NS of $S_0$; $S_3$ and $S_4$ are NS of $S_1$; and $S_5$ and $S_6$ are NS of $S_4$)

**III.** $(0, 1, 4, 6), (2, 3, 5)$

Figure 1-24(a) gives an assignment map, which satisfies the guidelines, and the corresponding transition table. Since state 001 is not used, the next state and outputs for this state are don't cares. The next state and output equations are derived from this table in Figure 1-25. Figure 1-26 shows the realization of the code converter using NAND gates and D flip-flops.

**FIGURE 1-24: State Assignment for BCD to Excess-3 Code Converter**

Assignment map (a):

| $Q_2Q_3$ \ $Q_1$ | 0 | 1 |
|---|---|---|
| 00 | S0 | S1 |
| 01 |  | S2 |
| 11 | S5 | S3 |
| 10 | S6 | S4 |

(a) Assignment map

Transition table (b):

| $Q_1Q_2Q_3$ | $Q_1^+ Q_2^+ Q_3^+$ X=0 | X=1 | z X=0 | X=1 |
|---|---|---|---|---|
| 000 | 100 | 101 | 1 | 0 |
| 100 | 111 | 110 | 1 | 0 |
| 101 | 110 | 110 | 0 | 1 |
| 111 | 011 | 011 | 0 | 1 |
| 110 | 011 | 010 | 1 | 0 |
| 011 | 000 | 000 | 0 | 1 |
| 010 | 000 | xxx | 1 | x |
| 001 | xxx | xxx | x | x |

(b) Transition table

FIGURE 1-25:
**Karnaugh Maps for Code Converter**

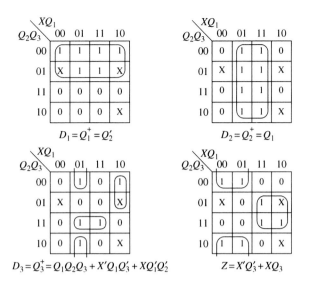

$$D_1 = Q_1^+ = Q_2'$$

$$D_2 = Q_2^+ = Q_1$$

$$D_3 = Q_3^+ = Q_1 Q_2 Q_3 + X'Q_1 Q_3' + XQ_1'Q_2'$$

$$Z = X'Q_3' + XQ_3$$

FIGURE 1-26:
**Realization of Code Converter**

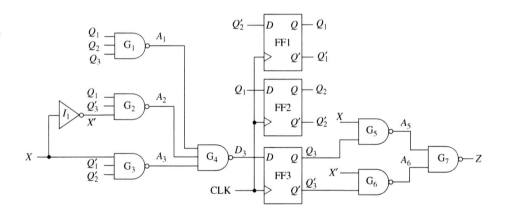

If J-K flip-flops are used instead of D flip-flops, the input equations for the J-K flip-flops can be derived from the next state maps. Given the present state flip-flop $(Q)$ and the desired next state $(Q^+)$, the $J$ and $K$ inputs can be determined from Table 1-6, also known as the excitation table. This table is derived from the truth table in Figure 1-12.

TABLE 1-6:
**Excitation Table for a J-K Flip-Flop**

| $Q$ | $Q^+$ | $J$ | $K$ | |
|---|---|---|---|---|
| 0 | 0 | 0 | X | (No change in $Q$; $J$ must be 0, $K$ may be 1 to reset $Q$ to 0.) |
| 0 | 1 | 1 | X | (Change to $Q = 1$; $J$ must be 1 to set or toggle.) |
| 1 | 0 | X | 1 | (Change to $Q = 0$; $K$ must be 1 to reset or toggle.) |
| 1 | 1 | X | 0 | (No change in $Q$; $K$ must be 0, $J$ may be 1 to set $Q$ to 1.) |

Figure 1-27 shows derivation of J-K flip-flop input equations for the state table of Figure 1-23 using the state assignment of Figure 1-24. First, we derive the J-K input equations for flip-flop $Q_1$ using the $Q_1^+$ map as the starting point. From the preceding table, whenever $Q_1$ is 0, $J = Q_1^+$ and $K = X$. So, we can fill in the $Q_1 = 0$ half of the $J_1$ map the same as $Q_1^+$ and the $Q_1 = 0$ half of the $K_1$ map as all X's. When $Q_1$ is 1, $J_1 = X$ and $K_1 = (Q_1^+)'$. So, we can fill in the $Q_1 = 1$ half of the $J_1$ map with X's and the $Q_1 = 1$ half of the $K_1$ map with the complement of the $Q_1^+$. Since half of every J and K map is don't cares, we can avoid drawing separate J and K maps and read the J's and K's directly from the $Q^+$ maps, as illustrated in Figure 1-27(b). This shortcut method is based on the following: If $Q = 0$, then $J = Q^+$, so loop the 1's on the $Q = 0$ half of the map to get J. If $Q = 1$, then $K = (Q^+)'$, so loop the 0's on the $Q = 1$ half of the map to get K. The J and K equations will be independent of Q, since Q is set to a constant value (0 or 1) when reading J and K. To make reading the J's and K's off the map easier, we cross off the Q values on each map. In effect, using the shortcut method is equivalent to splitting the four-variable $Q^+$ map into two three-variable maps, one for $Q = 0$ and one for $Q = 1$.

**FIGURE 1-27:**
**Derivation of J-K**
**Input Equations**

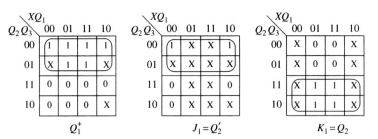

(a) Derivation using separate J–K maps

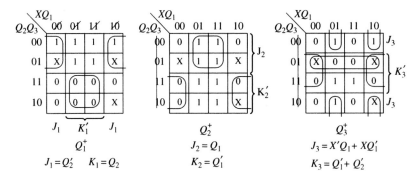

(b) Derivation using the shortcut method

The following summarizes the steps required to design a sequential circuit:

1. Given the design specifications, determine the required relationship between the input and output sequences. Then find a state graph and state table.
2. Reduce the table to a minimum number of states. First eliminate duplicate rows by row matching; then form an implication table and follow the procedure in Section 1.9.

3. If the reduced table has $m$ states ($2^{n-1} < m \leq 2^n$), $n$ flip-flops are required. Assign a unique combination of flip-flop states to correspond to each state in the reduced table. This is the encoded state assignment technique. Alternately, a one-hot assignment with m flip-flops can be used.
4. Form the transition table by substituting the assigned flip-flop states for each state in the reduced state tables. The resulting transition table specifies the next states of the flip-flops and the output in terms of the present states of the flip-flops and the input.
5. Plot next-state maps and input maps for each flip-flop and derive the flip-flop input equations. Derive the output functions.
6. Realize the flip-flop input equations and the output equations using the available logic gates.
7. Check your design using computer simulation or another method.

Steps 2 through 7 may be carried out using a suitable computer-aided design (CAD) program.

• • • • • • • • • • • • •

## 1.8 Moore Sequential Circuit Design

In a Moore circuit, the outputs depend only on the present state. Moore machines are typically easier to design and debug compared to Mealy machines, but they often contain more states than equivalent Mealy machines. In Moore machines, there are no outputs that happen during the transition. The outputs are associated entirely to the state.

### 1.8.1 Moore Machine Design Example 1: Sequence Detector

As an example, let us design the sequence detector of Section 1.7.1 using the Moore Method. The circuit will examine a string of 0's and 1's applied to the $X$ input and generate an output $Z = 1$ only when the input sequence ends in 101. The input $X$ can change only between clock pulses. The circuit does not reset when a 1 output occurs.

As in the Mealy machine example, we start in a reset state designated $S_0$ in Figure 1-28. If a 0 input is received, we can stay in state $S_0$ as the input sequence we are looking for does not start with 0. However, if a 1 is received, the circuit goes to a new state, $S_1$. When in $S_1$, if we receive a 0, the circuit must change to a new state ($S_2$) to remember that the first two inputs of the desired sequence (10) have been

FIGURE 1-28: **State Graph of the Moore Sequence Detector**

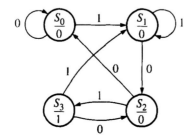

received. If a 1 is received in state $S_2$, the circuit should go to a new state to indicate that the desired input sequence is complete. Let us designate this new state as $S_3$. In state $S_3$, the output must have a value of 1. The outputs in states $S_0$, $S_1$ and $S_2$ must be 0's. The sequence 100 resets the circuit to $S_0$. A sequence 1010 takes the circuit back to $S_2$ because another 1 input should cause $Z$ to become 1 again.

The state table corresponding to the circuit is given by Table 1-7. Note that there is a single column for output because the output is determined by the present state and does not depend on $X$. Note that this sequence detector requires one more state than the Mealy sequence detector in Table 1-3, which detects the same input sequence.

**TABLE 1-7: State Table for Sequence Detector**

| Present State | Next State $X = 0$ | Next State $X = 1$ | Present Output ($Z$) |
|---|---|---|---|
| $S_0$ | $S_0$ | $S_1$ | 0 |
| $S_1$ | $S_2$ | $S_1$ | 0 |
| $S_2$ | $S_0$ | $S_3$ | 0 |
| $S_3$ | $S_2$ | $S_1$ | 1 |

Because there are four states, two flip-flops are required to realize the circuit. Using the state assignment $AB = 00$ for $S_0$, $AB = 01$ for $S_1$, $AB = 11$ for $S_2$, and $AB = 10$ for $S_3$, the transition table shown in Table 1-8 is obtained.

**TABLE 1-8: Transition Table for Moore Sequence Detector**

| $AB$ | $A^+B^+$ $X = 0$ | $A^+B^+$ $X = 1$ | $Z$ |
|---|---|---|---|
| 00 | 00 | 01 | 0 |
| 01 | 11 | 01 | 0 |
| 11 | 00 | 10 | 0 |
| 10 | 11 | 01 | 1 |

The output function $Z = AB'$. Note that $Z$ depends only on the flip-flop states and is independent of $X$, while for the corresponding Mealy machine, $Z$ was a function of $X$. (It was equal to $AX$ in Figure 1-21.) The transition table can be used to write the next state maps and inputs to the flip-flops can be derived.

### 1.8.2 Moore Machine Design Example 2: NRZ to Manchester Code Converter

As another example of designing a Moore sequential machine, we will design a converter for serial data. Binary data is frequently transmitted between computers as a serial stream of bits. Figure 1-29 shows three different coding schemes for serial data. The example shows transmission of the bit sequence 0, 1, 1, 1, 0, 0, 1, 0. With the NRZ (nonreturn-to-zero) code, each bit is transmitted for one bit time without any change. In contrast, for the RZ (return-to-zero) code, a 0 is transmitted as 0 for one full bit time, but a 1 is transmitted as a 1 for the first half of the bit time, and then

FIGURE 1-29:
**Coding Schemes
for Serial Data
Transmission**

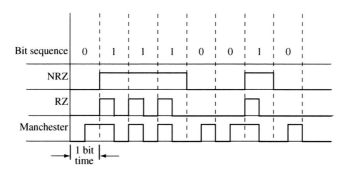

the signal returns to 0 for the second half. For the Manchester code, a 0 is transmitted as 0 for the first half of the bit time and a 1 for the second half, but a 1 is transmitted as a 1 for the first half and a 0 for the second half. Thus, the Manchester encoded bit always changes in the middle of the bit time.

We will design a Moore sequential circuit that converts an NRZ-coded bit stream to a Manchester-coded bit stream (Figure 1-30). In order to do this, we will use a clock (*CLOCK2*) that is twice the frequency of the basic bit clock. If the NRZ bit is 0, it will be 0 for two *CLOCK2* periods, and if it is 1, it will be 1 for two *CLOCK2* periods. Thus, starting in the reset state ($S_0$), the only two possible input sequences are 00 and 11, and the corresponding output sequences are 01 and 10. When a 0 is received, the circuit goes to $S_1$ and outputs a 0; when the second 0 is received, it goes to $S_2$ and outputs a 1. Starting in $S_0$, if a 1 is received, the circuit goes to $S_3$ and outputs a 1, and when the second 1 is received, it must go to a state with a 0 output. Going back to $S_0$ is appropriate since $S_0$ has a 0 output and the circuit is ready to receive another 00 or 11 sequence. When in $S_2$, if a 00 sequence is received, the circuit can go to $S_1$ and then back to $S_2$. If a 11 sequence is received in $S_2$, the circuit can go to $S_3$ and then back to $S_0$. The corresponding Moore state table has two don't cares, which correspond to input sequences that cannot occur.

FIGURE 1-30:
**Moore Circuit for
NRZ-to-Manchester
Conversion**

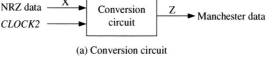

(a) Conversion circuit

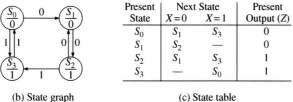

(b) State graph

| Present | Next State | | Present |
|---------|------------|--------|---------|
| State | $X=0$ | $X=1$ | Output ($Z$) |
| $S_0$ | $S_1$ | $S_3$ | 0 |
| $S_1$ | $S_2$ | — | 0 |
| $S_2$ | $S_1$ | $S_3$ | 1 |
| $S_3$ | — | $S_0$ | 1 |

(c) State table

Figure 1-31 shows the timing chart for the Moore circuit. Note that the Manchester output is shifted one clock time with respect to the NRZ input. This

FIGURE 1-31:
**Timing for Moore Circuit**

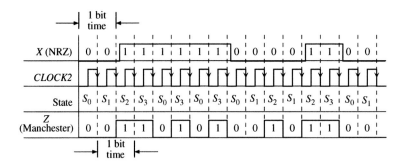

shift occurs because a Moore circuit cannot respond to an input until the active edge of the clock occurs. This is in contrast to a Mealy circuit, for which the output can change after the input changes and before the next clock.

# 1.9 Equivalent States and Reduction of State Tables

The concept of equivalent states is important for the design and testing of sequential circuits. It helps to reduce the hardware consumed by circuits. Two states in a sequential circuit are said to be *equivalent* if we cannot tell them apart by observing input and output sequences. Consider two sequential circuits, $N_1$ and $N_2$ (see Figure 1-32). $N_1$ and $N_2$ could be copies of the same circuit. $N_1$ is started in state $s_i$, and $N_2$ is started in state $s_j$. We apply the same input sequence, $\underline{X}$, to both circuits and observe the output sequences, $\underline{Z}_1$ and $\underline{Z}_2$. (The underscore notation indicates a sequence.) If $\underline{Z}_1$ and $\underline{Z}_2$ are the same, we reset the circuits to states $s_i$ and $s_j$, apply a different input sequence, and observe $\underline{Z}_1$ and $\underline{Z}_2$. If the output sequences are the same for all possible input sequences, we say the $s_i$ and $s_j$ are equivalent ($s_i \equiv s_j$). Formally, we can define equivalent states as follows: $s_i \equiv s_j$ if and only if, for every input sequence $\underline{X}$, the output sequences $\underline{Z}_1 = \lambda_1(s_i, \underline{X})$ and $\underline{Z}_2 = \lambda_2(s_j, \underline{X})$ are the same. This is not a very practical way to test for state equivalence since, at least in theory, it requires input sequences of infinite length. In practice, if we have a bound on number of states, then we can limit the length of the test sequences.

A more practical way to determine state equivalence uses the state equivalence theorem: $s_i \equiv s_j$ if and only if for every single input $X$, the outputs are the same and the

FIGURE 1-32:
**Sequential Circuits**

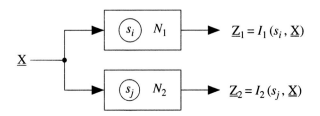

next states are equivalent. When using the definition of equivalence, we must consider all input sequences, but we do not need any information about the internal state of the system. When using the state equivalence theorem, we must look at both the output and next state, but we need to consider only single inputs rather than input sequences.

The table of Figure 1-33(a) can be reduced by eliminating equivalent states. First, observe that states $a$ and $h$ have the same next states and outputs when $X = 0$ and also when $X = 1$. Therefore, $a \equiv h$ so we can eliminate row $h$ and replace $h$ with $a$ in the table. To determine if any of the remaining states are equivalent, we will use the state equivalence theorem. From the table, since the outputs for states $a$ and $b$ are the same, $a \equiv b$ if and only if $c \equiv d$ and $e \equiv f$. We say that $c$-$d$ and $e$-$f$ are implied pairs for $a$-$b$. To keep track of the implied pairs, we make an *implication chart*, as shown in Figure 1-33(b). We place $c$-$d$ and $e$-$f$ in the square at the intersection of row $a$ and column $b$ to indicate the implication. Since states $d$ and $e$ have different outputs, we place an X in the $d$-$e$ square to indicate that $d \not\equiv e$. After completing the implication chart in this way, we make another pass through the chart. The $e$-$g$ square contains $c$-$e$ and $b$-$g$. Since the $c$-$e$ square has an X, $c \not\equiv e$, which implies $e \not\equiv g$, so we X out the $e$-$g$ square. Similarly, since $a \not\equiv g$, we X out the $f$-$g$ square. On the next pass through the chart, we X out all the squares that contain $e$-$g$ or $f$-$g$ as implied pairs (shown on the chart with dashed x's). In the next pass, no additional squares are X'ed out, so the process terminates. Since all the squares corresponding to non-equivalent states have been X'ed out, the coordinates of the remaining squares indicate equivalent state pairs. From the first column, $a \equiv b$; from third column, $c \equiv d$; and from the fifth column, $e \equiv f$.

The implication table method of determining state equivalence can be summarized as follows:

1. Construct a chart that contains a square for each pair of states.
2. Compare each pair of rows in the state table. If the outputs associated with states $i$ and $j$ are different, place an X in square $i$-$j$ to indicate that $i \not\equiv j$. If the outputs are the same, place the implied pairs in square $i$-$j$. (If the next states of $i$ and $j$ are $m$ and $n$ for some input $x$, then $m$-$n$ is an implied pair.) If the outputs and next states are the same (or if $i$-$j$ implies only itself), place a check ($\sqrt{}$) in square $i$-$j$ to indicate that $i \equiv j$.
3. Go through the table square by square. If square $i$-$j$ contains the implied pair $m$-$n$, and square $m$-$n$ contains an X, then $i \not\equiv j$, and an X should be placed in square $i$–$j$.
4. If any X's were added in step 3, repeat step 3 until no more X's are added.
5. For each square $i$-$j$ that does not contain an X, $i \equiv j$.

If desired, row matching can be used to partially reduce the state table before constructing the implication table. Although we have illustrated this procedure for a Mealy table, the same procedure applies to a Moore table.

Two sequential circuits are said to be equivalent if every state in the first circuit has an equivalent state in the second circuit, and vice versa.

Optimization techniques such as this are incorporated in CAD tools. The importance of state minimization has slightly diminished in recent years due to the abundance of transistors on chips; however, it is still important to do obvious state minimizations to reduce the circuit's area and power.

**FIGURE 1-33: State Table Reduction**

| Present State | Next State X=0 | 1 | Present Output X=0 | 1 |
|---|---|---|---|---|
| a | c | f | 0 | 0 |
| b | d | e | 0 | 0 |
| c | ~~h~~a | g | 0 | 0 |
| d | b | g | 0 | 0 |
| e | e | b | 0 | 1 |
| f | f | a | 0 | 1 |
| g | c | g | 0 | 1 |
| ~~h~~ | ~~e~~ | ~~f~~ | ~~0~~ | ~~0~~ |

(a) State table reduction by row matching

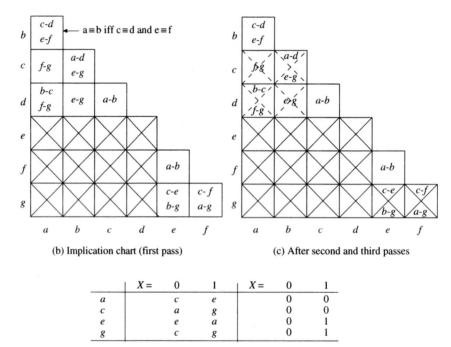

(b) Implication chart (first pass)          (c) After second and third passes

| | X = 0 | 1 | X = 0 | 1 |
|---|---|---|---|---|
| a | c | e | 0 | 0 |
| c | a | g | 0 | 0 |
| e | e | a | 0 | 1 |
| g | c | g | 0 | 1 |

(d) Final reduced table

• • • • • • • • • • • •

# 1.10  Sequential Circuit Timing

The correct functioning of sequential circuits involves several timing issues. Propagation delays of flip-flops, gates and wires, setup times and hold times of flip-flops, clock synchronization, clock skew, etc become important issues while designing sequential circuits. In this section, we look at various topics related to sequential circuit timing.

### 1.10.1 **Propagation Delays; Setup and Hold Times**

There is a certain amount of time, albeit small, that elapses from the time the clock changes to the time the $Q$ output changes. This time, called *propagation delay*, is indicated in Figure 1-34. The propagation delay can depend on whether the output is changing from high to low or vice versa. In the figure, the propagation delay for a low-to-high change in $Q$ is denoted by $t_{plh}$, and for a high-to-low change it is denoted by $t_{phl}$.

For an ideal D flip-flop, if the $D$ input changed at exactly the same time as the active edge of the clock, the flip-flop would operate correctly. However, for a real flip-flop, the $D$ input must be stable for a certain amount of time before the active edge of the clock. This interval is called the *setup time* $(t_{su})$. Furthermore, $D$ must be stable for a certain amount of time after the active edge of the clock. This interval is called the *hold time* $(t_h)$. Figure 1-34 illustrates setup and hold times for a $D$ flip-flop that changes state on the rising edge of the clock. $D$ can change at any time during the shaded region on the diagram, but it must be stable during the time interval $t_{su}$ before the active edge and for $t_h$ after the active edge. If $D$ changes at any time during the forbidden interval, it cannot be determined whether the flip-flop will change state. Even worse, the flip-flop may malfunction and output a short pulse or even go into oscillation. Minimum values for $t_{su}$ and $t_h$ and maximum values for $t_{plh}$ and $t_{phl}$ can be read from manufacturers' data sheets.

**FIGURE 1-34: Setup and Hold Times for D Flip-Flop**

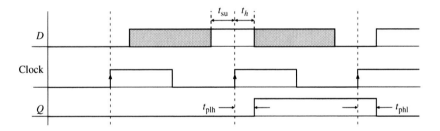

### 1.10.2 **Maximum Clock Frequency of Operation**

In a synchronous sequential circuit, state changes occur immediately following the active edge of the clock. The maximum clock frequency for a sequential circuit depends on several factors. The clock period must be long enough so that all flip-flop and register inputs will have time to stabilize before the next active edge of the clock. Propagation delays and setup and hold times create complications in sequential circuit timing.

Consider a simple circuit of the form of Figure 1-35(a). The output of a D flip-flop is fed back to its input through an inverter. Assume a clock as indicated by the waveform CLK in Figure 1-35(b). If the current output of the flip-flop is 1, a value of 0 will appear at the flip-flop's $D$ input after the propagation delay of the inverter. Assuming that the next active edge of the clock arrives after the setup time has elapsed, the output of the flip-flop will change to 0. This process will continue, yielding the output $Q$

FIGURE 1-35:
**Simple Frequency
Divider**

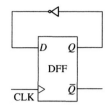

(a) A frequency divider

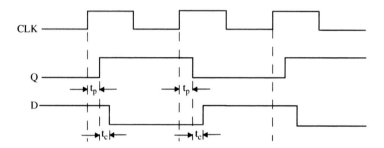

(b) Frequency divider timing diagram

of the flip-flop to be a waveform with twice the period of the clock. Essentially the circuit behaves as a frequency divider.

If we increase the frequency of the clock slightly, the circuit will still work yielding half of the increased frequency at the output. However, if we increase the frequency to be very high, the output of the inverter may not get enough time to stabilize and meet the setup time requirements. Similarly, if the inverter was very fast and fed the inverted output to the $D$ input extremely quickly, there will be timing problems because the hold time of the flip-flop may not be met. So we can easily see a variety of ways in which timing problems could arise from propagation delays and setup and hold time requirements.

### 1.10.3 Timing Conditions for Proper Operation

For a circuit of the general form of Figure 1-17, assume that the maximum propagation delay through the combinational circuit is $t_{cmax}$ and the maximum propagation delay from the time the clock changes to the time the flip-flop output changes is $t_{pmax}$, where $t_{pmax}$ is the maximum of $t_{plh}$ and $t_{phl}$. There are four conditions this circuit has to meet in order to ensure proper operation.

1. **Clock period should be long enough to satisfy flip-flop setup time.** The clock period should be long enough to allow the flip-flop outputs to change and the combinational circuitry to change while still leaving enough time to satisfy the setup time. Once the clock arrives, it could take a delay of up to $t_{pmax}$ before the flip-flop output changes. Then it could take a delay of up to $t_{cmax}$ before the output of the combinational circuitry changes. Thus the maximum time from the active edge of the clock to the time the change in $Q$ propagates back to the

D flip-flop inputs is $t_{pmax} + t_{cmax}$. In order to ensure proper flip-flop operation, the combinational circuit output must be stable $t_{su}$ before the end of the clock period. If the clock period is $t_{ck}$,

$$t_{ck} \geq t_{pmax} + t_{cmax} + t_{su}$$

The difference between $t_{ck}$ and $(t_{pmax} + t_{cmax} + t_{su})$ is referred to as the setup time margin.

2. **Propagation delays should be long enough to satisfy flip-flop hold time.** A hold-time violation could occur if the change in $Q$ was fed back through the combinational circuit and caused $D$ to change too soon after the clock edge. The hold time is satisfied if

$$t_{pmin} + t_{cmin} \geq t_h$$

When checking for hold-time violations, the worst case occurs when the timing parameters have their minimum values. Since $t_{pmin} > t_h$ for normal flip-flops, a hold-time violation due to $Q$ changing does not occur.

3. **External input changes to the circuit should satisfy flip-flop setup time.** A setup time violation could occur if the $X$ input to the circuit changes too close to the active edge of the clock. When the $X$ input to a sequential circuit changes, we must make sure that the input change propagates to the flip-flop inputs such that the setup time is satisfied before the active edge of the clock. If $X$ changes $t_x$ time units before the active edge of the clock (see Figure 1-36), then it could take up to the maximum propagation delay of the combinational circuit, before the change in $X$ propagates to the flip-flop input. There should still be a margin of $t_{su}$ left before the edge of the clock. Hence, the setup time is satisfied if

$$t_x \geq t_{cxmax} + t_{su}$$

where $t_{cxmax}$ is the maximum propagation delay from $X$ to the flip-flop input.

**FIGURE 1-36: Setup and Hold Timing for Changes in *X***

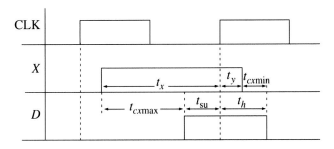

4. **External input changes to the circuit should satisfy flip-flop hold times.** In order to satisfy the hold time, we must make sure that $X$ does not change too soon after the clock. If a change in $X$ propagates to the flip-flop input in zero time, $X$ should not change for a duration of $t_h$ after the clock edge. Fortunately, it takes some positive propagation delay for the change in $X$

to reach the flip-flop. If $t_{cxmin}$ is the minimum propagation delay from $X$ to the flip-flop input, changes in $X$ will not reach the flip-flop input until at least a time of $t_{cxmin}$ has elapsed after the clock edge. So, if $X$ changes $t_y$ time units after the active edge of the clock, then the hold time is satisfied if

$$t_y \geq t_h - t_{cxmin}$$

If $t_y$ is negative, $X$ can change before the active clock edge and still satisfy the hold time.

Given a circuit, we can determine the safe frequency of operation and safe regions for input changes using the above principles. As an example, consider the frequency divider circuit in Figure 1-35(a). If the minimum and maximum delays of the inverter are 1 ns and 3 ns, and if $t_{pmin}$ and $t_{pmax}$ are 5 ns and 8 ns, the maximum frequency at which it can be clocked can be derived using requirement (1) above. Assume that the setup and hold times of the flip-flop are 4 ns and 2 ns. For proper operation, $t_{ck} \geq t_{pmax} + t_{cmax} + t_{su}$. In this example, $t_{pmax}$ for the flip-flops is 8 ns, $t_{cmax}$ is 3 ns, and $t_{su}$ is 4 ns. Hence

$$t_{ck} \geq 8 + 3 + 4 = 15 \text{ ns}$$

The maximum clock frequency is then $1/t_{ck}$ = 66.67 MHz. We should also make sure that the hold time requirement is satisfied. Hold time requirement means that the D input should not change before 2 ns after the clock edge. This will be satisfied if $t_{pmin} + t_{cmin} \geq 2$ ns. In this circuit, $t_{pmin}$ is 5 ns and $t_{cmin}$ is 1 ns. Thus the Q output is guaranteed to not change until 5 ns after the clock edge, and at least 1 ns more should elapse before the change can propagate through the inverter. Hence the D input will not change until 6 ns after the clock edge, which automatically satisfies the hold time requirements. Since there are no external inputs, these are the only timing constraints that we need to satisfy.

Now consider a circuit as in Figure 1-37(a). Assume that the delay of the combinational circuit is in the range 2 to 4 ns, the flip-flop propagation delays are in the range 5 to10 ns, the setup time is 8 ns, and hold time is 3 ns. In order to satisfy the setup time, the clock period has to be greater than $t_{pmax} + t_{cmax} + t_{su}$. So

$$t_{ck} \geq 10 + 4 + 8 = 22 \text{ ns}$$

The hold time requirement is satisfied if the output does not change until 3 ns after the clock. Here, the output is not expected to change until $t_{pmin} + t_{cmin}$. Since $t_{pmin}$ is 5 ns and $t_{cmin}$ is 2 ns, the output is not expected to change until 7 ns, which automatically satisfies the hold time requirement. This circuit has external inputs that allow us to identify safe regions where the input $X$ can change using requirements (3) and (4) above. The $X$ input should be stable for a duration of $t_{cxmax} + t_{su}$ (i.e., 4 ns + 8 ns) before the clock edge. Similarly, it should be stable for a duration of $t_h - t_{cxmin}$ (i.e., 3 ns − 2 ns) after the clock edge. Thus, the $X$ input should not change 12 ns before the clock edge and 1 ns after the clock edge. Although the hold time is 3 ns, we see that the input $X$ can change 1 ns after the clock edge, because it takes at least another 2 ns (minimum delay of combinational circuit) before the input change can propagate to the D input of the flip-flop. The shaded regions in

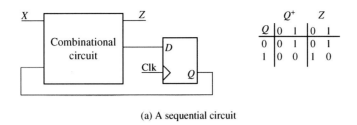

(a) A sequential circuit

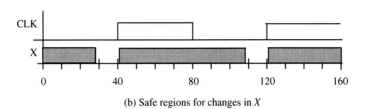

(b) Safe regions for changes in X

the waveform for $X$ indicate safe regions where the input signal $X$ may change without causing erroneous operation in the circuit.

### 1.10.4 Glitches in Sequential Circuits

Sequential circuits often have external inputs that are asynchronous. Input changes can cause temporary false values called glitches at the outputs and next states. For example, if the state table of Figure 1-23(b) is implemented in the form of Figure 1-17, the timing waveforms are as shown in Figure 1-38. Propagation delays in the flip-flop have been neglected; hence state changes are shown to coincide with clock edges. In this example, the input sequence is 0 0 1 0 1 0 0 1, and $X$ is assumed to change in the middle of the clock pulse. At any given time, the next state and $Z$ output can be read from the next state table. For example, at time $t_a$, State $= S_5$ and $X = 0$, so Next State $= S_0$ and $Z = 0$. At time $t_b$ following the rising edge of the clock, State $= S_0$ and $X$ is still 0, so Next State $= S_1$ and $Z = 1$. Then $X$ changes to 1, and at time $t_c$ Next State $= S_2$ and $Z = 0$. Note that there is a *glitch* (sometimes called a false output) at $t_b$. The $Z$ output

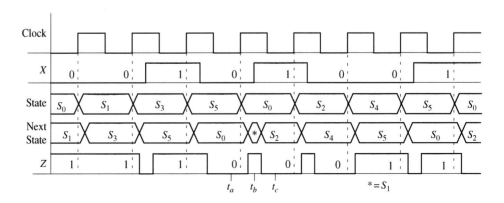

momentarily has an incorrect value at $t_b$, because the change in $X$ is not exactly synchronized with the active edge of the clock. The correct output sequence, as indicated on the waveform, is 1 1 1 0 0 0 1 1. Several glitches appear between the correct outputs; however, these are of no consequence if $Z$ is read at the right time. The glitch in the next state at $t_b$ ($S_1$) also does not cause a problem, because the next state has the correct value at the active edge of the clock.

The timing waveforms derived from the circuit of Figure 1-26 are shown in Figure 1-39. They are similar to the general timing waveforms given in Figure 1-38 except that State has been replaced with the states of the three flip-flops, and a propagation delay of 10 ns has been assumed for each gate and flip-flop.

**FIGURE 1-39:**
**Timing Diagram for Figure 1-26**

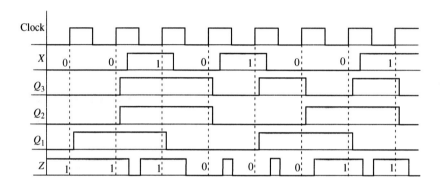

### 1.10.5 Synchronous Design

One of the most commonly used digital design techniques is *synchronous design*. In this type of design, a clock is used to synchronize the operation of all flip-flops, registers, and counters in the system. Synchronous circuits are more reliable compared to asynchronous circuits. In synchronous circuits, events are expected to occur immediately following the active edge of the clock. Outputs from one part have a full clock cycle to propagate to the next part of the circuit. Synchronous design philosophy makes design and debugging easier compared to asynchronous techniques.

Figure 1-40 illustrates a synchronous digital system. Assume that the system is built from several modules or devices. The devices could be flip-flops, registers, counters, adders, multipliers, and so on. All of the sequential devices are synchronized with respect to the same clock in a synchronous system. A traditional way to view a digital system is to consider it as a control section plus a data section. The various devices shown in Figure 1-40 are part of the data section. The control section is a sequential machine that generates control signals to control the operation of the data section. For example, if the data section contains a shift register, the control section may generate signals that determine when the register is to be loaded ($Ld$) and when it is to be shifted ($Sh$). A common clock synchronizes the operation of the control and data sections. The data section may generate status signals (not shown in this figure) that affect the control sequence. For example, if a data operation produces an arithmetic overflow, then the data section might generate a condition

FIGURE 1-40:
**A Synchronous Digital System**

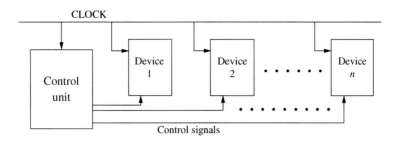

signal *V* to indicate an overflow. The control section is also called *controller* and the data section is often called *architecture* or *data path*.

In a synchronous digital system, we desire to see all changes happen immediately at the active edge of the clock, but that might not happen in a practical circuit. Modern integrated circuits (ICs) are fabricated at feature sizes such as or smaller than 0.1 microns. Modern microprocessors are clocked at several gigahertz. In these chips, wire delays are significant compared to the clock period. Even if two flip-flops are connected to the same clock, the clock edge might arrive at the two flip-flops at different times due to unequal wire delays. If unequal amounts of combinational circuitry (e.g., buffers or inverters) are used in the clock path to different devices, that also could result in unequal delays, making the clock reach different devices at slightly different times. This problem is called **clock skew**.

There are also problems that occur due to glitches in control signals. Consider Figure 1-41, which illustrates the operation of a digital system that uses devices that change state on the falling edge of the clock. Several flip-flops may change state in response to this falling edge. The time at which each flip-flop changes state is determined by the propagation delay for that flip-flop. The changes in flip-flop states in the control section will propagate through the combinational circuit that generates the control signals, and some of the control signals may change as a result. The exact times at which the control signals change depend on the propagation delays in the gate circuits that generate the signals as well as the flip-flop delays. Thus, after

FIGURE 1-41:
**Timing Chart for System with Falling-Edge Devices**

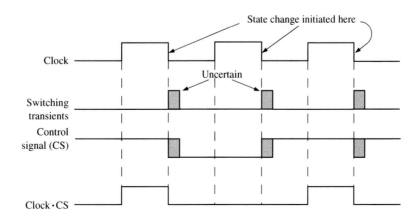

the falling edge of the clock, there is a period of uncertainty during which control signals may change. Glitches and spikes may occur in the control signals due to hazards. Furthermore, when signals are changing in one part of the circuit, noise may be induced in another part of the circuit. As indicated by the shading in Figure 1-41, there is a time interval after each falling edge of the clock in which there may be noise in a control signal (CS), and the exact time at which the control signal changes is not known.

If we want a device in the data section to change state on the falling edge of the clock only if the control signal CS = 1, we can AND the clock with CS, as shown in Figure 1-42(a). This technique is called clock gating. The transitions will occur in synchronization with the clock CLK except for a small delay in the AND gate. The gated CLK signal is clean because the clock is 0 during the time interval in which the switching transients occur in CS.

Gating the clock with the control signal, as illustrated in Figure 1-42(a), can solve some synchronization problems. However, clock gating can also lead to clock skew and additional timing problems in high-speed circuits. Instead of gating the clock with the control signal, it is more desirable to use devices with clock enable (CE) pins and feed the control signal to the enable pin, as illustrated in Figure 1-42(b). Many registers, counters, and other devices used in synchronous systems have an enable input. When enable = 1, the device changes state in response to the clock, and when enable = 0, no state change occurs. Use of the enable input eliminates the need for a gate on the clock input, and associated timing problems are avoided.

**FIGURE 1-42:**
**Techniques to**
**Synchronize**
**Control Signals**

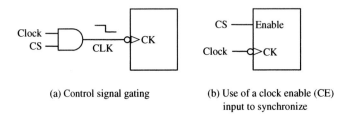

(a) Control signal gating

(b) Use of a clock enable (CE) input to synchronize

We discourage designers from gating clocks or feeding the output of combinational circuits to clock inputs. While clock skew from wire delays is unavoidable to some extent, clock skew due to combinational circuitry in the clock path can easily be avoided. Circuits as in Figure 1-43 should be avoided as much as possible to minimize timing problems.

**FIGURE 1-43:**
**Examples of**
**Circuits to Avoid**

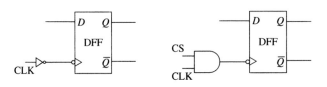

Due to wire delays or other unforeseen problems, at times we end up with circuits where the clock edge reaches different flip-flops at different times. Consider the circuit in Figure 1-44, where the clock reaches the two flip-flops at slightly different times. Proper synchronous operation means that both flip-flops operate as if they receive the same clock. Despite the delay in the clock to the second flip-flop, its state change must be triggered before the new value of $Q_1$ reaches $D_2$. The maximum clock frequency for synchronous operation should be decided considering the delay between the clocks as well.

FIGURE 1-44: **A Circuit with Clock Skew**

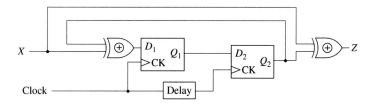

If devices do not have enables and synchronous operation cannot be obtained without clock gating, we should pay attention to gate the clocks correctly. A device with negative edge triggering can be made to function correctly by ANDing the clock signal with the control signal, as in Figure 1-42(a). In the following paragraphs, we describe issues associated with control signal gating for positive edge triggered devices.

Figure 1-45 illustrates the operation of a digital system that uses devices that change state on the rising edge of the clock. In this case, the switching transients that result in noise and uncertainty will occur following the rising edge of the clock. The shading indicates the time interval in which the control signal $CS$ may be noisy. If we want a device to change state on the rising edge of the clock when $CS = 1$, transition is expected at (a) and (c), but no change is expected at (b) since $CS = 0$ when the clock edge arrives. In order to create a gated control signal, it is tempting to

FIGURE 1-45: **Timing Chart for System with Rising-Edge Devices**

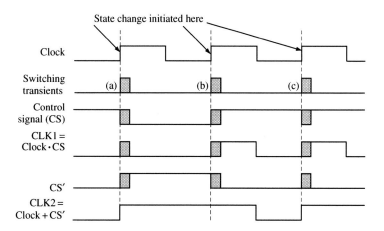

AND the clock with *CS*, as shown in Figure 1-46(a). The resulting signal, which goes to the *CK* input of the device, may be noisy and timed incorrectly. In particular, the *CLK1* pulse at (a) will be short and noisy. It may be too short to trigger the device, or it may be noisy and trigger the device more than once. In general, it will be out of synchronization with the clock, because the control signal does not change until after some of the flip-flops in the control circuit have changed state. The rising edge of the pulse at (b) again will be out of synch with the clock, and it may be noisy. But even worse, the device will trigger near point (b) when it should not trigger there at all. Since *CS* = 0 at the time of the rising edge of the clock, triggering should not occur until the next rising edge, when *CS* = 1.

**FIGURE 1-46:**
**Incorrect Clock**
**Gating for**
**Rising-Edge**
**Devices**

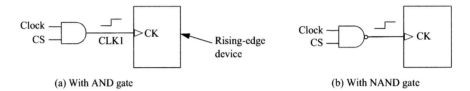

(a) With AND gate  (b) With NAND gate

For a rising-edge device, if we changed the AND gate in Figure 1-42 to NAND gate as in Figure 1-46(b), it would be incorrect because the synchronization will happen at the wrong edge. The correct way to gate the control signal will be as in Figure 1-47, which will result in the *CK* input to the device having a positive edge only when the control signal is positive and clock is going to have a positive edge. The *CK* input is then

$$CLK2 = (CS \cdot clock')' = CS' + clock$$

The last waveform in Figure 1-45 illustrates this gated control signal. While this circuit can solve the synchronization problem, we encourage designers to refrain from gating clocks at all if possible.

**FIGURE 1-47:**
**Correct Control**
**Signal Gating for**
**Rising-Edge Device**

In summary, synchronous design is based on the following principles:

- Method:      All clock inputs to flip-flops, registers, counters, and so on are driven directly from the system clock.
- Result:      All state changes occur immediately following the active edge of the clock signal.
- Advantage:   All switching transients, switching noise, and so on occur between clock pulses and have no effect on system performance.

Asynchronous design is generally more difficult than synchronous design. Since there is no clock to synchronize the state changes, problems may arise when several state variables must change at the same time. A race occurs if the final state depends on the order in which the variables change. Asynchronous design requires special techniques to eliminate problems with races and hazards. On the other hand, synchronous design has several disadvantages: In high-speed circuits where the propagation delay in the wiring is significant, the clock signal must be carefully routed so that it reaches all the clock inputs at essentially the same time (i.e., to minimize clock skew). The maximum clock rate is determined by the worst-case delay of the longest path. The system inputs may not be synchronized with the clock, so use of synchronizers may be required. Synchronous systems also consume more power than asynchronous systems. The clock distribution circuitry in synchronous chips often consumes a significant fraction of the chip power.

## 1.11 Tristate Logic and Busses

Normally, if we connect the outputs of two gates or flip-flops together, the circuit will not operate properly. It can also cause damage to the circuit. Hence, when we need to connect multiple gate outputs to the same wire or channel, one way to do that is by using tristate buffers. Tristate buffers are gates with a high impedance state (hi-Z) in addition to high and low logic states. The high impedance state is equivalent to an open circuit. In digital systems, transferring data back and forth between several system components is often necessary. Tristate busses can be used to facilitate data transfers between registers. When several gates are connected onto a wire, what we expect is that at any one point, one of the gates is going to actually drive the wire, and the other gates should behave as if they are not connected to the wire. The high impedance state achieves this.

Tristate buffers can be inverting or non-inverting. The control input can be active high or active low. Figure 1-48 shows four kinds of tristate buffers. $B$ is the control input used to enable or disable the buffer output. When a buffer is enabled, the output ($C$) is equal to the input ($A$) or its complement. However, we can connect two tristate buffer outputs, provided that only one output is enabled at a time.

FIGURE 1-48: **Four Kinds of Tristate Buffers**

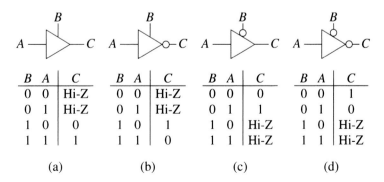

| $B$ | $A$ | $C$ |
|---|---|---|
| 0 | 0 | Hi-Z |
| 0 | 1 | Hi-Z |
| 1 | 0 | 0 |
| 1 | 1 | 1 |

(a)

| $B$ | $A$ | $C$ |
|---|---|---|
| 0 | 0 | Hi-Z |
| 0 | 1 | Hi-Z |
| 1 | 0 | 1 |
| 1 | 1 | 0 |

(b)

| $B$ | $A$ | $C$ |
|---|---|---|
| 0 | 0 | 0 |
| 0 | 1 | 1 |
| 1 | 0 | Hi-Z |
| 1 | 1 | Hi-Z |

(c)

| $B$ | $A$ | $C$ |
|---|---|---|
| 0 | 0 | 1 |
| 0 | 1 | 0 |
| 1 | 0 | Hi-Z |
| 1 | 1 | Hi-Z |

(d)

Figure 1-49 shows a system with three registers connected to a tristate bus. Each register is 8 bits wide, and the bus consists of 8 wires connected in parallel. Each tristate buffer symbol in the figure represents 8 buffers operating in parallel with a common enable input. Only one group of buffers is enabled at a time. For example, if $Enb = 1$, the register $B$ output is driven onto the bus. The data on the bus is routed to the inputs of register $A$, register $B$, and register $C$. However, data is loaded into a register only when its load input is 1 and the register is clocked. Thus, if $Enb = Ldc = 1$, the data in register $B$ will be copied into register $C$ when the active edge of the clock occurs. If $Eni = Lda = Ldb = 1$, the input data will be loaded in registers $A$ and $B$ when the registers are clocked.

**FIGURE 1-49: Data Transfer Using Tristate Bus**

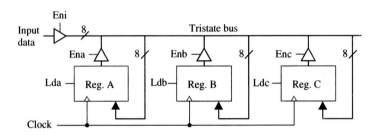

## Problems

**1.1** Write out the truth table for the following equation.

$$F = (A \oplus B) \cdot C + A' \cdot (B' \oplus C)$$

**1.2** A full subtracter computes the difference of three inputs $X$, $Y$, and $B_{in}$, where $Diff = X - Y - B_{in}$. When $X < (Y + B_{in})$, the borrow output $B_{out}$ is set. Fill in the truth table for the subtracter and derive the sum-of-products and product-of-sums equations for $Diff$ and $B_{out}$.

**1.3** Simplify $Z$ using a four-variable map with map-entered variables. $ABCD$ represents the state of a control circuit. Assume that the circuit can never be in state $0100, 0001$, or $1001$.

$$Z = BC'DE + ACDF' + ABCD'F' + ABC'D'G + B'CD + ABC'D'H'$$

**1.4** For the following functions, find the minimum sum of products using four-variable maps with map-entered variables. In (a) and (b), $m_i$ represents a minterm of variables $A$, $B$, $C$, and $D$.

**(a)** $F(A, B, C, D, E) = \Sigma m(0, 4, 6, 13, 14) + \Sigma d(2, 9) + E(m_1 + m_{12})$
**(b)** $Z(A, B, C, D, E, F, G) = \Sigma m(2, 5, 6, 9) + \Sigma d(1, 3, 4, 13, 14) + E(m_{11} + m_{12})$
$$+ F(m_{10}) + G(m_0)$$
**(c)** $H = A'B'CDF' + A'CD + A'B'CD'E + BCDF'$
**(d)** $G = C'E'F + DEF + AD'E'F' + BC'E'F + AD'EF'$

*Hint:* Which variables should be used for the map sides and which variables should be entered into the map?

1.5 Identify the static 1-hazards in the following circuit. State the condition under which each hazard can occur. Draw a timing diagram (similar to Figure 1–10(b)) that shows the sequence of events when a hazard occurs.

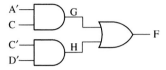

1.6 Find all of the 1-hazards in the given circuit. Indicate what changes are necessary to eliminate the hazards.

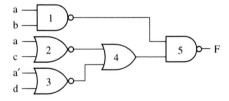

1.7 **(a)** Find all the static hazards in the following circuit. For each hazard, specify the values of the input variables and which variable is changing when the hazard occurs. For one of the hazards, specify the order in which the gate outputs must change.

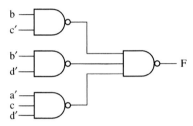

**(b)** Design a NAND-gate circuit that is free of static hazards to realize the same function.

1.8 **(a)** Find all the static hazards in the following circuit. State the condition under which each hazard can occur.
**(b)** Redesign the circuit so that it is free of static hazards. Use gates with at most three inputs.

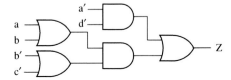

1.9 **(a)** Show how you can construct a T flip-flop using a J-K flip-flop.
**(b)** Show how you can construct a J-K flip-flop using a D flip-flop and gates.

1.10 Construct a clocked D flip-flop, triggered on the rising edge of $CLK$, using two transparent D latches and any necessary gates. Complete the following timing diagram, where $Q_1$ and $Q_2$ are latch outputs. Verify that the flip-flop output changes to $D$ after the rising edge of the clock.

1.11 A synchronous sequential circuit has one input and one output. If the input sequence 0101 or 0110 occurs, an output of two successive 1's will occur. The first of these 1's should occur coincident with the last input of the 0101 or 0110 sequence. The circuit should reset when the second 1 output occurs. For example,

$$\text{input sequence:} \quad X = 0\,1\,0\,0\,1\,1\,1\,0\,1\,0\,1\,0\,1\,0\,1\,1\,0\,1\ldots$$
$$\text{output sequence:} \quad Z = 0\,0\,0\,0\,0\,0\,0\,0\,0\,0\,1\,1\,0\,0\,0\,0\,1\,1\ldots$$

**(a)** Derive a Mealy state graph and table with a minimum number of states (six states).
**(b)** Try to choose a good state assignment. Realize the circuit using J-K flip-flops and NAND gates. Repeat using NOR gates. (Work this part by hand.)
**(c)** Check your answer to (b) using the *LogicAid* program. Also use the program to find the NAND solution for two other state assignments.

1.12 A sequential circuit has one input ($X$) and two outputs ($Z_1$ and $Z_2$). An output $Z_1 = 1$ occurs every time the input sequence 010 is completed provided that the sequence 100 has never occurred. An output $Z_2 = 1$ occurs every time the input sequence 100 is completed. Note that once a $Z_2 = 1$ output has occurred, $Z_1 = 1$ can never occur, but *not* vice versa.

**(a)** Derive a Mealy state graph and table with a minimum number of states (eight states).
**(b)** Try to choose a good state assignment. Realize the circuit using J-K flip-flops and NAND gates. Repeat using NOR gates. (Work this part by hand.)
**(c)** Check your answer to (b) using the *LogicAid* program. Also use the program to find the NAND solution for two other state assignments.

1.13 A sequential circuit has one input ($X$) and two outputs ($S$ and $V$). $X$ represents a 4-bit binary number $N$, which is input least significant bit first. $S$ represents a 4-bit binary number equal to $N + 2$, which is output least significant bit first. At the time the fourth input occurs, $V = 1$ if $N + 2$ is too large to be represented by 4 bits; otherwise, $V = 0$. The value of $S$ should be the proper value, not a don't care, in both cases. The circuit always resets after the fourth bit of $X$ is received.

    **(a)** Derive a Mealy state graph and table with a minimum number of states (six states).
    **(b)** Try to choose a good state assignment. Realize the circuit using D flip-flops and NAND gates. Repeat using NOR gates. (Work this part by hand.)
    **(c)** Check your answer to (b) using the *LogicAid* program. Also use the program to find the NAND solution for two other state assignments.

1.14 A sequential circuit has one input ($X$) and two outputs ($D$ and $B$). $X$ represents a 4-bit binary number $N$, which is input least significant bit first. $D$ represents a 4-bit binary number equal to $N - 2$, which is output least significant bit first. At the time the fourth input occurs, $B = 1$ if $N - 2$ is negative; otherwise, $B = 0$. The circuit always resets after the fourth bit of $X$ is received.

    **(a)** Derive a Mealy state graph and table with a minimum number of states (six states).
    **(b)** Try to choose a good state assignment. Realize the circuit using J-K flip-flops and NAND gates. Repeat using NOR gates. (Work this part by hand.)
    **(c)** Check your answer to (b) using the *LogicAid* program. Also use the program to find the NAND solution for two other state assignments.

1.15 A Moore sequential circuit has one input and one output. The output goes to 1 when the input sequence 111 has occurred and the output goes to 0 if the input sequence 000 occurs. At all other times, the output holds its value.
Example:

$$X = 0\ 1\ 0\ 1\ 1\ 1\ 0\ 1\ 0\ 0\ 0\ 1\ 1\ 1\ 0\ 0\ 1\ 0\ 0\ 0$$
$$Z = 0\ 0\ 0\ 0\ 0\ 0\ 1\ 1\ 1\ 1\ 0\ 0\ 0\ 1\ 1\ 1\ 1\ 1\ 1\ 0$$

Derive a Moore state graph and table for the circuit.

1.16 Derive the state transition table and flip-flop input equations for a modulo-6 counter that counts 000 through 101 and then repeats. Use J-K flip-flops.

1.17 Derive the state transition table and D flip-flop input equations for a counter that counts from 1 to 6 and then repeats.

1.18 Reduce the following state table to a minimum number of states.

| Present | Next State | | Output | |
|---|---|---|---|---|
| State | $X = 0$ | $X = 1$ | $X = 0$ | $X = 1$ |
| A | B | G | 0 | 1 |
| B | A | D | 1 | 1 |
| C | F | G | 0 | 1 |
| D | H | A | 0 | 0 |
| E | G | C | 0 | 0 |
| F | C | D | 1 | 1 |
| G | G | E | 0 | 0 |
| H | G | D | 0 | 0 |

1.19 A Mealy sequential circuit is implemented using the circuit shown in Figure 1-44. Assume that if the input $X$ changes, it changes at the same time as the falling edge of the clock.

(a) Complete the timing diagram below. Indicate the proper times to read the output ($Z$). Assume that "delay" is 0 ns and that the propagation delay for the flip-flop and XOR gate has a nominal value of 10 ns. The clock period is 100 ns.

Clock

X

$Q_1$

$Q_2$

Z

(b) Assume the following delays: XOR gate—10 to 20 ns, flip-flop propagation delay—5 to 10 ns, setup time—5 ns, and hold time—2 ns. Also assume that the "delay" is 0 ns. Determine the maximum clock rate for proper synchronous operation. Consider both the feedback path that includes the flip-flop propagation delay and the path starting when $X$ changes.

(c) Assume a clock period of 100 ns. Also assume the same timing parameters as in (b). What is the maximum value that "delay" can have and still achieve proper synchronous operation? That is, the state sequence must be the same as for no delay.

1.20 Two flip-flops are connected as shown below. The delay represents wiring delay between the two clock inputs, which results in clock skew. This can cause possible loss of synchronization. The flip-flop propagation delay from clock to $Q$ is 10 ns $<$ $t_p < 15$ ns; the setup and hold times are 4 ns and 2 ns, respectively.

**(a)** What is the maximum value that the delay can have and still achieve proper synchronous operation? Draw a timing diagram to justify your answer.

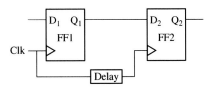

**(b)** Assuming that the delay is < 3 ns, what is the minimum allowable clock period?

1.21 A D flip-flop has a propagation delay from clock to Q of 7 ns. The setup time of the flip-flop is 10 ns and the hold time is 5 ns. A clock with a period of 50 ns (low until 25 ns, high from 25 to 50 ns, and so on) is fed to the clock input of the flip-flop. Assume a two-level AND-OR circuitry between the external input signals and the flip-flop inputs. Assume gate delays are between 2 and 4 ns. The flip-flop is positive edge triggered.

**(a)** Assume the D input equals 0 from $t = 0$ until $t = 10$ ns, 1 from 10 until 35, 0 from 35 to 70, and 1 thereafter. Draw timing diagrams illustrating the clock, D, and Q until 100 ns. If outputs cannot be determined (because of not satisfying setup and hold times), indicate this by XX in the region.

**(b)** The D input of the flip-flop should not change between __ ns before the clock edge and __ ns after the clock edge.

**(c)** External inputs should not change between __ ns before the clock edge and __ ns after the clock edge.

1.22 A sequential circuit consists of a PLA and a D flip-flop, as shown.

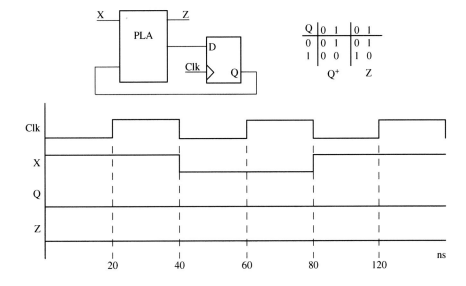

(a) Complete the timing diagram, assuming that the propagation delay for the PLA is in the range 5 to 10 ns, and the propagation delay from clock to output of the D flip-flop is 5 to 10 ns. Use cross-hatching on your timing diagram to indicate the intervals in which $Q$ and $Z$ can change, taking the range of propagation delays into account.

(b) Assuming that $X$ always changes at the same time as the falling edge of the clock, what is the maximum setup and hold time specification that the flip-flop can have and still maintain proper operation of the circuit?

1.23 A D flip-flop has a propagation delay from clock to $Q$ of 15 ns. The setup time of the flip-flop is 10 ns and the hold time is 2 ns. A clock with a period of 50 ns (low until 25 ns, high from 25 to 50 ns, and so on) is fed to the clock input of the flip-flop. The flip-flop is positive edge triggered. $D$ goes up at 20, down at 40, up at 60, down at 80, and so on. Draw timing diagrams illustrating the clock, $D$, and $Q$ until 100 ns. If outputs cannot be determined (because of not satisfying setup and hold times), indicate it by placing XX in that region.

1.24 A D flip-flop has a setup time of 5 ns, a hold time of 3 ns, and a propagation delay from the rising edge of the clock to the change in flip-flop output in the range of 6 to 12 ns. An OR gate delay is in the range of 1 to 4 ns.

(a) What is the minimum clock period for proper operation of the following circuit?

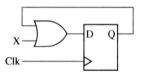

(b) What is the earliest time after the rising clock edge that $X$ is allowed to change?
(c) Show how you can construct a T flip-flop using a J-K flip-flop using a block diagram. Circuits inside the flip-flops are NOT to be shown.

1.25 In the following circuit, the XOR gate has a delay in the range of 2 to 16 ns. The D flip-flop has a propagation delay from clock to $Q$ in the range 12 to 24 ns. The setup time is 8 ns, and the hold time is 4 ns.

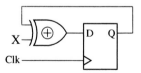

(a) What is the minimum clock period for proper operation of the circuit?
(b) What are the earliest and latest times after the rising clock edge that $X$ is allowed to change and still have proper synchronous operation? (Assume minimum clock period from (a).)

**1.26** A Mealy sequential machine has the following state table:

| PS | NS X = 0 | X = 1 | Z X = 0 | X = 1 |
|----|----------|-------|---------|-------|
| 1  | 2        | 3     | 0       | 1     |
| 2  | 3        | 1     | 1       | 0     |
| 3  | 2        | 2     | 1       | 0     |

Complete the following timing diagram. Clearly mark on the diagram the times at which you should read the values of Z. All state changes occur after the rising edge of the clock.

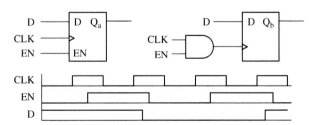

**1.27** **(a)** Do the following two circuits have essentially the same timing?
   **(b)** Draw the timing for $Q_a$ and $Q_b$ given the timing diagram.
   **(c)** If your answer to (a) is no, show what change(s) should be made in the second circuit so that the two circuits have essentially the same timing (do not change the flip-flop).

**1.28** A simple binary counter has only a clock input (*CK1*). The counter increments on the rising edge of *CK1*.

   **(a)** Show the proper connections for a signal *En* and the system clock (*CLK*), so that when *En* = 1, the counter increments on the rising edge of *CLK* and when *En* = 0, the counter does not change state.

**(b)** Complete the following timing diagram. Explain, in terms of your diagram, why the switching transients that occur on *En* after the rising edge of *CLK* do not affect the proper operation of the counter.

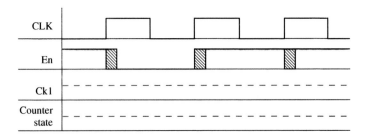

**1.29** Referring to Figure 1-49, specify the values of *Eni*, *Ena*, *Enb*, *Enc*, *Lda*, *Ldb*, and *Ldc* so that the data stored in Reg. C will be copied into Reg. A and Reg. B when the circuit is clocked.

# Introduction to VHDL

As integrated circuit technology has improved to allow more and more components on a chip, digital systems have continued to grow in complexity. While putting a few transistors on an integrated circuit (IC) was a miracle when it happened, technology improvements have advanced the **VLSI** (very large scale integration) field continually. The early integrated circuits belonged to **SSI** (small scale integration), **MSI** (medium scale integration), or **LSI** (large scale integration) categories depending on the density of integration. SSI referred to ICs with 1 to 20 gates, MSI referred to ICs with 20 to 200 gates, and LSI referred to devices with 200 to a few thousand gates. Many popular building blocks, such as adders, multiplexers, decoders, registers, and counters, are available as MSI standard parts. When the term *VLSI* was coined, devices with 10,000 gates were called VLSI chips. The boundaries between the different categories are fuzzy today. Many modern microprocessors contain more than 100 million transistors. Compared to what was referred to as VLSI in its initial days, modern integration capability could be described as ULSI (ultra large scale integration). Despite the changes in integration ability and the fuzzy definition, the term *VLSI* remains popular, while terms like *LSI* are not practically used any more.

As digital systems have become more complex, detailed design of the systems at the gate and flip-flop level has become very tedious and time-consuming. Two or three decades ago, digital systems were created using hand-drawn schematics, bread-boards, and wires that were connected to the bread-board. Now, hardware design often involves no hands-on tasks with bread-boards and wires.

In this chapter, first we present an introduction to computer-aided design. Then we present an introduction to hardware description languages. Basic features of VHDL are presented and examples are presented to illustrate how digital hardware is described, simulated, and synthesized using VHDL. Advanced features of VHDL are presented in Chapter 8.

## 2.1 Computer-Aided Design

**Computer-aided design (CAD)** tools have advanced significantly in the past decade, and nowadays, digital design is performed using a variety of software tools. Prototypes or even final designs can be created without discrete components and interconnection wires.

Figure 2-1 illustrates the steps in modern digital system design. Like any engineering design, the first step in the design flow is formulating the problem, stating the **design requirements** and arriving at the **design specification**. The next step is to **formulate the design** at a conceptual level, either at a block diagram level or at an algorithmic level.

**FIGURE 2-1: Design Flow in Modern Digital System Design**

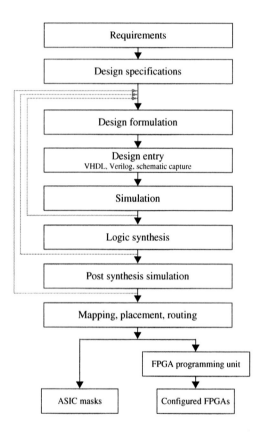

**Design entry** is the next step in the design flow. In olden days, this would have been a hand-drawn schematic or blueprint. Now with CAD tools, the design conceptualized in the previous step needs to be entered into the CAD system in an appropriate manner. Designs can be entered in multiple forms. A few years ago, CAD tools used to provide a graphical method to enter designs. This was called **schematic capture**. The schematic editors typically were supplemented with a library of standard digital building blocks like gates, flip-flops, multiplexers, decoders, counters, registers, and so on. ORCAD (a company that produced design automation tools) provided a very popular schematic editor. Nowadays, **hardware description languages** (HDLs) are used to enter designs. Two popular HDLs are VHDL and Verilog. The acronym VHDL stands for **VHSIC hardware description language**, and **VHSIC** in turn stands for **very high speed integrated circuit**.

A hardware description language allows a digital system to be designed and debugged at a higher level of abstraction than schematic capture with gates, flip-flops, and standard MSI building blocks. The details of the gates and flip-flops do not need to be handled during early phases of design. A design can be entered in what is called a **behavioral description** of the design. In a behavioral HDL description, one only specifies the general working of the design at a flow-chart or algorithmic level without associating to any specific physical parts, components, or implementations. Another method to enter a design in VHDL and Verilog is the **structural description** entry. In structural design, specific components or specific implementations of components are associated with the design. A structural VHDL or Verilog model of a design can be considered as a textual description of a schematic diagram that you would have drawn interconnecting specific gates and flip-flops.

Once the design has been entered, it is important to simulate it to confirm that the conceptualized design does function correctly. Initially, one should perform the **simulation** at the high-level behavioral model. This early simulation unveils problems in the initial design. If problems are discovered, the designer goes back and alters the design to meet the requirements.

Once the functionality of the design has been verified through simulation, the next step is **synthesis**. *Synthesis* means "conversion of the higher-level abstract description of the design to actual components at the gate and flip-flop level." Use of computer-aided design tools to do this conversion (a.k.a. synthesis) is becoming widespread. The output of the synthesis tool, consisting of a list of gates and a list of interconnections specifying how to interconnect them, is often referred to as a **netlist.** Synthesis is analogous to writing software programs in a high-level language such as C and then using a compiler to convert the programs to machine language. Just like a C compiler can generate optimized or unoptimized machine code, a synthesis tool can generate optimized or unoptimized hardware. The synthesis software generates different hardware implementations depending on algorithms embedded in the software to perform the translation and optimization techniques incorporated into the tool. A synthesis tool is nothing but a compiler to convert design descriptions to hardware, and it is not unusual to name synthesis packages with phrases similar to design compiler, silicon compiler, and so on.

The next step in the design flow is **post-synthesis simulation**. The earlier simulation at a higher level of abstraction does not take into account specific implementations of the hardware components that the design is using. If post-synthesis simulation unveils problems, one should go back and modify the design to meet timing requirements. Arriving at a proper design implementation is an iterative process.

Next, a designer moves into specific realizations of the design. A design can be implemented in several different target technologies. It could be a completely custom IC or it could be implemented in a standard part that is easily available from a vendor. The target technologies that are commonly available now are illustrated in Figure 2-2.

At the lowest level of sophistication and density is an old-fashioned printed circuit board with off-the-shelf gates, flip-flops, and other standard logic building blocks. Slightly higher in density are programmable logic arrays (PLAs), programmable array logic (PAL), and simple programmable logic devices (SPLDs). PLDs

FIGURE 2-2:
**Spectrum of Design
Technologies**

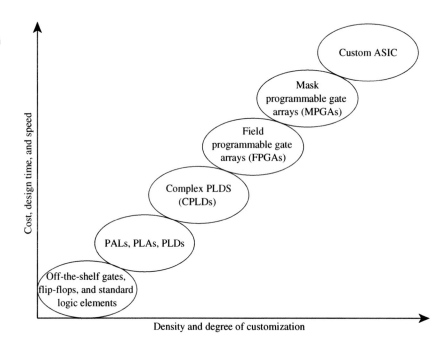

with higher density and gate count are called complex programmable logic devices (CPLDs). Then there are the popular field programmable gate arrays (FPGAs) and mask programmable gate arrays (MPGAs), or simply gate arrays. The highest level of density and performance is a fully custom application-specific integrated circuit (ASIC).

Two most common target technologies nowadays are FPGAs and ASICs. The initial steps in the design flow are largely the same for either realization. Toward the final stages in the design flow, different operations are performed depending on the target technology. This is indicated in Figure 2-1. The design is **mapped** into specific target technology and **placed** into specific parts in the target ASIC or FPGA. The paths taken by the connections between components are decided during the **routing**. If an ASIC is being designed, the routed design is used to generate a photomask that will be used in the IC manufacturing process. If a design is to be implemented in an FPGA, the design is translated to a format specifying what is to be done to various programmable points in the FPGA. In modern FPGAs, programming simply involves writing a sequence of 0's and 1's into the programmable cells in the FPGA, and no specific programming unit other than a personal computer (PC) is required.

● ● ● ● ● ● ● ● ● ● ●
## 2.2 Hardware Description Languages

**Hardware description languages** (HDLs) are a popular mode of design entry. As mentioned previously, two popular HDLs are VHDL and Verilog. This book uses VHDL for illustrating principles of modern digital system design.

VHDL is a hardware description language used to describe the behavior and structure of digital systems. VHDL is a general-purpose HDL that can be used to describe and simulate the operation of a wide variety of digital systems, ranging in complexity from a few gates to an interconnection of many complex integrated circuits. VHDL was originally developed under funding from the Department of Defense (DoD) to allow a uniform method for specifying digital systems. When VHDL was developed, the main purpose was to have a mechanism to describe and document hardware unambiguously. Synthesizing hardware from high-level descriptions was not one of the original purposes. The VHDL language has since become an IEEE (Institute of Electronic and Electrical Engineers) standard, and it is widely used in industry. IEEE created a VHDL standard in 1987 (VHDL-87) and later modified the standard in 1993 (VHDL-93). Further revisions were done to the standard in 2000 and 2002.

VHDL can describe a digital system at several different levels—**behavioral, data flow**, and **structural**. For example, a binary adder could be described at the behavioral level in terms of its function of adding two binary numbers without giving any implementation details. The same adder could be described at the data flow level by giving the logic equations for the adder. Finally, the adder could be described at the structural level by specifying the gates and the interconnections between the gates that comprise the adder.

VHDL leads naturally to a top-down design methodology, in which the system is first specified at a high level and tested using a simulator. After the system is debugged at this level, the design can gradually be refined, eventually leading to a structural description closely related to the actual hardware implementation. VHDL was designed to be technology independent. If a design is described in VHDL and implemented in today's technology, the same VHDL description could be used as a starting point for a design in some future technology. Although initially conceived as a hardware documentation language, most of VHDL can now be used for simulation and logic synthesis.

Verilog is another popular HDL. It was developed by the industry at about the same time the U.S. DoD was funding the creation of VHDL. Verilog was introduced by Gateway Design Automation in 1984 as a proprietary HDL. Synopsis created synthesis tools for Verilog around 1988. Verilog became an IEEE standard in 1995.

VHDL has its syntactic roots in ADA while Verilog has its syntactic roots in C. ADA was a general-purpose programming language, also sponsored by the Department of Defense. Due to the similarity with C, some find Verilog easier or less intimidating to learn. Many find VHDL to be excellent for supporting design and documentation of large systems. VHDL and Verilog enjoy approximately 50/50 market share. Both languages can accomplish most requirements for digital design rather easily. Often design companies continue to use what they are used to, and hence, Verilog users continue to use Verilog and VHDL users continue to use VHDL. If you know one of these languages, it is not difficult to transition to the other.

More recently, there also have been efforts in system design languages such as **System C, Handel-C**, and **System Verilog**. System C is created as an extension to C++, and hence some who are very comfortable with general-purpose software development find it less intimidating. These languages are primarily targeted at describing large digital systems at a high level of abstraction. They are primarily used for verification and validation. When different parts of a large system are designed by different teams, one team can use a system level behavioral description of the block being designed by the other team during initial design. Problems that might otherwise become obvious only during system integration may become evident in early stages reducing the design cycle for large systems. System-level simulation languages are used during design of large systems.

### 2.2.1 Learning a Language

There are several challenges when you learn a new language, whether it be a language for common communication (English, Spanish, French, etc.), a computer language like C, or a special-purpose language such as VHDL. If it is not your first language, you typically have a tendency to compare it to a language you know. In the case of VHDL, if you already know another hardware description language, it is good to compare it with VHDL, but you should be careful when comparing it with languages like C. VHDL and Verilog have a very different purpose than languages like C, and a comparison with C is not a meaningful activity. We will be describing the language assuming it is your first HDL; however, we will assume basic knowledge of computer languages like C and the basic compilation and execution flow.

When one learns a new language, one needs to study the alphabet of the new language, its vocabulary, grammar, syntax rules, and semantics of language descriptions. The process of learning VHDL is not much different. One needs to learn the alphabet, vocabulary or lexical elements of the language, syntax (grammar and rules), and semantics (meaning of descriptions). VHDL-87 uses the ASCII character set while VHDL-93 allows use of the full ISO character set. The ISO character set includes the ASCII characters and additionally includes accented characters. The ASCII character set only includes the first 128 characters of the ISO character set. The lexical elements of the language include various **identifiers, reserved words**, special symbols, and literals. We have listed these in Appendix A. The syntax or grammar determines what combinations of lexical elements can be combined to make valid VHDL descriptions. These are the rules that govern the use of different VHDL constructs. Then one needs to understand the semantics or meaning of VHDL descriptions. It is here that one understands what descriptions represent combinational hardware versus sequential hardware. And just like fluency in a natural language comes by speaking, reading, and writing the language, mastery of VHDL comes by repeated use of the language to create models for various digital systems.

Since VHDL is a hardware description language, it differs from an ordinary programming language in several ways. Most importantly, VHDL has statements that execute concurrently since they must model real hardware in which the components are all in operation at the same time. VHDL is popularly used for the purposes of describing, documenting, simulating, and automatically generating hardware. Hence, its constructs are tailored for these purposes. We will present the various methods to model different kinds of digital hardware using examples in the following sections.

| Common Abbreviations | |
|---|---|
| VHDL: | VHSIC hardware description language |
| VHSIC: | Very high speed integrated circuit |
| HDL: | Hardware description language |
| CAD: | Computer-aided design |
| EDA: | Electronic design automation |
| LSI: | Large scale integration |
| MSI: | Medium scale integration |
| SSI: | Small scale integration |
| VLSI: | Very large scale integration |
| ULSI: | Ultra large scale integration |
| ASCII: | American standard code for information interchange |
| ISO: | International Standards Organization |
| ASIC: | Application-specific integrated circuit |
| FPGA: | Field programmable gate array |
| PLA: | Programmable logic array |
| PAL: | Programmable array logic |
| PLD: | Programmable logic device |
| CPLD: | Complex programmable logic device |

## 2.3 VHDL Description of Combinational Circuits

The biggest difficulty in modeling hardware using a general-purpose computer language is representing concurrently operating hardware. Computer programs that you are normally accustomed to are sequences of instructions with a well-defined order. At any point of time during execution, the program is at a specific point in its flow and it encounters and executes different parts of the program sequentially. In order to model combinational circuits, which have several gates (all of which are working simultaneously), one needs to be able to "simulate" the execution of several parts of the circuit at the same time.

VHDL models combinational circuits by what are called **concurrent statements**. Concurrent statements are statements which are always ready to execute. These are statements which get evaluated any time and every time a signal on the right side of the statement changes.

We will start by describing a simple gate circuit in VHDL. If each gate in the circuit of Figure 2-3 has a 5-ns propagation delay, the circuit can be described by two VHDL statements as shown, where A, B, C, D, and E are signals. A signal in VHDL usually corresponds to a signal in a physical system. The symbol "<=" is the signal

FIGURE 2-3:
A Simple Gate
Circuit

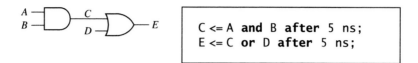

```
C <= A and B after 5 ns;
E <= C or D after 5 ns;
```

assignment operator, which indicates that the value computed on the right side is assigned to the signal on the left side. When the statements in Figure 2-3 are simulated, the first statement will be evaluated anytime $A$ or $B$ changes, and the second statement will be evaluated anytime $C$ or $D$ changes. Suppose that initially $A = 1$ and $B = C = D = E = 0$. If $B$ changes to 1 at time 0, $C$ will change to 1 at time $= 5$ ns. Then $E$ will change to 1 at time $= 10$ ns.

VHDL signal assignment statements, like the ones in the preceding example, are examples of concurrent statements. The VHDL simulator monitors the right side of each concurrent statement, and anytime a signal changes, the expression on the right side is immediately re-evaluated. The new value is assigned to the signal on the left side after an appropriate delay. This is exactly the way the hardware works. Anytime a gate input changes, the gate output is recomputed by the hardware, and the output changes after the gate delay. The location of the concurrent statement in the program is not important.

When we initially describe a circuit, we may not be concerned about propagation delays. If we write

```
C <= A and B;
E <= C or D;
```

this implies that the propagation delays are 0 ns. In this case, the simulator will assume an infinitesimal delay referred to as $\Delta$ (delta). Assume that initially $A = 1$ and $B = C = D = E = 0$. If $B$ is changed to 1 at time $= 1$ ns, then $C$ will change at time $1 + \Delta$ and $E$ will change at time $1 + 2\Delta$.

Unlike a sequential program, the order of the preceding concurrent statements is unimportant. If we write

```
E <= C or D;
C <= A and B;
```

the simulation results would be exactly the same as before.

In general, a signal assignment statement has the form

```
signal_name <= expression [after delay];
```

The expression is evaluated when the statement is executed, and the signal on the left side is scheduled to change after **delay**. The square brackets indicate that **after delay** is optional; they are not part of the statement. If **after delay** is omitted, then the signal is scheduled to be updated after a delta delay. Note that the time at which the statement executes and the time at which the signal is updated are not the same.

Even if a VHDL program has no explicit loops, concurrent statements may execute repeatedly as if they were in a loop. Figure 2-4 shows an inverter with the output connected back to the input. If the output is '0', then this '0' feeds back to the input and the inverter output changes to '1' after the inverter delay, assumed

FIGURE 2-4:
Inverter with
Feedback

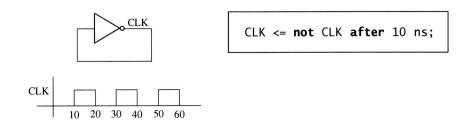

to be 10 ns. Then the '1' feeds back to the input and the output changes to '0' after the inverter delay. The signal *CLK* will continue to oscillate between '0' and '1' as shown in the waveform. The corresponding concurrent VHDL statement will produce the same result. If *CLK* is initialized to '0', the statement executes and *CLK* changes to '1' after 10 ns. Since *CLK* has changed, the statement executes again, and *CLK* will change back to '0' after another 10 ns. This process will continue indefinitely.

The statement in Figure 2-4 generates a clock waveform with a half period of 10 ns. On the other hand, the concurrent statement

```
CLK <= not CLK;
```

will cause a run-time error during simulation. Since there is 0 delay, the value of CLK will change at times $0 + \Delta$, $0 + 2\Delta$, $0 + 3\Delta$, and so on. Since $\Delta$ is an infinitesimal time, time will never advance to 1 ns.

In general, **VHDL is not case sensitive**; that is, uppercase and lowercase letters are treated the same by the compiler and by the simulator. Thus, the statements

```
Clk <= NOT clk After 10 ns;
```

and

```
CLK <= not CLK after 10 ns;
```

would be treated exactly the same. Signal names and other **VHDL identifiers** may contain letters, numbers, and the underscore character (_). An identifier must start with a letter, and it cannot end with an underscore. Thus C123 and ab_23 are legal identifiers, but 1ABC and ABC_ are not. Every VHDL statement must be terminated with a semicolon. Spaces, tabs, and carriage returns are treated in the same way. This means that a VHDL statement can be continued over several lines, or several statements can be placed on one line. In a line of VHDL code, anything following a double dash (--) is treated as a comment. Words such as **and, or**, and **after** are reserved words (or keywords) which have a special meaning to the VHDL compiler. In this text, we will put all reserved words in boldface type.

Figure 2-5 shows three gates that have the signal *A* as a common input and the corresponding VHDL code. The three concurrent statements execute simultaneously whenever *A* changes, just as the three gates start processing the signal change at the same time. However, if the gates have different delays, the gate outputs can change at different times. If the gates have delays of 2 ns, 1 ns, and 3 ns, respectively, and *A* changes at time 5 ns, then the gate outputs *D, E,* and *F* can change at times 7 ns, 6 ns,

FIGURE 2-5: **Three Gates with a Common Input and Different Delays**

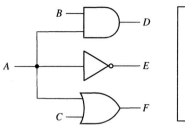

```
-- when A changes, these concurrent
-- statements all execute at the
-- same time
   D <= A and B after 2 ns;
   E <= not A after 1 ns;
   F <= A or C after 3 ns;
```

and 8 ns, respectively. The VHDL statements work in the same way. Even though the statements execute simultaneously, the signals *D*, *E*, and *F* are updated at times 7 ns, 6 ns, and 8 ns. However, if no delays were specified, then *D*, *E*, and *F* would all be updated at time $5 + \Delta$.

In the preceding examples, every signal is of type bit, which means it can have a value of '0' or '1'. (Bit values in VHDL are enclosed in single quotes to distinguish them from integer values.)

In digital design, we often need to perform the same operation on a group of signals. A one-dimensional array of bit signals is referred to as a **bit-vector**. If a 4-bit vector named *B* has an index range 0 through 3, then the four elements of the bit-vector are designated $B(0)$, $B(1)$, $B(2)$, and $B(3)$. One can declare a bit-vector using a statement such as:

```
B:  in bit_vector(3 downto 0);
```

The statement B <= "1100" assigns '1' to $B(3)$, '1' to $B(2)$, '0' to $B(1)$, and '0' to $B(0)$.

Figure 2-6 shows an array of four AND gates. The inputs are represented by bit-vectors *A* and *B*, and the output by bit-vector *C*. Although we can write four VHDL statements to represent the four gates, it is much more efficient to write a single VHDL statement that performs the **and** operation on the bit-vectors *A* and *B*. When applied to bit-vectors, the **and** operator performs the **and** operation on corresponding pairs of elements.

FIGURE 2-6: **Array of AND Gates**

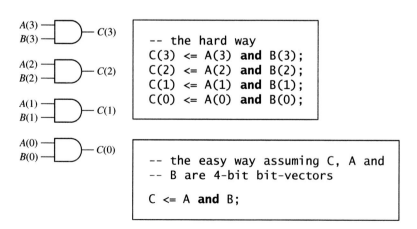

```
-- the hard way
C(3) <= A(3) and B(3);
C(2) <= A(2) and B(2);
C(1) <= A(1) and B(1);
C(0) <= A(0) and B(0);
```

```
-- the easy way assuming C, A and
-- B are 4-bit bit-vectors

C <= A and B;
```

• • • • • • • • • • •
# 2.4 VHDL Modules

The general structure of a VHDL module is an **entity** description and an **architecture** description. The **entity** description declares the input and output signals, and the **architecture** description specifies the internal operation of the module. As an example, consider Figure 2-7. The **entity** declaration gives the name two_gates to the module. The **port** declaration specifies the inputs and outputs to the module. *A, B,* and *D* are input signals of type bit, and *E* is an output signal of type bit. The architecture is named gates. The signal *C* is declared within the architecture since it is an internal signal. The two concurrent statements that describe the gates are placed between the keywords **begin** and **end**.

**FIGURE 2-7: VHDL Module with Two Gates**

```
entity two_gates is
   port(A, B, D: in bit; E: out bit);
end two_gates;

architecture gates of two_gates is
signal C: bit;
begin
   C <= A and B;  -- concurrent
   E <= C or D;   -- statements
end gates;
```

The **entity** description can be considered as the black box picture of the module being designed and its external interface (i.e., it represents the interconnections from this module to the external world, as in Figure 2-8).

**FIGURE 2-8: Black Box View of the Two-Gate Module**

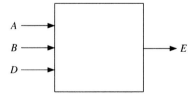

Just as in the preceding simple example, when we describe a system in VHDL, we must specify an entity and architecture at the top level and also specify an entity and architecture for each of the component modules that are part of the system (see Figure 2-9). Each entity declaration includes a list of interface signals that can be used to connect to other modules or to the outside world. We will use entity declarations of the form

```
entity entity-name is
   [port(interface-signal-declaration);]
end [entity] [entity-name];
```

The items enclosed in square brackets are optional. The `interface-signal-declaration` normally has the following form:

```
list-of-interface-signals: mode type [:= initial-value]
{; list-of-interface-signals: mode type [:= initial-value]};
```

The curly brackets indicate zero or more repetitions of the enclosed clause. **Mode** indicates the direction of information; whether information is flowing into the port or out of it. Input port signals are of mode **in**, output port signals are of mode **out**, and bidirectional signals are of mode **inout**. **Type** specifies the data type or kind of information that can be communicated. So far, we have only used type bit and bit-vector; other types are described in Section 2.10. The optional `initial-value` is used to initialize the signals on the associated list; otherwise, the default initial value is used for the specified type. For example, the port declaration

```
port(A, B: in integer := 2; C, D: out bit);
```

indicates that A and B are input signals of type integer that are initially set to 2, and C and D are output signals of type bit that are initialized by default to '0'. These initial values are significant only for simulation and not for synthesis.

In addition to **in, out** and **inout** modes, there are two other modes: **buffer** and **linkage**. The **buffer** mode is similar to **inout** mode, in that it can be read and written into in the entity. The **buffer** mode is useful if a signal is truly an output, but we would like to read the ports internally as well. A linkage port is useful when VHDL entities are connected to non-VHDL entities. Both of these modes involve several restrictions and we generally restrict ourselves to **in, out** and **inout** modes.

**FIGURE 2-9: VHDL Program Structure**

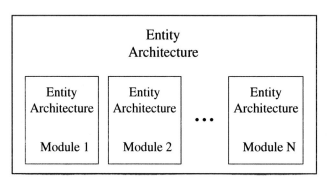

Associated with each **entity** is one or more **architecture** declarations of the form

```
architecture architecture-name of entity-name is
  [declarations]
begin
  architecture body
end [architecture] [architecture-name];
```

In the `declarations` section, we can declare signals and components that are used within the architecture. The architecture body contains statements that describe the operation of the module.

Next, we will write the entity and architecture for a full adder module. A full adder adds 2 bits and a carry input to generate a sum bit and a carry output bit. The entity specifies the inputs and outputs of the adder module as shown in Figure 2-10. The port declaration specifies that $X$, $Y$, and $C_{in}$ are input signals of type bit, and that $C_{out}$ and *Sum* are output signals of type bit.

**FIGURE 2-10: Entity Declaration for a Full Adder Module**

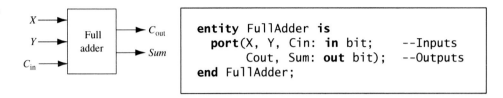

```
entity FullAdder is
    port(X, Y, Cin: in bit;       --Inputs
         Cout, Sum: out bit);     --Outputs
end FullAdder;
```

The operation of the full adder is specified by an architecture declaration:

```
architecture Equations of FullAdder is
begin          -- concurrent assignment statements
   Sum  <= X xor Y xor Cin after 10 ns;
   Cout <= (X and Y) or (X and Cin) or (Y and Cin) after 10 ns;
end Equations;
```

In this example, the architecture name **(Equations)** is arbitrary, but the entity name **(FullAdder)** must match the name used in the associated entity declaration. The VHDL assignment statements for *Sum* and $C_{out}$ represent the logic equations for the full adder. Several other architectural descriptions, such as a truth table or an interconnection of gates, could have been used instead. In the $C_{out}$ equation, parentheses are required around **(X and Y)** since VHDL does not specify an order of precedence for the logic operators except the NOT operator.

### 2.4.1 Four-Bit Full Adder

Next, we will show how to use the **FullAdder** module defined above as a **component** in a system, which consists of four full adders connected to form a 4-bit binary adder (see Figure 2-11). We first declare the 4-bit adder as an entity (see Figure 2-12). Since the inputs and the sum output are 4 bits wide, we declare them as bit-vectors which are dimensioned **3 downto** 0. (We could have used a range **1 to** 4 instead).

**FIGURE 2-11: Four-Bit Binary Adder**

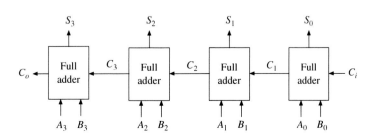

Next, we specify the `FullAdder` as a component within the architecture of `Adder4` (Figure 2-12). The **component** specification is very similar to the **entity** declaration for the full adder, and the input and output port signals correspond to those declared for the full adder. Anytime a module created in one part of the code has to be used in another part, a component declaration needs to be used. The component declaration does not need to be in the same file where you are using the component. It can be where the component entity and architecture are defined. It is typical to create libraries of components for reuse in code, and typically the component declarations are placed in the library file.

Following the component statement, we declare a 3-bit internal carry signal $C$. In the body of the architecture, we create several instances of the `FullAdder` component. (In CAD jargon, we "instantiate" four copies of the `FullAdder`.) Each copy of `FullAdder` has a name (such as FA0) and a port map. The signal names following the port map correspond one-to-one with the signals in the component port. Thus, $A(0)$, $B(0)$, and $C_i$ correspond to the inputs $X$, $Y$, and $C_{in}$, respectively. $C(1)$ and $S(0)$ correspond to the $C_{out}$ and $Sum$ outputs. Note that the order of the signals in the port map must be the same as the order of the signals in the port of the component declaration.

FIGURE 2-12: **Structural Description of a 4-Bit Adder**

```
entity Adder4 is
  port(A, B: in bit_vector(3 downto 0); Ci: in bit; -- Inputs
       S: out bit_vector(3 downto 0); Co: out bit); -- Outputs
end Adder4;
architecture Structure of Adder4 is
component FullAdder
  port (X, Y, Cin: in bit;        -- Inputs
        Cout, Sum: out bit);      -- Outputs
end component;
signal C: bit_vector(3 downto 1); -- C is an internal signal
begin      --instantiate four copies of the FullAdder
  FA0: FullAdder port map (A(0), B(0), Ci, C(1), S(0));
  FA1: FullAdder port map (A(1), B(1), C(1), C(2), S(1));
  FA2: FullAdder port map (A(2), B(2), C(2), C(3), S(2));
  FA3: FullAdder port map (A(3), B(3), C(3), Co, S(3));
end Structure;
```

In preparation for simulation, we can place the entity and architecture for the `FullAdder` and for `Adder4` together in one file and compile. Alternatively, we could compile the `FullAdder` separately and place the resulting code in a library which is linked in when we compile `Adder4`.

All of the simulation examples in this text use the **ModelSim VHDL simulator** from Mentor Graphics. Most other VHDL simulators use similar command files and can produce output in a similar format. We will use the following simulator commands to test `Adder4`:

```
add list A B Co C Ci S   -- put these signals on the output list
force A 1111             -- set the A inputs to 1111
force B 0001             -- set the B inputs to 0001
force Ci 1               -- set Ci to 1
run 50 ns                -- run the simulation for 50 ns
force Ci 0
force A 0101
force B 1110
run 50 ns
```

We have chosen to run the simulation for 50 ns since this is more than enough time for the carry to propagate through all of the full adders. The simulation results for the preceding command list are as follows:

| ns | delta | a | b | co | c | ci | s |
|----|-------|------|------|----|-----|----|------|
| 0 | +0 | 0000 | 0000 | 0 | 000 | 0 | 0000 |
| 0 | +1 | 1111 | 0001 | 0 | 000 | 1 | 0000 |
| 10 | +0 | 1111 | 0001 | 0 | 001 | 1 | 1111 |
| 20 | +0 | 1111 | 0001 | 0 | 011 | 1 | 1101 |
| 30 | +0 | 1111 | 0001 | 0 | 111 | 1 | 1001 |
| 40 | +0 | 1111 | 0001 | 1 | 111 | 1 | 0001 |
| 50 | +0 | 0101 | 1110 | 1 | 111 | 0 | 0001 |
| 60 | +0 | 0101 | 1110 | 1 | 110 | 0 | 0101 |
| 70 | +0 | 0101 | 1110 | 1 | 100 | 0 | 0111 |
| 80 | +0 | 0101 | 1110 | 1 | 100 | 0 | 0011 |

The listing shows how the carry propagates one position every 10 ns. The full adder inputs change at time = $\Delta$:

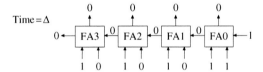

The sum and carry are computed by each FA and appear at the FA outputs 10 ns later:

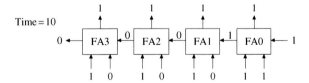

Since the inputs to FA1 have changed, the outputs change 10 ns later:

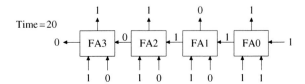

The final simulation results are

$$1111 + 0001 + 1 = 0001 \text{ with a carry of 1 (at time} = 40 \text{ ns) and}$$
$$0101 + 1110 + 0 = 0011 \text{ with a carry of 1 (at time} = 80 \text{ ns)}$$

The simulation stops at 80 ns since no further changes occur after that time.

In this section we have shown how to construct a VHDL module using an entity-architecture pair. The 4-bit adder module demonstrates the use of VHDL components to write structural VHDL code. Components used within the architecture are declared at the start of the architecture using a component declaration of the form

```
component component-name
   port(list-of-interface-signals-and-their-types);
end component;
```

The port clause used in the component declaration has the same form as the port clause used in an entity declaration. The connections to each component used in a circuit are specified using a component instantiation statement of the form

```
label: component-name port map (list-of-actual-signals);
```

The list of actual signals must correspond one-to-one to the list of interface signals specified in the component declaration.

### 2.4.2 Use of "Buffer" Mode

Let us consider the example in Figure 2-13. Assume that all variables are 0 at 0 ns, but $A$ changes to 1 at 10 ns.

FIGURE 2-13: **VHDL Code Which Will Not Compile**

```
entity gates is
  port(A, B, C: in bit; D, E: out bit);
end gates;

architecture example of gates is
begin
  D <= A or B after 5 ns;   -- statement 1
  E <= C or D after 5 ns;   -- statement 2
end example;
```

The code in Figure 2-13 will not actually compile, simulate, or synthesize in most tools because $D$ is declared only as an output. Statement 2 uses $D$ on the right side of the assignment. Hence, $D$ should be either **inout** or **buffer** mode as in Figure 2-14. Use of inout mode results in the synthesis tools creating a truly bidirectional signal. In actuality, $D$ is not an external input to the circuit, and hence the mode **buffer** is more appropriate. The mode **buffer** indicates a signal that is an output to the external world; however, its value can also be read inside the entity's architecture. The following code uses buffer mode for signal $D$ instead of out mode.

FIGURE 2-14: **VHDL Code Illustrating Use of Mode Buffer**

```
entity gates is
  port(A, B, C: in bit; D: buffer bit; E: out bit);
end gates;

architecture example of gates is
begin
  D <= A or B after 5 ns;   -- statement 1
  E <= C or D after 5 ns;   -- statement 2
end example;
```

All signals remain at '0' until time 10 ns. The change in $A$ at 10 ns results in statement 1 reevaluating. The value of $D$ becomes '1' at time equal to 15 ns. The change in $D$ at time 15 ns results in statement 2 reevaluating. Signal $E$ changes to '1' at time 20 ns. The description represents two gates, each with a delay of 5 ns.

● ● ● ● ● ● ● ● ● ● ● ●
## 2.5 Sequential Statements and VHDL Processes

The concurrent statements from the previous section are useful in modeling combinational logic. Combinational logic constantly reacts to input changes. In contrast, synchronous sequential logic responds to changes dependent on the clock. Many input changes might be ignored since output and state changes occur only at valid conditions of the clock. Modeling sequential logic requires primitives to model selective activity conditional on clock, edge-triggered devices, sequence of operations, and so on. In this unit, we will learn VHDL processes which help to model sequential logic.

A VHDL process has the following basic form:

```
process(sensitivity-list)
begin
   sequential-statements
end process;
```

When a process is used, the statements between the **begin** and the **end** are executed sequentially. The expression in parentheses after the word **process** is called a sensitivity list, and the process executes whenever any signal in the sensitivity list changes. For example, if the process begins with **process**(A, B, C), then the process executes whenever any one of $A$, $B$, or $C$ changes. Whenever one of the signals in the sensitivity list changes, the sequential statements in the process body are executed in sequence one time. When a process finishes executing, it goes back to the beginning and waits for a signal on the sensitivity list to change again.

When the concurrent statements

```
C <= A and B;   -- concurrent
E <= C or D;    -- statements
```

are used in a process, they become sequential statements executed in the order in which they appear in the process. Remember that when they were concurrent statements outside a process, their sequence did not matter. But, if they are in a process, the sequence determines the order of execution.

```
process(A, B, C, D)
begin
  C <= A and B;   -- sequential
  E <= C or D;    -- statements
end process;
```

The process executes once when any of the signals $A$, $B$, $C$, or $D$ changes. If $C$ changes when the process executes, then the process will execute a second time because $C$ is on the sensitivity list.

VHDL processes can be used for modeling combinational logic and sequential logic; however, processes are not necessary for modeling combinational logic. They are, however, required for modeling sequential logic. One should be very careful when using processes to represent combinational logic. Consider the code in Figure 2-15, where a process is used. One may write this code thinking of two cascaded gates; however, it does not actually represent such a circuit.

FIGURE 2-15: **VHDL Code with a Process**

```
entity nogates is
  port(A, B, C: in bit;
       D: buffer bit;
       E: out bit);
end nogates;

architecture behave of nogates is
begin
  process(A, B, C)
  begin
    D <= A or B after 5 ns;   -- statement 1
    E <= C or D after 5 ns;   -- statement 2
  end process;
end behave;
```

The sensitivity list of the process only includes $A$, $B$, and $C$, the only external inputs to the circuit. Let us assume that all variables are '0' at 0 ns. Then $A$ changes to '1' at 10 ns. That causes the process to execute. Both statements inside the process execute once sequentially, but the change in $D$ does not happen right at execution. Hence, execution of statement 2 is with the value of $D$ at the beginning of the process. $D$ becomes '1' at 15 ns, but $E$ stays at '0'. Since the change in $D$ does not propagate to signal $E$, this VHDL model is not equivalent to two gates. If $D$ was included in the sensitivity list of the process, the process would execute again making $E$ change at 20 ns. This would result in simulation outputs matching a circuitry with cascaded gates, but it is preferable to realize gates using concurrent statements.

Understanding sequential statements and operation of processes will take several more examples. In the next section, we explain how simple flip-flops can be modeled using processes, and then we explain the basics of the VHDL simulation process. After that, we present more examples illustrating the working of processes and the simulation process.

● ● ● ● ● ● ● ● ● ● ● ● ●

## 2.6 Modeling Flip-Flops Using VHDL Processes

A flip-flop can change state either on the rising or on the falling edge of the clock input. This type of behavior is modeled in VHDL by a process. For a simple D flip-flop with a $Q$ output that changes on the rising edge of CLK, the corresponding process is given in Figure 2-16.

FIGURE 2-16: **VHDL Code for a Simple D Flip-Flop**

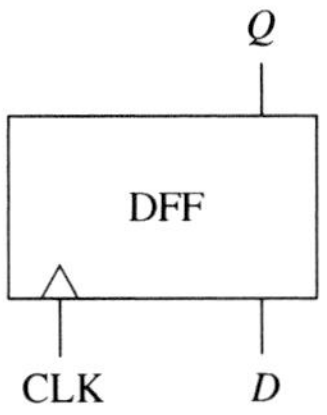

```
process(CLK)
begin
  if CLK'event and CLK = '1'  -- rising edge of CLK
    then Q <= D;
  end if;
end process;
```

In Figure 2-16, whenever $CLK$ changes, the process executes once through and then waits at the start of the process until $CLK$ changes again. The **if** statement tests for a rising edge of the clock, and $Q$ is set equal to $D$ when a rising edge occurs. The expression CLK'event is used to accomplish the functionality of an edge-triggered device. The expression 'event is a predefined attribute for any signal. There are two types of signal attributes in VHDL, those that return values and those that return signals. The 'event attribute returns a value. The expression CLK'event (read as "clock tick event") is TRUE whenever the signal $CLK$ changes. If CLK = '1' is also TRUE, this means that the change was from '0' to '1', which is a rising edge.

If VHDL is used only for simulation purposes, one might use a statement such as

```
if CLK = '1'
      ...
```

and obtain action corresponding to rising edge. However, when VHDL code is used to synthesize hardware, this statement will result in latches, whereas the expression CLK'event results in edge-triggered devices.

If the flip-flop has a delay of 5 ns between the rising edge of the clock and the change in the $Q$ output, we would replace the statement Q <= D; with Q <= D after 5 ns; in the preceding process.

The statements between **begin** and **end** in a process operate as sequential statements. In the preceding process, Q <= D; is a sequential statement that only executes following the rising edge of *CLK*. In contrast, the concurrent statement Q <= D; executes whenever *D* changes. If we synthesize the above process, the synthesizer infers that *Q* must be a flip-flop since it only changes on the rising edge of *CLK*. If we synthesize the concurrent statement Q <= D;, the synthesizer will simply connect *D* to *Q* with a wire or a buffer.

In Figure 2-16, note that *D* is not on the sensitivity list because changing *D* will not cause the flip-flop to change state. Figure 2-17 shows a transparent latch and its VHDL representation. Both *G* and *D* are on the sensitivity list since if *G* = '1', a change in *D* causes *Q* to change. If *G* changes to '0', the process executes, but *Q* does not change.

**FIGURE 2-17:**
**VHDL Code for a**
**Transparent Latch**

```
process(G, D)
begin
   if G = '1' then Q <= D; end if;
end process;
```

If a flip-flop has an active-low asynchronous clear input (*ClrN*) that resets the flip-flop independently of the clock, then we must modify the process of Figure 2-16 so that it executes when either *CLK* or *ClrN* changes. To do this, we add *ClrN* to the sensitivity list. The VHDL code for a D flip-flop with asynchronous clear is given in Figure 2-18. Since the asynchronous *ClrN* signal overrides *CLK*, *ClrN* is tested first and the flip-flop is cleared if *ClrN* is '0'. Otherwise, *CLK* is tested, and *Q* is updated if a rising edge has occurred.

**FIGURE 2-18:**
**VHDL Code for a**
**D Flip-Flop with**
**Asynchronous**
**Clear**

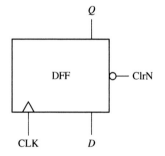

```
process(CLK, ClrN)
begin
   if CLRn = '0' then Q <= '0';
   else if CLK'event and CLK = '1'
      then Q <= D;
   end if;
   end if;
end process;
```

In the preceding examples, we have used two types of sequential statements—signal assignment statements and **if** statements. The basic **if** statement has the form

```
if condition then
   sequential statements1
else sequential statements2
end if;
```

The condition is a Boolean expression which evaluates to TRUE or FALSE. If it is TRUE, `sequential statements1` are executed; otherwise, `sequential statements2` are executed.

VHDL **if** statements are sequential statements that can be used within a process, but they cannot be used as concurrent statements outside of a process.

The most general form of the **if** statement is

```
if condition then
   sequential statements
{elsif condition then
   sequential statements}
   -- 0 or more elsif clauses may be included
[else sequential statements]
end if;
```

The curly brackets indicate that any number of **elsif** clauses may be included, and the square brackets indicate that the **else** clause is optional. The example of Figure 2-19 shows how a flow chart can be represented using nested **if**s or the equivalent using **elsif**s. In this example, C1, C2, and C3 represent conditions that can be true or false, and S1, S2, ... , S8 represent sequential statements. Each **if** requires a corresponding **end if**, but **elsif**s do not.

FIGURE 2-19:
**Equivalent Representations of a Flow Chart Using Nested Ifs and Elsifs**

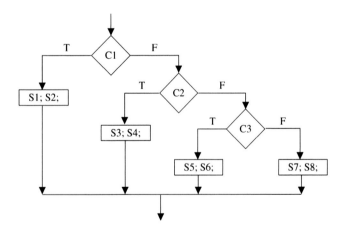

```
if (C1) then S1; S2;
   else if (C2) then S3; S4;
      else if (C3) then S5; S6;
         else S7; S8;
         end if;
      end if;
end if;
```

```
if (C1) then S1; S2;
   elsif (C2) then S3; S4;
   elsif (C3) then S5; S6;
   else S7; S8;
end if;
```

Next, we will write a VHDL module for a J-K flip-flop (Figure 2-20). This flip-flop has active-low asynchronous preset (*SN*) and clear (*RN*) inputs. State changes related to *J* and *K* occur on the falling edge of the clock. In this chapter, we use a suffix N to indicate an active-low (negative-logic) signal. For simplicity, we will assume that the condition $SN = RN = 0$ does not occur.

FIGURE 2-20:
J-K Flip-Flop

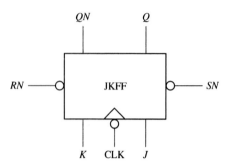

The VHDL code for the J-K flip-flop is given in Figure 2-21. The port declaration in the entity defines the input and output signals. Within the architecture we define a signal $Q_{int}$ that represents the state of the flip-flop internal to the module. The two concurrent statements after **begin** transmit this internal signal to the *Q* and *QN* outputs of the flip-flop. We do it this way because an output signal in a port cannot appear on the right side of an assignment statement within the architecture. This is another solution to the problem presented in Figure 2-13. The flip-flop can change state in response to changes in *SN*, *RN*, and *CLK*, so these three signals are in the sensitivity list of the process. Since *RN* and *SN* reset and set the flip-flop independently of the clock, they are tested first. If *RN* and *SN* are both '1', then we test for the falling edge of the clock. The condition (CLK'event and CLK = '0') is TRUE only if *CLK* has just changed from '1' to '0'. The next state of the flip-flop is determined by its characteristic equation:

$$Q^+ = JQ' + K'Q$$

FIGURE 2-21: **J-K Flip-Flop Model**

```
entity JKFF is
  port(SN, RN, J, K, CLK: in bit; -- inputs
       Q, QN: out bit);
end JKFF;

architecture JKFF1 of JKFF is
signal Qint: bit;                 -- Qint can be used as input or output
begin
  Q <= Qint;                      -- output Q and QN to port
  QN <= not Qint;                 -- combinational output
                                  -- outside process
```

```
   process(SN, RN, CLK)
   begin
     if RN = '0' then Qint <= '0' after 8 ns; -- RN = '0' will clear the FF
     elsif SN = '0' then Qint <= '1' after 8 ns; -- SN='0' will set the FF
     elsif CLK'event and CLK = '0' then        -- falling edge of CLK
       Qint <= (J and not Qint) or (not K and Qint) after 10 ns;
     end if;
   end process;
end JKFF1;
```

The 8-ns delay represents the time it takes to set or clear the flip-flop output after *SN* or *RN* changes to '0'. The 10-ns delay represents the time it takes for *Q* to change after the falling edge of the clock.

● ● ● ● ● ● ● ● ● ● ● ●
## 2.7 Processes Using Wait Statements

An alternative form for a process uses wait statements instead of a sensitivity list. A process cannot have both wait statements and a sensitivity list. A process with wait statements may have the form

```
process
begin
  sequential-statements
  wait-statement
  sequential-statements
  wait-statement
  . . .
end process;
```

This process will execute the `sequential-statements` until a wait statement is encountered. Then it will wait until the specified wait condition is satisfied. It will then execute the next set of `sequential-statements` until another wait is encountered. It will continue in this manner until the end of the process is reached. Then it will start over again at the beginning of the process.

Wait statements can be of three different forms:

```
wait on sensitivity-list;
wait for time-expression;
wait until Boolean-expression;
```

The first form waits until one of the signals on the sensitivity-list changes. For example, **wait on** A, B, C; waits until *A, B,* or *C* changes and then execution proceeds. The second form waits until the time specified by `time-expression` has lapsed. If **wait for** 5 ns is used, the process waits for 5 ns before continuing. If **wait for** 0 ns is used, the wait is for one delta time. Wait statements of the form **wait for** xxx ns are useful for writing VHDL code for simulation; however, they should not be used when writing VHDL code for synthesis since they are not synthesizable. For the third form of wait statement, the `Boolean-expression` is evaluated whenever one of the signals in the expression changes, and the process continues execution when the expression evaluates to TRUE. For example,

```
wait until A = B;
```

will wait until either *A* or *B* changes. Then A = B is evaluated and if the result is TRUE, the process will continue; otherwise, the process will continue to wait until *A* or *B* changes again and A = B is TRUE.

A process cannot have both wait statements and a sensitivity list. It is not acceptable to have some of the signals to be in a sensitivity list and others in wait statements.

After a VHDL simulator is initialized, it executes each process with a sensitivity list one time through, and then waits at the beginning of the process for a change in one of the signals on the sensitivity list. If a process has a wait statement, it will initially execute until a wait statement is encountered. The following two processes are equivalent:

```
process(A, B, C, D)            process
begin                          begin
  C <= A and B after 5 ns;       C <= A and B after 5 ns;
  E <= C or D after 5 ns;        E <= C or D after 5 ns;
end process;                     wait on A, B, C, D;
                               end process;
```

The wait statement at the end of the process replaces the sensitivity list at the beginning. In this way, both processes will initially execute the sequential statements one time and then wait until *A*, *B*, *C*, or *D* changes.

The order in which sequential statements execute in a process is not necessarily the order in which the signals are updated. Consider the following example:

```
process
begin
  wait until clk'event and clk = '1';
  A <= E after 10 ns;        -- (1)
  B <= F after 5 ns;         -- (2)
  C <= G;                    -- (3)
  D <= H after 5 ns;         -- (4)
end process;
```

This process waits for a rising clock edge. Suppose the clock rises at time = 20 ns. Statements (1), (2), (3), (4) immediately execute in sequence. *A* is scheduled to change to *E* at time = 30 ns; *B* is scheduled to change to *F* at time = 25 ns; *C* is scheduled to change to *G* at time = 20 + delta; and *D* is scheduled to change to *H* at time 25 ns. As the simulated time advances, first *C* changes. Then *B* and *D* change at time = 25 ns, and finally *A* changes at time = 30 ns. When *clk* changes to '0', the wait statement is reevaluated, but it keeps waiting until *clk* changes to '1', and then the remaining statements execute again.

If several VHDL statements in a process update the same signal at a given time, the last value overrides. For example,

```
process(CLK)
begin
  if CLK'event and CLK = '0' then
    Q <= A; Q <= B; Q <= C;
  end if;
end process;
```

Every time *CLK* changes from '1' to '0', after delta time, *Q* will change to *C*.

A process must have either a sensitivity list or wait statements. The VHDL code in Figure 2-22 will not simulate because there is no sensitivity list or wait statement.

FIGURE 2-22:  **Example of VHDL Code That Will Not Simulate**

```
entity gates is
  port(A, B, C: in bit; D, E: out bit);
end gates;

architecture exam of gates is
begin
  process
  begin
    D <= A or B after 2 ns;
    E <= not C and A;
  end process;
end exam;
```

In this section, we have introduced processes with sensitivity lists and processes with wait statements. The statements within a process are called sequential statements because they execute in sequence, in contrast with concurrent statements that execute only when a signal on the right-hand-side changes. Signal assignment statements can be either concurrent or sequential. However, **if** statements are always sequential.

● ● ● ● ● ● ● ● ● ● ● ●

## 2.8   Two Types of VHDL Delays: Transport and Inertial Delays

In one of the initial examples in this chapter, we used the statement

```
C <= A and B after 5 ns;
```

to model an AND gate with a propagation delay of 5 ns. The preceding statement will model the AND gate's delay; however, it also introduces some complication, which many readers will not normally expect. If you simulate this AND gate with inputs that change very often in comparison to the gate delay (e.g., at 1 ns, 2 ns, 3 ns, etc.), the simulation output will not show the changes. This is due to how VHDL delays work.

VHDL provides two types of delays—transport delays and inertial delays. The default delay is inertial delay; hence, the after clause in the preceding statement represents an **inertial delay**. Inertial delays are slightly different from simple delays that readers normally assume.

Inertial delay is intended to model gates and other devices that do not propagate short pulses from the input to the output. If a gate has an ideal inertial delay $T$, in addition to delaying the input signals by time $T$, any pulse with a width less than $T$ is rejected. For example, if a gate has an inertial delay of 5 ns, a pulse of width 5 ns would pass through, but a pulse of width 4.999 ns would be rejected. Real devices do not behave in this way. Perhaps they would reject very narrow spurious pulses, but it might be unreasonable to assume that all pulses narrower than the delay duration will be rejected. VHDL does allow one to model devices which reject only very narrow pulses. Rejection of pulses of any arbitrary duration up to the specified inertial delay can be modeled by adding a reject clause to the assignment statement. A statement of the form

```
signal_name <= reject pulse-width after delay-time
```

evaluates the expression, rejects any pulses whose width is less than pulse-width, and then sets the signal equal to the result after a delay of delay-time. In statements of this type, the rejection pulse width must be less than the delay time.

The second type of VHDL delay is **transport delay**, which is intended to model the delay introduced by wiring, simply delays an input signal by the specified delay time. In order to model this delay, the key word **transport** must be specified in the code. Figure 2-23 illustrates the difference between transport and inertial delays. Consider the following VHDL statements:

```
Z1 <= transport X after 10 ns;   -- transport delay
Z2 <= X after 10 ns;             -- inertial delay
Z3 <= reject 4 ns X after 10 ns;  -- delay with specified
                                 -- rejection pulse width
```

FIGURE 2-23:
**Transport and Inertial Delays**

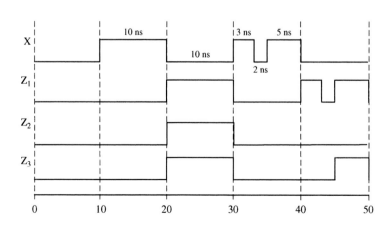

$Z_1$ is the same as $X$, except that it is shifted 10 ns in time. $Z_2$ is similar to $Z_1$, except the pulses in $X$ shorter than 10 ns are filtered out and do not appear in $Z_2$. $Z_3$ is the same as $Z_2$, except that only the pulses of width less than 4 ns have been rejected.

In general, using **reject** is equivalent to using a combination of an inertial delay and a transport delay. The statement for $Z_3$ given here could be replaced with the concurrent statements:

```
Zm <= X after 4 ns; -- inertial delay rejects short pulses
Z3 <= transport Zm after 6 ns;   -- total delay is 10 ns
```

Note that these delays are relevant only for simulation. Understanding how inertial delay works can remove a lot of frustration in your initial experience with VHDL simulation. The pulse rejection associated with inertial delay can inhibit many output changes. In simulations with basic gates and simple circuits, one should make sure that test sequences that you apply are wider than the inertial delays of the modeled devices.

● ● ● ● ● ● ● ● ● ● ● ● ●
## 2.9 Compilation, Simulation, and Synthesis of VHDL Code

After describing a digital system in VHDL, simulation of the VHDL code is important for two reasons. First, we need to verify the VHDL code correctly implements the intended design, and second, we need to verify that the design meets its specifications. We first simulate the design and then synthesize it to the target technology (e.g., FPGA or custom ASIC). In this section, first we describe steps in simulation and then introduce synthesis. As illustrated in Figure 2-24, there are three phases in the simulation of VHDL code: **analysis (compilation), elaboration**, and **simulation**.

FIGURE 2-24:
**Compilation, Elaboration, and Simulation of VHDL Code**

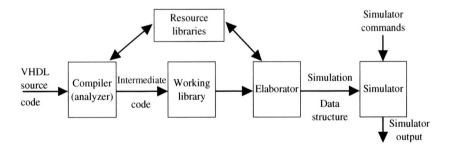

Before the VHDL model of a digital system can be simulated, the VHDL code must first be compiled. The VHDL compiler, also called an **analyzer**, first checks the VHDL source code to see that it conforms to the syntax and semantic rules of VHDL. If there is a syntax error, such as a missing semicolon, or if there is a semantic error, such as trying to add two signals of incompatible types, the compiler will output an error message. The compiler also checks to see that references to libraries are correct. If the VHDL code conforms to all of the rules, the compiler generates intermediate code, which can be used by a simulator or by a synthesizer.

In preparation for simulation, the VHDL intermediate code must be converted to a form which can be used by the simulator. This step is referred to as **elaboration**. During elaboration, a *driver* is created for each signal. Each driver holds the current value of a signal and a queue of future signal values. Each time a signal is scheduled to change in the future, the new value is placed in the queue along with the time at which the change is scheduled. In addition, ports are created for each instance of a component; memory storage is allocated for the required signals; the interconnections among the port signals are specified; and a mechanism is established for executing the VHDL statements in the proper sequence. The resulting data structure represents the digital system being simulated.

The simulation process consists of an **initialization phase** and actual **simulation**. The simulator accepts simulation commands, which control the simulation of the digital system and which specify the desired simulator output. VHDL simulation uses what is known as **discrete event simulation**. The passage of time is simulated in discrete steps in this method of simulation. The initialization phase is used to give an initial value to the signal. During simulation, the VHDL statements are executed and corresponding actions are scheduled. These actions are called transactions, and the process is called **scheduling a transaction**. The scheduled action happens, not necessarily when the statement executes, but when the scheduled time has been reached. A transaction does not mean that there is a change in the value of a signal. The new value for the signal after the transaction may be the same as the old value. If a change in the value occurs, we say that an **event** has taken place.

To facilitate correct initialization, the initial value can be specified in the VHDL model. In the absence of any specifications of the initial values, some simulator packages will assign an initial value depending on the type of the signal. Please note that this initialization is only for simulation and not for synthesis. During initialization, simulation time is set to zero and each process is activated. The process "executes," scheduling corresponding transactions; however, the scheduled transactions do not happen until one reaches the time at which the scheduled transaction is to occur. Execution of a process happens once, and then the process waits for a signal in the sensitivity list to change.

Understanding the role of the delta (Δ) time delays is important when interpreting output from a VHDL simulator. Although the delta delays do not show up on waveform outputs from the simulator, they show up on listing outputs. The simulator uses delta delays to make sure that signals are processed in the proper sequence. Basically, the simulator works as follows: Whenever a component input changes, the output is scheduled to change after the specified delay, or after Δ if no delay is specified. When all input changes have been processed, simulated time is advanced to the next time at which an output change is specified. When time is advanced by a finite amount (1 ns for example), the Δ counter is reset and simulation resumes. Real time does not advance again until all Δ delays associated with the current simulation time have been processed.

The following example illustrates how the simulator works for the circuit of Figure 2-25. Suppose that *A* changes at time = 3 ns. Statement 1 executes and *B* is

**FIGURE 2-25:**
**Illustration of Delta Delays during Simulation of Concurrent Statements**

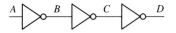

| ns | delta | A | B | C | D |
|----|-------|---|---|---|---|
| 0 | +0 | 0 | 1 | 0 | 1 |
| 3 | +0 | 1 | 1 | 0 | 1 |
| 3 | +1 | 1 | 0 | 0 | 1 |
| 3 | +2 | 1 | 0 | 1 | 1 |
| 8 | +0 | 1 | 0 | 1 | 0 |

```
1  B <= not A;
2  C <= not B;
3  D <= not C after 5 ns;
```

scheduled to change at time $3 + \Delta$. Then time advances to $3 + \Delta$, and statement 2 executes. $C$ is scheduled to change at time $3 + 2\Delta$. Time advances to $3 + 2\Delta$, and statement 3 executes. $D$ is then scheduled to change at 8 ns. You might think the change should occur at $(3 + 2\Delta + 5)$ ns. However, when time advances a finite amount (as opposed to $\Delta$, which is infinitesimal), the $\Delta$ counter is reset. For this reason, when events are scheduled a finite time in the future, the $\Delta$'s are ignored. Since no further changes are scheduled after 8 ns, the simulator goes to an idle mode and waits for another input change. The table gives the simulator output listing.

### 2.9.1 Simulation with Multiple Processes

If a model contains more than one process, all processes execute concurrently with other processes. If there are concurrent statements outside processes, they also execute concurrently. Statements inside of each process execute sequentially. A process takes no time to execute unless it has wait statements in it. (Examples: wait for 10 ns, wait for 0 ns, and wait on E.) Signals take delta time to update when no delay is specified.

As an example of simulation of multiple processes, we trace execution of the VHDL code shown in Figure 2-26. The keyword **transport** specifies the type of delay as transport delay.

**FIGURE 2-26: VHDL Code to Illustrate Process Simulation**

```
entity simulation_example is
end simulation_example;

architecture test1 of simulation_example is
signal A,B: bit;
begin
  P1: process(B)
  begin
    A <= '1';
    A <= transport '0' after 5 ns;
  end process P1;

  P2: process(A)
  begin
    if A = '1' then B <= not B after 10 ns; end if;
  end process P2;
end test1;
```

Figure 2-27 shows the drivers for the signals $A$ and $B$ as the simulation progresses. After elaboration is finished, each driver holds '0', since this is the default initial value for a bit. When simulation begins, initialization takes place. Both processes are executed simultaneously one time through, and then the processes wait until a signal on the sensitivity list changes. When process $P_1$ executes at zero time, two changes in $A$ are scheduled ($A$ changes to '1' at time $\Delta$ and back to '0' at time = 5 ns). Meanwhile, process $P_2$ executes at zero time, but no change in $B$ occurs, since $A$ is still '0' during execution at time 0 ns. Time advances to $\Delta$, and $A$ changes to '1'. The change in $A$ causes process $P_2$ to execute, and since $A$ = '1', $B$ is scheduled to change to '1' at time 10 ns. The next scheduled change occurs at time = 5 ns, when $A$ changes

**FIGURE 2-27:**
**Signal Drivers**
**for Simulation**
**Example**

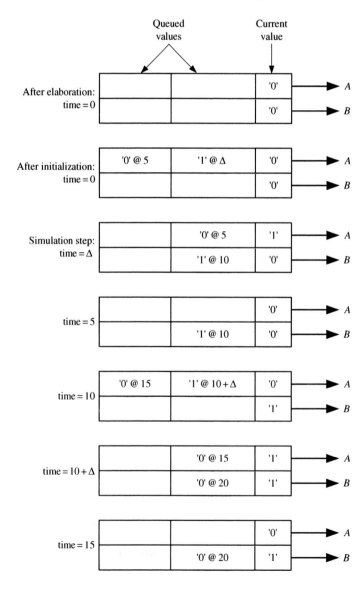

to '0'. This change causes $P_2$ to execute, but $B$ does not change. $B$ changes to '1' at time = 10 ns. The change in $B$ causes $P_1$ to execute, and 2 changes in $A$ are scheduled. When $A$ changes to '1' at time $10 + \Delta$, process $P_2$ executes, and $B$ is scheduled to change at time 20 ns. Then $A$ changes at time 15 ns, and the simulation continues in this manner until the run-time limit is reached. It should be understood that $A$ changes at 15 ns and not at $15 + \Delta$. The $\Delta$ delay comes into the picture only when no time delay is specified.

VHDL simulators use event-driven simulation, as illustrated in the preceding example. A change in a signal is referred to as an *event*. Each time an event occurs, any processes that have been waiting on the event are executed in zero time, and any resulting signal changes are queued up to occur at some future time. When all the active processes are finished executing, simulation time is advanced to the time for which the next event is scheduled, and the simulator processes that event. This continues until either no more events have been scheduled or the simulation time limit is reached.

When VHDL was originally created, simulation was the primary purpose; however, nowadays, one of the most important uses of VHDL is to synthesize or automatically create hardware from a VHDL description. The **synthesis** software for VHDL translates the VHDL code to a circuit description that specifies the needed components and the connections between the components. The initial steps (analysis and elaboration) in Figure 2-24 are common whether VHDL is used for simulation or synthesis. The simulation and synthesis processes are shown in Figure 2-28.

FIGURE 2-28:
**Compilation, Simulation, and Synthesis of VHDL Code**

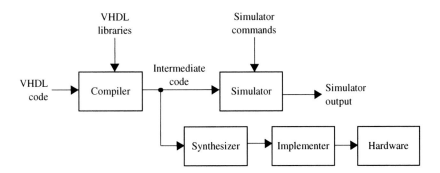

Although synthesis can be done in parallel to simulation, synthesis follows simulation because designers would normally want to catch errors before attempting to synthesize. After the VHDL code for a digital system has been simulated to verify that it works correctly, the VHDL code can be synthesized to produce a list of required components and their interconnections. The synthesizer output can then be used to implement the digital system using specific hardware, such as a CPLD or FPGA, or an ASIC. The CAD software used for implementation generates the necessary information to program the CPLD or FPGA hardware. In the case of an ASIC, it generates the mask required to create the ASIC. Synthesis and implementation of digital logic from VHDL code is discussed in more detail later.

• • • • • • • • • • • •
# 2.10 VHDL Data Types and Operators

### 2.10.1 Data Types

VHDL has several predefined data types. Signals can have these predefined data types, or they can have a user-defined type. Some of the predefined types are as follows:

| | |
|---|---|
| bit | '0' or '1' |
| boolean | FALSE or TRUE |
| integer | an integer in the range $-(2^{31} - 1)$ to $+(2^{31} - 1)$ (some implementations support a wider range) |
| real | floating-point number in the range $-1.0E38$ to $+1.0E38$ |
| character | any legal VHDL character including upper- and lowercase letters, digits, and special characters (each printable character must be enclosed in single quotes; e.g., 'd', '7', '+') |
| time | an integer with units fs, ps, ns, us, ms, sec, min, or hr |

Note that the integer range for VHDL is symmetrical, even though the range for a 32-bit 2's complement integer is $-2^{31}$ to $+(2^{31} - 1)$.

Users can define and create their own data types. A common user-defined type is the **enumeration** type in which all of the values are enumerated. For example, the declarations

```
type state_type is (S0, S1, S2, S3, S4, S5);
signal state: state_type := S1;
```

define a signal called *state* that can have any one of the values $S_0, S_1, S_2, S_3, S_4,$ or $S_5$ and is initialized to $S_1$. If no initialization is given, the default initialization is the leftmost element in the enumeration list, $S_0$ in this example.

VHDL is a **strongly typed language**, so signals and variables of different types generally cannot be mixed in the same assignment statement, and no automatic type conversion is performed. Thus, the statement

```
A <= B or C;
```

is valid only if $A$, $B$, and $C$ all have the same type or closely related types. If types do not match, explicit type conversions should be performed, or "overloaded operators" should be created. Operator overloading is described in Sections 2.13 and 8.4. The overloaded operators in the IEEE packages are presented in Section 2.13.

### 2.10.2 VHDL Operators

Predefined VHDL operators can be grouped into seven classes:

**1.** Binary logical operators: **and or nand nor xor xnor**
**2.** Relational operators: = /= < <= > >=
**3.** Shift operators: **sll srl sla sra rol ror**
**4.** Adding operators: + − & (concatenation)

**5.** Unary sign operators: $+$ $-$
**6.** Multiplying operators: $*$ $/$ **mod rem**
**7.** Miscellaneous operators: **not abs** $**$

When parentheses are not used, operators in class 7 have highest precedence and are applied first, followed by class 6, then class 5, and so on. Class 1 operators have lowest precedence and are applied last. Operators in the same class have the same precedence and are applied from left to right in an expression. The precedence order can be changed by using parentheses. Consider the following expression, where *A, B, C*, and *D* are bit_vectors:

```
(A & not B or C ror 2 and D) = "110010"
```

Note that this is a relational expression performing an equality test; it is not an assignment statement.

To evaluate the expression, the operators are applied in the order

**not, &, ror, or, and, =**

If $A$ = "110", $B$ = "111", $C$ = "011000", and $D$ = "111011", the computation proceeds as follows:

```
not B = "000" (bit-by-bit complement)
A & not B = "110000" (concatenation)
C ror 2 = "000110" (rotate right 2 places)
(A & not B) or (C ror 2) = "110110" (bit-by-bit or)
(A & not B or C ror 2) and D = "110010" (bit-by-bit and)
[(A & not B or C ror 2 and D) = "110010"] = TRUE (the parentheses
force the equality test to be done last and the result is TRUE)
```

The binary logical operators (class 1) as well as **not** can be applied to bits, booleans, bit_vectors, and boolean_vectors. The class 1 operators require 2 operands of the same type, and the result is of that type.

The result of applying a relational operator (class 2) is always a Boolean (FALSE or TRUE). Equals ($=$) and not equals ($/=$) can be applied to almost any type. The other relational operators can be applied to any numeric or enumerated type as well as to some array types. For example, if $A$ = 5, $B$ = 4, and $C$ = 3, the expression (A <= B) **and** (B <= C) evaluates to FALSE.

The shift operators can be applied to any bit_vector or boolean_vector. In the following examples, *A* is a bit_vector equal to "10010101":

```
A sll 2 is "01010100"    (shift left logical, filled with '0')
A srl 3 is "00010010"    (shift right logical, filled with '0')
A sla 3 is "10101111"    (shift left arithmetic, filled with
                          right bit)
A sra 2 is "11100101"    (shift right arithmetic, filled with
                          left bit)
A rol 3 is "10101100"    (rotate left)
A ror 5 is "10101100"    (rotate right)
```

The + and − operators can be applied to integer or real numeric operands. The + and − operators are not defined for bits or bit-vectors. That is why we had to make a full adder by specifically creating carry and sum bits for each bit (Figure 2-12). However, several standard libraries do provide functions for + and − that can work on bit-vectors. If we use such a library, we can perform addition using the statement $C <= A + B$. Some of the popular libraries are described in Section 2.13.

The & operator can be used to concatenate two vectors (or an element and a vector, or two elements) to form a longer vector. For example, "010" & '1' is "0101" and "ABC" & "DEF" is "ABCDEF".

The * and / operators perform multiplication and division on integer or floating-point operands. The rem and mod operators calculate the remainder and modulus for integer operands. The ** operator raises an integer or floating-point number to an integer power, and abs finds the absolute value of a numeric operand.

●●●●●●●●●●●●●

## 2.11 Simple Synthesis Examples

Synthesis tools try to infer the hardware components needed by "looking" at the VHDL code. In order for code to synthesize correctly, certain conventions must be followed. When writing VHDL code, you should always keep in mind that you are designing hardware, not simply writing a computer program. Each VHDL statement implies certain hardware requirements. So, poorly written VHDL code may result in poorly designed hardware. Even if VHDL code gives the correct result when simulated, it may not result in hardware that works correctly when synthesized. Timing problems may prevent the hardware from working properly even though the simulation results are correct.

Consider the VHDL code in Figure 2-29. (Note that $B$ is missing from the process sensitivity list.) This code will simulate as follows: Whenever $A$ changes, it will cause the process to execute once. The value of $C$ will reflect the values of $A$ and $B$ when the process began. If $B$ changes now, that will not cause the process to execute.

FIGURE 2-29: **VHDL Code Example where Simulation and Synthesis Results in Different Outputs**

```
entity Q1 is
  port(A, B: in bit;
       C: out bit);
end Q1;

architecture circuit of Q1 is
begin
  process(A)
  begin
    C <= A or B after 5 ns;
  end process;
end circuit;
```

If this code is synthesized, most synthesizers will output an OR gate as in Figure 2-30. The synthesizer will warn you that $B$ is missing from the sensitivity list, but will go ahead and synthesize the code properly. The synthesizer will also ignore the 5-ns delay on the above statement. If you want to model an exact 5-ns delay, you will have to use counters. The simulator output will not match the synthesizer's output since the process will not execute when $B$ changes. This is an example of where the synthesizer guessed a little more than what you wrote; it assumed that you probably meant an OR gate and created that circuit (accompanied by a warning). But this circuit functions differently from what simulated before synthesis. It is important that you always check for synthesizer warnings of missing signals in the sensitivity list. Perhaps the synthesizer helped you; perhaps it created hardware that you did not intend to.

**FIGURE 2-30:**
**Synthesizer**
**Output for Code**
**in Figure 2-29**

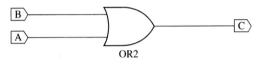

OR2

Now, consider the VHDL code in Figure 2-31. What hardware will you get if you synthesized this code?

FIGURE 2-31: **Example VHDL Code**

```
entity Q3 is
  port(A,B,F, CLK: in bit;
       G: out bit);
end Q3;

architecture circuit of Q3 is
signal C: bit;
begin
  process(Clk)
  begin
    if (Clk = '1' and Clk'event) then
      C <= A and B; -- statement 1
      G <= C or F; -- statement 2
    end if;
  end process;
end circuit;
```

Let us think about the block diagram of the circuit represented by this code without worrying about the details inside. The block diagram is as shown in Figure 2-32. The ability to hide details and use abstractions is an important part of good system design.

Note that $C$ is an internal signal, and therefore it does not show up in the block diagram.

FIGURE 2-32: **Block Diagram for VHDL Code in Figure 2-31**

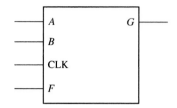

Now, let us think about the details of the circuit inside this block. This circuit is not two cascaded gates; the signal assignment statements are in a process. An edge-triggered clock is implied by the use of clk'event in the clock statement preceding the signal assignment. Since the values of $C$ and $G$ need to be retained after the clock edge, flip-flops are required for both $C$ and $G$. Please note that a change in the value of $C$ from statement 1 will not be considered during the execution of statement 2 in that pass of the process. It will be considered only in the next pass, and the flip-flop for $C$ makes this happen in the hardware also. Hence the code implies hardware shown in Figure 2-33.

FIGURE 2-33: **Hardware Corresponding to VHDL Code in Figure 2-31**

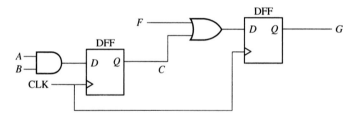

We saw earlier that the following code represents a D-latch:

```
process(G, D)
begin
   if G = '1' then Q <= D; end if;
end process;
```

Let us understand why this code does not represent an AND gate with $G$ and $D$ as inputs. If $G$ = '1', an AND gate will result in the correct output to match the **if** statement. However, what happens if currently $Q$ = '1' and then $G$ changes to '0'? When $G$ changes to '0', an AND gate would propagate that to the output; however, the device we have modeled here should not. It is expected to make no changes to the output if $G$ is not equal to '1'. Hence, it is clear that this device has to be a D-latch and not an AND gate.

In order to infer flip-flops or registers that change state on the rising edge of a clock signal, an **if**-clause of the form

```
if clock'event and clock = '1' then . . . end if;
```

is required by most synthesizers. For every assignment statement between **then** and **end if** above, a signal on the left side of the assignment will cause creation of a register or flip-flop. The moral to this story is, if you don't want to create unnecessary flip-flops, don't put the signal assignments in a clocked process. If `clock'event` is omitted, the synthesizer may produce latches instead of flip-flops.

Now consider the VHDL code in Figure 2-34. If you attempt to synthesize this code, the synthesizer will generate an empty block diagram. This is because $D$, the output of the above block, is never assigned. It will generate warnings that

```
Input <CLK> is never used.
Input <A> is never used.
Input <B> is never used.
Output <D> is never assigned.
```

FIGURE 2-34: **Example VHDL Code That Will Not Synthesize**

```vhdl
entity no_syn is
  port(A,B, CLK: in bit;
       D: out bit);
end no_syn;

architecture no_synthesis of no_syn is
  signal C: bit;
begin
  process(Clk)
  begin
    if (Clk='1' and Clk'event) then
      C <= A and B;
    end if;
  end process;
end no_synthesis;
```

●  ●  ●  ●  ●  ●  ●  ●  ●  ●  ●  ●

## 2.12  VHDL Models for Multiplexers

A multiplexer is a combinational circuit and can be modeled using concurrent statements only or using processes. A conditional signal assignment statement such as **when** or a selective signal assignment statement using **with select** can be used to model a multiplexer without processes. A case statement within a process can also be used to make a model for a multiplexer.

### 2.12.1  Using Concurrent Statements

Figure 2-35 shows a 2-to-1 multiplexer (MUX) with two data inputs and one control input. The MUX output is $F = A' \cdot I_0 + A \cdot I_1$. The corresponding VHDL statement is

```vhdl
F <= (not A and I0) or (A and I1);
```

Here, the MUX can be modeled as a single concurrent signal assignment statement. Alternatively, we can represent the MUX by a conditional signal assignment statement as shown in Figure 2-35. This statement executes whenever $A$, $I_0$, or $I_1$

FIGURE 2-35: **2-to-1 Multiplexer**

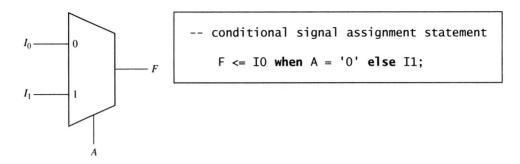

```
-- conditional signal assignment statement

F <= I0 when A = '0' else I1;
```

changes. The MUX output is $I_0$ when $A = $ '0', and otherwise it is $I_1$. In the conditional statement, $I_0$, $I_1$, and $F$ can either be bits or bit-vectors.

The general form of a conditional signal assignment statement is

```
signal_name <= expression1 when condition1
          else expression2 when condition2
          [else expressionN];
```

This concurrent statement is executed whenever a change occurs in a signal used in one of the expressions or conditions. If `condition1` is true, `signal_name` is set equal to the value of `expression1`, otherwise if `condition2` is true, `signal_name` is set equal to the value of `expression2`, and so on. The line in square brackets is optional. Figure 2-36 shows how two cascaded MUXes can be represented by a conditional signal assignment statement. The output MUX selects $A$ when $E = $ '1'; otherwise, it selects the output of the first MUX, which is $B$ when $D = $ '1', or it is $C$.

FIGURE 2-36: **Cascaded 2-to-1 MUXes**

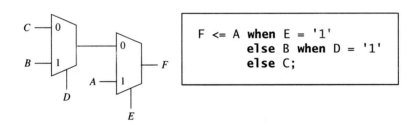

```
F <= A when E = '1'
      else B when D = '1'
      else C;
```

Figure 2-37 shows a 4-to-1 multiplexer (MUX) with four data inputs and two control inputs, $A$ and $B$. The control inputs select which one of the data inputs is transmitted to the output. The logic equation for the 4-to-1 MUX is

$$F = A'B'I_0 + A'BI_1 + A B'I_2 + A B I_3$$

Thus, one way to model the MUX is with the VHDL statement

```
F <= (not A and not B and I0) or (not A and B and I1) or
      (A and not B and I2) or (A and B and I3);
```

Another way to model the 4-to-1 MUX is to use a conditional assignment statement:

```
F <= I0 when A&B = "00"
else I1 when A&B = "01"
else I2 when A&B = "10"
else I3;
```

The expression A&B means that $A$ is concatenated with $B$; that is, the two bits $A$ and $B$ are merged together to form a 2-bit vector. This bit-vector is tested and the appropriate MUX input is selected. For example, if $A =$ '1' and $B =$ '0', A&B = "10" and $I_2$ is selected. Instead of concatenating $A$ and $B$, we could use a more complex condition:

```
F <= I0 when A = '0' and B = '0'
else I1 when A = '0' and B = '1'
else I2 when A = '1' and B = '0'
else I3;
```

A third way to model the MUX is to use a selected signal assignment statement, as shown in Figure 2-37. A&B cannot be used in this type of statement, so we concatenate $A$ and $B$ to create sel. The value of sel then selects the MUX input that is assigned to $F$.

FIGURE 2-37: **4-to-1 Multiplexer**

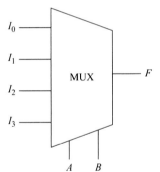

```
sel <= A&B;
--selected signal assignment statement
with sel select
  F <= I0 when "00",
       I1 when "01",
       I2 when "10",
       I3 when "11",
```

The general form of a selected signal assignment statement is

```
with expression_s select
  signal_s <= expression1 [after delay-time] when choice1,
              expression2 [after delay-time] when choice2,
              . . .
              [expression_n [after delay-time] when others];
```

This concurrent statement executes whenever a signal changes in any of the expressions. First, expression_s is evaluated. If it equals choice1, signal_s is set equal to expression1; if it equals choice2, signal_s is set equal to expression2; and so on. If all possible choices for the value of expression_s are given, the last line should be omitted; otherwise, the last line is required. When it is present, if `expression_s` is not equal to any of the enumerated choices, `signal_s` is set equal to `expression_n`. Then `signal_s` is updated after the specified `delay-time`, or after Δ if the **after** `delay-time` is omitted.

### 2.12.2 Using Processes

If a MUX model is used inside a process, a concurrent statement cannot be used. As an alternative, the MUX can be modeled using a **case** statement:

```
case Sel is
  when 0 => F <= I0;
  when 1 => F <= I1;
  when 2 => F <= I2;
  when 3 => F <= I3;
end case;
```

The case statement has the general form

```
case expression is
  when choice1 => sequential statements1
  when choice2 => sequential statements2
  . . .
  [when others => sequential statements]
end case;
```

The `expression` is evaluated first. If it is equal to `choice1`, then `sequential statements1` are executed; if it is equal to `choice2`, then `sequential statements2` are executed; and so on. All possible values of the expression must be included in the choices. If all values are not explicitly given, a **when others** clause is required in the **case** statement.

One might notice that combinational circuits can be described using concurrent or sequential statements. Sequential circuits generally require a **process** statement. **Process** statements can be used to make sequential or combinational circuits.

• • • • • • • • • • • •
## 2.13 VHDL Libraries

VHDL libraries and packages are used to extend the functionality of VHDL by defining types, functions, components, and overloaded operators. In standard VHDL, some operations are valid only for certain data types. If those operations are desired for other data types, one has to use function "overloading" to create an "overloaded" operator. The concept of "function overloading" exists in many general-purpose languages. It means that two or more functions may have the same name, so long as the parameter types are sufficiently different enough to distinguish which function is actually intended. Overloaded functions can also be created to handle operations involving heterogeneous data types.

In the initial days of CAD, every tool vendor used to create its own libraries and packages. Porting designs from one environment to another became a problem under those conditions. The IEEE has developed standard libraries and packages to make design portability easier. The original VHDL standard only defines 2-valued logic (bits and bit-vectors). One of the earliest extensions was to define multivalued logic as an IEEE standard. The package IEEE.std_logic_1164 defines a std_logic type that has nine values, including '0', '1', 'X' (unknown), and 'Z' (high impedance). The package also defines std_logic_vectors, which are vectors of the std_logic type. This standard defines logic operations and other functions for working with std_logic and std_logic_vectors, but it does not provide for arithmetic operations. The std_logic_1164 package and its use for simulation and synthesis will be described in more detail in Chapter 8.

When VHDL became more widely used for synthesis, the IEEE introduced two packages to facilitate writing synthesizable code: IEEE.numeric_bit and IEEE.numeric_std. The former uses bit_vectors to represent unsigned and signed binary numbers, and the latter uses std_logic_vectors. Both packages define overloaded logic and arithmetic operators for unsigned and signed numbers. Prior to Chapter 8, we will use the numeric_bit package and unsigned numbers for arithmetic operations.

To access functions and components from a library, you need a library statement and a use statement. The statement

```
library IEEE;
```

allows your design to access all packages in the IEEE library. The statement

```
use IEEE.numeric_bit.all;
```

allows your design to use the entire numeric_bit package, which is found in the IEEE library. Whenever a package is used in a module, the library and use statements must be placed before the entity in that module period.

The numeric_bit package defines unsigned and signed types as unconstrained arrays of bits:

```
type unsigned is array (natural range <>) of bit;
type   signed is array (natural range <>) of bit;
```

Signed numbers are represented in 2's complement form. The package contains overloaded operators for arithmetic, relational, logical, and shifting operations on unsigned and signed numbers.

Unsigned and signed types are basically bit-vectors. However, overloaded operators are defined for these types and not for bit-vectors. The statement

```
C <= A + B;
```

will cause a compiler error if $A$, $B$, and $C$ are bit_vectors. If these signals are of type unsigned or signed, the compiler will invoke the appropriate overloaded operator to carry out the addition.

The numeric_bit package defines the following overloaded operators:

arithmetic: $+, -, *, /,$ rem, mod
relational:: $=, /=, >, <, >=, <=$
logical: not, and, or, nand, nor, xor, xnor
shifting: shift_left, shift_right, rotate_left, rotate_right, sll, srl, rol, ror

The arithmetic, relational, and logical operators (except not) each require a left operand and a right operand. For arithmetic and relational operators, the following left and right operand pairs are acceptable: unsigned and unsigned, unsigned and natural, natural and unsigned, signed and signed, signed and integer, integer and signed. For logical operators (except not), left and right operands must either both be unsigned or both signed. When the $+$ and $-$ operators are used with unsigned operands of different lengths, the shortest operand will be extended by filing in 0's on the left. Any carry is discarded so that the result has the same number of bits as the longest operand. For example, when working with unsigned numbers

"1011" + "110" = "1011" + "0110" = "0001" and the carry is discarded.

The numeric_bit package provides an overloaded operator to add an integer to an unsigned, but not to add a bit to an unsigned type. Thus, if $A$ and $B$ are unsigned, A+B+1 is allowed, but a statement of the form

```
Sum <= A + B + carry;
```

is not allowed when carry is of type bit. The carry must be converted to unsigned before it can be added to the unsigned vector A+B. The notation **unsigned'(0 => carry)** will accomplish the necessary conversion.

Figure 2-38 shows behavioral VHDL code that uses overloaded operators from the numeric_bit package to describe a 4-bit adder with a carry input. The entity declaration is the same as in Figure 2-12, except type unsigned is used instead of

FIGURE 2-38: **VHDL Code for 4-Bit Adder Using Unsigned Vectors**

```
library IEEE;
use IEEE.numeric_bit.all;

entity Adder4 is
  port(A, B: in unsigned(3 downto 0); Ci: in bit;  -- Inputs
       S: out unsigned(3 downto 0); Co: out bit);  -- Outputs
end Adder4;

architecture overload of Adder4 is
signal Sum5: unsigned(4 downto 0);
begin
  Sum5 <= '0' & A + B + unsigned'(0=>Ci);  -- adder
  S <= Sum5(3 downto 0);
  Co <= Sum5(4);
end overload;
```

bit_vector. Because adding two 4-bit numbers produces a 5-bit sum, a 5-bit signal (Sum5) is declared within the architecture. If we compute $A + B$, the result is only 4 bits. Since we want a 5-bit result, we must extend $A$ to 5 bits by concatenating '0' and $A$. ($B$ will automatically be extended to match.) After Sum5 is calculated using the overloaded operators from the numeric_bit package, it is split into a 4-bit sum ($S$) and a carry ($C_o$). Most synthesis tools will implement the code of Figure 2-38 as an adder with a carry input and output. One version of the Xilinx synthesizer produces the result shown in Figure 2-39.

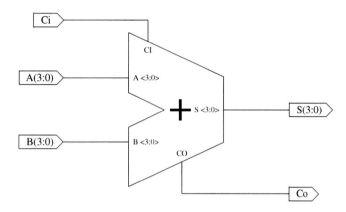

**FIGURE 2-39:**
**Synthesizer Output for VHDL Code of Figure 2-38**

Useful conversion functions found in the numeric_bit package include the following:

`TO_INTEGER(A):` converts an unsigned vector $A$ to an integer
`TO_UNSIGNED(B, N):` converts an integer to an unsigned vector of length $N$
`UNSIGNED(A):` causes the compiler to treat a bit_vector $A$ as an unsigned vector
`BIT_VECTOR(B):` causes the compiler to treat an unsigned vector $B$ as a bit_vector

If multivalued logic is desired, one can use the IEEE standard numeric_std package instead of the numeric_bit package. The numeric_std package defines unsigned and signed types as std_logic vectors instead of bit_vectors. Three statements are required to use this package:

```
library IEEE;
use IEEE.std_logic_1164.all;
use IEEE.numeric_std.all;
```

This package defines the same set of overloaded operators and functions on unsigned and signed numbers as the numeric_bit package.

Another popular VHDL package used for simulation and synthesis with multivalued logic is the std_logic_arith package developed by Synopsis. This package defines unsigned and signed types and overloaded operators similarly to the IEEE numeric_std package; however, the conversion functions have different names and there are some other differences. A major deficiency of the std_logic_arith package is that it does not define logic operations for unsigned or signed vectors. This package is not an IEEE standard even though it is commonly placed in the IEEE library.

Yet another option is to use the std_logic_unsigned package, also developed by Synopsis. This package does not define unsigned types, but instead it defines some overloaded arithmetic operators for std_logic_vectors. These operators treat std_logic_vectors as if they were unsigned numbers. When used in conjunction with the std_logic_1164 package, both arithmetic and logic operations can be performed on std_logic_vectors because the 1164 package defines the logic operations. The std_logic_unsigned package is not an IEEE standard even though it is commonly placed in the IEEE library. The VHDL code for the 4-bit adder of Figure 2-38 is rewritten in Figure 2-40 using the std_logic_unsigned package. Because the package provides an overloaded operator to add a std_logic bit to a std_logic_vector, type conversion is not needed. The result of synthesizing this code is the same as that for Figure 2-38.

**FIGURE 2-40: VHDL Code for 4-Bit Adder Using the std_logic_unsigned Package**

```
library IEEE;
use IEEE.std_logic_1164.all;
use IEEE.std_logic_unsigned.all;
entity Adder4 is
  port(A, B: in std_logic_vector(3 downto 0); Ci: in std_logic;  --Inputs
       S: out std_logic_vector(3 downto 0); Co: out std_logic); --Outputs
end Adder4;

architecture overload of Adder4 is
signal Sum5: std_logic_vector(4 downto 0);
begin
  Sum5 <= '0' & A + B + Ci;   --adder
  S <= Sum5(3 downto 0);
  Co <= Sum5(4);
end overload;
```

In this section, we have discussed four different packages, which provide overloaded operators for arithmetic and relational operations. We will initially use the numeric_bit package because it is easiest to use and it is an IEEE standard. Starting in Chapter 8, we will use the IEEE numeric_std package because it is an IEEE standard, provides multivalued signals, and is similar in functionality to the numeric_bit package. We have chosen not to use the std_logic_arith and std_logic_unsigned packages because they are not IEEE standards and they have less functionality than the IEEE numeric_std package.

## 2.14 Modeling Registers and Counters Using VHDL Processes

When several flip-flops change state on the same clock edge, statements representing these flip-flops can be placed in the same clocked process. Figure 2-41 shows three flip-flops connected as a cyclic shift register. These flip-flops all change state following the rising edge of the clock. We have assumed a 5-ns propagation delay between the clock edge and the output change. Immediately following the clock edge, the three statements in the process execute in sequence with no delay. The new values of the Q's are then scheduled to change after 5 ns. If we omit the delay and replace the sequential statements with

Q1 <= Q3;     Q2 <= Q1;     Q3 <= Q2;

the operation is basically the same. The three statements execute in sequence in zero time, and then the Q's values change after a delta delay. In both cases, the old values of $Q_1$, $Q_2$, and $Q_3$ are used to compute the new values. This may seem strange at first, but that is the way the hardware works. At the rising edge of the clock, all of the $D$ inputs are loaded into the flip-flops, but the state change does not occur until after a propagation delay.

FIGURE 2-41: **Cyclic Shift Register**

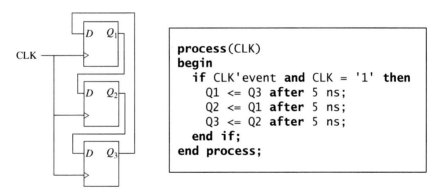

```
process(CLK)
begin
  if CLK'event and CLK = '1' then
    Q1 <= Q3 after 5 ns;
    Q2 <= Q1 after 5 ns;
    Q3 <= Q2 after 5 ns;
  end if;
end process;
```

Figure 2-42 shows a simple register that can be loaded or cleared on the rising edge of the clock. If $CLR$ = '1', the register is cleared, and if $Ld$ = '1', the $D$ inputs are loaded into the register. This register is fully synchronous so that the $Q$ outputs only change in response to the clock edge and not in response to a change in $Ld$ or $CLR$. In the VHDL code for the register, $Q$ and $D$ are bit-vectors dimensioned 3 **downto** 0. Since the register outputs can only change on the rising edge of the clock, $CLR$ is not on the sensitivity list. It is tested after the rising edge of the clock. If $CLR$ = $Ld$ = '0', no change of $Q$ occurs. Since $CLR$ is tested before $Ld$, if $CLR$ = '1', the **elsif** prevents $Ld$ from being tested and $CLR$ overrides $Ld$.

FIGURE 2-42:
**Register with
Synchronous Clear
and Load**

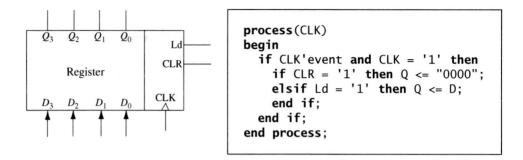

```
process(CLK)
begin
  if CLK'event and CLK = '1' then
    if CLR = '1' then Q <= "0000";
    elsif Ld = '1' then Q <= D;
    end if;
  end if;
end process;
```

Next, we will model a left shift register using a VHDL process. The register in Figure 2-43 is similar to that in Figure 2-42, except that we have added a left shift control input ($LS$). When $LS$ is '1', the contents of the register are shifted left and the rightmost bit is set equal to $R_{in}$. The shifting is accomplished by taking the rightmost 3 bits of Q, Q(2 downto 0), and concatenating them with $R_{in}$. For example, if $Q$ = "1101" and $R_{in}$ = '0', then Q(2 downto 0) & Rin = "1010", and this value is loaded back into the $Q$ register on the rising edge of $CLK$. The code implies that if $CLR = Ld = LS = $ '0', then $Q$ remains unchanged.

FIGURE 2-43: **Left Shift Register with Synchronous Clear and Load**

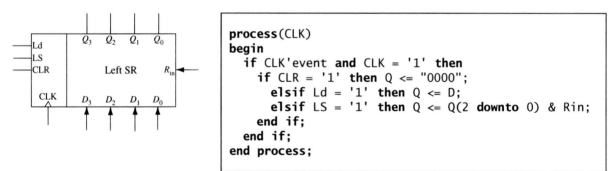

```
process(CLK)
begin
  if CLK'event and CLK = '1' then
    if CLR = '1' then Q <= "0000";
      elsif Ld = '1' then Q <= D;
      elsif LS = '1' then Q <= Q(2 downto 0) & Rin;
    end if;
  end if;
end process;
```

Figure 2-44 shows a simple synchronous counter. On the rising edge of the clock, the counter is cleared when $ClrN$ = '0', and it is incremented when $ClrN = En$ = '1'. In this example, the signal $Q$ represents the 4-bit value stored in the counter. Since addition is not defined for bit-vectors, we have declared $Q$ to be of type unsigned. Then we can increment the counter using the overloaded " + " operator that is defined in the ieee.numeric_bit package. The statement Q <= Q + 1; increments the counter. When the counter is in state "1111", the next increment takes it back to state "0000".

Now, let us create a VHDL model for a generic counter, the 74163. It is a 4-bit fully synchronous binary counter, which is available in both TTL and CMOS logic families. Although rarely used in new designs at present, it represents a general type of counter that is found in many CAD design libraries. In addition to performing the

**FIGURE 2-44: VHDL Code for a Simple Synchronous Counter**

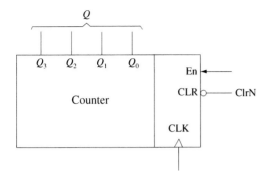

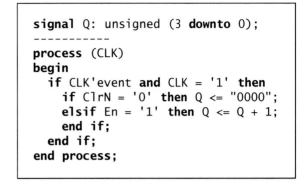

```
signal Q: unsigned (3 downto 0);
-----------
process (CLK)
begin
  if CLK'event and CLK = '1' then
    if ClrN = '0' then Q <= "0000";
    elsif En = '1' then Q <= Q + 1;
    end if;
  end if;
end process;
```

counting function, it can be cleared or loaded in parallel. All operations are synchronized by the clock, and all state changes take place following the rising edge of the clock input. A block diagram of the counter is provided in Figure 2-45.

This counter has four control inputs—$ClrN$, $LdN$, $P$, and $T$. $P$ and $T$ are used to enable the counting function. Operation of the counter is as follows:

**1.** If $ClrN = $ '0', all flip-flops are set to '0' following the rising clock edge.
**2.** If $ClrN = $ '1' and $LdN = $ '0', the $D$ inputs are transferred in parallel to the flip-flops following the rising clock edge.
**3.** If $ClrN = LdN = $ '1' and $P = T = $ '1', the count is enabled and the counter state will be incremented by 1 following the rising clock edge.

If $T = $ '1', the counter generates a carry ($C_{out}$) in state 15, so

$$C_{out} = Q_3\,Q_2\,Q_1\,Q_0\,T$$

The truth table in Figure 2-45 summarizes the operation of the counter. Note that $ClrN$ overrides the load and count functions in the sense that when $ClrN = $ '0',

**FIGURE 2-45: 74163 Counter Operation**

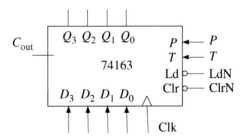

| Control Signals | | | Next State | | | | |
|---|---|---|---|---|---|---|---|
| ClrN | LdN | PT | $Q_3^+$ | $Q_2^+$ | $Q_1^+$ | $Q_0^+$ | |
| 0 | X | X | 0 | 0 | 0 | 0 | (clear) |
| 1 | 0 | X | $D_3$ | $D_2$ | $D_1$ | $D_0$ | (parallel load) |
| 1 | 1 | 0 | $Q_3$ | $Q_2$ | $Q_1$ | $Q_0$ | (no change) |
| 1 | 1 | 1 | present state + 1 | | | | (increment count) |

clearing occurs regardless of the values of *LdN*, *P*, and *T*. Similarly, *LdN* overrides the count function. The *ClrN* input on the 74163 is referred to as a *synchronous* clear input because it clears the counter in synchronization with the clock, and no clearing can occur if no clock pulse is present.

The VHDL description of the counter is shown in Figure 2-46. Q represents the four flip-flops that comprise the counter. The counter output, $Q_{out}$, changes whenever Q changes. The carry output is computed whenever Q or T changes. The first if statement in the process tests for a rising edge of Clk. Since clear overrides load and count, the next **if** statement tests *ClrN* first. Since load overrides count, *LdN* is tested next. Finally, the counter is incremented if both *P* and *T* are '1'. Since Q is of type unsigned, we can use the overloaded "+" operator from the ieee.numeric_bit package to add 1 to increment the counter. The expression Q+1 would not be legal if *Q* were a bit-vector since addition is not defined for bit-vectors.

FIGURE 2-46: **74163 Counter Model**

```
-- 74163 FULLY SYNCHRONOUS COUNTER

library IEEE;
use IEEE.numeric_bit.all;

entity c74163 is
  port(LdN, ClrN, P, T, Clk: in bit;
       D: in unsigned(3 downto 0);
       Cout: out bit; Qout: out unsigned(3 downto 0));
end c74163;

architecture b74163 of c74163 is
signal Q: unsigned(3 downto 0);    -- Q is the counter register
begin
  Qout <= Q;
  Cout <= Q(3) and Q(2) and Q(1) and Q(0) and T;
  process(Clk)
  begin
    if Clk'event and Clk = '1' then    -- change state on rising edge
      if ClrN = '0' then Q <= "0000";
      elsif LdN = '0' then Q <= D;
      elsif (P and T) = '1' then Q <= Q + 1;
      end if;
    end if;
  end process;
end b74163;
```

To test the counter, we have cascaded two 74163's to form an 8-bit counter (Figure 2-47). When the counter on the right is in state 1111 and $T_1$ = '1', *Carry1* = '1'. Then for the left counter, *PT* = '1' if *P* = '1'. If *PT* = '1', on the next clock the right counter is incremented to 0000 at the same time the left counter is incremented.

FIGURE 2-47: **Two 74163 Counters Cascaded to Form an 8-Bit Counter**

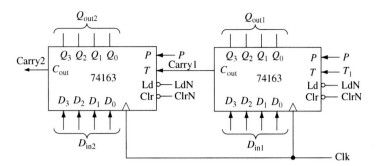

Figure 2-48 shows the VHDL code for the 8-bit counter. In this code we have used the c74163 model as a component and instantiated two copies of it. For convenience in reading the output, we have defined a signal *Count*, which is the integer equivalent of the 8-bit counter value. The function **to_integer** converts an unsigned vector to an integer.

Let us now synthesize the VHDL code for a left shift register (Figure 2-43). Before synthesis is started, we must specify a target device (e.g., a particular FPGA

FIGURE 2-48: **VHDL for 8-Bit Counter**

```
--Test module for 74163 counter

library IEEE;
use IEEE.numeric_bit.ALL;

entity eight_bit_counter is
  port(ClrN, LdN, P, T1, Clk: in bit;
       Din1, Din2: in unsigned(3 downto 0);
       Count: out integer range 0 to 255;
       Carry2: out bit);
end eight_bit_counter;

architecture cascaded_counter of eight_bit_counter is
component c74163
  port(LdN, ClrN, P, T, Clk: in bit;
       D: in unsigned(3 downto 0);
       Cout: out bit; Qout: out unsigned(3 downto 0));
end component;

signal Carry1: bit;
signal Qout1, Qout2: unsigned(3 downto 0);
begin
  ct1: c74163 port map (LdN, ClrN, P, T1, Clk, Din1, Carry1, Qout1);
  ct2: c74163 port map (LdN, ClrN, P, Carry1, Clk, Din2, Carry2, Qout2);
  Count <= to_integer(Qout2 & Qout1);
end cascaded_counter;
```

or CPLD) so that the synthesizer knows what components are available. Let us assume that the target is a CPLD or FPGA that has D flip-flops with clock enable (D-CE flip-flops). $Q$ and $D$ are 4-bit vectors. Because updates to $Q$ follow `"CLK'event and CLK = '1' then"`, this infers that $Q$ must be a register composed of four flip-flops, which we will label $Q_3$, $Q_2$, $Q_1$, and $Q_0$. Since the flip-flops can change state when $Clr$, $Ld$, or $Ls$ is '1', we connect the clock enables to an OR gate whose output is $Clr + Ld + Ls$. Then we connect gates to the $D$ inputs to select the data to be loaded into the flip-flops. If $Clr$ = '0' and $Ld$ = '1', $D$ is loaded into the register on the rising clock edge. If $Clr = Ld$ = '0' and $Ls$ = '1', then $Q_2$ is loaded into $Q_3$, $Q_1$ is loaded into $Q_2$, and so on. Figure 2-49 shows the logic circuit for the first two flip-flops. If $Clr$ = '1', the D flip-flop inputs are '0' and the register is cleared.

**FIGURE 2-49:**
**Synthesis of VHDL**
**Code for Left Shift**
**Register from**
**Figure 2-43**

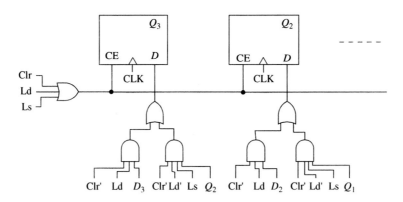

A VHDL synthesizer cannot synthesize delays. Clauses of the form `"after time-expression"` will be ignored by most synthesizers, but some synthesizers require that **after** clauses be removed. Although initial values for signals may be specified in port and signal declarations, these initial values are ignored by the synthesizer. A reset signal should be provided if the hardware must be set to a specific initial state. Otherwise, the initial state of the hardware may be unknown and the hardware may malfunction. When an integer signal is synthesized, the integer is represented in hardware by its binary equivalent. If the range of an integer is not specified, the synthesizer will assume the maximum number of bits, usually 32. Thus

```
signal count: integer range 0 to 7;
```

would result in a 3-bit counter, but

```
signal count: integer;
```

could result in a 32-bit counter.

VHDL signals retain their current values until they are changed. This can result in creation of unwanted latches when the code is synthesized. For example, in a combinational process, the statement

```
if X = '1' then B <= 1; end if;
```

would create latches to hold the value of $B$ when $X$ changes to '0'. To avoid creation of unwanted latches in a combinational process, always include an **else** clause in every **if** statement. For example,

```
if X = '1' then B <= 1 else B <= 0; end if;
```

would create a MUX to switch the value of $B$ from 1 to 0.

## 2.15 Behavioral and Structural VHDL

Any circuit or device can be represented in multiple forms of abstraction. Consider the different representations for a NAND gate, as illustrated in Figure 2-50. When hearing the term NAND, different designers, depending on the domain of their design level, think of these different representations of the same NAND device.

**FIGURE 2-50:**
**Different Levels of Abstraction of a NAND Device**

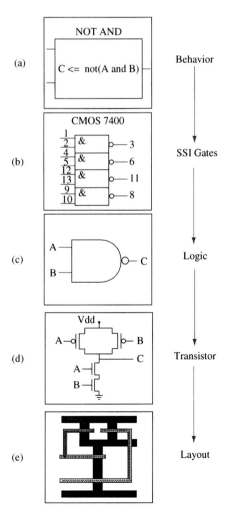

Some would think of just a block representing the behavior of a NAND operator, as illustrated in Figure 2-50(a). Some others might think of the four gates in a CMOS 7400 chip, as in Figure 2-50(b). For designers who work at the logic level, they think of the logic symbol for a NAND gate, as in Figure 2-50(c). Transistor-level circuit designers think of the transistor-level circuit to achieve the NAND functionality, as in Figure 2-50(d). What passes through the mind of a physical level designer is the layout of a NAND gate, as in Figure 2-50(e). All of the figures represent the same device, but they differ in the amount of detail provided in the description.

Just as a NAND gate can be described in different ways, any logic circuit can be described with different levels of detail. Figure 2-51 indicates a behavioral level representation of the logic function $F = ab + bc$, whereas Figures 2-52 represents 2 equivalent structural representations. The functionality specified in the abstract description in Figure 2-51 can be achieved in different ways, two examples of which are by using two AND gates and one OR gate or three NAND gates. A structural description gives different descriptions for Figures 2-52(a) and 2-52(b), whereas the same behavioral description could result in either of these two representations. A structural description specifies more details, whereas the behavioral level description only specifies the behavior at a higher level of abstraction.

**FIGURE 2-51: A Block Diagram with A, B, C as Inputs and F = AB + BC as Output**

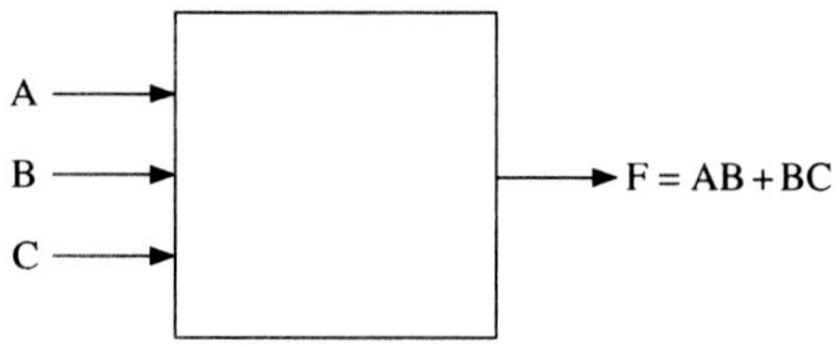

A → 
B → 
C → 
→ F = AB + BC

**FIGURE 2-52: Two Implementations of F = AB + BC**

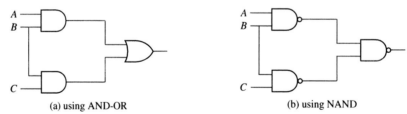

(a) using AND-OR

(b) using NAND

You noticed that the same circuit can be described in different ways. Similarly, VHDL allows you to create design descriptions at multiple levels of abstraction. The most common ones are **behavioral models, dataflow (register transfer language [RTL]) models**, and **structural models**. Behavioral VHDL models describe the circuit or system at a high level of abstraction without implying any particular structure or technology. Only the overall behavior is specified. In contrast, in structural models, the components used and the structure of the interconnection between the components are clearly specified. Structural models may be detailed enough to specify use of particular gates and flip-flops from specific libraries/packages. The structural VHDL model is at a low level of abstraction. VHDL code can be written at an intermediate level of abstraction, at the dataflow level or RTL level, in addition to pure behavioral level or structural level. Register transfer languages have been used for decades to describe the behavior of synchronous systems where a system is viewed as registers plus control logic required to perform loading and manipulation of registers. In the dataflow model, data path and control signals are specified. The working of the system is described in terms of the data transfer between registers.

If designs are specified at higher levels of abstraction, they need to get converted to the lower levels in order to get implemented. In the early days of design automation, there were not enough automatic software tools to perform this conversion; hence, designs needed to be specified at the lower levels of abstraction. Designs were entered using schematic capture or lower levels of abstraction. Nowadays, synthesis tools perform very efficient conversion of behavioral level designs into target technologies.

Behavioral and structural design techniques are often combined. Different parts of the design are often done with different techniques. State-of-the-art design automation tools generate efficient hardware for logic and arithmetic circuits; hence, a large part of those designs is done at the behavioral level. However, memory structures often need manual optimizations and are done by custom design, as opposed to automatic synthesis.

### 2.15.1 Modeling a Sequential Machine

In this section, we discuss several ways of writing VHDL descriptions for sequential machines. Let us assume that we have to write a **behavioral** model for a Mealy sequential circuit represented by the state table in Figure 2-53 (note that this is the BCD to excess-3 code converter designed in Chapter 1). A block diagram of this state machine is also shown in Figure 2-53. This view of the circuit can be used to write its entity description. Please note that the current state and next state are not visible externally.

**FIGURE 2-53:**
**State Table and Block Diagram of Sequential Machine**

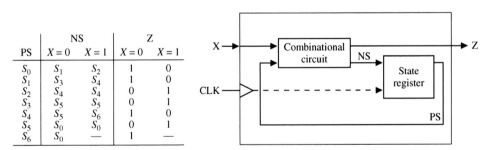

| | NS | | Z | |
|---|---|---|---|---|
| PS | $X = 0$ | $X = 1$ | $X = 0$ | $X = 1$ |
| $S_0$ | $S_1$ | $S_2$ | 1 | 0 |
| $S_1$ | $S_3$ | $S_4$ | 1 | 0 |
| $S_2$ | $S_4$ | $S_4$ | 0 | 1 |
| $S_3$ | $S_5$ | $S_5$ | 0 | 1 |
| $S_4$ | $S_5$ | $S_6$ | 1 | 0 |
| $S_5$ | $S_0$ | $S_0$ | 0 | 1 |
| $S_6$ | $S_0$ | — | 1 | — |

There are several ways to model this sequential machine. One approach would be to use two processes to represent the two parts of the circuit. One process models the combinational part of the circuit and generates the next state information and outputs. The other process models the state register and updates the state at the appropriate edge of the clock. Figure 2-54 illustrates such a model for this Mealy machine. The first process represents the combinational circuit. At the behavioral level, we will represent the state and next state of the circuit by integer signals initialized to 0. Please remember that this initialization is meaningful only for simulations. Since the circuit outputs, *Z* and *Nextstate*, can change when either the *State* or *X* changes, the sensitivity list includes both *State* and *X*. The case statement tests the value of *State*, and depending on the value of *X, Z* and *Nextstate* are assigned new values. The second process represents the state register. Whenever the rising edge of the clock occurs, *State* is updated to the value of *Nextstate*, so *CLK* appears in the sensitivity list. The second process will simulate correctly if written as

```
process(CLK)                    -- State Register
begin
  if CLK = '1' then             -- rising edge of clock (simulation)
    State <= Nextstate;
  end if;
end process;
```

but in order to synthesize with edge-triggered flip-flops, the `clk'event` attribute must be used, as in

```
process(CLK)                          -- State Register
begin                                 -- (synthesis)
  if CLK'event and CLK = '1' then     -- rising edge of clock
    State <= Nextstate;
  end if;
end process;
```

In Figure 2-54, *State* is an integer with range 0 to 6. The statement **when others => null** is not actually needed here because the outputs and next states of all possible values of *State* are explicitly specified; however, it should be included whenever the **else** clause of any **if** statement is omitted or when actions for all possible values of *State* are not specified. The **null** implies no action, which is appropriate since the other values of *State* should never occur. If else clauses are omitted or actions for any conditions are unspecified, synthesis typically results in creation of latches.

FIGURE 2-54: **Behavioral Model for Excess-3 Code Converter**

```
-- This is a behavioral model of a Mealy state machine (Figure 2-53)
-- based on its state table. The output (Z) and next state are
-- computed before the active edge of the clock. The state change
-- occurs on the rising edge of the clock.

entity Code_Converter is
  port(X, CLK: in bit;
       Z: out bit);
end Code_Converter;

architecture Behavioral of Code_Converter is
signal State, Nextstate: integer range 0 to 6;
begin
  process(State, X)              -- Combinational Circuit
  begin
    case State is
      when 0 =>
        if X = '0' then Z <= '1'; Nextstate <= 1;
```

```
            else Z <= '0'; Nextstate <= 2; end if;
        when 1 =>
          if X = '0' then Z <= '1'; Nextstate <= 3;
          else Z <= '0'; Nextstate <= 4; end if;
        when 2 =>
          if X = '0' then Z <= '0'; Nextstate <= 4;
          else Z <= '1'; Nextstate <= 4; end if;
        when 3 =>
          if X = '0' then Z <= '0'; Nextstate <= 5;
          else Z <= '1'; Nextstate <= 5; end if;
        when 4 =>
          if X = '0' then Z <= '1'; Nextstate <= 5;
          else Z <= '0'; Nextstate <= 6; end if;
        when 5 =>
          if X = '0' then Z <= '0'; Nextstate <= 0;
          else Z <= '1'; Nextstate <= 0; end if;
        when 6 =>
          if X = '0' then Z <= '1'; Nextstate <= 0;
          else Z <= '0'; Nextstate <= 0; end if;
        when others => null;           -- should not occur
      end case;
    end process;

    process(CLK)                       -- State Register
    begin
      if CLK'EVENT and CLK = '1' then -- rising edge of clock
        State <= Nextstate;
      end if;
    end process;
  end Behavioral;
```

A simulator command file that can be used to test Figure 2-54 is as follows:

```
add wave CLK X State NextState Z
force CLK 0 0, 1 100 -repeat 200
force X 0 0, 1 350, 0 550, 1 750, 0 950, 1 1350
run 1600
```

The first command specifies the signals that are to be included in the waveform output. The next command defines a clock with a period of 200 ns. *CLK* is '0' at time 0 ns, is '1' at time 100 ns, and repeats every 200 ns. In a command of the form

```
force signal_name v1 t1, v2 t2, . . .
```

signal_name gets the value v1 at time t1, the value v2 at time t2, and so on. *X* is '0' at time 0 ns, changes to '1' at time 350 ns, changes to '0' at time 550 ns, and so on. The *X* input corresponds to the sequence 0010 1001, and only the times at which *X* changes are specified. Execution of the preceding command file produces the waveforms shown in Figure 2-55.

In Chapter 1, we manually designed this state machine (Figure 1-26). This circuitry contained three flip-flops, four 3-input NAND gates, two 3-input NAND

FIGURE 2-55:
Simulator Output
for Excess-3 Code
Converter

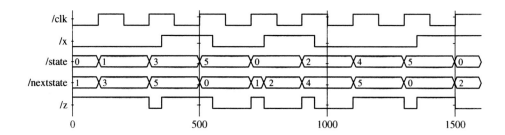

gates, and one inverter. The behavioral model of Figure 2-54 may not result in exactly that circuit. In fact, when we synthesized it using Xilinx ISE tools, we got a circuit that contains seven D-flip-flops, fifteen 2-input AND gates, three 2-input OR gates, and one 7-input OR gate. Apparently, the Xilinx synthesis tool may be using one-hot design by default, instead of encoded design. One-hot design is a popular approach for FPGAs, where flip-flops are abundant.

Figure 2-56 shows an alternative behavioral model for the code converter that uses a single process instead of two processes. The next state is not computed explicitly, but instead the state register is updated directly to the proper next state value on the rising edge of the clock. Since $Z$ can change whenever State or $X$ changes, $Z$ should not be computed in the clocked process. Instead, we have used a conditional assignment statement to compute $Z$. If $Z$ were updated in the clocked process, then a flip-flop would be created to store $Z$ and $Z$ would be updated at the wrong time. In general, the two-process model for a state machine is preferable to the one-process model, since the former corresponds more closely to the hardware implementation which uses a combinational circuit and a state register.

FIGURE 2-56: **Behavioral Model for Code Converter Using a Single Process**

```
-- This is a behavioral model of the Mealy state machine for BCD to
-- Excess-3 Code Converter based on its state table. The state change
-- occurs on the rising edge of the clock. The output is computed by a
-- conditional assignment statement whenever State or Z changes.

entity Code_Converter is
  port(X, CLK: in bit;
       Z: out bit);
end Code_Converter;

architecture one_process of Code_Converter is
signal State: integer range 0 to 6 := 0;
begin
  process(CLK)
  begin
    if CLK'event and CLK = '1' then
      case State is
        when 0 =>
          if X = '0' then State <= 1; else State <= 2; end if;
```

```
      when 1 =>
        if X = '0' then State <= 3; else State <= 4; end if;
      when 2 =>
        State <= 4;
      when 3 =>
        State <= 5;
      when 4 =>
        if X = '0' then State <= 5; else State <= 6; end if;
      when 5 =>
        State <= 0;
      when 6 =>
        State <= 0;
    end case;
  end if;
end process;
Z <= '1' when (State = 0 and X = '0') or (State = 1 and X = '0')
           or (State = 2 and X = '1') or (State = 3 and X = '1')
           or (State = 4 and X = '0') or (State = 5 and X = '1')
           or State = 6
      else '0';
end one_process;
```

Another way to model this Mealy machine is using the **dataflow** approach (i.e., using equations). The dataflow VHDL model of Figure 2-57 is based on the next state and output equations, which are derived in Chapter 1 (Figure 1-25). The flip-flops are updated in a process that is sensitive to *CLK*. When the rising edge of the clock occurs, $Q_1, Q_2$, and $Q_3$ are all assigned new values. A 10-ns delay is included to represent the propagation delay between the active edge of the clock and the change of the flip-flop outputs. Even though the assignment statements in the process are executed sequentially, $Q_1, Q_2$, and $Q_3$ are all scheduled to be updated at the same time, $T + \Delta$, where $T$ is the time at which the rising edge of the clock occurred. Thus,

FIGURE 2-57: **Sequential Machine Model Using Equations**

```
-- The following is a description of the sequential machine of
-- the BCD to Excess-3 code converter in terms of its next state
-- equations. The following state assignment was used:
-- S0-->0; S1-->4; S2-->5; S3-->7; S4-->6; S5-->3; S6-->2

entity Code_Converter is
  port(X, CLK: in bit;
       Z: out bit);
end Code_Converter;

architecture Equations of Code_Converter is
signal Q1, Q2, Q3: bit;
begin
  process(CLK)
```

```
  begin
    if CLK = '1' and CLK'event then      -- rising edge of clock
      Q1 <= not Q2 after 10 ns;
      Q2 <= Q1 after 10 ns;
      Q3 <= (Q1 and Q2 and Q3) or (not X and Q1 and not Q3) or
            (X and not Q1 and not Q2) after 10 ns;
    end if;
  end process;
  Z <= (not X and not Q3) or (X and Q3) after 20 ns;
end Equations;
```

the old value of $Q_1$ is used to compute $Q_2^+$, and the old values of $Q_1, Q_2$, and $Q_3$ are used to compute $Q_3^+$. The concurrent assignment statement for $Z$ causes $Z$ to be updated whenever a change in $X$ or $Q_3$ occurs. The 20-ns delay represents two gate delays. Note that in order to do VHDL modeling at this level, we need to perform state assignments, derive next state equations, and so on. In contrast, at the behavioral level, the state table was sufficient to create the VHDL model.

Yet another approach to creating a VHDL model of the aforementioned Mealy machine is to create a **structural** model describing the gates and flip-flops in the circuit. Figure 2-58 shows a structural VHDL representation of the circuit of Figure 1-20. Note that the designer had to manually perform the design and obtain the gate level circuitry here in order to create a model as in Figure 2-58. Seven NAND gates, three D flip-flops, and one inverter are used in the design presented in Chapter 1. When primitive components like gates and flip-flops are required, each of these components can be defined in a separate VHDL module. Depending on which CAD tools are used, the component modules can be included in the same file as the main VHDL description, or they be inserted as separate files in a VHDL project. The code in Figure 2-58 requires component modules DFF, Nand3, Nand2, and Inverter. CAD tools might include packages with similar components. If such packages are used, one should use the exact component names and port-map statements that match the input-output signals of the component in the package. The DFF module is as follows:

```
--D Flip-Flop
entity DFF is
  port(D, CLK: in bit;
       Q: out bit; QN: out bit := '1');
-- initialize QN to '1' since bit signals are defaulted to '0'
end DFF;
architecture SIMPLE of DFF is
begin
  process(CLK)              -- process is executed when CLK changes
  begin
    if CLK'event and CLK = '1' then      -- rising edge of clock
      Q <= D after 10 ns;
      QN <= not D after 10 ns;
    end if;
  end process;
end SIMPLE;
```

FIGURE 2-58: **Structural Model of Sequential Machine**

```
-- The following is a STRUCTURAL VHDL description of
-- the circuit to realize the BCD to Excess-3 code Converter.
-- This circuit was illustrated in Figure 1-20.
-- Uses components NAND3, NAND2, INVERTER and DFF
-- The component modules can be included in the same file
-- or they can be inserted as separate files.

entity Code_Converter is
  port(X,CLK: in bit;
       Z: out bit);
end Code_Converter;

architecture Structure of Code_Converter is
component DFF
  port(D, CLK: in bit; Q: out bit; QN: out bit := '1');
end component;
component Nand2
  port(A1, A2: in bit; Z: out bit);
end component;
component Nand3
  port(A1, A2, A3: in bit; Z: out bit);
end component;
component Inverter
  port(A: in bit; Z: out bit);
end component;
signal A1, A2, A3, A5, A6, D3: bit;
signal Q1, Q2, Q3: bit;
signal Q1N, Q2N, Q3N, XN: bit;
begin
  I1:  Inverter port map (X, XN);
  G1:  Nand3 port map (Q1, Q2, Q3, A1);
  G2:  Nand3 port map (Q1, Q3N, XN, A2);
  G3:  Nand3 port map (X, Q1N, Q2N, A3);
  G4:  Nand3 port map (A1, A2, A3, D3);
  FF1: DFF port map (Q2N, CLK, Q1, Q1N);
  FF2: DFF port map (Q1, CLK, Q2, Q2N);
  FF3: DFF port map (D3, CLK, Q3, Q3N);
  G5:  Nand2 port map (X, Q3, A5);
  G6:  Nand2 port map (XN, Q3N, A6);
  G7:  Nand2 port map (A5, A6, Z);
end Structure;
```

The Nand3 module is as follows:

```
--3 input NAND gate
entity Nand3 is
  port(A1, A2, A3: in bit; Z: out bit);
end Nand3;
```

```
architecture concur of Nand3 is
begin
    Z <= not (A1 and A2 and A3) after 10 ns;
end concur;
```

The Nand2 and Inverter modules are similar except for the number of inputs. We have assumed a 10-ns delay in each component, and this can easily be changed to reflect the actual delays in the hardware being used.

Since $Q_1$, $Q_2$, and $Q_3$ are initialized to '0', the complementary flip-flop outputs ($Q_1N$, $Q_2N$, and $Q_3N$) are initialized to '1'. $G_1$ is a three-input NAND gate with inputs $Q_1, Q_2, Q_3$, and output $A_1$. $FF_1$ is a D flip-flop with the $D$ input connected to $Q_2N$. Executing the simulator command file given next produces the waveforms of Figure 2-59, which are very similar to Figure 1-39.

```
add wave CLK X Q1 Q2 Q3 Z
force CLK 0 0, 1 100 -repeat 200
force X 0 0, 1 350, 0 550, 1 750, 0 950, 1 1350
run 1600
```

FIGURE 2-59:
**Waveforms for Code Converter**

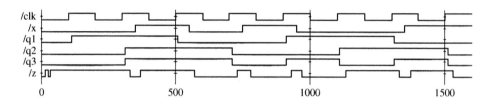

If we synthesized this structural description, we would get exactly the same circuit that we had in mind. Now the circuit includes only three D-flip-flops, three 2-input NAND gates, and four 3-input NAND gates. Compare it against the seven D-flip-flops, fifteen 2-input AND gates, three 2-input OR gates, and one 7-input OR gate generated when Figure 2-54 was synthesized. When the designer specified all components and their interconnections, the synthesizer tool did not have to infer or "guess."

Those who have developed C code with assembly inlining may feel some similarity to the phenomenon occurring here. By inlining the assembly code, you can precisely describe what microprocessor instruction sequence you want to be used, and the compiler gives you that. In a similar way, the synthesizer does not actually have to translate any structural descriptions that the designer wrote; it simply gives the hardware that the designer specified in a structural fashion. Some optimizing tools are capable of optimizing imperfect circuits that you might have specified. In general, you have more control of the generated circuitry when you use structural coding. However, it takes a lot more effort to produce a structural model because one needs to perform state assignments, derive next-state equations, and so on. **Time-to-market** is an important criterion for success in the IC market, and hence designers often use behavioral design in order to achieve quick time-to-market. Additionally, CAD tools have matured significantly during the past decade, and most synthesis tools are capable of producing efficient hardware for arithmetic and logic circuits.

● ● ● ● ● ● ● ● ● ● ● ●

# 2.16 Variables, Signals, and Constants

So far, we have used only signals in the VHDL code and have not used variables. VHDL also provides variables as in other general-purpose high-level languages. Variables may be used for local storage in processes. They can also be used in procedures and functions (which are yet to be introduced). A large part of what is described in this section is relevant only for simulation.

A variable declaration has the form

```
variable list_of_variable_names: type_name [ := initial_value];
```

Variables must be declared within the process in which they are used and are local to that process. (An exception to this rule is *shared* variables, which are not discussed in this text.) Signals, on the other hand, must be declared outside of a process. Signals declared at the start of an architecture can be used anywhere within that architecture. A signal declaration has the form

```
signal list_of_signal_names: type_name [ := initial_value];
```

Variables are updated using a variable assignment statement of the form

```
variable_name := expression;
```

When this statement is executed, the variable is instantaneously updated with no delay, not even a delta delay. In contrast, consider a signal assignment of the form

```
signal_name <= expression [after delay];
```

The expression is evaluated when this statement is executed, and the signal is scheduled to change after delay. If no delay is specified, then the signal is scheduled to be updated after a delta delay.

It is incorrect to use

```
variable_name <= expression [after delay];
```

---

**When to Use a Signal versus Variable:** If whatever you are modeling actually corresponds to some physical signal in your circuit, you should use a signal. If whatever you are modeling is simply a temporary value that you are using for convenience of programming, a variable will be sufficient. Values represented using variables will not appear on any physical wire in the implied circuit. If you would like them to appear, you should use signals.

---

The examples in Figures 2-60 and 2-61 illustrate the difference between using variables and signals in a process. The variables must be declared and initialized inside the process, whereas the signals must be declared and initialized outside the process. In Figure 2-60, if *trigger* changes at time = 10 ns, Var1, Var2, and Var3 are computed sequentially and updated instantly, and then Sum is computed using the

FIGURE 2-60: Process Using Variables and Corresponding Simulation Output

```
entity dummy is
end dummy;

architecture var of dummy is
signal trigger, sum: integer:=0;
begin
  process
  variable var1: integer:=1;
  variable var2: integer:=2;
  variable var3: integer:=3;
  begin
    wait on trigger;
    var1 := var2 + var3;
    var2 := var1;
    var3 := var2;
    sum <= var1 + var2 + var3;
  end process;
end var;
```

Simulation Output of 2-60

| ns | delta | trigger | Var1 | Var2 | Var3 | Sum |
|----|-------|---------|------|------|------|-----|
| 0  | +0    | 0       | 1    | 2    | 3    | 0   |
| 0  | +1    | 0       | 1    | 2    | 3    | 0   |
| 10 | +0    | 1       | 5    | 5    | 5    | 0   |
| 10 | +1    | 1       | 5    | 5    | 5    | 15  |

new variable values. The sequence is Var1 = 2 + 3 = 5, Var2 = 5, Var3 = 5. Then Sum = 5 + 5 + 5 is computed. Since Sum is a signal, it is updated $\Delta$ time later, so Sum = 15 at time = 10 + $\Delta$. In summary, variables work just as variables you are used to in another language, whereas signals get updated with time delays. In Figure 2-61, if *trigger* changes at time = 10 ns, signals Sig1, Sig2, Sig3, and Sum are all computed at time 10 ns, but the signals are not updated until time 10 + $\Delta$. The old values of Sig1 and Sig2 are used to compute Sig2 and Sig3. Therefore, at time = 10 + $\Delta$, Sig1 = 5, Sig2 = 1, Sig3 = 2, and Sum = 6.

FIGURE 2-61: Process Using Signals and Corresponding Simulation Output

```
entity dummy is
end dummy;

architecture sig of dummy is
signal trigger, sum: integer:=0;
signal sig1: integer:=1;
signal sig2: integer:=2;
signal sig3: integer:=3;
begin
  process
  begin
    wait on trigger;
    sig1 <= sig2 + sig3;
    sig2 <= sig1;
    sig3 <= sig2;
    sum <= sig1 + sig2 + sig3;
  end process;
end sig;
```

Simulation Output of 2-61

| ns | delta | trigger | Sig1 | Sig2 | Sig3 | Sum |
|----|-------|---------|------|------|------|-----|
| 0  | +0    | 0       | 1    | 2    | 3    | 0   |
| 0  | +1    | 0       | 1    | 2    | 3    | 0   |
| 10 | +0    | 1       | 1    | 2    | 3    | 0   |
| 10 | +1    | 1       | 5    | 1    | 2    | 6   |

During simulation, initialization makes the process execute once, and it stops when wait statements are encountered. Hence, simulation outputs can vary depending on whether the wait statements are put at the beginning of the process, end of the process, or whether a sensitivity list is used. Figures 2-62 and 2-63 illustrate various possibilities. Please remember that these differences are not important when VHDL is used for synthesis of hardware. These are subtle differences that only affect simulation of behavioral VHDL.

**FIGURE 2-62:** **Process Using Variables and Corresponding Simulation Output**

```
entity dummy is
end dummy;

architecture var of dummy is
signal trigger, sum: integer:=0;
begin
  process(trigger)
  variable var1: integer:=1;
  variable var2: integer:=2;
  variable var3: integer:=3;
  begin
    var1 := var2 + var3;
    var2 := var1;
    var3 := var2;
    sum <= var1 + var2 + var3;
  end process;
end var;
```

Simulation Output of 2-62

| ns | delta | trigger | Var1 | Var2 | Var3 | Sum |
|----|-------|---------|------|------|------|-----|
| 0  | +0    | 0       | 1    | 2    | 3    | 0   |
| 0  | +1    | 0       | 5    | 5    | 5    | 15  |
| 10 | +0    | 1       | 10   | 10   | 10   | 15  |
| 10 | +1    | 1       | 10   | 10   | 10   | 30  |

**FIGURE 2-63:** **Process Using Signals and Corresponding Simulation Output**

```
entity dummy is
end dummy;

architecture sig of dummy is
signal trigger, sum: integer:=0;
signal sig1: integer:=1;
signal sig2: integer:=2;
signal sig3: integer:=3;
begin
  process(trigger)
  begin
    sig1 <= sig2 + sig3;
    sig2 <= sig1;
    sig3 <= sig2;
    sum <= sig1 + sig2 + sig3;
  end process;
end sig;
```

Simulation Output of 2-63

| ns | delta | trigger | Sig1 | Sig2 | Sig3 | Sum |
|----|-------|---------|------|------|------|-----|
| 0  | +0    | 0       | 1    | 2    | 3    | 0   |
| 0  | +1    | 0       | 5    | 1    | 2    | 6   |
| 10 | +0    | 1       | 5    | 1    | 2    | 6   |
| 10 | +1    | 1       | 3    | 5    | 1    | 8   |

### 2.16.1 **Constants**

Like variables, constants are also used for convenience of programming.

A common form of constant declaration is

**constant** constant_name: type_name := constant_value;

A constant *delay1* of type time, having the value of 5 ns, can be defined as

**constant** delay1: time := 5 ns;

Constants declared at the start of an architecture can be used anywhere within that architecture, but constants declared within a process are local to that process.

Variables, signals, and constants can have any one of the predefined VHDL types, or they can have a user-defined type.

## 2.17 **Arrays**

Digital systems often use memory arrays. VHDL arrays can be used to specify the values to be stored in these arrays. A key feature of VLSI circuits is the repeated use of similar structures. Arrays in VHDL can be used while modeling the repetition.

In order to use an array in VHDL, we must first declare an array type and then declare an array object. For example, the following declaration defines a one-dimensional array type named SHORT_WORD:

**type** SHORT_WORD **is array** (15 **downto** 0) **of** bit;

An array of this type has an integer index with a range from **15 downto** 0, and each element of the array is of type bit. SHORT_WORD is the name of the newly created data type. We may note that SHORT_WORD is nothing but a bit_vector of size 16.

Now, we can declare array objects of type SHORT_WORD as follows:

```
signal    DATA_WORD:   SHORT_WORD;
variable  ALT_WORD:    SHORT_WORD := "0101010101010101";
constant  ONE_WORD:    SHORT_WORD := (others => '1');
```

Three different arrays are defined by the preceding statements. *DATA_WORD* is a signal array of 16 bits, indexed **15 downto** 0, which is initialized (by default) to all '0' bits. *ALT_WORD* is a variable array of 16 bits, which is initialized to alternating 0's and 1's. *ONE_WORD* is a constant array of 16 bits; all bits are set to 1 by (others => '1').

We can reference individual elements of the defined array by specifying an index value. For example, *ALT_WORD(0)* accesses the rightmost bit of *ALT_WORD*. We can also specify a portion of the array by specifying an index range: *ALT_WORD(5 downto 0)* accesses the low-order 6 bits of *ALT_WORD*, which have an initial value of "010101".

The array type and array object declarations illustrated here have the general forms

```
type array_type_name is array index_range of element_type;
signal array_name: array_type_name [ := initial_values];
```

In the preceding declaration, **signal** may be replaced with **variable** or **constant**.

### 2.17.1 Matrices

Multidimensional array types may also be defined with two or more dimensions. The following example defines a two-dimensional array variable, which is a matrix of integers with four rows and three columns:

```
type matrix4x3 is array (1 to 4, 1 to 3) of integer;
variable matrixA: matrix4x3 := ((1, 2, 3), (4, 5, 6), (7, 8, 9),
        (10, 11, 12));
```

The variable *matrixA* will be initialized to

$$
\begin{array}{ccc}
1 & 2 & 3 \\
4 & 5 & 6 \\
7 & 8 & 9 \\
10 & 11 & 12
\end{array}
$$

The array element *matrixA(3, 2)* references the element in the third row and second column, which has a value of 8.

When an array type is declared, the dimensions of the array may be left undefined. This is referred to as an *unconstrained array type*. For example,

```
type intvec is array (natural range <>) of integer;
```

declares intvec as an array type that defines a one-dimensional array of integers with an unconstrained index range of natural numbers. The default type for array indices is integer, but another type may be specified. Since the index range is not specified in the unconstrained array type, the range must be specified when the array object is declared. For example,

```
signal intvec5: intvec(1 to 5) := (3, 2, 6, 8, 1);
```

defines a signal array named *intvec5* with an index range of 1 to 5 that is initialized to 3, 2, 6, 8, 1. The following declaration defines matrix as a two-dimensional array type with unconstrained row and column index ranges:

```
type matrix is array (natural range <>, natural range <>) of
        integer;
```

*Example*

Parity bits are often used in digital communication for error detection and correction. The simplest of these involve transmitting one additional bit with the data, a parity bit. Use VHDL arrays to represent a parity generator that generates a 5-bit-odd-parity generation for a 4-bit input number using the look-up table (LUT) method.

### Answer

The input word is a 4-bit binary number. A 5-bit odd-parity representation will contain exactly an odd number of 1's in the output word. This can be accomplished by the read-only memory (ROM) method using a look-up table of size 16 entries × 5 bits. The look-up table is indicated in Figure 2-64.

**FIGURE 2-64: LUT Contents for a Parity Code Generator**

| Input (LUT Address) | | | | Output (LUT Data) | | | | |
|---|---|---|---|---|---|---|---|---|
| A | B | C | D | P | Q | R | S | T |
| 0 | 0 | 0 | 0 | 0 | 0 | 0 | 0 | 1 |
| 0 | 0 | 0 | 1 | 0 | 0 | 0 | 1 | 0 |
| 0 | 0 | 1 | 0 | 0 | 0 | 1 | 0 | 0 |
| 0 | 0 | 1 | 1 | 0 | 0 | 1 | 1 | 1 |
| 0 | 1 | 0 | 0 | 0 | 1 | 0 | 0 | 0 |
| 0 | 1 | 0 | 1 | 0 | 1 | 0 | 1 | 1 |
| 0 | 1 | 1 | 0 | 0 | 1 | 1 | 0 | 1 |
| 0 | 1 | 1 | 1 | 0 | 1 | 1 | 1 | 0 |
| 1 | 0 | 0 | 0 | 1 | 0 | 0 | 0 | 0 |
| 1 | 0 | 0 | 1 | 1 | 0 | 0 | 1 | 1 |
| 1 | 0 | 1 | 0 | 1 | 0 | 1 | 0 | 1 |
| 1 | 0 | 1 | 1 | 1 | 0 | 1 | 1 | 0 |
| 1 | 1 | 0 | 0 | 1 | 1 | 0 | 0 | 1 |
| 1 | 1 | 0 | 1 | 1 | 1 | 0 | 1 | 0 |
| 1 | 1 | 1 | 0 | 1 | 1 | 1 | 0 | 0 |
| 1 | 1 | 1 | 1 | 1 | 1 | 1 | 1 | 1 |

The VHDL code for the parity generator is illustrated in Figure 2-65. The IEEE numeric bit package is used here. $X$ and $Y$ are defined to be unsigned vectors. The first four bits of the output are identical to the input. Hence, instead of storing all five bits of the output, we might store only the parity bit and then concatenate it to the input bits. In the VHDL code (Figure 2-65), a new data type OutTable is defined to be an array of 16 bits. A constant table of type OutTable is defined using the following statement:

```
type OutTable is array(0 to 15) of bit;
```

The index of this array is an integer in the range 0 **to** 15. Hence, unsigned vector $X$ needs to be converted to an integer first, which can be done using the to_integer function defined in the library.

**FIGURE 2-65: Parity Code Generator Using the LUT Method**

```
library IEEE;
use IEEE.numeric_bit.all;

entity parity_gen is
  port(X: in unsigned(3 downto 0);
       Y: out unsigned(4 downto 0));
end parity_gen;

architecture Table of parity_gen is
type OutTable is array(0 to 15) of bit;
signal ParityBit: bit;
```

```
constant OT: OutTable := ('1','0','0','1','0','1','1','0',
                          '0','1','1','0','1','0','0','1');
begin
  ParityBit <= OT(to_integer(X));
  Y <= X & ParityBit;
end Table;
```

Predefined unconstrained array types in VHDL include bit_vector and string, which are defined as follows:

```
type bit_vector is array (natural range <>) of bit;
type string is array (positive range <>) of character;
```

The characters in a string literal must be enclosed in double quotes. For example, "This is a string." is a string literal. The following example declares a constant *string1* of type string:

```
constant string1: string(1 to 29) :=
    "This string is 29 characters."
```

A bit_vector literal may be written either as a list of bits separated by commas or as a string. For example, ('1','0','1','1','0') and "10110" are equivalent forms. The following declares a constant *A* that is a bit_vector with a range 0 **to** 5:

```
constant A: bit_vector(0 to 5) := "101011";
```

After a type has been declared, a related subtype can be declared to include a subset of the values specified by the type. For example, the type SHORT_WORD, which was defined at the start of this section, could have been defined as a subtype of bit_vector:

```
subtype SHORT_WORD is bit_vector (15 downto 0);
```

Two predefined subtypes of type integer are POSITIVE, which includes all positive integers, and NATURAL, which includes all positive integers and 0.

● ● ● ● ● ● ● ● ● ● ●
## 2.18 **Loops in VHDL**

Often, we encounter systems where some activity is happening in a repetitive fashion. VHDL loop statements can be used to express this behavior. A loop statement is a sequential statement. VHDL has several kinds of loop statements including **for** loops and **while** loops.

### 1. infinite loop

Infinite loops are undesirable in common computer languages, but they can be useful in hardware modeling where a device works continuously and continues to work until the power is off.

The general form for an infinite loop is

```
[loop-label:] loop
   sequential statements
end loop [loop-label];
```

An exit statement of the form

```
exit; or exit when condition;
```

may be included in the loop. The loop will terminate when the exit statement is executed, provided that the condition is TRUE.

2. **for loop**

One way to augment the basic loop is the **for** loop, where the number of invocations of the loop can be specified.

The general form of a **for** loop is

```
[loop-label:] for loop-index in range loop
   sequential statements
end loop [loop-label];
```

The `loop-index` is automatically defined when the loop is entered, and it should not explicitly be declared. It is initialized to the first value in the range and then the `sequential statements` are executed. The range is specified, for example as 0 **to** n, where $n$ can be a constant or variable. The `loop-index` can be used within the `sequential statements` inside the loop, but it cannot be changed within the loop. When the end of the loop is reached, the `loop-index` is set to the next value in the `range` and the `sequential statements` are executed again. This process continues until the loop has been executed for every value in the `range`, and then the loop terminates. After the loop terminates, the `loop-index` is no longer available.

We could use this type of a loop in behavioral models. The following excerpt models a 4-bit adder. The loop index ($i$) will be initialized to 0 when the **for** loop is entered, and the sequential statements will be executed. Execution will be repeated for $i = 1$, $i = 2$, and $i = 3$; then the loop will terminate. The carry out from one iteration (*cout*) is copied to the carry in (*cin*) before the end of the loop. Since variables are used for the sum and carry bits, the update of carry out happens instantaneously. Code like this often appears in VHDL functions and procedures (described in Chapter 8):

```
loop1: for i in 0 to 3 loop
   cout := (A(i) and B(i)) or (A(i) and cin) or (B(i) and cin);
   sum(i) := A(i) xor B(i) xor cin;
   cin := cout;
end loop loop1;
```

You could also use the for loop construct to create multiple copies of a basic cell. When the preceding code is synthesized, the synthesizer typically provides four copies of a 1-bit adder connected in a ripple carry fashion.

3. **while loop**

In the **for** loop, the loop index cannot be changed by the programmer. However, in the **while** loop, the loop index can be manipulated by the programmer. So incrementing the loop index by 2 can be done in the **while** loop. As in **while** loops in most languages, a condition is tested before each iteration. The loop is terminated if the condition is false. The general form of a **while** loop is

```
[loop-label:] while condition loop
   sequential statements
end loop [loop-label];
```

This construct is primarily for simulation.

Figure 2-66 illustrates a **while** loop that models a down counter. We use the **while** statement to continue the decrementing process until the stop is encountered or the counter reaches 0. The counter is decremented on every rising edge of clk until either the count is 0 or stop is 1.

FIGURE 2-66: **Use of While Loop**

```
while stop = '0' and count /= 0 loop
  wait until clk'event and clk = '1';
    count <= count - 1 ;
  wait for 0 ns;
end loop;
```

● ● ● ● ● ● ● ● ● ● ● ●
## 2.19   **Assert and Report Statements**

Once a VHDL model for a system is made, the next step is to test it. A model must be tested and validated before it can be successfully used. VHDL provides some special statements, such as **assert, report**, and **severity**, to aid in the testing and validation process.

The **assert** statement checks to see if a certain condition is true, and, if not, it causes an error message to be displayed. One form of the assert statement is

```
assert boolean-expression
  report string-expression
  [severity severity-level];
```

The **assert** statement specifies a Boolean expression which indicates the condition to be met. If the condition has not been met, an assertion violation has occurred. If an assertion violation occurs during simulation, the simulator reports it with the string-expression provided in the **report** clause. If the boolean-expression is false, then the string-expression is displayed on the monitor along with the severity-level. If the boolean-expression is true, no message is displayed.

There are four possible `severity-levels`: note, warning, error, and failure. We can include one of these to indicate the degree to which the violation of the particular assertion affects the operation of the model. For instance, a serious violation may have to be flagged as a failure, whereas some minor violation only needs to be flagged as a note or warning. The action taken for these severity-levels depends on the simulator. The severity-level is optional.

If the **assert** clause is omitted, then the report is always made. Thus, the statement

> **report** "ALL IS WELL";

will display the message "ALL IS WELL" whenever the statement is executed.

**Assert** and **report** statements are very useful for creation of **test benches**. A test bench is a piece of VHDL code that can provide input combinations to test a VHDL model for the system under test. It provides stimuli to the system/circuit under test. Test benches are frequently used during simulation to provide sequences of inputs to the circuit/VHDL model under test. Figure 2-67 shows a test-bench for testing the 4-bit binary adder that we created earlier in this chapter. The adder we are testing will be treated as a component and embedded in the test bench program. The signals generated within the test bench are interfaced to the adder, as shown in Figure 2-67. The test bench code in Figure 2-68 uses constant arrays to define the test inputs for the adder and the expected outputs. It uses a **for** loop to select the inputs from the

**FIGURE 2-67:**
**Interfacing of Signals while Using a Test Bench to Test a 4-Bit Adder**

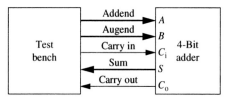

**FIGURE 2-68: Test Bench for 4-Bit Adder**

```
entity TestAdder is
end TestAdder;

architecture test1 of TestAdder is
component Adder4
  port(A, B: in bit_vector(3 downto 0); Ci: in bit;
       S: out bit_vector(3 downto 0); Co: out bit);
end component;
constant N: integer := 11;
type bv_arr is array(1 to N) of bit_vector(3 downto 0);
type bit_arr is array(1 to N) of bit;
constant addend_array: bv_arr := ("0111", "1101", "0101", "1101",
        "0111", "1000", "0111", "1000", "0000", "1111", "0000");
constant augend_array: bv_arr := ("0101", "0101", "1101", "1101",
        "0111", "0111", "1000", "1000", "1101", "1111", "0000");
constant cin_array: bit_arr := ('0', '0', '0', '0', '1', '0', '0',
        '0', '1', '1', '0');
```

```
constant sum_array: bv_arr := ("1100", "0010", "0010", "1010",
        "1111", "1111", "1111", "0000", "1110", "1111", "0000");
constant cout_array: bit_arr := ('0', '1', '1', '1', '0', '0', '0',
        '1', '0', '1', '0');
signal addend, augend, sum: bit_vector(3 downto 0);
signal cin, cout: bit;
begin
  process
  begin
    for i in 1 to N loop
      addend <= addend_array(i);
      augend <= augend_array(i);
      cin <= cin_array(i);
      wait for 40 ns;
      assert (sum = sum_array(i) and cout = cout_array(i))
        report "Wrong Answer"
        severity error;
    end loop;
    report "Test Finished";
  end process;
  add1: adder4 port map (addend, augend, cin, sum, cout);
end test1;
```

arrays. It uses **assert** and **report** statements to check the outputs and report whether the output matched the expected output for the particular combination of inputs. The **assert** statement is meaningful only for simulation. During synthesis, the synthesizer may simply assume that the assertion violation does not exist.

We will provide another example to illustrate how a waveform input can be provided in a test bench. In earlier examples in this chapter, we used simulator commands to test VHDL models. Figure 2-69 illustrates a piece of VHDL code that

FIGURE 2-69: Generating a Test Sequence for Testing VHDL Model for Code Converter

```
entity test_code_conv is
end test_code_conv;

architecture tester of test_code_conv is
signal X, CLK, Z: bit;
component Code_Converter is
  port(X, CLK: in bit;
       Z: out bit);
end component;
begin
  clk <= not clk after 100 ns;
  X <= '0', '1' after 350 ns, '0' after 550 ns, '1' after
       750 ns, '0' after 950 ns, '1' after 1350 ns;
  CC: Code_Converter port map (X, clk, Z);
end tester;
```

accomplishes exactly the same testing that was done using simulator commands in Figure 2-55. A time-varying signal is provided to input $X$ using the statement

```
X <= '0', '1' after 350 ns, '0' after 550ns, '1' after 750 ns, '0'
     after 950 ns, '1' after 1350 ns;
```

In this chapter, we have covered the basics of VHDL. We have shown how to use VHDL to model combinational logic and sequential machines. Since VHDL is a hardware description language, it differs from an ordinary programming language in several ways. Most importantly, VHDL statements execute concurrently, since they must model real hardware in which the components are all in operation at the same time. Statements within a process execute sequentially, but the processes themselves operate concurrently. VHDL signals model actual signals in the hardware, but variables may be used for internal computation that is local to processes, procedures, and functions. We will cover more advanced features of VHDL in Chapter 8.

● ● ● ● ● ● ● ● ● ● ● ● ● ●

# Problems

2.1 **(a)** What do the acronyms VHDL and VHSIC stand for?
   **(b)** How does a hardware description language like VHDL differ from an ordinary programming language?
   **(c)** What are the advantages of using a hardware description language as compared with schematic capture in the design process?

2.2 **(a)** Which of the following are legal VHDL identifiers? `123A`, `A_123`, `_A123`, `A123_`, `c1__c2`, `and`, `and1`
   **(b)** Which of the following identifiers are equivalent? `aBC`, `ABC`, `Abc`, `abc`

2.3 Given the concurrent VHDL statements:

```
B <= A and C after 3ns;
C <= not B after 2ns;
```

   **(a)** Draw the circuit the statements represent.
   **(b)** Draw a timing diagram if initially $A = B = $ '0' and $C = $ '1', and $A$ changes to '1' at time 5 ns.

2.4 Write a VHDL description of the following combinational circuit using concurrent statements. Each gate has a 5-ns delay, excluding the inverter, which has a 2-ns delay.

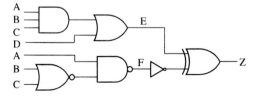

2.5 **(a)** Write VHDL code for a full subtracter using logic equations.

**(b)** Write VHDL code for a 4-bit subtracter using the module defined in (a) as a component.

2.6 Write VHDL code for the following circuit. Assume that the gate delays are negligible.

**(a)** Use concurrent statements.

**(b)** Use a process with sequential statements.

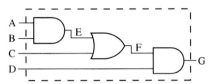

2.7 In the following VHDL code, $A$, $B$, $C$, and $D$ are integers that are 0 at time 10 ns. If $D$ changes to 1 at 20 ns, specify the times at which $A$, $B$, and $C$ will change and the values they will take.

```
process(D)
begin
  A <= 1 after 5 ns;
  B <= A + 1;            -- executes before A changes
  C <= B after 10 ns;    -- executes before B changes
end process;
```

2.8 **(a)** What device does the following VHDL code represent?

```
process(CLK, Clr, Set)
begin
  if Clr = '1' then Q <= '0';
  elsif Set = '1' then Q <= '1';
  elsif CLK'event and CLK <= '0' then
    Q <= D;
  end if;
end process;
```

**(b)** What happens if $Clr = Set = $ '1' in the device in part (a)?

2.9 Write a VHDL description of an S-R latch using a process.

2.10 An M-N flip-flop responds to the falling clock edge as follows:

If $M = N = $ '0', the flip-flop changes state.
If $M = $ '0' and $N = $ '1', the flip-flop output is set to '1'.
If $M = $ '1' and $N = $ '0', the flip-flop output is set to '0'.
If $M = N = $ '1', no change of flip-flop state occurs.
The flip-flop is cleared asynchronously if $CLRn = $ '0'.

Write a complete VHDL module that implements an M-N flip-flop.

2.11 A DD flip-flop is similar to a D flip-flop, except that the flip-flop can change state ($Q^+ = D$) on both the rising edge and falling edge of the clock input. The flip-flop has a direct reset input, $R$, and $R = $ '0' resets the flip-flop to $Q = $ '0' independent of the clock. Similarly, it has a direct set input, $S$, that sets the flip-flop to '1' independent of the clock. Write a VHDL description of a DD flip-flop.

2.12 An inhibited toggle flip-flop has inputs $I0$, $I1$, $T$, and *Reset*, and outputs $Q$ and $QN$. *Reset* is active high and overrides the action of the other inputs. The flip-flop works as follows. If $I0 = $ '1', the flip-flop changes state on the rising edge of $T$; if $I1 = $ '1', the flip-flop changes state on the falling edge of $T$. If $I0 = I1 = $ '0', no state change occurs (except on reset). Assume the propagation delay from $T$ to output is 8 ns and from reset to output is 5 ns.

**(a)** Write a complete VHDL description of this flip-flop.
**(b)** Write a sequence of simulator commands that will test the flip-flop for the input sequence $I1 = $ '1', toggle $T$ twice, $I1 = $ '0', $I0 = $ '1', toggle $T$ twice.

2.13 In the following VHDL process $A$, $B$, $C$, and $D$ are all integers that have a value of 0 at time $= 10$ ns. If $E$ changes from '0' to '1' at time $= 20$ ns, specify the time(s) at which each signal will change and the value to which it will change. List these changes in chronological order $(20, 20 + \Delta, 20 + 2\Delta, \text{etc.})$.

```
p1: process
begin
  wait on E;
  A <= 1 after 5 ns;
  B <= A + 1;
  C <= B after 10 ns;
  wait for 0 ns;
  D <= B after 3 ns;
  A <= A + 5 after 15 ns;
  B <= B + 7;
end process p1;
```

2.14 In the following VHDL process $A$, $B$, $C$, and $D$ are all integers that have a value of 0 at time $= 10$ ns. If $E$ changes from '0' to '1' at time $= 20$ ns, specify the time(s) at which each signal will change and the value to which it will change. List these changes in chronological order $(20, 20 + \Delta, 20 + 2\Delta, \text{etc.})$.

```
p2: process(E)
begin
  A <= 1 after 5 ns;
  B <= A + 1;
  C <= B after 10 ns;

  D <= B after 3 ns;
  A <= A + 5 after 15 ns;
  B <= B + 7;
end process p2;
```

2.15  For the following VHDL code, assume that $D$ changes to '1' at time 5 ns. Give the values of $A$, $B$, $C$, $D$, $E$, and $F$ each time a change occurs. That is, give the values at time 5 ns, $5 + \Delta$, $5 + 2\Delta$, and so on. Carry this out until either 20 steps have occurred, until no further change occurs, or until a repetitive pattern emerges.

```
entity prob is
  port(D: inout bit);
end prob;

architecture q1 of prob is
  signal A, B, C, E, F: bit;
begin
  C <= A;
  A <= (B and not E) or D;
  P1: process (A)
  begin
    B <= A;
  end process P1;
  P2: process
  begin
    wait until A = '1';
    wait for 0 ns;
    E <= B after 5 ns;
    D <= '0';
    F <= E;
  end process P2;
end architecture q1;
```

2.16  Assuming $B$ is driven by the simulator command

```
force B 0 0, 1 10, 0 15, 1 20, 0 30, 1 35
```

draw a timing diagram illustrating $A$, $B$, and $C$ if the following concurrent statements are executed:

```
A <= transport B after 5 ns;
C <= B after 8 ns;
```

2.17  Assuming $B$ is driven by the simulator command

```
force B 0 0, 1 4, 0 10, 1 15, 0 20, 1 30, 0 40
```

draw a timing diagram illustrating $A$, $B$, and $C$ if the following concurrent statements are executed:

```
A <= transport B after 5 ns;
C <= B after 5 ns;
```

2.18  In the following VHDL code, $A$, $B$, $C$, and $D$ are bit signals that are '0' at time = 4 ns. If $A$ changes to 1 at time 5 ns, make a table showing the values of $A$, $B$, $C$, and $D$ as a function of time until time = 18 ns. Include deltas. Indicate the times at which each process begins executing.

```
P1: process(A)
begin
   B <= A after 5 ns;
   C <= B after 2 ns;
end process;
P2: process
begin
   wait on B;
   A <= not B;
   D <= not A xor B;
end process;
```

2.19 If A = "101", B = "011", and C = "010", what are the values of the following statements?

(a) (A & B) or (B & C)
(b) A ror 2
(c) A sla 2
(d) A & not B = "111110"
(e) A or B and C

2.20 Consider the following VHDL code:

```
entity Q3 is
   port(A, B, C, F, Clk: in bit;
        E: out bit);
end Q3;

architecture Qint of Q3 is
   signal D, G: bit;
begin
   process(Clk)
   begin
      if Clk'event and Clk = '1' then
         D <= A and B and C;
         G <= not A and not B;
         E <= D or G or F;
      end if;
   end process;
end Qint;
```

(a) Draw a block diagram for the circuit (no gates—at block level only).
(b) Give the circuit generated by the preceding code (at the gate level)

2.21 Implement the following VHDL code using these components: D flip-flops with clock enable, a multiplexer, an adder, and any necessary gates. Assume that *Ad* and *Ora* will never be '1' at the same time, and only enable the flip-flops when *Ad* or *Ora* is '1'.

```
library IEEE;
use IEEE.numeric_bit.all;

entity module1 is
port(A, B: in unsigned (2 downto 0);
     Ad, Ora, clk: in bit;
     C: out unsigned (2 downto 0));
end module1;

architecture RT of module1 is
begin
  process(clk)
  begin
    if clk = '1' and clk'event then
      if Ad = '1' then C <= A + B; end if;
      if Ora = '1' then C <= A or B; end if;
    end if;
  end process;
end RT;
```

2.22 Draw the circuit represented by the following VHDL process. Use only two gates.

```
process(clk, clr)
begin
  if clr = '1' then Q <= '0';
  elsif clk'event and clk = '0' and CE = '1' then
    if C = '0' then Q <= A and B;
    else Q <= A or B; end if;
  end if;
end process;
```

Why is *clr* on the sensitivity list but *C* is not?

2.23 **(a)** Write a selected signal assignment statement to represent the 4-to-1 MUX shown below. Assume that there is an inherent delay in the MUX that causes the change in output to occur 10 ns after a change in input.
   **(b)** Repeat (a) using a conditional signal assignment statement.
   **(c)** Repeat (a) using a process and a case statement.

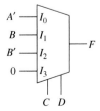

2.24 **(a)** Write a VHDL process that is equivalent to the following concurrent statement:

    A <= B1 **when** C = 1 **else** B2 **when** C = 2 **else** B3 **when** C = 3 **else** 0;

**(b)** Draw a circuit to implement the following VHDL statement:

    A <= B1 **when** C1 = '1' **else** B2 **when** C2 = '1' **else**
         B3 **when** C3 = '1' **else** '0';

where all signals are of type bit.

2.25 Write a VHDL description of an SR latch.

**(a)** Use a conditional assignment statement.
**(b)** Use the characteristic equation.
**(c)** Use logic gates.

2.26 For the VHDL code of Figure 2-38, what will be the values of $S$ and $Co$ if $A$ = "1101", $B$ = "111", and $Ci$ = '1'?

2.27 Write VHDL code to add a positive integer $B$ ($B < 16$) to a 4-bit bit-vector $A$ to produce a 5-bit bit-vector as a result. Use an overloaded operator in the IEEE numeric bit package to do the addition. Use calls to conversion functions as needed. The final result should be a bit-vector, not an unsigned vector.

2.28 A 4-bit magnitude comparator chip (e.g., 74LS85) compares two 4-bit numbers $A$ and $B$ and produces outputs to indicate whether $A < B, A = B$, or $A > B$. There are three output signals to indicate each of the above conditions. Note that exactly one of the output lines will be high and the other two lines will be low at any time. The chip is a cascadable chip and has three inputs, $A > B.IN$, $A = B.IN$, and $A < B.IN$, in order to allow cascading the chip to make 8-bit or bigger magnitude comparators.

**(a)** Draw block diagram of a 4-bit magnitude comparator
**(b)** Draw a block diagram to indicate how you can construct an 8-bit magnitude comparator using two 4-bit magnitude comparators.
**(c)** Write behavioral VHDL description for the 4-bit comparator.
**(d)** Write VHDL code for the 8-bit comparator using two 4-bit comparators as components.

2.29 Write a VHDL module that describes a 16-bit serial-in, serial-out shift register with inputs $SI$ (serial input), $EN$ (enable), and $CK$ (clock, shifts on rising edge) and a serial output ($SO$).

2.30 A description of a 74194 four-bit bidirectional shift register follows:
The *CLRb* input is asynchronous and active low and overrides all the other control inputs. All other state changes occur following the rising edge of the clock. If the control inputs $S_1 = S_0 = 1$, the register is loaded in parallel. If $S_1 = 1$ and $S_0 = 0$, the register is shifted right and *SDR* (serial data right) is shifted into $Q_3$. If $S_1 = 0$ and $S_0 = 1$, the register is shifted left and *SDL* is shifted into $Q_0$. If $S_1 = S_0 = 0$, no action occurs.

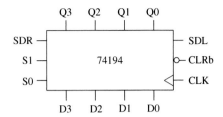

(a) Write a behavioral-level VHDL model for the 74194.
(b) Draw a block diagram and write a VHDL description of an 8-bit bidirectional shift register that uses two 74194's as components. The parallel inputs and outputs to the 8-bit register should be $X(7 \textbf{ downto } 0)$ and $Y(7 \textbf{ downto } 0)$. The serial inputs should be $RSD$ and $LSD$.

2.31  A synchronous (4-bit) up/down decade counter with output $Q$ works as follows: All state changes occur on the rising edge of the $CLK$ input, except the asynchronous clear $(CLR)$. When $CLR = 0$, the counter is reset regardless of the values of the other inputs.

If the $LOAD$ input is 0, the data input $D$ is loaded into the counter.
If $LOAD = ENT = ENP = UP = 1$, the counter is incremented.
If $LOAD = ENT = ENP = 1$ and $UP = 0$, the counter is decremented.
If $ENT = UP = 1$, the carry output $(CO) = 1$ when the counter is in state 9.
If $ENT = 1$ and $UP = 0$, the carry output $(CO) = 1$ when the counter is in state 0.

(a) Write a VHDL description of the counter.
(b) Draw a block diagram and write a VHDL description of a decimal counter that uses two of the above counters to form a two-decade decimal up/down counter that counts up from 00 to 99 or down from 99 to 00.
(c) Simulate for the following sequence: load counter with 98, increment three times, do nothing for two clocks, decrement four times, and clear.

2.32  Write a VHDL model for a 74HC192 synchronous 4-bit up/down counter. Ignore all timing data. Your code should contain a statement of the form **process**(DOWN, UP, CLR, LOADB)

2.33  Consider the following 8-bit bi-directional synchronous shift register with parallel load capability. The notation used to represent the input/output pins is explained below.

| | |
|---|---|
| $CLR$ | Asynchronous Clear, overrides all other inputs |
| $Q(7:0)$ | 8-bit output |
| $D(7:0)$ | 8-bit input |
| $S0, S1$ | mode control inputs |
| $LSI$ | serial input for left shift |
| $RSI$ | serial input for right shift |

The mode control inputs work as follows:

| S0 | S1 | Action |
|----|----|--------|
| 0 | 0 | No action |
| 0 | 1 | Right shift |
| 1 | 0 | Left shift |
| 1 | 1 | Load parallel data (i.e., $Q = D$) |

**(a)** Write an entity description for this shift register.

**(b)** Write an architecture description of this shift register.

**(c)** Draw a block diagram illustrating how two of these can be connected to form a 16-bit cyclic shift register, which is controlled by signals $L$ and $R$. If $L = $ '1' and $R = $ '0', then the 16-bit register is cycled left. If $L = $ '0' and $R = $ '1', the register is cycled right. If $L = R = $ '1', the 16-bit register is loaded from $X(15:0)$. If $L = R = $ '0', the register is unchanged.

**(d)** Write an entity description for the module in part (c).

**(e)** Write an architecture description using the module from parts (a) and (b).

2.34 Complete the following VHDL code to implement a counter that counts in the following sequence: $Q = $ 1000, 0111, 0110, 0101, 0100, 0011, 1000, 0111, 0110, 0101, 0100, 0011, . . . (repeats). The counter is synchronously loaded with 1000 when $Ld8 = $ '1'. It goes through the prescribed sequence when $Enable = $ '1'. The counter outputs $S5 = $ '1' whenever it is in state 0101. Do not change the entity in any way. Your code must be synthesizable.

```
library IEEE;
use IEEE.numeric_bit.all;

entity countQ1 is
    port(clk, Ld8, Enable: in bit; S5: out bit;
         Q: out unsigned(3 downto 0));
end countQ1;
```

2.35 A synchronous 4-bit UP/DOWN binary counter has a synchronous clear signal $CLR$ and a synchronous load signal $LD$. $CLR$ has higher priority than $LD$. Both $CLR$ and $LD$ are active high. $D$ is a 4-bit input to the counter and $Q$ is the 4-bit output from the counter. $UP$ is a signal that controls the direction of counting. If $CLR$ and $LD$ are not active and $UP = 1$, the counter increments. If $CLR$ and $LD$ are not active and $UP = 0$, the counter decrements. All changes occur on the falling edge of the clock.

**(a)** Write a behavioral VHDL description of the counter.

**(b)** Use the above UP/DOWN counter to implement a synchronous modulo 6 counter that counts from 1 to 6. This modulo 6 counter has an external reset which, if applied, makes the count $= 1$. A count enable signal $CNT$ makes it count in the sequence 1, 2, 3, 4, 5, 6, 1, 2, . . . incrementing once for each clock pulse. You should use any necessary logic to make the counter go to count $= 1$ after count $= 6$. The modulo 6 counter only counts in the UP sequence. Provide a textual/pictorial description of your approach.

**(c)** Write a behavioral VHDL description for the modulo-6 counter in part (b).

2.36  Examine the following VHDL code and complete the following exercises:

```
entity Problem
  port(X, CLK: in bit;
       Z1, Z2: out bit);
end Problem;

architecture Table of Problem is
  signal State, Nextstate: integer range 0 to 3 := 0;
begin
  process(State, X)        --Combinational Circuit
  begin
    case State is
      when 0 =>
        if X = '0' then Z1 < = '1'; Z2 <= '0'; Nextstate < = 0;
        else Z1 < = '0'; Z2 < = '0'; Nextstate < = 1; end if;
      when 1 =>
        if X = '0' then Z1 < = '0'; Z2 <= '1'; Nextstate < = 1;
        else Z1 < = '0'; Z2 < = '1'; Nextstate < = 2; end if;
      when 2 =>
        if X = '0' then Z1 < = '0'; Z2 <= '1'; Nextstate < = 2;
        else Z1 < = '0'; Z2 < = '1'; Nextstate < = 3; end if;
      when 3 =>
        if X = '0' then Z1 < = '0'; Z2 <= '0'; Nextstate < = 0;
        else Z1 < = '1'; Z2 < = '0'; Nextstate < = 1; end if;
    end case;
  end process;
  process(CLK)             --State Register
  begin
    if CLK'event and CLK = '1' then     --rising edge of clock
      State <= Nextstate;
    end if;
  end process;
end Table;
```

(a) Draw a block diagram of the circuit implemented by this code.

(b) Write the state table that is implemented by this code.

2.37 (a) Write a behavioral VHDL description of the state machine you designed in Problem 1.13. Assume that state changes occur on the falling edge of the clock pulse. Instead of using if-then-else statements, represent the state table and output table by arrays. Compile and simulate your code using the following test sequence:

$$X = 1101\ 1110\ 1111$$

$X$ should change 1/4 clock period after the rising edge of the clock.

(b) Write a data flow VHDL description using the next state and output equations to describe the state machine. Indicate on your simulation output at which times $S$ and $V$ are to be read.

**(c)** Write a structural model of the state machine in VHDL that contains the inter-connection of the gates and D flip-flops.

2.38 **(a)** Write a behavioral VHDL description of the state machine that you designed in Problem 1.14. Assume that state changes occur on the falling edge of the clock pulse. Use a case statement together with if-then-else statements to represent the state table. Compile and simulate your code using the following test sequence:

$$X = 1011\ 0111\ 1000$$

$X$ should change 1/4 clock period after the falling edge of the clock.
**(b)** Write a data flow VHDL description using the next state and output equations to describe the state machine. Indicate on your simulation output at which times $D$ and $B$ should be read.
**(c)** Write a structural model of the state machine in VHDL that contains the inter-connection of the gates and J-K flip-flops.

2.39 A Moore sequential machine with two inputs ($X_1$ and $X_2$) and one output ($Z$) has the following state table:

| Present State | Next State $X_1X_2 = 00$ | 01 | 10 | 11 | Output (Z) |
|---|---|---|---|---|---|
| 1 | 1 | 2 | 2 | 1 | 0 |
| 2 | 2 | 1 | 2 | 1 | 1 |

Write VHDL code that describes the machine at the behavioral level. Assume that state changes occur 10 ns after the falling edge of the clock, and output changes occur 10 ns after the state changes.

2.40 Write VHDL code to implement the following state table. Use two processes. State changes should occur on the falling edge of the clock. Implement the $Z_1$ and $Z_2$ outputs using concurrent conditional statements. Assume that the combinational part of the sequential circuit has a propagation delay of 10 ns, and the propagation delay between the rising-edge of the clock and the state register output is 5 ns.

| Present State | Next state $X_1X_2 = 00$ | 01 | 11 | Output $(Z_1Z_2)$ |
|---|---|---|---|---|
| 1 | 3 | 2 | 1 | 00 |
| 2 | 2 | 1 | 3 | 10 |
| 3 | 1 | 2 | 3 | 01 |

2.41 In the following code, *state* and *nextstate* are integers with a range of 0 to 2.

```
process(state, X)
begin
  case state is
    when 0 => if X = '1' then nextstate <= 1;
    when 1 => if X = '0' then nextstate <= 2;
```

```
   when 2 => if X = '1' then nextstate <= 0;
 end case;
end process;
```

**(a)** Explain why a latch would be created when the code is synthesized.
**(b)** What signal would appear at the latch output?
**(c)** Make changes in the code which would eliminate the latch.

2.42 For the process given below, $A$, $B$, $C$, and $D$ are all integers that have a value of 0 at time = 10 ns. If $E$ changes from '0' to '1' at time 20 ns, specify all resulting changes. Indicate the time at which each change will occur, the signal/variable affected, and the value to which it will change.

```
process
  variable F: integer: =1; variable A: integer: =0;
begin
  wait on E;
  A := 1;
  F := A + 5;
  B <= F + 1 after 5 ns;
  C <= B + 2 after 10 ns;
  D <= C + 5 after 15 ns;
  A := A + 5;
end process;
```

2.43 What is wrong with the following model of a 4-to-1 MUX? (It is not a syntax error.)

```
architecture mux_behavioral of Fourto1mux is
signal sel: integer range 0 to 3;
begin
  process(A, B, I0, I1, I2, I3)
  begin
    sel <= 0;
    if A = '1' then sel <= sel + 1; end if;
    if B = '1' then sel <= sel + 2; end if;
    case sel is
      when 0 => F <= I0;
      when 1 => F <= I1;
      when 2 => F <= I2;
      when 3 => F <= I3;
    end case;
  end process;
end mux_behavioral;
```

2.44 When the following VHDL code is simulated, $A$ is changed to '1' at time 5 ns. Make a table that shows all changes in $A$, $B$, and $D$ and the times at which they occur through time = 40 ns.

```
entity Q1F00 is
  port(A: inout bit);
end Q1F00;
```

```
architecture Q1F00 of Q1F00 is
  signal B, D: bit;
begin
  D <= A xor B after 10 ns;
  process(D)
    variable C: bit;
  begin
    C := not D;
    if C = '1' then
      A <= not A after 15 ns;
    end if;
    B <= D;
  end process;
end Q1F00;
```

2.45 What device does the following VHDL code represent?

```
process(CLK, RST)
  variable Qtmp: bit;
begin
  if RST '1' then Qtmp := '0';
  elsif CLK'event and CLK = '1' then
    if T = '1' then
      Qtmp := not Qtmp;
    end if;
  end if;
  Q <= Qtmp;
end process;
```

2.46 **(a)** Write a VHDL module for a LUT with four inputs and three outputs. The 3-bit output should be a binary number equal to the number of 1's in the LUT input.

**(b)** Write a VHDL module for a circuit that counts the number of 1's in a 12-bit number. Use three of the modules from (a) along with overloaded addition operators.

**(c)** Simulate your code and test if for the following data inputs:

$$111111111111, 010110101101, 100001011100$$

2.47 Implement a 3-to-8 decoder using a LUT. Give the LUT truth table and write the VHDL code. The inputs should be $A$, $B$, and $C$ and the output should be an 8-bit unsigned vector.

2.48 $A(1 \text{ to } 20)$ is an array of 20 integers. Write VHDL code that finds the largest integer in the array

**(a)** Using a **for** loop

**(b)** Using a **while** loop

2.49 Write VHDL code to test a Mealy sequential circuit with one input ($X$) and one output ($Z$). The code should include the Mealy circuit as a component. Assume the Mealy circuit changes state on the rising edge of *CLK*. Your test code should generate a clock with 100 ns period. The code should apply the following test sequence:

$$X = 0, 1, 1, 0, 1, 1, 0, 1, 1, 1, 0, 0$$

$X$ should change 10 ns after the rising edge of CLK. Your test code should read $Z$ at an appropriate time and verify that the following output sequence was generated:

$$Z = 1, 0, 0, 1, 1, 0, 1, 1, 0, 1, 1, 0$$

Report an error if the output sequence from the Mealy circuit is incorrect; otherwise, report "sequence correct." Complete the following architecture for the tester:

```
architecture test1 of tester is
  component Mealy
    -- sequential circuit to be tested; assume this component
    -- is available in your design; do NOT write code for the
    -- component
    port(X, CLK: in bit; Z: out bit);
  end component;
  signal XA: bit_vector(0 to 11) := "011011011100";
  signal ZA: bit_vector(0 to 11) := "100110110110";
```

2.50 Write a VHDL test bench that will test the VHDL code for the sequential circuit of Figure 2-58. Your test bench should generate all ten possible input sequences (0000, 1000, 0100, 1100, . . . ) and verify that the output sequences are correct. Remember that the components have a 10-ns delay. The input should be changed 1/4 of a clock period after the rising edge of the clock and the output should be read at the appropriate time. Report "Pass" if all sequences are correct; otherwise, report "Fail."

2.51 Write a test bench to test the counter of Problem 2.34. The test bench should generate a clock with a 100-ns period. The counter should be loaded on the first clock; then it should count for five clocks; then it should do nothing for two clocks; then it should continue counting for ten clocks. The test bench port should output the current time (in time units, not the count) whenever $S5 = `1$'. Use only concurrent statements in your test bench.

2.52 Complete the following VHDL code to implement a test bench for the sequential circuit SMQ1. Assume that the VHDL code for the SMQ1 sequential circuit module is already available. Use a clock with a 50-ns half-period. Your test bench should test the circuit for the input sequence $X = 1, 0, 0, 1, 1$. Assume that the correct output sequence for this input sequence is $1, 1, 0, 1, 0$. Use a single concurrent statement to generate the $X$ sequence. The test bench should read the values of output $Z$ at the proper times and compare them with the correct values of $Z$. The correct answer is stored as a bit-vector constant:

$$answer(1 \text{ } to \text{ } 5) = \text{"11010"};$$

The port signal *correct* should be set to TRUE if the answer is correct; otherwise, it should be set to FALSE. Make sure that your read $Z$ at the correct time. Use wait statements in your test bench.

```
entity testSMQ1 is
  port(correct: out Boolean);
end testSMQ1;
architecture testSM of testSMQ1 is
  component SMQ1  -- the sequential circuit module
    port(X, CLK: in bit; Z: out bit);
  end component;
  constant answer: bit_vector(1 to 5) := "11010";
begin
```

# Introduction to Programmable Logic Devices

Chapter 1 illustrated how the same digital circuit can be implemented using a variety of standard building blocks. If we can put several of these building blocks into an integrated circuit (IC) and provide the user with mechanisms to modify the configuration, we can implement almost any circuit within a chip. This is the general principle of programmable logic devices.

This chapter introduces the use of programmable logic devices in digital design. Read-only memories (ROMs), programmable logic arrays (PLAs), and programmable array logic (PAL) devices are discussed first. Then complex programmable logic devices (CPLDs) and field programmable gate arrays (FPGAs) are introduced. Use of these devices allows us to implement complex logic functions, which require many gates and flip-flops, with a single IC. Although FPGAs are introduced, only an overview is provided in this chapter. A detailed treatment of FPGAs is provided in Chapter 6.

## 3.1 Brief Overview of Programmable Logic Devices

Designers have always liked programmable logic devices such as PALs and FPGAs for implementation of digital circuits. First, there is reasonable integration ability, allowing implementation of a significant amount of functionality into one physical chip. Programmable logic devices remove the use of multiple off-the-shelf devices and the inconvenience and unreliability associated with external wires. Second, there is the increased ability to change designs. Many of the programmable devices allow easy reprogramming. In general, it is easier to change the design in case of errors or changes in design specifications. Nowadays, programmable logic comes in different types: devices that can be programmed only once and those that can be reprogrammed many times.

Figure 3-1 illustrates a classification of popular programmable logic devices. Programmable logic can be considered to fall into field programmable logic and factory programmable logic. The term *field* indicates that this type of device is programmed in the user's "field" rather than in a semiconductor fab. Often, many may refer to programmable logic to mean devices that are field programmable.

FIGURE 3-1: **Major Programmable Logic Devices**

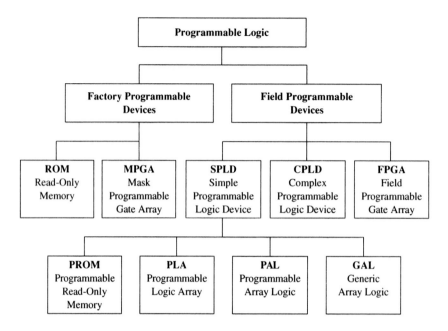

However, there are factory programmable devices, too. These are generic devices which can be programmed at the factory to meet customers' requirements. The programming technology uses an irreversible process; hence, programming can be done only once. Examples of factory programmable logic are mask programmable gate arrays (MPGAs) and read-only memories (ROMs). The earliest generations of many programmable devices were programmable only at the factory.

Read-only memories can be considered as an early form of programmable logic. While primarily meant for use as memory, ROMs can be used to implement any combinational circuitry. This will be illustrated later in Section 3.2.1. MPGAs are traditional gate arrays, which require a mask to be designed. MPGAs are often simply called **gate arrays** and have been a popular technology for creating application-specific integrated circuits (ASICs).

User programmable logic in the form of AND-OR circuits was developed at the beginning of the 1970s. By 1972–1973, one-time field programmable logic arrays that permitted instant customizations by designers were available. Some referred to these devices as field programmable logic arrays or FPLAs. Monolithic Memories Inc. (MMI), a company that was bought by Advanced Micro Devices (AMD), created integrated circuits called programmable logic arrays (PLAs) in 20- and 24-pin packages that could yield the same functionality as 5 to 20 off-the-shelf chips. A similar device is the programmable array logic or PAL.

PALs and PLAs contain arrays of gates. In the PLA, there is a programmable AND array and a programmable OR array, allowing users to implement combinational functions in two levels of gates. The PAL is a special case of a PLA, in that the OR array is fixed and only the AND array is programmable. Many PALs also contain flip-flops.

In the 1970s and 1980s, PALs and PLAs were very popular. Part of the popularity was due to the ease of design. MMI and Advanced Micro Devices created a simple programming language, called PALASM, to easily convert Boolean equations into PLA configurations. PALASM made programming PALs and PLAs relatively simple.

The early programmable devices allowed only one-time programming. The next technological innovation that helped programmable logic was advancement in erasure of programmable devices. In early days, erasure of programmable logic used ultraviolet light. With ultraviolet light, erasing the configuration of a device meant removing the device from the circuit and placing it in an ultraviolet environment. Hence, in-circuit erasure was not possible. Ultraviolet erasers were slow; typically 10 or 15 minutes were required to perform erasures. Then electrically erasable technology came along. This led to the creation of field programmable logic arrays that can be easily and quickly erased and reprogrammed without removing the chip from the board.

The early PALs and PLAs were soon followed by CMOS electrically erasable programmable logic devices (PLDs). While the term *PLDs* can be used to refer to any programmable logic devices, there are a set of devices, including the popular PALCE22V10, that are often referred to as PLDs. PLDs contain macroblocks with arrays of gates, multiplexers, flip-flops, or other standard building blocks. Several of these macroblocks appear in a PLD. Lattice Semiconductor created similar devices with easy reprogrammability and called its line of devices **GALs** or **generic array logic**.

Now, many refer to PLAs, PALs, GALs, PLDs, and PROMs collectively as simple PLDs (SPLDs) in contrast to another type of product that has come on the market, complex PLDs (CPLDs). As the name suggests, CPLDs have more integration capability than SPLDs. They come in sizes ranging from 500 to 16,000 gates. CPLDs essentially put multiple PLDs into the same chip with some kind of an interconnection circuit, typically a crossbar switch.

In the late 1980s, Xilinx started using static random-access memory (RAM) storage elements to hold configuration information for programmable devices and created devices called FPGAs that can integrate a fairly large amount of logic. Contrary to their names, the basic building blocks in these devices were not arrays of gates but were bigger and complex blocks containing static RAMs and multiplexers. Several PLD vendors and gate array companies soon jumped into the market, creating a variety of FPGA architectures, some of which used reprogrammable technologies and others of which used one-time programmable fuse technologies. The FPGA technology has continually improved in the last 15 years. Now, there are FPGAs that can contain more than 5 million gates.

Programmable logic devices basically contain an array of basic building blocks which can be used to implement whatever functionality one desires. Different programmable devices differ in the building blocks or the amount of programmability they provide. Table 3-1 illustrates a comparison of various programmable logic devices. FPGAs are bigger and more complex than CPLDs. The routing resources in FPGAs are more complex than those in simple programmable devices. The variety of alternate routes that can be taken causes the paths taken by signals to be

TABLE 3-1:
A Comparison of
Programmable
Devices

|  | SPLD | CPLD | FPGA |
|---|---|---|---|
| **Density** | Low<br>Few hundred gates | Low to Medium<br>500 to 12,000 gates | Medium to High 3,000<br>to 5,000,000 gates |
| **Timing** | Predictable | Predictable | Unpredictable |
| **Cost** | Low | Low to Medium | Medium to High |
| **Major Vendors** | Lattice Semiconductor<br>Cypress<br>AMD | Xilinx<br>Altera | Xilinx<br>Altera<br>Lattice Semiconductor<br>Actel |
| **Example Device Families** | **Lattice Semiconductor**<br>GAL16LV8<br>GAL22V10<br><br>**Cypress**<br>PALCE16V8<br><br>**AMD**<br>22V10 | **Xilinx**<br>CoolRunner<br>XC9500<br><br>**Altera**<br>MAX | **Xilinx**<br>Virtex<br>Spartan<br><br>**Altera**<br>Stratix<br><br>**Lattice**<br>Mach<br>ECP<br><br>**Actel**<br>Accelerator |

unpredictable. FPGAs are more expensive than CPLDs and SPLDs. They contain more overhead for programming. In this chapter, we describe various programmable devices, including SPLDs, CPLDs, and FPGAs.

Many names and abbreviations in this field have historically been used to refer to specific types of programmable devices; however, one may not find the name to be meaningful. Consider PALs and PLAs. Both are arrays of logic. The fact that PLAs contain programmable AND and OR arrays and PALs contain only programmable AND arrays is due to nothing but historical reasons. PALs and PLAs could very well be named the other way around. But it is important for students to understand what these names popularly refer to because they will need to communicate with fellow designers and other design teams. Conventions are important in facilitating communication.

● ● ● ● ● ● ● ● ● ● ● ●
## 3.2  Simple Programmable Logic Devices

With the advent of CPLDs and FPGAs, the early generation programmable logic devices, such as ROMs, PALs, PLAs, and PLDs, can be collectively called simple programmable logic devices (SPLDs). In this section, we describe the implementation of digital circuits in simple PLDs.

### 3.2.1 **Read-Only Memories**

A read-only memory (ROM) consists of an array of semiconductor devices that are interconnected to store an array of binary data. Once binary data is stored in the ROM, it can be read out whenever desired, but the data that is stored cannot be changed under normal operating conditions. Figure 3-2(a) shows a ROM that has three input lines and four output lines. Figure 3-2(b) shows a typical truth table, which relates the ROM inputs and outputs. For each combination of input values on the three input lines, the corresponding pattern of 0's and 1's appears on the ROM output lines. For example, if the combination $ABC = 010$ is applied to the input lines, the pattern $F_0F_1F_2F_3 = 0111$ appears on the output lines. Each of the output patterns that is stored in the ROM is called a *word*. Since the ROM has three input lines, we have $2^3 = 8$ different combinations of input values. Each input combination serves as an *address*, which can select one of the eight words stored in the memory. Since there are four output lines, each word is four bits long, and the size of this ROM is 8 words $\times$ 4 bits.

FIGURE 3-2:
**An 8-Word × 4-Bit ROM**

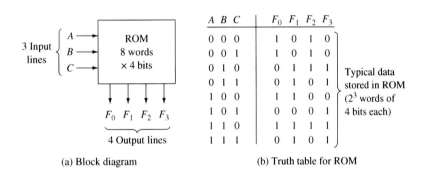

(a) Block diagram          (b) Truth table for ROM

A ROM which has $n$ input lines and $m$ output lines (Figure 3-3) contains an array of $2^n$ words, and each word is $m$ bits long. The input lines serve as an address to select one of the $2^n$ words. When an input combination is applied to the ROM, the pattern of 0's and 1's stored in the corresponding word in the memory appears at the output lines. For the example in Figure 3-3, if $00\ldots11$ is applied to the input (address lines) of the ROM, the word $110\ldots010$ will be selected and transferred to

FIGURE 3-3:
**Read-Only Memory with *n* Inputs and *m* Outputs**

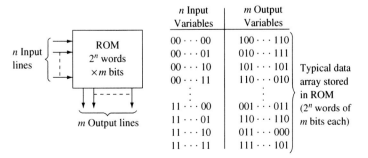

the output lines. A $2^n \times m$ ROM can realize $m$ functions of $n$ variables since it can store a truth table with $2^n$ rows and $m$ columns. Typical sizes for commercially available ROMs range from 32 words $\times$ 4 bits to 512K words $\times$ 8 bits, or larger.

A ROM basically consists of a decoder and a memory array. When a pattern of $n$ 0's and 1's is applied to the decoder inputs, exactly one of the $2^n$ decoder outputs is 1. This decoder output line selects one of the words in the memory array, and the bit pattern stored in this word is transferred to the memory output lines.

Basic types of ROMs include mask programmable ROMs, user programmable ROMs (PROMs), erasable programmable ROMs (usually called EPROMs), electrically erasable and programmable ROMs (EEPROMs), and flash memories. In the mask programmable ROM, the data array is permanently stored at the time of manufacture. This is accomplished by selectively including or omitting the switching elements at the row-column intersections of the memory array. This requires preparation of a special "mask," which is used during fabrication of the integrated circuit. Preparation of this mask is expensive, so use of mask programmable ROMs is economically feasible only if a large quantity (typically several thousand or more) is required with the same data array. There are also one-time user programmable ROMs or PROMs.

Modification of the data stored in a ROM is often necessary during the developmental phases of a digital system, so EPROMs are used instead of mask programmable ROMs. EPROMs use a special charge-storage mechanism to enable or disable the switching elements in the memory array. An EPROM programmer is used to provide appropriate voltage pulses to store electronic charges in the memory array locations. The data stored in this manner is generally permanent until erased using ultraviolet light. After erasure, a new set of data can be stored in the EPROM.

The EEPROM is similar to an EPROM, except that erasure is accomplished using electrical pulses instead of ultraviolet light. A traditional EEPROM can be erased and reprogrammed only a limited number of times, typically 100 to 1000 times. **Flash memories** are similar to EEPROMs, except that they use a different charge-storage mechanism. They usually have built-in programming and erasure capability so that data can be written to the flash memory while it is in a circuit without the need for a separate programmer.

A ROM can implement any combinational circuit. Essentially, if the outputs for all combinations of inputs are stored in the ROM, the outputs can be "looked up" in the table stored in the ROM. The ROM method is also called the **look-up table (LUT)** method for this reason.

Consider the implementation of a 2-bit adder in a ROM. This adder must add two 2-bit numbers. Since the maximum value of a 2-bit number is 3, the maximum sum is 6, necessitating 3 bits for the sum. The truth table for such an adder is illustrated in Figure 3-4. We could also design a 2-bit full adder assuming a carry input in addition to the two 2-bit numbers.

This 2-bit adder can be implemented with a $16 \times 3$ ROM. The input numbers ($X$ and $Y$) must be connected to the four address lines, and the three data lines will produce the sum bits.

**FIGURE 3-4: Block Diagram and Truth Table of a 2-Bit Adder**

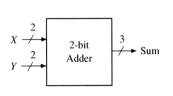

| $X_1$ | $X_0$ | $Y_1$ | $Y_0$ | $S_2$ | $S_1$ | $S_0$ |
|---|---|---|---|---|---|---|
| 0 | 0 | 0 | 0 | 0 | 0 | 0 |
| 0 | 0 | 0 | 1 | 0 | 0 | 1 |
| 0 | 0 | 1 | 0 | 0 | 1 | 0 |
| 0 | 0 | 1 | 1 | 0 | 1 | 1 |
| 0 | 1 | 0 | 0 | 0 | 0 | 1 |
| 0 | 1 | 0 | 1 | 0 | 1 | 0 |
| 0 | 1 | 1 | 0 | 0 | 1 | 1 |
| 0 | 1 | 1 | 1 | 1 | 0 | 0 |
| 1 | 0 | 0 | 0 | 0 | 1 | 0 |
| 1 | 0 | 0 | 1 | 0 | 1 | 1 |
| 1 | 0 | 1 | 0 | 1 | 0 | 0 |
| 1 | 0 | 1 | 1 | 1 | 0 | 1 |
| 1 | 1 | 0 | 0 | 0 | 1 | 1 |
| 1 | 1 | 0 | 1 | 1 | 0 | 0 |
| 1 | 1 | 1 | 0 | 1 | 0 | 1 |
| 1 | 1 | 1 | 1 | 1 | 1 | 0 |

Figure 3-5 illustrates the ROM implementation of this 2-bit full adder. Assuming the connections that are shown, the contents of the ROM in its 16 locations should be 0, 1, 2, 3, 1, 2, 3, 4, 2, 3, 4, 5, 3, 4, 5, and 6, respectively (representing the digits in decimal). The LSB of the sum will come from the LSB of the data bus.

**FIGURE 3-5: ROM Implementation of a 2-Bit Full Adder**

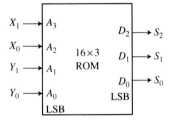

**Example**

Compute the size of the ROM required to implement an 8-to-3 priority encoder.

### Solution

An encoder performs the inverse function of a decoder. An 8-to-3 priority encoder is illustrated in Figure 3-6. If input $y_i$ is 1 and the other inputs are 0, then the $abc$ outputs represent a binary number equal to $i$. An additional output $d$ is used to indicate invalid outputs. A value of 1 on bit $d$ indicates that the output bits $a$, $b$, and $c$ are valid. If more than one input is 1 in a priority encoder, the highest numbered input determines the output. The truth table in Figure 3-6 illustrates the output combinations for each input combination. The X's in the truth table indicate "don't cares." As illustrated, the 8-to-3 priority encoder has eight inputs and four outputs. Hence, it needs a $2^8 \times 4$ bit ROM.

### Comment

There will be 256 entries in this ROM. When all the "don't cares" in the truth table in Figure 3-6 are expanded, it does result in 256 entries.

FIGURE 3-6: **8-to-3 Priority Encoder**

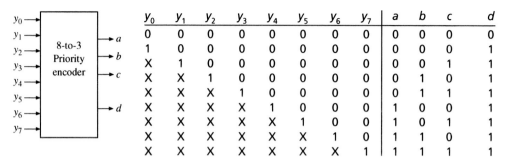

| $y_0$ | $y_1$ | $y_2$ | $y_3$ | $y_4$ | $y_5$ | $y_6$ | $y_7$ | a | b | c | d |
|---|---|---|---|---|---|---|---|---|---|---|---|
| 0 | 0 | 0 | 0 | 0 | 0 | 0 | 0 | 0 | 0 | 0 | 0 |
| 1 | 0 | 0 | 0 | 0 | 0 | 0 | 0 | 0 | 0 | 0 | 1 |
| X | 1 | 0 | 0 | 0 | 0 | 0 | 0 | 0 | 0 | 1 | 1 |
| X | X | 1 | 0 | 0 | 0 | 0 | 0 | 0 | 1 | 0 | 1 |
| X | X | X | 1 | 0 | 0 | 0 | 0 | 0 | 1 | 1 | 1 |
| X | X | X | X | 1 | 0 | 0 | 0 | 1 | 0 | 0 | 1 |
| X | X | X | X | X | 1 | 0 | 0 | 1 | 0 | 1 | 1 |
| X | X | X | X | X | X | 1 | 0 | 1 | 1 | 0 | 1 |
| X | X | X | X | X | X | X | 1 | 1 | 1 | 1 | 1 |

**Example**

Implement, in ROM, a sequential machine whose state table is given in Figure 3-7. You may note that this is the BCD to excess-3 code converter that we designed in Chapter 1.

FIGURE 3-7:
**State Table for a Sequential Circuit**

| PS | NS $X=0$ | NS $X=1$ | Z $X=0$ | Z $X=1$ |
|---|---|---|---|---|
| $S_0$ | $S_1$ | $S_2$ | 1 | 0 |
| $S_1$ | $S_3$ | $S_4$ | 1 | 0 |
| $S_2$ | $S_4$ | $S_4$ | 0 | 1 |
| $S_3$ | $S_5$ | $S_5$ | 0 | 1 |
| $S_4$ | $S_5$ | $S_6$ | 1 | 0 |
| $S_5$ | $S_0$ | $S_0$ | 0 | 1 |
| $S_6$ | $S_0$ | — | 1 | — |

## Solution

A sequential circuit can easily be designed using a ROM and flip-flops. The combinational part of the sequential circuit can be realized using the ROM. The ROM can be used to realize the output functions and the next state functions. The state of the circuit can then be stored in a register of D flip-flops and fed back to the input of the ROM. Use of D flip-flops is preferable to J-K flip-flops since using 2-input flip-flops would require increasing the number of inputs for the flip-flops (which are outputs from the ROM). The fact that the D flip-flop input equations would generally require more gates than the J-K equations is of no consequence since the size of the ROM depends only on the number of inputs and outputs and not on the complexity of the equations being realized. For this reason, the state assignment used is also of little importance, and, generally, a state assignment in straight binary order is as good as any.

In order to realize the above sequential machine, a ROM and three D flip-flops are necessary. The ROM will generate the next state equations and output $Z$ from the present states and input $X$. Hence, the ROM needs four address lines (three coming from flip-flops and one for $X$) and it should provide four outputs (three next state bits and output $Z$). Figure 3-8 illustrates the general organization of the implementation. Since the ROM has four inputs, it contains $2^4 = 16$ words. In general, a Mealy sequential circuit with $i$ inputs, $j$ outputs, and $k$ state variables can be realized using $k$ D flip-flops and a ROM with $i + k$ inputs ($2^{i+k}$ words) and $j + k$ outputs.

FIGURE 3-8:
**Realization of a
Mealy Sequential
Circuit with a ROM**

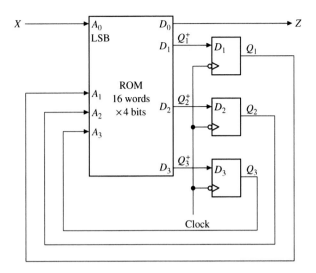

Now, let us derive the contents of the ROM. Table 3-2 gives the truth table for the sequential circuit, which implements the state table of Figure 3-7 with the "don't cares" replaced by 0's and using a straight binary state assignment.

TABLE 3-2:
**ROM Truth Table**

| $Q_3$ | $Q_2$ | $Q_1$ | $X$ | $Q_3^+$ | $Q_2^+$ | $Q_1^+$ | $Z$ |
|---|---|---|---|---|---|---|---|
| 0 | 0 | 0 | 0 | 0 | 0 | 1 | 1 |
| 0 | 0 | 0 | 1 | 0 | 1 | 0 | 0 |
| 0 | 0 | 1 | 0 | 0 | 1 | 1 | 1 |
| 0 | 0 | 1 | 1 | 1 | 0 | 0 | 0 |
| 0 | 1 | 0 | 0 | 1 | 0 | 0 | 0 |
| 0 | 1 | 0 | 1 | 1 | 0 | 0 | 1 |
| 0 | 1 | 1 | 0 | 1 | 0 | 1 | 0 |
| 0 | 1 | 1 | 1 | 1 | 0 | 1 | 1 |
| 1 | 0 | 0 | 0 | 1 | 0 | 1 | 1 |
| 1 | 0 | 0 | 1 | 1 | 1 | 0 | 0 |
| 1 | 0 | 1 | 0 | 0 | 0 | 0 | 0 |
| 1 | 0 | 1 | 1 | 0 | 0 | 0 | 1 |
| 1 | 1 | 0 | 0 | 0 | 0 | 0 | 1 |
| 1 | 1 | 0 | 1 | 0 | 0 | 0 | 0 |
| 1 | 1 | 1 | 0 | 0 | 0 | 0 | 0 |
| 1 | 1 | 1 | 1 | 0 | 0 | 0 | 0 |

Assuming that $Q_3$, $Q_2$, $Q_1$, and $X$ are connected to the address lines in that order, with $X$ connected to the LSB, the contents of the ROM to implement this sequential machine are 3, 4, 7, 8, 8, 9, A, B, B, C, 0, 1, 1, 0, 0, and 0 (in hexadecimal representation). The hexadecimal (hex) representation is a concise and convenient way to represent the outputs. The output $Z$ will come from the LSB of the data lines. The next state information will be available from the three MSBs of the ROM data lines.

### 3.2.2 **Programmable Logic Arrays**

A programmable logic array (PLA) performs the same basic function as a ROM. A PLA with $n$ inputs and $m$ outputs (Figure 3-9) can realize $m$ functions of $n$ variables. The internal organization of the PLA is different from that of the ROM. The decoder is replaced with an AND array that realizes selected product terms of the input variables. The OR array OR's together the product terms needed to form the output functions.

**FIGURE 3-9:**
**Programmable Logic Array Structure**

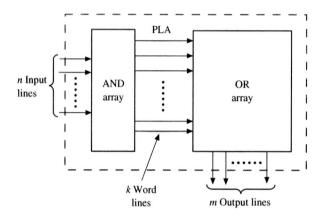

Figure 3-10 shows a PLA that realizes the following functions:

$$F_0 = \Sigma m(0, 1, 4, 6) = A'B' + AC' \tag{3-1}$$
$$F_1 = \Sigma m(2, 3, 4, 6, 7) = B + AC'$$
$$F_2 = \Sigma m(0, 1, 2, 6) = A'B' + BC'$$
$$F_3 = \Sigma m(2, 3, 5, 6, 7) = AC + B$$

The above logic functions contain three variables. In a PLA implementation, each product term in the equation is created first, and then required product terms are OR'ed using the OR gate. Hence, product terms can be shared while using the PLA. Instead of minimizing each function separately, we want to minimize the total number of product terms. There are five distinct product terms in the above four equations. Figure 3-10 illustrates a PLA with three inputs, five product terms, and four outputs, implementing the above four equations. It should be noted that the number of terms in each equation is not important, as long as there are AND gates to generate all product terms required for all outputs together.

Internally, the PLA may use NOR-NOR logic instead of AND-OR logic. The array shown in Figure 3-10 is thus equivalent to the nMOS PLA structure of Figure 3-11. Logic gates are formed in the array by connecting nMOS switching transistors between the column lines and the row lines.

**FIGURE 3-10:**
**PLA with Three Inputs, Five Product Terms, and Four Outputs (Logic Level)**

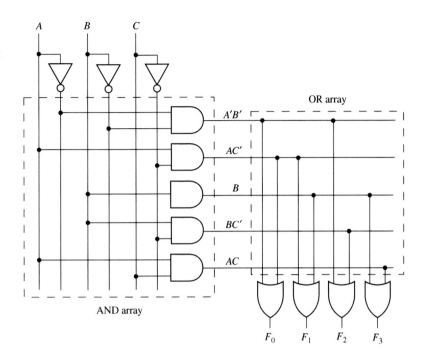

**FIGURE 3-11:**
**PLA with Three Inputs, Five Product Terms, and Four Outputs (Transistor Level)**

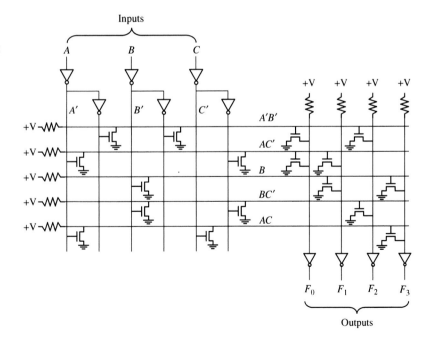

> Source, drain, and gate are the names of the three terminals of the metal oxide semiconductor (MOS) transistor. The gate is the one that is used to control the ON/OFF action. There are two types of MOS transistors, *n-channel MOS (nMOS) and p-channel MOS (pMOS)*. The illustrations in this section use nMOS transistors. A popular technology since the 1990s is complementary MOS (CMOS), where nMOS and pMOS transistors are used together in a complementary fashion.

Figure 3-12 shows the implementation of a two-input NOR gate using nMOS transistors. The transistors act as switches, so if the gate input is a logic 0, the transistor is off. If the gate input is a logic 1, the transistor provides a conducting path to ground. If $X_1 = X_2 = 0$, both transistors are off, and the pull-up resistor brings the $Z$ output to a logic 1 level $(+V)$. If either $X_1$ or $X_2$ is 1, the corresponding transistor is turned on, and $Z = 0$. Thus, $Z = (X_1 + X_2)' = X_1'X_2'$, which corresponds to a NOR gate. The part of the PLA array that realizes $F_0$ is equivalent to the NOR-NOR gate structure shown in Figure 3-13. After canceling the extra inversions, this reduces to an AND-OR structure.

**FIGURE 3-12:**
**nMOS NOR Gate**

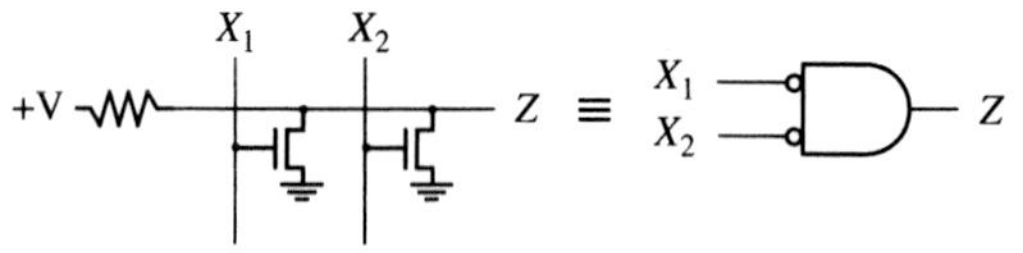

**FIGURE 3-13:**
**Conversion of NOR-NOR to AND-OR**

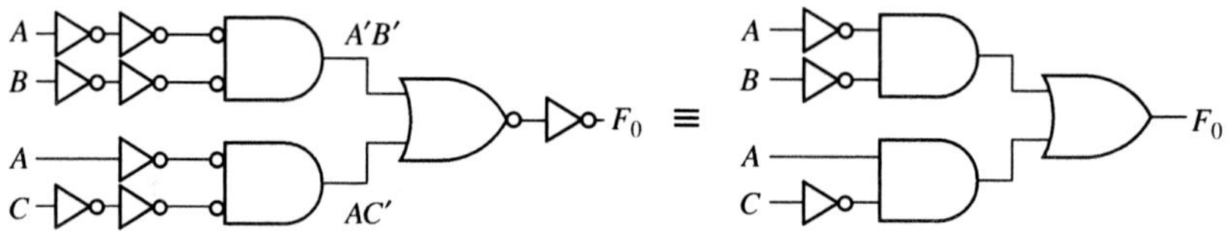

The contents of a PLA can be specified by a modified truth table. Table 3-3 specifies the PLA in Figure 3-10. The input side of the table specifies the product terms. The symbols 0, 1, and – indicate whether a variable is complemented, not complemented, or not present in the corresponding product term. The output side of the table specifies which product terms appear in each output function. A 1 or 0 indicates whether a given product term is present or not present in the corresponding output function. Thus, the first row of Table 3-3 indicates that the term $A'B'$ is present in output functions $F_0$ and $F_2$, and the second row indicates that $AC'$ is present in $F_0$ and $F_1$.

Next, we will realize the following functions using a PLA:

$$F_1 = \Sigma m(2, 3, 5, 7, 8, 9, 10, 11, 13, 15) \tag{3-2}$$
$$F_2 = \Sigma m(2, 3, 5, 6, 7, 10, 11, 14, 15)$$
$$F_3 = \Sigma m(6, 7, 8, 9, 13, 14, 15)$$

TABLE 3-3:
**PLA Table for Equations 3-1**

| Product Term | Inputs | | | Outputs | | | |
|---|---|---|---|---|---|---|---|
| | A | B | C | $F_0$ | $F_1$ | $F_2$ | $F_3$ |
| $A'B'$ | 0 | 0 | — | 1 | 0 | 1 | 0 |
| $AC'$ | 1 | — | 0 | 1 | 1 | 0 | 0 |
| $B$ | — | 1 | — | 0 | 1 | 0 | 1 |
| $BC'$ | — | 1 | 0 | 0 | 0 | 1 | 0 |
| $AC$ | 1 | — | 1 | 0 | 0 | 0 | 1 |

If we minimize each function separately, the result is

$$F_1 = bd + b'c + ab'$$ (3-3)
$$F_2 = c + a'bd$$
$$F_3 = bc + ab'c' + abd$$

If we implement these reduced equations in a PLA, a total of eight different product terms (including $c$) are required.

Instead of minimizing each function separately, we want to minimize the total number of rows in the PLA table. In this case, the number of terms in each equation is not important, since the size of the PLA does not depend on the number of terms within an equation. Equations (3-3) are plotted on the Karnaugh maps shown in Figure 3-14. Since the term $ab'c'$ is already needed for $F_3$, we can use it in $F_1$ instead of $ab'$. The other two 1's in $ab'$ are covered by the $b'c$ term. This eliminates the need to use a row of the PLA table for $ab'$. Since the terms $a'bd$ and $abd$ are needed in $F_2$ and $F_3$, respectively, we can replace $bd$ in $F_1$ with $a'bd + abd$. This eliminates the need for a row to implement $bd$. Since $b'c$ and $bc$ are used in $F_1$ and $F_3$, respectively, we can replace $c$ in $F_2$ with $b'c + bc$. The resulting Equations (3-4) correspond to the reduced PLA table (Table 3-4). Instead of using Karnaugh maps to reduce the number of rows

FIGURE 3-14:
**Multiple-Output Karnaugh Maps**

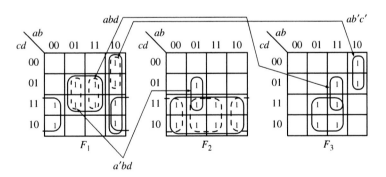

TABLE 3-4:
**Reduced PLA Table**

| a | b | c | d | $F_1$ | $F_2$ | $F_3$ |
|---|---|---|---|---|---|---|
| 0 | 1 | — | 1 | 1 | 1 | 0 |
| 1 | 1 | — | 1 | 1 | 0 | 1 |
| 1 | 0 | 0 | — | 1 | 0 | 1 |
| — | 0 | 1 | — | 1 | 1 | 0 |
| — | 1 | 1 | — | 0 | 1 | 1 |

$$F_1 = a'bd + abd + ab'c' + b'c$$ (3-4)
$$F_2 = a'bd + b'c + bc$$
$$F_3 = abd + ab'c' + bc$$

in the PLA, the Espresso algorithm can be used. This complex algorithm is described in *Logic Minimization Algorithms for VLSI Synthesis* by Brayton [12].

Equations (3-4) have only five different product terms, so the PLA table has only five rows. This is a significant improvement over Equations (3-3), which require eight product terms. Figure 3-15 shows the corresponding PLA structure, which has four inputs, five product terms, and three outputs. A dot at the intersection of a word line and an input or output line indicates the presence of a switching element in the array.

**FIGURE 3-15:**
**PLA Realization of**
**Equations (3-4)**

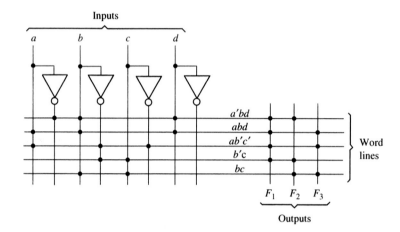

A PLA table is significantly different than a truth table for a ROM. In a truth table, each row represents a minterm; therefore, exactly one row will be selected by each combination of input values. The 0's and 1's of the output portion of the selected row determine the corresponding output values. On the other hand, each row in a PLA table represents a general product term. Therefore, zero, one, or more rows may be selected by each combination of input values. To determine the value of $F$ for a given input combination, the values of $F$ in the selected rows of the PLA table must be OR'ed together. The following examples refer to the PLA table of Table 3-4. If $abcd = 0001$, no rows are selected, and all $F_i$'s are 0. If $abcd = 1001$, only the third row is selected, and $F_1F_2F_3 = 101$. If $abcd = 0111$, the first and fifth rows are selected. Therefore, $F_1 = 1 + 0 = 1$, $F_2 = 1 + 1 = 1$, and $F_3 = 0 + 1 = 1$.

Next, we realize the sequential machine BCD to excess-3 code converter of Figure 1-23 using a PLA and three D flip-flops. The circuit structure is the same as Figure 3-8, except that the ROM is replaced by a PLA. The required PLA table, based on the equations given in Figure 1-25, is shown in Table 3-5.

| | Product Term | $Q_1$ | $Q_2$ | $Q_3$ | $X$ | $Q_1^+$ | $Q_2^+$ | $Q_3^+$ | $Z$ |
|---|---|---|---|---|---|---|---|---|---|
| **TABLE 3-5:** | $Q_2'$ | — | 0 | — | — | 1 | 0 | 0 | 0 |
| **PLA Table** | $Q_1$ | 1 | — | — | — | 0 | 1 | 0 | 0 |
| | $Q_1Q_2Q_3$ | 1 | 1 | 1 | — | 0 | 0 | 1 | 0 |
| | $Q_1Q_3'X'$ | 1 | — | 0 | 0 | 0 | 0 | 1 | 0 |
| | $Q_1'Q_2'X$ | 0 | 0 | — | 1 | 0 | 0 | 1 | 0 |
| | $Q_3'X'$ | — | — | 0 | 0 | 0 | 0 | 0 | 1 |
| | $Q_3X$ | — | — | 1 | 1 | 0 | 0 | 0 | 1 |

### 3.2.3 Programmable Array Logic

The PAL (programmable array logic) is a special case of the programmable logic array in which the AND array is programmable and the OR array is fixed. The basic structure of the PAL is the same as the PLA shown in Figure 3-9. Because only the AND array is programmable, the PAL is less expensive than the more general PLA, and the PAL is easier to program. For this reason, logic designers frequently use PALs to replace individual logic gates when several logic functions must be realized.

Figure 3-16(a) represents a segment of an unprogrammed PAL. The symbol

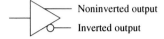

represents an input buffer, which is logically equivalent to

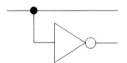

A buffer is used since each PAL input must drive many AND gate inputs. When the PAL is programmed, some of the interconnection points are programmed to make the desired connections to the AND gate inputs. Connections to the AND gate inputs in a PAL are represented by X's as shown in the following diagram:

As an example, we will use the PAL segment of Figure 3-16(a) to realize the function $I_1I_2' + I_1'I_2$. The X's in Figure 3-16(b) indicate that $I_1$ and $I_2'$ lines are connected to the first AND gate, and the $I_1'$ and $I_2$ lines are connected to the other gate.

When designing with PALs, we must simplify our logic equations and try to fit them into one (or more) of the available PALs. Unlike the more general PLA, the AND terms cannot be shared among two or more OR gates; therefore, each function to be realized can be simplified by itself without regard to common terms. For a given type of PAL, the number of AND terms that feed each output OR gate is fixed and limited. If the number of AND terms in a simplified function is too large, we may be forced to choose a PAL with more gate inputs and fewer outputs.

As an example of programming a PAL, we will implement a full adder. The logic equations for the full adder are

$$Sum = X'Y'C_{in} + X'YC_{in}' + XY'C_{in}' + XYC_{in}$$
$$C_{out} = XC_{in} + YC_{in} + XY$$

FIGURE 3-16:
**PAL Segment**

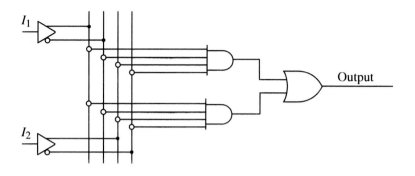

(a) Unprogrammed

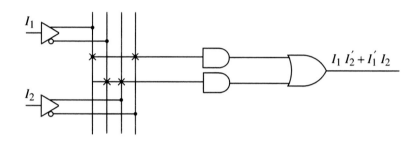

(b) Programmed

Figure 3-17 shows a section of a PAL where each OR gate is driven by four AND gates. The X's on the diagram show the connections that are programmed into the PAL to implement the full adder equations. For example, the first row of X's implements the product term $X'Y'C_{in}$.

FIGURE 3-17:
**Implementation of a Full Adder Using a PAL**

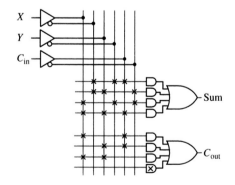

Typical combinational PALs have from 10 to 20 inputs and from 2 to 10 outputs, with 2 to 8 AND gates driving each OR gate. PALs are also available that contain D flip-flops with inputs driven from the programmable array logic. Such PALs are called sequential PALs. They provide a convenient way of realizing sequential circuits. Figure 3-18 shows a segment of a sequential PAL. The D flip-flop is driven from an OR gate, which is fed by two AND gates. The flip-flop output is fed back to the programmable AND array through a buffer. Thus, the AND gate inputs can be connected to $A$, $A'$, $B$, $B'$, $Q$, or $Q'$. The diagram shows the realization of the next state equation:

$$Q^+ = D = A'BQ' + AB'Q$$

The flip-flop output is connected to an inverting tristate buffer, which is enabled when EN = 1.

**FIGURE 3-18:**
**Segment of a**
**Sequential PAL**

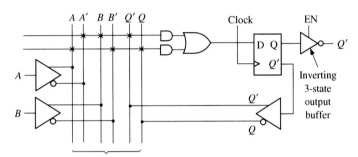

A few decades ago, PALs were very popular among digital system designers. A very popular PAL was the 16R4. This PAL has an AND gate array with 16 input variables, and it has four D flip-flops. Nowadays, several other programmable devices, such as GALs (described in the next section), CPLDs, and FPGAs, have arrived. PALs have practically disappeared; hence, we do not describe further any of the traditional PAL devices.

### 3.2.4 Programmable Logic Devices/Generic Array Logic

PALs and PLAs have been very popular for implementing small circuitry and interface logic often needed by designers. As integrated circuit technology has improved, a wide variety of other programmable logic devices have become available. Traditional PALs are not reprogrammable. However, there are flash erasable/reprogrammable PALs now. Often, these are referred to as PLDs.

The 22CEV10 (Figure 3-19) is a CMOS electrically erasable PLD that can be used to realize both combinational and sequential circuits. The abbreviation PLD has been used as a generic term for all programmable logic devices and also refers to specific devices such as the 22CEV10. In addition to the AND-OR arrays that the PALs have, most PLDs have some type of a macroblock that contains some multiplexers and some additional programmability. These PLDs are named with reference to their input and output capability. For instance, the 22CEV10 has 12 dedicated input pins and 10 pins that can be programmed as either inputs or outputs. It contains 10 D flip-flops and 10 OR gates. The number of AND gates that feeds each OR gate ranges from 8 through 16. Each OR gate

FIGURE 3-19: **Block Diagram for 22V10**

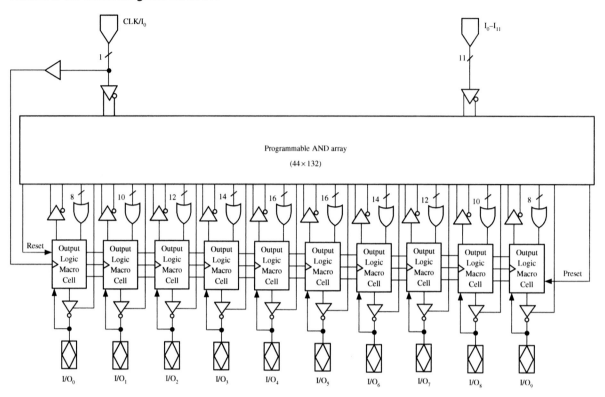

drives an *output logic macrocell*. Each macrocell contains one of the 10 D flip-flops. The flip-flops have a common clock, a common asynchronous reset (AR) input, and a common synchronous preset (SP) input. The name 22V10 indicates a versatile PAL with a total of 22 input and output pins, 10 of which are bidirectional I/O (input/output) pins.

Figure 3-20 shows the details of a 22CEV10 output macrocell. The connections to the output pins are controlled by programming this macrocell. The output MUX control inputs $S_1$ and $S_0$ select one of the data inputs. For example, $S_1 S_0 = 10$ selects data input 2. Each macrocell has two programmable interconnect bits. $S_1$ or $S_0$ is connected to ground (logic 0) when the corresponding bit is programmed. Erasing a bit disconnects the control line ($S_1$ or $S_0$) from ground and allows it to float to logic 1. When $S_1 = 1$, the flip-flop is bypassed, and the output is from the OR gate. The OR gate output is connected to the I/O pin through the multiplexer and the output buffer. The OR gate is also fed back so that it can be used as an input to the AND gate array. If $S_1 = 0$, then the flip-flop output is connected to the output pin, and it is also fed back so that it can be used for AND gate inputs. When $S_0 = 1$, the output is not inverted, so it is an active high. When $S_0 = 0$, the output is inverted, so it is an active low. The output pin is driven by a tristate inverting buffer. When the buffer output is in a high-impedance state, the

OR gate and flip-flop are disconnected from the output pin, and the pin can be used as an input. The dashed lines in Figure 3-20(a) show the path when both $S_1$ and $S_0$ are 0, and the dashed lines in Figure 3-20(b) show the path when both $S_1$ and $S_0$ are 1. Note that in the first case, the flip-flop output $Q$ is inverted by the output buffer, and in the second case the OR gate output is inverted twice, so there is no net inversion.

**FIGURE 3-20: PLD Output Macrocell**

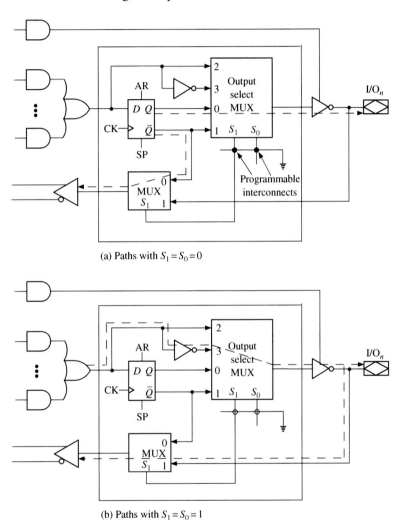

(a) Paths with $S_1 = S_0 = 0$

(b) Paths with $S_1 = S_0 = 1$

Several PLDs similar to the 22V10 have been popular. Typically these PLDs had 8 to 12 I/O pins. Each output pin is typically connected to an output macrocell, and each macrocell has a D flip-flop. The I/O pins can be programmed so that they act as inputs or as combinational or flip-flop outputs. Some of the PLDs have a dedicated clock input, and the others have a dual-purpose pin that can be used either as a clock or as an input. All the PLDs typically have tristate buffers at the outputs, and some of them have a dedicated output enable ($\overline{OE}$).

Lattice Semiconductor created similar devices which are in-circuit programmable and called them generic array logic (GAL). GALs are perfect for implementing small amounts of interface logic, often called "glue" logic. Most of the common PLDs, like the PALCE22V10, PALCE20V8, and so on, have GAL equivalents, called GAL22V10, GAL20V8, and so on.

### Design Flow for PLDs

Computer-aided design programs for PALs and PLDs are widely available. Such programs accept logic equations, truth tables, state graphs, or state tables as inputs and automatically generate the required bit patterns. These patterns can then be downloaded into a PLD programmer, which will create the necessary connections and verify the operation of the PAL. Many of the newer types of PLDs are erasable and reprogrammable in a manner similar to EPROMs and EEPROMs. Hence, in these newer devices, bit patterns corresponding to the required EEPROM content will be generated by the software.

**PALASM** and **ABEL** are examples of two languages that were popularly used with PALs and PLDs. PALASM is a PLD design language from MMI and AMD. ABEL is a PLD design language from DATA I/O. Intel used to manufacture PLDs and had a PLD language called PLDShell. While PALASM and ABEL can still be used, nowadays designs for GALs can be done using hardware description languages such as VHDL or Verilog.

● ● ● ● ● ● ● ● ● ● ● ● ●
## 3.3    Complex Programmable Logic Devices

Improvements in integrated circuit technology have made it possible to create programmable ICs equivalent to several PLDs in the same chip. These chips are called complex programmable logic devices (CPLDs). When storage elements such as flip-flops are also included on the same IC, a small digital system can be implemented with a single CPLD.

CPLDs are an extension of the PAL concept. In general, a CPLD is an IC that consists of a number of PAL-like logic blocks together with a programmable interconnect matrix. CPLDs typically contain 500 to 10,000 logic gates. Essentially, several PLDs are interconnected using a crossbar-like switch and fabricated inside the same IC. An $N \times M$ crossbar switch is one in which each of the $N$ input lines can be connected to any of the $M$ output lines simultaneously. It is expensive to build these switches; however, use of such a switch results in predictable timing. Many CPLDs are electronically erasable and reprogrammable and are sometimes referred to as EPLDs (erasable PLDs).

A typical CPLD contains a number of macrocells that are grouped into function blocks. Connections between the function blocks are made through an interconnection array. Each macrocell contains a flip-flop and an OR gate, which has its inputs connected to an AND gate array. Some CPLDs are based on PALs, in which case each OR gate has a fixed set of AND gates associated with it. Other CPLDs are based on PLAs, in which case any AND gate output within a function block can be connected to any OR gate input in that block.

Xilinx, Altera, Lattice Semiconductor, Cypress, and Atmel are the major CPLD manufacturers in the market today. The major products available on the market are listed in Table 3-6. Some vendors specify their gate capacities in usable gates, and some specify it in terms of logic elements.

TABLE 3-6:
**Major CPLDs and their Approximate Capacity**

| Vendor | CPLD family | Gate Count |
|---|---|---|
| Xilinx | CoolRunner-II | 750 to 12K |
| | CoolRunner XPLA3 | 750 to 12K |
| | XC9500XV | 800 to 6400 |
| | XC9500 | 800 to 6400 |
| | XC9500XL | 800 to 6400 |
| Atmel | CPLD ATF15 | 750 to 3000 usable gates |
| | CPLD-2 22V10 | 500 usable gates |
| Cypress | Delta39K | 30K to 200K |
| | Flash370i | 800 to 3200 |
| | Quantum38K | 30K to 100K |
| | Ultra37000 | 960 to 7700 |
| | MAX340 high-density EPLDs | 600 to 3750 |
| Lattice Semiconductor | ispXPLD 5000MX | 75K to 300K |
| | ispMACH 4000B/C/V/Z | 640 to 10,240 |
| Altera | MAX II | 240 to 2210 logic elements |
| | MAX3000 | 600 to 10K usable gates |
| | MAX7000 | 600 to 10K usable gates |

### 3.3.1 **An Example CPLD: The Xilinx CoolRunner**

Xilinx has two major series of CPLDs, the CoolRunner and the XC9500. Figure 3-21 shows the basic architecture of a CoolRunner family CPLD, the Xilinx XCR3064XL. This CPLD has four function blocks, and each block has 16 associated **macrocells**

FIGURE 3-21: **Architecture of Xilinx CoolRunner XCR3064XL CPLD**

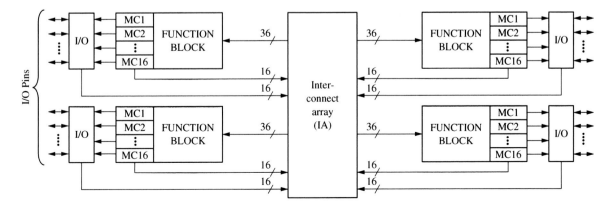

(MC1, MC2, ...). Each function block is a programmable AND-OR array that is configured as a PLA. Each macrocell contains a flip-flop and multiplexers that route signals from the function block to the input/output (I/O) block or to the **interconnect array (IA)**. The interconnect array selects signals from the macrocell outputs or I/O blocks and connects them back to function block inputs. Thus, a signal generated in one function block can be used as an input to any other function block. The I/O blocks provide an interface between the bidirectional I/O pins on the IC and the interior of the CPLD.

Figure 3-22 shows how a signal generated in the PLA (function block) is routed to an I/O pin through a macrocell. Any of the 36 inputs from the IA (or their complements) can be connected to any inputs of the 48 AND gates. Each OR gate can accept up to 48 product term inputs from the AND array. The macrocell logic in this diagram is a simplified version of the actual logic. The first mux (1) can be programmed to select the OR gate output or its complement. The mux (2) at the output of the macrocell can be programmed to select either the combinational output ($G$) or the flip-flop output ($Q$). This output goes to the interconnect array and to the output cell. The output cell includes a three-state buffer (3) to drive the I/O pin. The buffer enable input can be programmed from several sources. When the I/O pin is used as an input, the buffer must be disabled.

**FIGURE 3-22: CPLD Function Block and Macrocell (Simplified Version of XCR3064XL)**

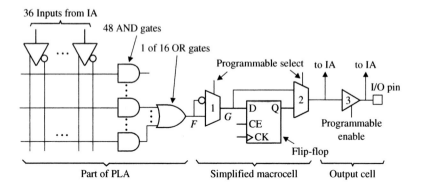

Figure 3-23 shows how a Mealy sequential machine with two inputs, two outputs, and two flip-flops can be implemented by a CPLD. Four macrocells are required, two to generate the D inputs to the flip-flops and two to generate the $Z$ outputs. The flip-flop outputs are fed back to the AND array inputs via the interconnection matrix (not shown). The number of product terms required depends on the complexity of the equations for the $D$'s and the $Z$'s.

## CPLD Implementation of a Parallel Adder with Accumulator

Assume that we need to implement an adder with an accumulator, as in Figure 3-24, in a CPLD. The accumulator register needs one flip-flop for each bit. Each bit also needs to generate the sum and carry bits corresponding to that bit.

FIGURE 3-23: **CPLD Implementation of a Mealy Machine**

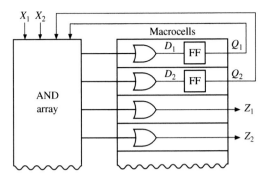

FIGURE 3-24: **N-Bit Parallel Adder with Accumulator**

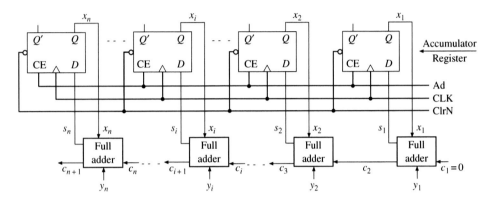

Figure 3-25 shows how three bits of such a parallel adder with an accumulator can be implemented using a CPLD. Each bit of the adder requires two macrocells. One of the macrocells implements the sum function and an accumulator flip-flop. The other macrocell implements the carry, which is fed back into the AND array. The $Ad$ signal can be connected to the enable input ($CE$) of each flip-flop via an

FIGURE 3-25: **CPLD Implementation of a Parallel Adder with Accumulator**

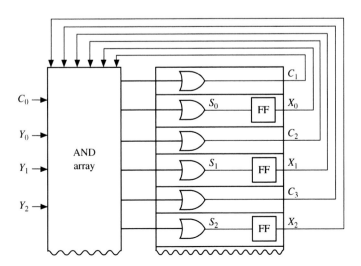

AND gate (not shown). Each bit of the adder requires eight product terms (four for the sum, three for the carry, and one for $CE$). For each accumulator flip-flop,

$$D_i = X_i^+ = S_i = X_i \oplus Y_i \oplus C_i$$

If the flip-flops are programmed as T flip-flops, then the logic for the sum can be simplified. For each accumulator flip-flop

$$X_i^+ = X_i \oplus Y_i \oplus C_i$$

Therefore, the T input is

$$T_i = X_i^+ \oplus X_i = Y_i \oplus C_i$$

The add signal can be AND'ed with the $T_i$ input so that the flip-flop state only can change when $Ad = 1$:

$$T_i = Ad(Y_i \oplus C_i) = Ad\, Y_i\, C_i' + Ad\, Y_i'\, C_i$$

The equation for carry is

$$C_{i+1} = X_i Y_i + X_i C_i + Y_i C_i$$

● ● ● ● ● ● ● ● ● ● ● ●
## 3.4 Field Programmable Gate Arrays

In this section, we introduce field programmable gate arrays (FPGAs). FPGAs are ICs that contain an array of identical logic blocks with programmable interconnections. The user can program the functions realized by each logic block and the connections between the blocks. FPGAs have revolutionized the way prototyping and designing are done. The flexibility offered by reprogrammable FPGAs has enhanced the design process. While different kinds of programmable devices had been around, when Xilinx used static RAM (SRAM) storage elements to create programmable logic blocks and introduced its family of XC2000 devices in 1985, the world received a totally new and powerful technology. There are a variety of FPGA products available in the market now. Xilinx, Altera, Lattice Semiconductor, Actel, Cypress, QuickLogic, and Atmel are examples of companies that design and sell FPGAs.

FPGAs provide several advantages over traditional gate arrays or mask programmable gate arrays (MPGAs). A traditional gate array can be used to implement any circuit but is programmable only in the factory. A specific mask to match the particular circuit is created in order to fabricate the gate array. The design time of a gate-array-based IC is a few months. FPGAs are standard off-the-shelf products. Manufacturing time reduces from months to hours as one adopts FPGAs instead of MPGAs. Design iterations become easier with FPGAs. This is a tremendous advantage when it comes to time-to-market. It becomes easy to correct mistakes that creep into designs. Mistakes and design specification changes become less costly. Prototyping cost is reduced. At low volumes, FPGAs are cheaper than MPGAs.

FPGAs have disadvantages, too. FPGAs are less dense than traditional gate arrays (MPGAs). In FPGAs, a lot of resources are spent to merely achieve the programmability. MPGAs have better performance than FPGAs. Programmable points have resistance and capacitance. They slow down signals, so FPGAs are slower than traditional gate arrays. Also, interconnection delays are unpredictable in FPGAs. PLDs, like PALs and GALs, are simple and inexpensive. CPLDs are faster than FPGAs and are cheaper. The overhead for programmability is fairly low in PALs and CPLDs. The main advantage of CPLDs over FPGAs is the lower cost and predictability in timing.

Several commercial FPGAs are listed in Table 3-7. As we notice, some of these chips contain logic equivalent to 5 million gates. The capacity of some FPGAs is specified in number of look-up tables (LUTs). Due to the large capacity, it is possible to prototype or even manufacture large systems in a single FPGA. In this chapter, we describe the basic organization of FPGAs. Design examples with FPGAs are presented in Chapter 6.

**TABLE 3-7:**
**Examples of**
**Commercial FPGAs**

| Vendor | FPGA Product | Capacity (Approx) in Gates/LUTs |
|---|---|---|
| Xilinx | Spartan-II | 15K to 200K |
| | Spartan-IIE | 50K to 600K |
| | Spartan-3 | 50K to 5M |
| | Virtex-5 | 19,200 to 207,360 LUTs |
| | Virtex | 57,906 to 1,124,022 |
| | Virtex-E | 71,693 to 4,074,387 |
| | Virtex-II | 40K to 8M |
| Altera | ACEX 1K | 56K to 257K |
| | APEX II | 1.9M to 5.25M |
| | FLEX 10K | 10K to 50K |
| | Stratix/Stratix II | 10,570 to 132,540 logic elements |
| Lattice Semiconductor | LatticeECP2 | 6K to 68K LUTs |
| | Lattice SC | 15.2K to 115.2K LUTs |
| | ispXPGA | 139K to 1.25M |
| | MachXO | 256 to 2280 LUTs |
| | LatticeECP | 6.1K to 32.8K LUTs |
| Actel | Axcelerator | 125K To 2M |
| | eX | 3K to 12K |
| | ProASIC3 | 30K to 3M |
| | MX | 3K to 54K |
| Quick Logic | Eclipse/EclipsePlus | 248K to 662K |
| | Quick RAM | 45K to 176K |
| | pASIC 3 | 5K to 75K |
| Atmel | AT40K | 5K to 40K |
| | AT40KAL | 5K to 50K |

### 3.4.1 Organization of FPGAs

Figure 3-26 shows the layout of a typical FPGA. The interior of FPGAs typically contains three elements that are programmable:

Programmable logic blocks
Programmable input/output blocks
Programmable routing resources

FIGURE 3-26: **Layout of a Typical FPGA**

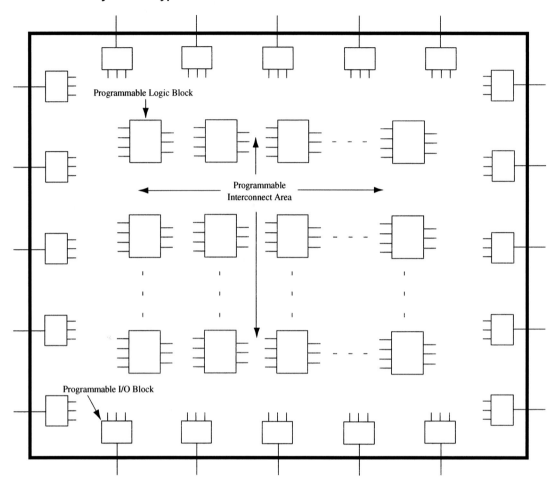

Arrays of programmable logic blocks are distributed within the FPGA. These logic blocks are surrounded by input/output (I/O) interface blocks. These I/O blocks can be considered to be on the periphery of the chip. They connect the logic signals to FPGA pins. The space between the logic blocks is used to route connections between the logic blocks.

The "field" programmability in FPGAs is achieved by reconfigurable elements, which can be programmed or reconfigured by the user. As mentioned, there are three major programmable elements in FPGAs: the logic block, the interconnect, and the input/output block. Programmable logic blocks are created by using multiplexers, look-up tables, and AND-OR or NAND-NAND arrays. "Programming" them means changing the input or control signals to the multiplexers, changing the look-up table contents, or selecting/not selecting particular gates in AND-OR gate blocks. For a programmable interconnect, "programming" means making or breaking specific connections. This is required to interconnect various blocks in the chip and to connect specific I/O pins to specific logic blocks. Programmable I/O blocks denote blocks which can be programmed to be input, output, or bidirectional lines. Typically, they can also be "programmed" to adjust the properties of their buffers such as inverting/noninverting, tristate, passive pull-up, or even to adjust the **slew rate**, which is the rate of change of signals on that pin.

What makes an FPGA distinct from a CPLD is the flexible general-purpose interconnect. In a CPLD, the interconnect is fairly restricted. The general-purpose interconnect in an FPGA gives it a lot of flexibility, but it also has the disadvantage of being slow. A connection from one part of the chip to another part might have to travel through several programmable interconnect points, resulting in large and unpredictable signal delays.

While Figure 3-26 was used to illustrate the general structure of an FPGA, not all FPGAs look like that. Commercial FPGAs use a variety of architectures. The FPGA architecture or organization refers to the manner or topology in which the logic blocks and interconnect resources are distributed inside the FPGA. The organization that is presented in Figure 3-26 is often referred to as symmetrical array architecture. If we examine the various FPGAs that have been on the market since their inception in the late 1980s, we could classify them into four different basic architectures or topologies:

Matrix-based (symmetrical array) architectures
Row-based architectures
Hierarchical PLD architectures
Sea-of-gates architecture

These architectures are illustrated in Figure 3-27.

### Matrix-Based (Symmetrical Array) Architectures

The logic blocks in this type of FPGA are organized in a matrix-like fashion as illustrated in Figure 3-27(a). Most Xilinx FPGAs belong to this category. The logic blocks in these architectures are typically of a large granularity (capable of implementing four-variable functions or more). These architectures typically contain $8 \times 8$ arrays in the smaller chips and $100 \times 100$ or larger arrays in the bigger chips. The routing resources are interspersed between the logic blocks. The routing in these architectures is often called two-dimensional channeled routing since routing resources are generally available in horizontal and vertical directions.

### Row-Based Architectures

These architectures were inspired by traditional gate arrays. The logic blocks in this architecture are organized in rows, as illustrated in Figure 3-27(b). Thus, there are rows of logic blocks and routing resources. The routing resources interspersed between the rows can be used to interconnect the various logic blocks. Traditional mask programmable gate arrays use very similar architectures. The routing in these architectures is often called one-dimensional channeled routing because the routing resources are located as a channel in between rows of logic resources. Some Actel FPGAs employ this architecture.

FIGURE 3-27: **Typical Architectures for FPGAs**

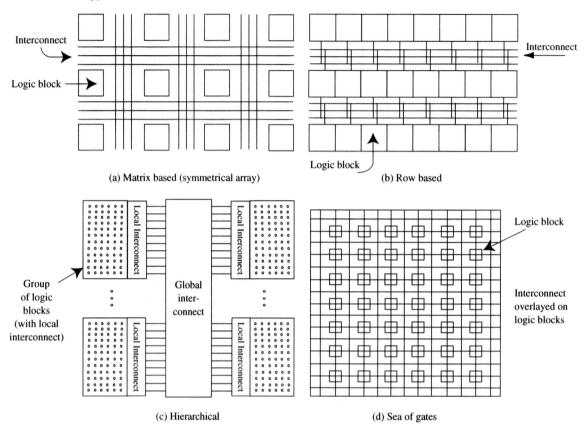

(a) Matrix based (symmetrical array)

(b) Row based

(c) Hierarchical

(d) Sea of gates

## Hierarchical Architectures

In some FPGAs, blocks of logic cells are grouped together by a local interconnect and several such groups are interconnected by another level of interconnect. For instance, in Altera APEX20 and APEX II FPGAs, 10 or so logic elements are connected to form what Altera calls a logic array block (LAB), and then several LABs are connected to form a MEGALAB. Thus, there is a hierarchy in the organization of these FPGAs. These FPGAs contain clusters of logic blocks with localized resources for interconnection. The global interconnect network is used for the interconnections between the clusters of logic blocks in these FPGAs.

## Sea-of-Gates Architecture

The sea-of-gates architecture is yet another manner to organize the logic blocks and interconnect in an FPGA. The general FPGA fabric consists of a large number of gates, and then there is an interconnect superimposed on the sea of gates as illustrated in Figure 3-27(d). Plessey, a manufacturer who was in the FPGA market in the mid-1990s, made FPGAs of this architecture. The basic cell they used was a NAND gate, in contrast to the larger basic cells used by manufacturers like Xilinx. While the terminology *sea of gates* is the most popular, there are also terminologies like **sea of cells** and **sea of tiles** to indicate the topology of FPGAs with a large number of fine-grain logic cells. The Actel Fusion FPGAs contain a sea of tiles, where each tile can be configured as a three-input logic function or a flip-flop/latch.

### 3.4.2 FPGA Programming Technologies

FPGAs consist of a large number of logic blocks interspersed with a programmable interconnect. The logic block is programmable in the sense that the same building block can be "programmed" or "configured" to create any desired circuitry. There is also programmability in the interconnections between the logic blocks.

Several techniques have been used to achieve the programmable interconnections between FPGAs. The term **programming technology** is used here to denote the technology by which the programmability in an FPGA is achieved. In some devices, the reconfigurability is achieved by changing the contents of static RAM cells. In some devices, it is achieved by using flash memory cells. In others, it is achieved by fusing metal links. In general, FPGAs use one of the following programming methods:

> StaticRAM programming technology
> EPROM/EEPROM/flash programming technology
> Antifuse programming technology

#### The SRAM Programming Technology

The SRAM programming technology involves creating reconfigurability by bits stored in static RAM (SRAM) cells. The logic blocks, I/O blocks, and interconnect can be made programmable by using configuration bits stored in SRAM. Reconfigurable logic blocks can easily be implemented as LUTs, which is the same approach as the ROM method described in Section 3.2.1. Sixteen SRAM cells can implement any function of four variables. The programmable interconnect can also be achieved by SRAM. The key idea is to use pass transistors to create switches and then control them using the SRAM content. Consider the arrangement in Figure 3-28(a). The SRAM cell is connected to the gate of the pass transistor. When the SRAM cell content is 0, the pass transistor is OFF, and hence no connection exists between points

**FIGURE 3-28:**
**Routing with Static RAM Programming Technology**

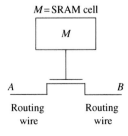

(a) Pass transistor connecting two points

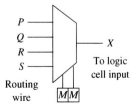

(b) Multiplexer controlled by two memory cells

*A* and *B*. A closed path can be achieved by turning the pass transistor ON by making the SRAM cell content 1. SRAM bits can be used to construct routing matrices by using multiplexers as in Figure 3-28(b). Changing the contents of the SRAM in the arrangement in Figure 3-28(b) will allow the designer to change what is connected to point *X*. The bits that are stored in the SRAM for deciding the LUT functionality or interconnection are called **configuration bits**.

A SRAM cell usually takes six transistors, as illustrated in Figure 3-29. Four cross-coupled transistors are required to create a latch, and two additional transistors are used to control passing data bits into the latch. When the *Word Line* is set

**FIGURE 3-29:**
**Typical Six-**
**Transistor SRAM**
**Cell**

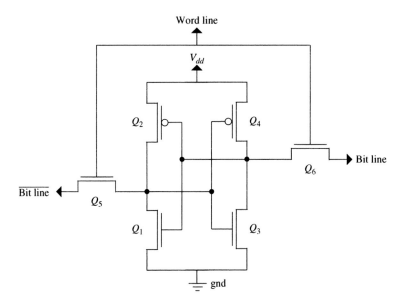

to high, the values on the *Bit Line* will be latched into the cell. This is the write operation. The read operation is performed by precharging the *Bit Line* and $\overline{Bit\ Line}$ to a logic 1 and then setting *Word Line* to high. The contents stored in the cell will then appear on the *Bit Line*. Some SRAM cell implementations only use five transistors. One advantage of using static RAM is that it is volatile and you can write new contents again and again. This provides flexibility during prototyping and development. Another advantage is that the fabrication steps for making SRAM cells are not different from the steps for making logic. The major disadvantage of the SRAM programming technology is that five or six transistors are used for every SRAM cell. This adds a tremendous cost to the chip. For example, if an FPGA has 1 million programmable points, it means that approximately 5 or 6 million transistors are spent in achieving this programmability.

Being volatile can become a disadvantage when an FPGA is used in the final product. Hence, when SRAM FPGAs are used, a nonvolatile device such as an EPROM should be used to permanently store the configuration bits. Typically, what is done is to use the EPROM as a "boot ROM." The EPROM contents are transferred to the SRAM when power comes up.

Xilinx FPGAs were the first FPGAs to use SRAM as the programming technology. In fact, it is the flexibility and reprogrammability of SRAM FPGAs that caused FPGAs to become widely popular. Now, many companies use the SRAM programming technology for their FPGAs.

### EPROM/EEPROM Programming Technology

In the EPROM/EEPROM programming technology, EPROM cells are used to control programmable connections. Assume that EPROM/EEPROM cells are used instead of the SRAM cells in Figure 3-28. A transistor with two gates, a floating gate and a control gate, is used to create an EPROM cell. Figure 3-30 illustrates an EPROM cell. The pull-up resistor connects the drain of the transistor to the power supply (labeled $V_{dd}$ in the figure). To turn the transistor off, charge can be injected on the floating gate using a high voltage between the control gate and the drain of the transistor. This charge increases the threshold voltage of the transistor and turns it off. The charge can be removed by exposing the floating gate to ultraviolet light. This lowers the threshold voltage of the transistor and makes it function normally.

FIGURE 3-30:
**The EPROM Programming Technology**

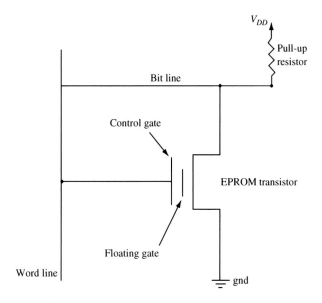

EPROMs are slower than SRAM; hence, SRAM-based FPGAs can be programmed faster. EPROMs also require more processing steps than SRAM. EPROM-based switches have high ON resistance and high static power consumption. The EEPROM is similar to EPROM, but removal of the gate charge can be done electrically.

Flash memory is a form of EEPROM that allows multiple locations to be erased in one operation. Flash memory stores information in floating-gate transistors as in traditional EPROM. The floating gate is isolated by an insulating oxide layer, and hence any electrons placed there are trapped. The cell is read by placing a specific voltage on the control gate. When the voltage to read is placed, electrical current will or will not flow depending on the threshold voltage of the cell, which is controlled by the number of electrons trapped in the floating gate. In some devices, the information is stored as absence or presence of current. In some advanced devices, the amount of current flow is sensed, and hence multiple bits of information can be stored in a cell. To erase, a large voltage differential is placed between the control gate and source, which pulls electrons off. Flash memory is erased in segments/sectors; all cells in a block are erased at the same time.

### The Antifuse Programming Technology

In some FPGAs, the programmable connections between different points are achieved by what is called an "antifuse." Contrary to fuse wires that blow open when high current passes through them, the "antifuse" programming element changes from high resistance (open) to low resistance (closed) when a high voltage is applied to it. Antifuses are often built using dielectric layers between N+ diffusion and polysilicon layers or by amorphous silicon between metal layers. Antifuses are normally OFF; permanently connected links are created when they are programmed. The process is irreversible, and hence antifuse FPGAs are only one-time programmable. Programming an antifuse requires applying a high voltage and currents in excess of normal currents. Special programming transistors larger than normal transistors are incorporated into the device in order to accomplish the programming. There are different antifuse technologies; a popular one is the Via antifuse technology.

Antifuse technology has the advantage that the area consumed by the programmable switch is small. Another advantage is that antifuse-based connections are faster than SRAM- and EEPROM-based switches. The disadvantage of the antifuse technology is that it is not reprogrammable. It is a permanent connection; if an error or design change necessitates reprogramming, a new device is required.

### Comparison of Programming Technologies

Table 3-8 compares the characteristics of the major programming technologies used by FPGAs. Only the SRAM and EEPROM programming technologies allow in-circuit programmability. In-circuit programmability means that an FPGA can be reprogrammed without removing it from the board in which it is used. In-circuit programmability is not possible in traditional EPROM-based devices, but EEPROM/flash technologies allow in-circuit reprogrammability.

TABLE 3-8: **Characteristics of the Major FPGA Programming Technologies**

| Programming Technology | Volatility | Programmability | Area Overhead | Resistance | Capacitance |
|---|---|---|---|---|---|
| SRAM | Volatile | In-circuit reprogrammable | Large | Medium to high | High |
| EPROM | Nonvolatile | Out-of-circuit reprogrammable | Small | High | High |
| EEPROM | Nonvolatile | In-circuit reprogrammable | Medium to high | High | High |
| Antifuse | Nonvolatile | Not reprogrammable | Small | Small | Small |

SRAM FPGAs have several disadvantages: high area overhead, large delays, volatility, and so on. However, the in-circuit programmability and fast programmability have made them very popular. SRAM FPGAs are more expensive than other types of FPGAs because each programmable point uses six transistors. This extra hardware contributes only to the reprogrammability but not to the actual circuitry realized with the FPGA. EEPROM/flash-based FPGAs are comparable to SRAM FPGAs in many aspects; however, they are not as fast as SRAM FPGAs.

### 3.4.3 Programmable Logic Block Architectures

FPGAs in the past have employed different kinds of programmable logic blocks as the basic building block. In this section, we present some generalized versions of typical building blocks in commercial FPGAs.

The logic blocks vary in the basic components they use. For instance, some FPGAs use LUT-based logic blocks, while others use multiplexers and logic gates to build their logic blocks. There also have been FPGAs where logic blocks simply consisted of transistor pairs (e.g., crosspoint FPGAs). Logic building blocks in early Altera FPGAs were PLD blocks. There were also FPGAs that used NAND gates as the building block (e.g., Plessey).

The logic blocks also vary in their architecture and size. Some FPGAs use large basic blocks, which can implement large functions (several five-variable or four-variable functions) and have several flip-flops in each basic block. In contrast, there are FPGA building blocks which only allow a three-variable function or a flip-flop in one block. Some FPGAs allow choices as to whether latched/unlatched or both kinds of outputs can be brought out. Some FPGAs allow one to control the type of flip-flop that is realized. Some allow positive edge/negative edge clock, direct set/reset inputs to the flip-flop, and so on. Different FPGA manufacturers use different names (often trademarked) to denote their logic blocks. In the Xilinx literature, a programmable logic block is called a **Configurable Logic Block (CLB)**. Altera calls their basic blocks **Logic Elements (LE)** and a collection of 8 or 10 of them **Logic Array Blocks (LABs)**. The basic cells in Actel Fusion FPGAs are referred to as **VersaTiles**.

### Look-Up Table–Based Programmable Logic Blocks

Many LUT-based FPGAs use a four-variable look-up table plus a flip-flop as the basic element and then combine several of them in various topologies. Consider the structure in Figure 3-31. There are two four-variable look-up tables (often denoted by the short form **LUT4**) and two flip-flops in this programmable logic block. The LUT4 can also be called a **four-variable function generator** since it can generate any function of four variables. The two LUT4s can generate any two functions of four variables. The inputs to the $X$-function generator are called $X_1$, $X_2$, $X_3$, and $X_4$, and the inputs to the $Y$-function generator are called $Y_1$, $Y_2$, $Y_3$, and $Y_4$. The functions can be steered to the output of the block ($X$ and $Y$) in combinational or latched form. There are two D flip-flops in the logic block. The D flip-flops are versatile in the sense that they have clock enable, direct set, and direct reset inputs. A multiplexer selects between the combinatorial output and the latched version of the output. The little box with "M" in it (beneath the multiplexer) indicates a memory cell

FIGURE 3-31:
**A Look-Up
Table–Based
Programmable
Logic Block**

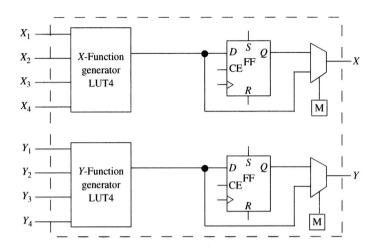

that is required to provide appropriate select signals to select between the latched and unlatched form of the function. An early Xilinx FPGA, the XC3000, used building blocks very similar to this structure.

Let us assume that we want to implement the function $F_1 = A'B'C + A'BC' + AB$ using an FPGA with programmable logic blocks as in Figure 3-31. Since this is a three-variable function, a four-input LUT is more than sufficient to implement the function. The path highlighted in Figure 3-32 assumes that the $X$-function generator (top LUT) is used. Let us assume that $X_1$ is the LSB and $X_4$ is the MSB to the LUT. Since function $F_1$ only uses three variables, the $X_4$ input is not used. A truth table can be constructed to represent the function, and the LUT contents can be derived.

The LUT contents to implement function $F_1$ will be 0, 1, 1, 0, 0, 0, 1, 1, 0, 1, 1, 0, 0, 0, 1, 1. The first 8 bits in the LUT reflect the truth table outputs when the function is represented in a truth table form. Since input $X_4$ is not grounded, the first 8 bits are repeated to take care of the possibility that the $X_4$ input might stay at a logic 1 when it is unused. Since the functions are stored in LUT form, the number of terms

FIGURE 3-32:
**Highlighting Paths
for Function $F_1$**

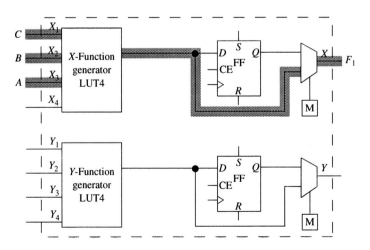

in the function is not important. Common minimizations to reduce the number of terms are not relevant. The number of variables is what is important.

Many commercial FPGAs use LUTs. Examples are the Xilinx Spartan/Virtex, Altera Cyclone II/APEX II, QuickLogic Eclipse/PolarPro, and Lattice Semiconductor ECP. Many of these FPGAs put two or more four-input LUTs into a block in various topologies. Some FPGAs also provide multiplexers in addition to look-up tables.

### Logic Blocks Based on Multiplexers and Gates

Some FPGAs use multiplexers as the basic building block. As you know, any combinational function can be implemented using multiplexers alone. In the most naïve method, a 4-to-1 multiplexer can generate any two-input function. If inverted inputs can be provided, a 4-to-1 multiplexer can generate any three-input function. Examples of multiplexer-based basic blocks are given in Figure 3-33. Logic blocks similar to these were used in early Actel FPGAs such as the ACT I and ACT II.

FIGURE 3-33:
**Multiplexer-Based Logic Blocks in FPGAs**

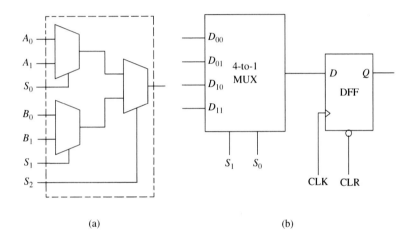

(a)                              (b)

Let us assume that we want to implement the function $F_1 = A'B'C + A'BC' + AB$ using an FPGA with programmable logic blocks consisting of 4-to-1 multiplexers. Two of the three-input variables can be connected to the multiplexer select lines. Then we have to provide appropriate signals to the multiplexer data input lines in order to realize the function. To derive these inputs, we will first construct a truth table of the function as shown below:

| A | B | C | F | Mux Input in Terms of {0, 1, C, C'} |
|---|---|---|---|---|
| 0 | 0 | 0 | 0 | } C |
| 0 | 0 | 1 | 1 | |
| 0 | 1 | 0 | 1 | } C' |
| 0 | 1 | 1 | 0 | |
| 1 | 0 | 0 | 0 | } 0 |
| 1 | 0 | 1 | 0 | |
| 1 | 1 | 0 | 1 | } 1 |
| 1 | 1 | 1 | 1 | |

Let us assume that $A$ and $B$ are connected to the select inputs of the multiplexer. Next, we will derive values of inputs to provide to the multiplexer input lines in terms of the third variable in the function. The third variable is $C$, and by providing one of the four values $\{C, C', 0, 1\}$, any three-variable function can be expressed. Considering the first two rows of the truth table, it can be seen that $F = C$ when $AB = 00$. Similarly, considering the third and fourth rows of the truth table, $F = C'$ when $AB = 01$. When $AB = 10$, $F = 0$ irrespective of the value of $C$. Similarly, when $AB = 11$, the value of the function equals 1. The last column in the truth table presents the required multiplexer inputs. Hence, one 4-to-1 multiplexer with the connections shown in Figure 3-34 can implement function $F_1$.

**FIGURE 3-34:**
**Multiplexer**
**Implementing**
**Function $F_1$**

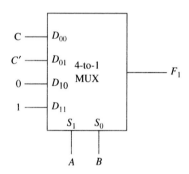

In the past three sections, we have provided an overview of the general architecture, logic block types, and programming technologies that can be used to build FPGAs. The general architecture, programming technology, and logic block types of

**TABLE 3-9: Architecture, Technology, and Logic Block Types of Commercial FPGAs**

| Company | Device Names | General Architecture | Logic Block Type | Programming Technology |
|---------|-------------|---------------------|------------------|------------------------|
| Actel | ProASIC/ProASIC3/ ProASICplus | Sea of Tiles | Multiplexers & Basic Gates | SRAM |
| | SX/SXA/eX/MX | Sea of Modules | Multiplexers & Basic Gates | Antifuse |
| | Accelerator | Sea of Modules | Multiplexers & Basic Gates | SRAM |
| | Fusion | Sea of Tiles | Multiplexers & Basic Gates | Flash, SRAM |
| Xilinx | Virtex | Symmetrical Array | LUT | SRAM |
| | Spartan | Symmetrical Array | LUT | SRAM |
| Atmel | AT40KAL | Cell Based | Multiplexers & Basic Gates | SRAM |
| QuickLogic | Eclipse II | Flexible Clock | LUT | SRAM |
| | PolarPro | Cell Based | LUT | SRAM |
| Altera | Cyclone II | Two-Dimensional Row and Column Based | LUT | SRAM |
| | Stratix II | Two-Dimensional Row and Column Based | LUT | SRAM |
| | APEX II | Row and Column, but Hierarchical Interconnect | LUT | SRAM |

several example commercial FPGAs are summarized in Table 3-9. LUT-based FPGAs are very common, especially for Xilinx and Altera. Actel is the manufacturer of multiplexer-based FPGAs. SRAM programming technology, while expensive, is also common.

### 3.4.4 Programmable Interconnects

A key element of an FPGA is the general-purpose programmable interconnect interspersed between the programmable logic blocks. There are different types of interconnection resources in all commercial FPGAs. Every vendor has its own specific names for the different types of interconnects in its FPGA.

#### Interconnects in Symmetric Array FPGAs

In this section, we discuss some of the basic elements used for interconnection in symmetric array FPGAs.

**General-Purpose Interconnect:**  Many FPGAs use switch matrices that provide interconnections between routing wires connected to the switch matrix. Figure 3-35(a) illustrates interconnecting logic blocks in an FPGA using switch matrices. Many FPGAs use this type of interconnect. A typical switch matrix is illustrated in Figure 3-35(b), where there is a switch at each intersection (i.e., wherever the lines cross). A switch matrix that supports every possible connection from every wire to every other wire is very expensive. The connectivity is often limited to some subset of a full crossbar connection; moreover, not all connections might be possible simultaneously. In the switch matrix illustrated in Figure 3-35(b), each wire from a side of the switch can be routed to other wires using some combination of the switches. In order to support this type of a connection, each cross point in the switch matrix must support six possible interconnections as marked in Figure 3-35(c).

Depending on the programming technology, SRAM cells, flash memory cells, or antifuse connections control the configuration of the switches. The switch matrices interspersed between the logic blocks in an FPGA allow general-purpose interconnectivity between arbitrary points in the chip. However, the switch matrices are expensive in area and time (delay). If a signal passes through several of these switch matrices, it could contribute to a significant signal delay. Moreover, the delays are variable and unpredictable depending on the number of the switch matrices involved in each signal. In contrast, the interconnection resources in a CPLD are more restricted. However, interconnections in CPLDs result in smaller and more predictable delays.

**Direct Interconnects:**  Many FPGAs provide special connections between adjacent logic blocks. These interconnects are fast because they do not go through the routing matrix. Many FPGAs provide direct interconnections to the four nearest neighbors: top, bottom, left, and right. Figure 3-36 illustrates examples of direct connections. In some cases, there are special interconnections to eight neighboring blocks, including the diagonally located logic blocks (Figure 3-36(b)). The direct interconnections do not go through the switch matrix but are implemented with

FIGURE 3-35:
**Routing Matrix for
General-Purpose
Interconnection in
an FPGA**

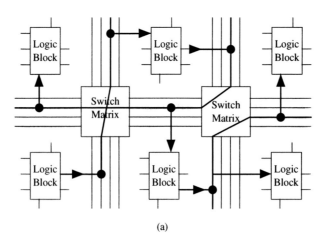

(a)

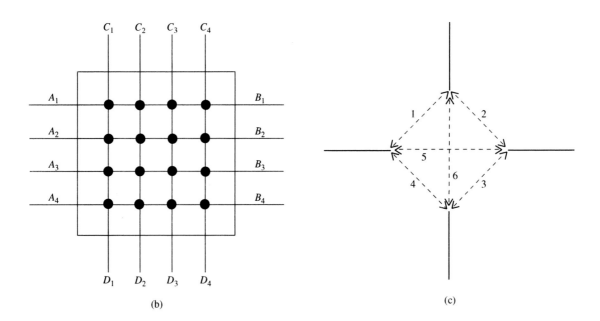

(b)

(c)

dedicated switches, resulting in smaller delays. These types of direct interconnects are used in some Xilinx FPGAs.

**Global Lines:** For purposes like high fan-out and low-skew clock distribution, most FPGAs provide routing lines that span the entire width of the device/height of the device. A limited number (two or four) of such global lines is provided by many FPGAs in the horizontal and vertical directions. Figure 3-37 illustrates **horizontal long lines** (global lines) in an example FPGA. The logic blocks often have tristate buffers to connect to the global lines.

**FIGURE 3-36: Direct Interconnects between Neighboring Logic Blocks**

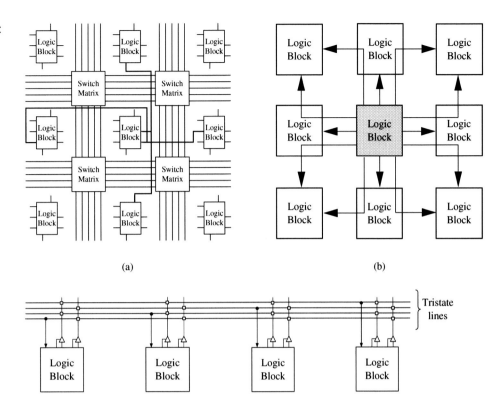

(a)                                                    (b)

**FIGURE 3-37: Global Lines**

Tristate lines

Logic Block    Logic Block    Logic Block    Logic Block

**Clock Skew**

There are several million gates in modern FPGA chips. When a clock is distributed to various parts of such a large chip, the delays in the wire carrying the clock can result in the clock edge arriving at different times at different parts. This difference in the actual edge of the clock as it arrives at different flip-flops or other devices is called clock skew. Clock skew is a problem in large systems, including modern microprocessors. Carefully planned clock distribution circuits are implemented in most systems in order to minimize the effect of clock skew. Modern FPGAs provide specialized clock distribution circuitry in order to create a clock of sufficient strength and low skew.

## Interconnects in Row-Based FPGAs

Many of the interconnect resources mentioned previously are very characteristic of symmetric array devices with a two-dimensional array of logic blocks (e.g., Xilinx). In devices that are row based, there are rows of logic blocks, and there are channels of switches to enable connections between the logic blocks. Several switches are used to route a signal from a logic block in one row to another logic block elsewhere in the chip. There are arrays of switches in the routing channel between the rows of logic. The routing resources in these FPGAs are very similar to routing in traditional gate arrays.

The interconnects in row-based channeled architecture can be classified into two categories: nonsegmented routing and segmented routing. In order to understand different types of channel routing, consider the connections $x$, $y$, and $z$ in Figure 3-38(a). Figure 3-38(b) indicates what is called as a **nonsegmented channel routing** architecture. There are three horizontal rows or tracks in this figure. There are several vertical wires and switches at the crosspoints. The switches technically can use any programming technology (SRAM, EPROM, or antifuse), although FPGAs that use this type of routing are typically antifuse FPGAs. Desired connectivity is obtained by programming the appropriate switches. Connectivity between the points marked $x$ is obtained by the two switches at row 1, columns 1 and 4. Typically this is called net $x$. Net $x$ simply means a wire that is named $x$. The connectivity for net $y$ is obtained by programming the switches at row 2, columns 3 and 8. It may be noticed that row 1 cannot be used for any other connections other than net $x$. Similarly, row 2 is exclusively used for net $y$. Thus, a problem with this type of interconnect resource is that a full-length track (i.e., an entire row) is used even for a short net. The area overhead of this type of routing is very high for this reason.

FIGURE 3-38:
**Typical Routing Resources in a Row-Based FPGA**

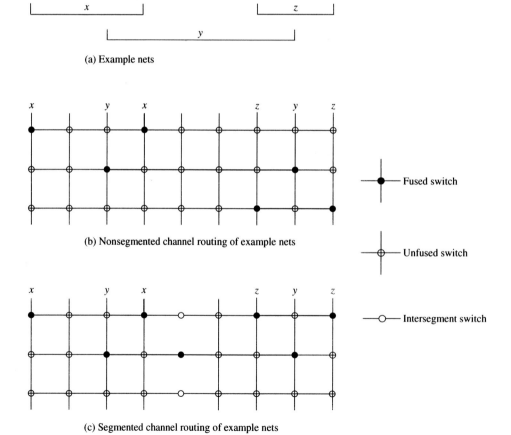

(a) Example nets

(b) Nonsegmented channel routing of example nets

(c) Segmented channel routing of example nets

Fused switch

Unfused switch

Intersegment switch

In order to reduce the area overhead associated with using full-length tracks for each net, we can use **segmented tracks**, as in Figure 3-38(c). Instead of being full length, a track is divided into segments. If a track in row 1 is segmented into two segments, we could use the same track for one more net. For example, nets $x$ and $z$ can both be routed on row 1 in Figure 3-38(c). That is the principle of segmented track routing. More nets can be routed using the same number of tracks; however, when long nets are desired, intersegment switches must be used to join the segments. These switches introduce more resistance and capacitance into the net. However, the overall routing resource area will reduce with segmented routing.

### 3.4.5 Programmable I/O Blocks in FPGAs

The I/O pads on an FPGA are connected to programmable input/output blocks, which facilitate connecting the signals from FPGA logic blocks to the external world in desired forms and formats. I/O blocks on modern FPGAs allow use of the pin as input and/or output, in direct (combinational) or latched forms, in tristate true or inverted forms, and with a variety of I/O standards.

Figure 3-39 shows an example configurable input/output block (I/OB). Each I/OB has a number of I/O options, which can be selected by configuration memory cells, indicated by boxes with an M. The I/O pad can be programmed to be an output or an input. To use the cell as an output, the tristate buffer must be enabled. To use the cell as an input, the tristate control must be set to place the tristate buffer, which drives the output pin, in the high-impedance state.

Flip-flops are provided so that input and output values can be stored within the I/O block. The flip-flops are bypassed when direct input or output is desired. The input flip-flop on many FPGAs can be programmed to act as an edge-triggered D flip-flop or as a transparent latch. Even if the I/O pin is not used, the I/O flip-flops can still be used to store data.

The configuration memory cells (marked M) allow control of various aspects associated with the I/O block. An output signal can be inverted by the I/O block if desired. The inversion is done using an XOR gate. The output signal goes through an exclusive-OR gate, where it is either complemented or not, depending on the contents of the configuration bit in the OUT-INVERT cell. The 3-STATE INVERT configuration bit allows one to create an active high or active low tristate control signal. If the *3-STATE* signal is 1 and the *3-STATE INVERT* bit is 0 (or if the *3-STATE* signal is 0 and the *3 STATE INVERT* bit is 1), the output buffer has a high-impedance output. Otherwise, the buffer drives the output signal to the I/O pad. When the I/O pad is used as an input, the output buffer must be in the high-impedance state. An external signal coming into the I/O pad goes through a buffer and then to the input of a D flip-flop. The buffer output provides a *DIRECT IN* signal to the logic array. Alternatively, the input signal can be stored in the D flip-flop, which provides the *LATCHED IN* signal to the logic array.

The *LATCHED OUTPUT* configuration bit allows one to provide the output in latched or combinational form. Depending on how the *LATCHED OUTPUT* bit is programmed, either the *OUT* signal or the flip-flop output goes to the output buffer. The *SLEW RATE* bit controls the rate at which the output signal can change. When the output drives an external device, reduction of the slew rate is desirable to reduce the

FIGURE 3-39: **Programmable I/O Block for an FPGA**

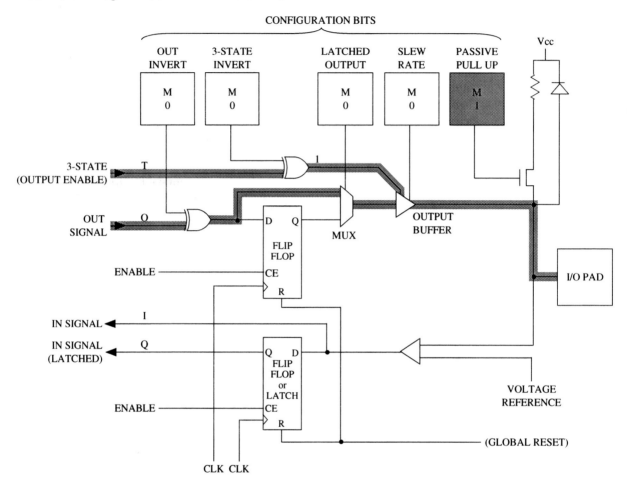

induced noise that can occur when the output changes rapidly. When the *PASSIVE PULL-UP* bit is set, a pull-up resistor is connected to the I/O pad. This internal pull-up resistor can be used to avoid floating inputs. The highlighted path indicates the I/O block in an output configuration, with tristate enabled and with a passive pull-up resistor.

---

**I/O Standards**

Early FPGAs provided TTL and CMOS signal compatibility, but nowadays there are many more standards for input/output signals. I/O blocks on modern FPGAs allow transforming signals to a variety of I/O signal standards, some of which are as follows:

LVTTL:      low-voltage transistor-transistor logic
PCI:          peripheral component interconnect

LVCMOS:   low-voltage complementary metal-oxide semiconductor
LVPECL:   low-voltage positive emitter-coupled logic
SSTL:     stub-series terminated logic
AGP:      advanced graphics port
CTT:      center tap terminated
GTL:      gunning transceiver logic
HSTL:     high-speed transceiver logic

Some of these standards use 5 volts whereas some use 3.3 volts or even 1.5 volts. The LVTTL is an example of a 3.3-V standard that can tolerate 5-V signals. The LVCMOS2 is a 2.5-V signal standard which can tolerate 5-V signals. The PCI standard has 5-V and 3.3-V versions. Some standards need an input voltage reference.

### 3.4.6 Dedicated Specialized Components in FPGAs

In the early days, FPGAs were simply logic blocks of medium or low complexity, integrated with programmable I/O and interconnect. More recently, FPGA vendors have incorporated embedded processors, digital signal processing (DSP) processors, dedicated multipliers, dedicated memory, analog-to-digital (A/D) converters, and so on into FPGAs. These specialized components help to efficiently achieve the provided special-purpose functionality. For instance, if dedicated multipliers are not provided, we will have to implement multipliers using general-purpose logic blocks, albeit in an inefficient manner.

### Dedicated Memory

A key feature of modern FPGAs is the embedding of dedicated memory blocks (RAM) onto the chip. The embedded RAM can be used to implement the memory needs of the circuit being designed. It could be a table storing constants/coefficients during processing, or it could be implementing memory for an embedded processor that you are designing using the FPGA. Modern FPGAs include 16K to 10M bits of memory. The width of the embedded RAM often can be adjusted. Let us assume that there are 32K of SRAM bits provided as blocks of RAM. This RAM can be used as 32K × 1, 16K × 2, 8K × 4, or 4K × 8. Essentially there are several tiles or blocks of memory. They can be placed in different ways to achieve different aspect ratios. The number of address lines and data lines get adjusted according to the aspect ratio, as illustrated in Table 3-10.

### Dedicated Arithmetic Units

Many users of FPGAs use them to implement arithmetic logic. When logic is implemented in FPGA logic blocks, the implementation generally takes more area and power and is slower than custom implementations. Hence, if most of the target users use arithmetic units such as adders and multipliers, it is beneficial to provide support for such dedicated operations inside the chip. Most FPGAs provide dedicated

**TABLE 3-10:**
**Variable-Width**
**RAM Aspect Ratios**

| Width | Depth | Addr Bus | Data Bus |
|---|---|---|---|
| 1 | 32K | 15 bits | 1 bit |
| 2 | 16K | 14 bits | 2 bits |
| 4 | 8K | 13 bits | 4 bits |
| 8 | 4K | 12 bits | 8 bits |
| 16 | 2K | 11 bits | 16 bits |

fast-carry logic to create fast adders. Nowadays, many FPGAs also contain dedicated multipliers (see Table 3-11). Thus, instead of mapping a multiplier into several logic blocks, dedicated multipliers provided on the FPGA fabric can be used. These dedicated multipliers are more efficient than a multiplier we could implement using the programmable logic in the FPGA. As indicated in Table 3-11, many Xilinx and Altera FPGAs provide 18 bit × 18 bit multipliers.

**TABLE 3-11:**
**Examples of FPGAs**
**with Dedicated**
**Multipliers**

| FPGA | Dedicated Multipliers |
|---|---|
| **Xilinx** Virtex-4, Virtex-II Pro/X, Spartan-3E, Spartan 3/3L | 18 × 18 multipliers |
| **Altera** Stratix II Cyclone II | 18 × 18 multipliers |

### Digital Signal Processing Blocks

Multiplication is a common operation in DSP. Hence the dedicated multipliers help DSP applications. Similar to multipliers, an FPGA vendor can provide DSP building blocks such as hardware for fast Fourier transforms (FFTs), finite impulse response (FIR) filters, infinite impulse response (IIR) filters, and so on. Encryption/decryption, compression/decompression, and security functions can also be provided. Once a large amount of specialized components are provided, a large part of an FPGA may be unused in applications that do not warrant such specialized components. In some FPGAs, DSP support is limited to the dedicated multipliers.

### Embedded Processors

Many modern FPGAs contain an entire processor core (see Table 3-12). This is extremely useful when designers use hybrid solutions, where part of a system is in a programmable processor, but part of the system is implemented in hardware. Circuitry that needs a large amount of flexibility can be implemented in the microprocessor, but circuit parts that need better performance than that of a programmable processor can be implemented in the FPGA logic blocks. Some FPGAs include the core of a small MIPS processor such as the MIPS R 4000, and some include an embedded version of the IBM PowerPC processor. Some FPGAs include custom processors designed by the FPGA vendors such as the MicroBlaze from Xilinx and the Nios processor from Altera.

TABLE 3-12:
**Examples of FPGAs with Embedded Microprocessors**

| FPGA | Embedded Processor |
|---|---|
| **Xilinx**<br>Virtex-4,<br>Virtex-II Pro/X | IBM 400 MHz<br>PowerPC |
| **Xilinx**<br>Spartan-3E,<br>Spartan 3/3I | MicroBlaze<br>PicoBlaze |
| **Altera**<br>Stratix II<br>Cyclone II | Nios II |
| **Altera**<br>APEX<br>APEX II | ARM,<br>MIPS,<br>Nios |
| **Altera**<br>Excalibur | ARM 9 |
| **Actel**<br>Fusion | ARM7 |

**Content Addressable Memories**

In some FPGAs, the memory blocks can be used as **content addressable memories** (CAMs). The general concept of a memory is that the user provides a memory address and the memory unit responds with the content. A CAM is a special kind of memory in which the content, not the address, is used to search the memory. We provide a data element, and the CAM responds with addresses where that data was found. CAMs contain more logic than RAMs because all locations of the memory have to be searched simultaneously to see whether the particular content is in any of the locations. Some FPGAs allow embedded CAM (e.g., Altera APEX II).

---

**The Actel Fusion architecture,** shown in Figure 3-40, provides several specialized components, including embedded RAM, decryption, and A/D converters. At the core of the chip are tiles of logic blocks (VersaTiles in Actel terminology). The embedded RAM is in the form of rows of SRAM blocks above and below the tiles of logic blocks. Several specialized components appear below the SRAM blocks in the bottom. There is a dedicated decryption unit that implements the AES decryption algorithm. (AES stands for Advanced Encryption Standard, which has been the cryptograhic standard for the U.S. government since 2001.) There is an analog-to-digital converter (ADC) that accepts inputs from several analog quads, which are circuitry to condition analog signals received by the FPGA. The analog quads contain circuitry to monitor and condition signals according to voltage, current, and temperature.

---

FIGURE 3-40: **Overview of the Actel Fusion Chip (© 2006 Actel Corporation)**

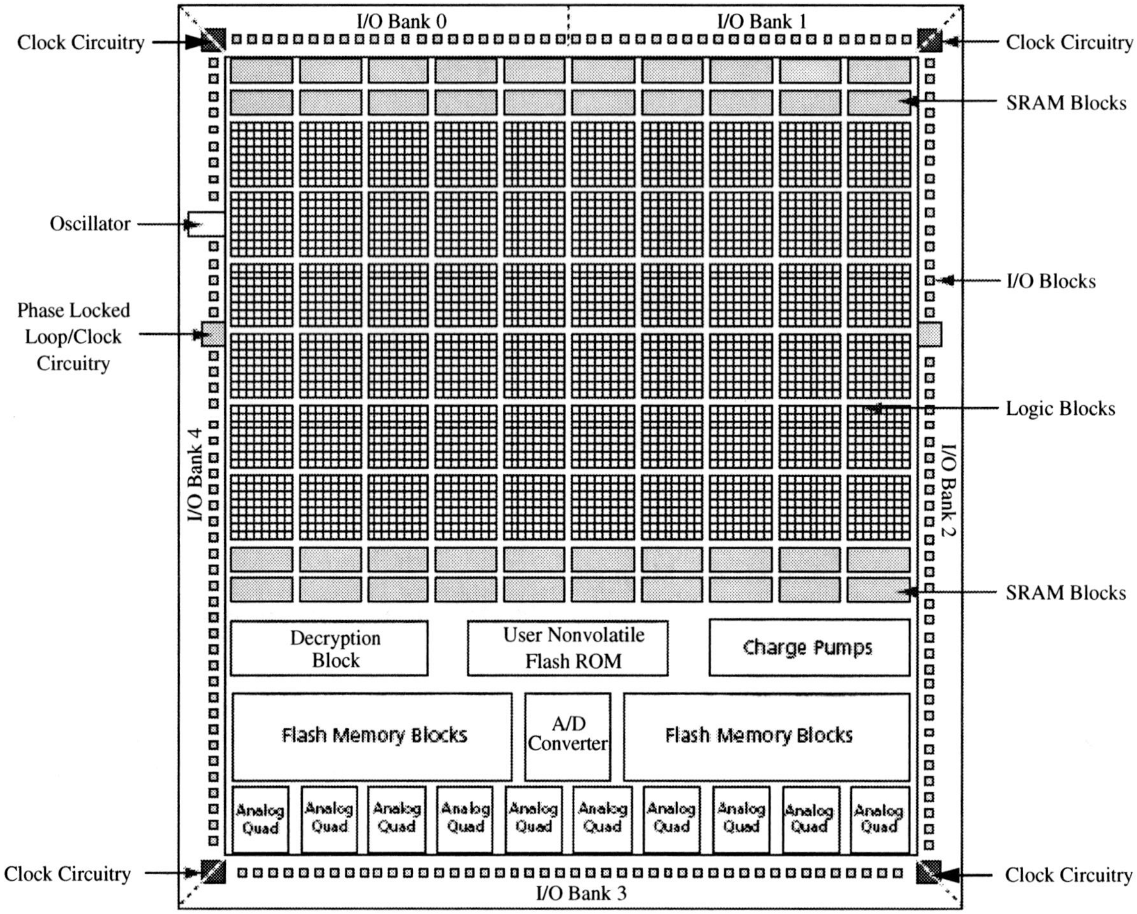

### 3.4.7 Applications of FPGAs

FPGAs have become a popular mode of circuit implementation for various applications:

#### Rapid Prototyping

FPGAs are very useful for building rapid prototypes of large systems. A designer can build proof-of-concept systems very quickly using field programmable gate arrays. Since FPGAs are large enough to contain 5 million or more gates, many large real-world systems can be prototyped using a single FPGA. If a single FPGA will not suffice, multiple FPGAs can be interconnected to realize large systems. Rapid prototyping of large systems is done by using boards with multiple FPGAs and plugging multiple boards into a backplane (motherboard).

## As Final Product in Medium-Speed Systems

Circuits realized using FPGAs typically operate in the 150–200-MHz clock rate. For applications where this speed is sufficient, FPGAs can be used for the final product itself as opposed to the prototype. When an FPGA is used as the final product, enhancements to the system can be done as software updates rather than hardware changes. Modern FPGA speeds are adequate for many applications.

## Reconfigurable Circuits and Systems

The reprogrammability of FPGAs lends itself to building dynamically reconfigurable circuits and systems. SRAM-based FPGAs make it possible to implement "soft" hardware. FPGAs have been used to design circuits and systems that need multiple functionalities at various times.

As an example, consider a reprogrammable Tomahawk missile that the Navy designed using FPGAs. [46] The conventional Tomahawk is a long-range Navy cruise missile designed to perform a variety of missions. The Navy designed a reconfigurable Tomahawk, which can operate in one of two modes, depending on the mission at hand. Rather than designing separate logic for each mode, the missile designers used FPGAs so that the configuration for each mode can be kept on-board in ROM. Depending on the mode of operation, the FPGA could be configured in midflight.

## Glue Logic

FPGAs have become the medium of choice for implementing interface or glue logic between modules and components. Small changes in interface protocols or formats would conventionally necessitate building new interface logic. With SRAM FPGAs, the new interface logic can be implemented on the same FPGA as in a software update.

## Hardware Accelerators/Coprocessors

A software application running on a conventional system can be accelerated if a coprocessor/accelerator can implement some key routines/kernels from the application in hardware. An FPGA can be used to implement the key kernel. A SRAM-based, reconfigurable FPGA is well suited for this type of use because depending on the application running, different kernels can be dynamically programmed into the FPGA. This approach has been demonstrated for applications, such as pattern matching. FPGA-based hardware is used for several applications, including computer architecture simulator acceleration, emulation boards, hardware test/verification, and so on.

### 3.4.8 Design Flow for FPGAs

Sophisticated CAD tools are available to assist with the design of systems using programmable gate arrays. Designs can be entered in many ways.

In the early days of FPGAs, designs were entered using schematic entry or even lower levels of design entry tools. Low-level design entry means less abstraction, whereas high-level means entering designs at a higher level of abstraction (e.g., behavioral VHDL/Verilog description). Early FPGA tools allowed low-level utilities to enter logic equations, Karnaugh maps, and so on into specific logic blocks in the FPGA. Schematic capture technique means that the designer develops a schematic of the design. Schematic diagrams utilizing standard hardware components are created and entered into the CAD software.

Nowadays, automatic synthesis tools are available that will take a VHDL description of the system as an input and generate an interconnection of gates and flip-flops to realize the system. Behavioral models can be translated into design implementations reasonably efficiently. Synthesis tools have advanced significantly in the last decade.

One method of designing a digital system with an FPGA uses the following steps:

**1.** Create a behavioral, register-transfer level (RTL), or structural model of the design in a hardware description language such as VHDL or Verilog.
**2.** Simulate and debug the design.
**3.** Synthesize the design targeting the desired device.
**4.** Run a mapping/partitioning program. This program will break the logic diagram into pieces that will fit into the configurable logic blocks.
**5.** Run an automatic place and route program. This will place the logic blocks in appropriate places in the FPGA and then route the interconnections between the logic blocks.
**6.** Run a program that will generate the bit pattern necessary to program the FPGA.
**7.** Download the bit pattern into the internal configuration cells in the FPGA, and test the operation of the FPGA.

Steps 3, 4, and 5 are often integrated in modern CAD tools. However, the processes mentioned in the steps are happening whether presented as one step or several steps. This is analogous to how general-purpose compilers have integrated compiling and assembling steps. In the early days of high-level language compilers, the term *compiling* only meant translation into an assembly language format. Converting from assembly language to machine language code was considered the assembler's job. Nowadays, the steps are integrated in most high-level language compilation environments.

In SRAM-based FPGAs, when the final system is built, the bit pattern for programming the FPGA is normally stored in an EPROM and automatically loaded into the FPGA when the power is turned on. The EPROM is connected to the FPGA, as shown in Figure 3-41. The FPGA resets itself after the power has been applied. Then it reads the configuration data from the EPROM by supplying a sequence of addresses to the EPROM inputs and storing the EPROM output data in the FPGA internal configuration memory cells. This is not required in flash memory based FPGAs because the flash technology is nonvolatile. In antifuse FPGAs, the configuration bits permanently alter the switches.

**FIGURE 3-41:**
**EPROM**
**Connections for**
**SRAM FPGA**
**Initialization**

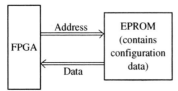

In this chapter we have introduced several different types of programmable logic devices and used them for designing circuits. The technology underlying early programmable logic devices, such as ROMs, PALs, and PLAs, was presented first. Simple PLDs and GALs were presented next. Examples were presented to illustrate implementations of simple logic functions in these devices. CPLDs and FPGAs were presented next. The discussion on FPGAs was limited to an overview of the general technology underlying this class of devices. General organization of FPGAs, general structure of logic blocks, typical programming techniques, and so on were discussed. More details on FPGAs will be presented in Chapter 6.

• • • • • • • • • • • •

# Problems

3.1 What is the size of the smallest ROM that is needed to implement the following?

**(a)** An 8-bit full adder (assume carry-in and carry-out)
**(b)** A BCD-to-binary converter (2 BCD digits)
**(c)** A 4-to-1 MUX
**(d)** A 32-bit adder (adds two 32-bit numbers to give a 33-bit sum)
**(e)** A 3-to-8 decoder
**(f)** A 32-bit adder (no carry in or carry out)
**(g)** A $16 \times 16$ bit multiplier
**(h)** A 16-bit full adder (with carry-in and carry-out)
**(i)** An 8-to-3 priority encoder
**(j)** A 10-to-4 priority encoder
**(k)** An 8-to-1 multiplexer

3.2 Given $F = A'B' + BC'$ and $G = AC + B'$, write a complete VHDL module that realizes the functions $F$ and $G$ using an 8-word $\times$ 2-bit ROM. Include the array type declaration and the constant declaration that defines the contents of the ROM.

3.3 Implement the following state table using a ROM and two D flip-flops. Use a straight binary state assignment.

**(a)** Show the block diagram and the ROM truth table. Truth table column headings should be in the order $Q_1 Q_0 X D_1 D_0 Z$.
**(b)** Write VHDL code for the implementation. Use an array to represent the ROM table, and use two processes.

| Present State | Next State X = 0 | Next State X = 1 | Output (Z) X = 0 | Output (Z) X = 1 |
|---|---|---|---|---|
| $S_0$ | $S_0$ | $S_1$ | 0 | 1 |
| $S_1$ | $S_2$ | $S_3$ | 1 | 0 |
| $S_2$ | $S_1$ | $S_3$ | 1 | 0 |
| $S_3$ | $S_3$ | $S_2$ | 0 | 1 |

3.4 The following state table is implemented using a ROM and two D flip-flops (falling edge triggered):

| $Q_1Q_2$ | $Q_1^+Q_2^+$ X = 0 | X = 1 | Z X = 0 | X = 1 |
|---|---|---|---|---|
| 00 | 01 | 10 | 0 | 1 |
| 01 | 10 | 00 | 1 | 1 |
| 10 | 00 | 01 | 1 | 0 |

**(a)** Draw the block diagram.
**(b)** Write VHDL code that describes the system. Assume that the ROM has a delay of 10 ns, and each flip-flop has a propagation delay of 15 ns.

3.5 Find a minimum-row PLA to implement the following three functions:

$$f(A, B, C, D) = \Sigma m(3, 6, 7, 11, 15)$$
$$g(A, B, C, D) = \Sigma m(1, 3, 4, 7, 9, 13)$$
$$h(A, B, C, D) = \Sigma m(4, 6, 8, 10, 11, 12, 14, 15)$$

**(a)** Use Karnaugh maps to find common terms. Give the logic equations with common terms underlined, the PLA table, and also a PLA diagram similar to Figure 3-15.
**(b)** Use the Espresso multiple-output simplification routine that is in *LogicAid*. Compare the *LogicAid* results with part (a). They might not be exactly the same since *LogicAid* Espresso only finds minimum row tables; it does not necessarily minimize the number of variables in each AND term. *Note*: Enter the variable names $A, B, C, D, F, G$, and $H$ in *LogicAid*. Printouts with variable names $X1, X2, X3, X4$, and so on are not acceptable.

3.6 Find a minimum-row PLA table to implement the following sets of functions.

**(a)** $f_1 (A, B, C, D) = \Sigma m(0, 2, 3, 6, 7, 8, 9, 11, 13)$,
$f_2 (A, B, C, D) = \Sigma m(3, 7, 8, 9, 13)$,
$f_3 (A, B, C, D) = \Sigma m(0, 2, 4, 6, 8, 12, 13)$
**(b)** $f_1 (A, B, C, D) = cd + ad + a'bc'd'$
$f_2 (A, B, C, D) = bc'd' + ac' + ad'$

3.7 **(a)** Find a minimum-row PLA table to implement the following equations:

$$x (A, B, C, D) = \Sigma m(0, 1, 4, 5, 6, 7, 8, 9, 11, 12, 14, 15)$$
$$y (A, B, C, D) = \Sigma m(0, 1, 4, 5, 8, 10, 11, 12, 14, 15)$$
$$z (A, B, C, D) = \Sigma m(0, 1, 3, 4, 5, 7, 9, 11, 15)$$

**(b)** Indicate the connections that will be made to program a PLA to implement your solution to part (a) on a diagram similar to Figure 3-15.

3.8 Write VHDL code that describes the output macrocell of a 22V10 (the part enclosed by a box on Figure 3-20). The entity should include $S_1$ and $S_0$. Note that the flip-flop has an asynchronous reset $(AR)$ and a synchronous preset $(SP)$.

3.9  An $N$-bit bidirectional shift register has $N$ parallel data inputs, $N$ outputs, a left serial input ($LSI$), a right serial input ($RSI$), a clock input, and the following control signals:

 $Load$:  Load the parallel data into the register (load overrides shift).
 $Rsh$:  Shift the register right ($LSI$ goes into the left end).
 $Lsh$:  Shift the register left ($RSI$ goes into the right end).

 **(a)** If the register is implemented using a 22V10, what is the maximum value of $N$?
 **(b)** Give equations for the rightmost two cells.

3.10  Show how the left shift register of Figure 2-43 could be implemented using a CPLD. Draw a diagram similar to Figure 3-25. Give the equations for the flip-flop D inputs.

3.11  A Mealy sequential circuit with four output variables is realized using a 22V10. What is the maximum number of input variables it can have? What is the maximum number of states? Can any Mealy circuit with these numbers of inputs and outputs be realized with a 22V10? Explain.

3.12  **(a)** What is the difference between a traditional gate array and an FPGA?
 **(b)** What are the different types of FPGAs based on architecture (organization)?
 **(c)** What are the different programming technologies for FPGAs?
 **(d)** What is the main advantage of SRAM FPGAs?
 **(e)** What is the main advantage of antifuse FPGAs?
 **(f)** What are the major programmable elements in an FPGA?
 **(g)** What are the disadvantages of SRAM FPGAs?
 **(h)** What are the disadvantages of antifuse FPGAs?
 **(i)** How many transistors are typically required to make an SRAM cell?
 **(j)** What is an MPGA?
 **(k)** What is the difference between a CPLD and an FPGA?
 **(l)** What is an advantage of a CPLD over an FPGA?
 **(m)** What is the advantage of an FPGA over a CPLD?
 **(n)** Name three vendors of CPLDs.
 **(o)** Name three vendors of FPGAs.

3.13  **(a)** In what type of applications should a designer use a CPLD rather than an FPGA?
 **(b)** In what type of applications should a designer use an MPGA rather than an FPGA?
 **(c)** In what type of applications should a designer use an FPGA rather than an MPGA?
 **(d)** A company is designing an experimental product, which is in version 1 now. It is expected that the product will undergo several revisions. The company's plan is to use an FPGA for the actual design. What type of FPGA (SRAM or anti-fuse) should be used?
 **(e)** A company is designing a product using an FPGA. The company's plan is to use an FPGA for the actual design. The product has undergone several revisions and is fairly stable. Minimizing area, power, and cost is important for the company. What type of FPGA (SRAM or antifuse) should be used?

(f) A company is designing a product. It expects to sell 1000 copies of it. Should the company use an MPGA or FPGA for this product?

(g) A company is designing a product. It expects to sell 100 million copies of it. Should the company use an MPGA or an FPGA for this product?

3.14 (a) Implement the function $F_1 = A'BC + B'C + AB$ using an FPGA with programmable logic blocks consisting of 4-to-1 multiplexers. Assume inputs and their complements are available as in Figure 3-34.

(b) Implement the function $F_1 = A'B + AB' + AC' + A'C$ using a multiplexer. What is the size of the smallest multiplexer needed, assuming inputs and their complements are available?

3.15 (a) Route the 'w', 'x', 'y', and 'z' nets on the nonsegmented tracks shown below. Use the minimum number of tracks possible.

(b) Route the 'w', 'x', 'y', and 'z' nets on the segmented tracks shown below. Use the minimum number of tracks possible.

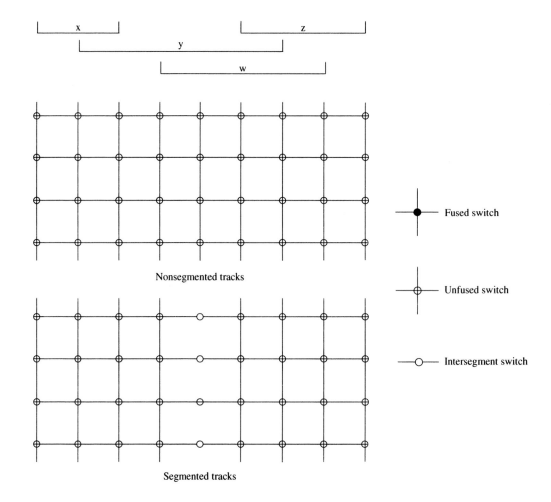

Nonsegmented tracks

Segmented tracks

Fused switch

Unfused switch

Intersegment switch

3.16 Consider the following programmable I/O block:

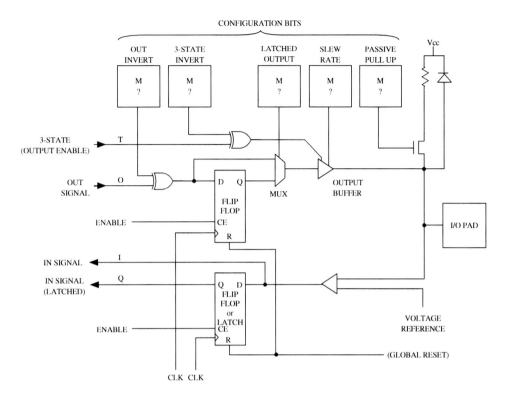

Highlight the connections to configure this I/O block as an **INPUT** pin. Specify the five configuration bits and the value of T.

# CHAPTER 4

# Design Examples

In this chapter, we present several VHDL design examples to illustrate the design of small digital systems. We present the concept of dividing a design into a controller and a data path and using the control circuit to control the sequence of operations in a digital system. We use VHDL to describe a digital system at the behavioral level so that we can simulate the system to test the algorithms used. We also show how designs have to be coded structurally if specific hardware structures are to be generated.

In any design, first you should understand the problem and the design specifications clearly. If the problem has not been stated clearly, try to get the features of the design clarified. In real-world designs, if another team or a client company is providing your team with the specifications, getting the design specifications clarified properly can save you a lot of grief later. Good design starts with a clear specification document.

Once the problem has been stated clearly, often designers start thinking about the basic blocks necessary to accomplish what is specified. Designers often think of standard building blocks, such as adders, shift registers, counters, and so on. Traditional design methodology splits a design into a "data path" and a "controller." The term *data path* refers to the hardware that actually performs the data processing. The controller sends control signals or commands to the data path, as in Figure 4-1. The controller can obtain feedback in the form of status signals from the data path.

In the context of a microprocessor, the data path is the *arithmatic logic unit* (ALU) that performs the core of the processing. The controller is the control logic that sends appropriate control signals to the data path, instructing it to perform addition, multiplication, shifting, or whatever action is called for by the instruction.

FIGURE 4-1:
**Separation of a Design into Data Path and Controller**

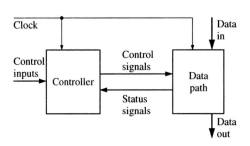

Many have a tendency to consider the term *data path* to be synonymous with the data bus, but *data path* in traditional design terminology refers to the actual data processing unit.

Maintaining a distinction between data path and controller helps in debugging (i.e., finding errors in the design). It also helps while modifying the design. Many modifications can be accomplished by changing only the control path because the same data path can support the new requirements. The controller can generate the new sequence of control signals to accomplish the functionality of the modified design. Design often involves refining the data path and controller in iterations.

In this chapter, we will discuss various design examples. Several arithmetic and nonarithmetic examples are presented. Nonarithmetic examples include a seven-segment decoder, a traffic light, a scoreboard, and a keypad scanner. Arithmetic circuits such as adders, multipliers, and dividers are presented.

● ● ● ● ● ● ● ● ● ● ● ● ●
## 4.1 BCD to Seven-Segment Display Decoder

Seven-segment displays are often used to display digits in digital counters, watches, and clocks. A digital watch displays time by turning on a combination of the segments on a seven-segment display. For this example, the segments are labeled as follows, and the digits have the forms as indicated in Figure 4-2.

**FIGURE 4-2:**
**Seven-Segment**
**Display**

Let us design a BCD to seven-segment display decoder. BCD stands for binary-coded decimal. In this format, each digit of a decimal number is encoded into 4-bit binary representation. This decoder is a purely combinational circuit, and hence no state machine is involved here. A block diagram of the decoder is shown in Figure 4-3. The decoder for one BCD digit is presented.

**FIGURE 4-3: Block Diagram of a BCD to Seven-Segment Display Decoder**

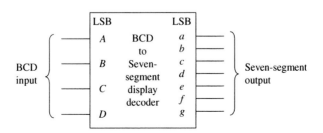

We will create a behavioral VHDL architectural description of this BCD to seven-segment decoder by using a single process with a case statement to model this combinational circuit, as in Figure 4-4. The sensitivity list of the process consists of the BCD number (4 bits).

FIGURE 4-4: **Behavioral VHDL Code for BCD to Seven-Segment Decoder**

```
entity bcd_seven is
  port(bcd: in bit_vector(3 downto 0);
       seven: out bit_vector(7 downto 1));
               -- LSB is segment a of the display. MSB is segment g
end bcd_seven;

architecture behavioral of bcd_seven is
begin
  process (bcd)
  begin
    case bcd is
      when "0000" => seven <= "0111111";
      when "0001" => seven <= "0000110";
      when "0010" => seven <= "1011011";
      when "0011" => seven <= "1001111";
      when "0100" => seven <= "1100110";
      when "0101" => seven <= "1101101";
      when "0110" => seven <= "1111101";
      when "0111" => seven <= "0000111";
      when "1000" => seven <= "1111111";
      when "1001" => seven <= "1101111";
      when others => null;
    end case;
  end process;
end behavioral;
```

● ● ● ● ● ● ● ● ● ● ● ●
## 4.2  A BCD Adder

In this example, we design a two-digit BCD adder, which will add two BCD numbers and produce the sum in BCD format. In BCD representation, each decimal digit is encoded into binary. For instance, decimal number 97 will be represented as 1001 0111 in the BCD format, where the first 4 bits represent digit 9 and the next 4 bits represent digit 7. Note that the BCD representation is different from the binary representation of 97, which is 1100001. It takes 8 bits to represent 97 in BCD, whereas the binary representation of 97 (1100001) only requires 7 bits. The 4-bit binary combinations 1010, 1011, 1100, 1101, 1110, and 1111 corresponding to hexadecimal numbers A to F are not used in the BCD representation. Since 6 out of 16 representations possible with 4 binary bits are skipped, a BCD number will take more bits than the corresponding binary representation.

When BCD numbers are added, each sum digit should be adjusted to skip the six unused codes. For instance, if 6 is added with 8, the sum is 14 in decimal form. A binary adder would yield 1110, but the lowest digit of the BCD sum should read 4. In order to obtain the correct BCD digit, 6 should be added to the sum whenever it is greater than 9. Figure 4-5 illustrates the hardware that will be required to perform the addition of two BCD digits. A binary adder adds the least significant digits. If the sum is greater than 9, an adder adds 6 to yield the correct sum digit and a carry digit to be added with the next digit. The addition of the higher digits is performed in a similar fashion.

FIGURE 4-5:
**Addition of Two
BCD Numbers**

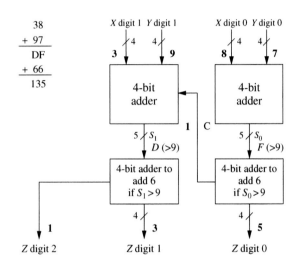

The VHDL code for the BCD adder is shown in Figure 4-6. The input BCD numbers are represented by $X$ and $Y$. The BCD sum of two 2-digit BCD numbers can exceed two digits, and hence three BCD digits are provided for the sum, which is represented by $Z$. The unsigned type from the IEEE numeric_bit library is used to represent $X$, $Y$, and $Z$. Aliases are defined to denote each digit of each BCD number. For example, the upper digit of $X$ can be denoted by $Xdig1$ by using the VHDL statement

```
alias Xdig1: unsigned(3 downto 0) is X(7 downto 4);
```

This statement allows us to use the name $Xdig1$ whenever we wish to refer to the upper digit of $X$. If BCD numbers 97 and 38 are added, the sum is 135, and hence, $Zdig2$ equals 1, $Zdig1$ equals 3 and $Zdig0$ equals 5.

The overloaded '+' operator from the IEEE numeric_bit library is used for adding each BCD digit. Adding two 4-bit vectors can result in a 5-bit sum. The sums are temporarily stored in $S0$ and $S1$, which are declared to be 5-bit numbers. Since we want a 5-bit result, we must extend $Xdig0$ to 5 bits by concatenating '0' and $Xdig0$. ($Ydig0$ will automatically be extended to match.) Hence

```
S0 <= '0' & Xdig0 + Ydig0;
```

FIGURE 4-6: **VHDL Code for BCD Adder**

```vhdl
library IEEE;
use IEEE.numeric_bit.all;

entity BCD_Adder is
  port(X, Y: in unsigned(7 downto 0);
       Z: out unsigned(11 downto 0));
end BCD_Adder;

architecture BCDadd of BCD_Adder is
alias Xdig1: unsigned(3 downto 0) is X(7 downto 4);
alias Xdig0: unsigned(3 downto 0) is X(3 downto 0);
alias Ydig1: unsigned(3 downto 0) is Y(7 downto 4);
alias Ydig0: unsigned(3 downto 0) is Y(3 downto 0);
alias Zdig2: unsigned(3 downto 0) is Z(11 downto 8);
alias Zdig1: unsigned(3 downto 0) is Z(7 downto 4);
alias Zdig0: unsigned(3 downto 0) is Z(3 downto 0);
signal S0, S1: unsigned(4 downto 0);
signal C: bit;
begin
  S0 <= '0' & Xdig0 + Ydig0; -- overloaded +
  Zdig0 <= S0(3 downto 0) + 6 when S0 > 9
      else S0(3 downto 0); -- add 6 if needed
  C <= '1' when S0 > 9 else '0';
  S1 <= '0' & Xdig1 + Ydig1 + unsigned'(0=>C);
                          -- type conversion done on C before adding
  Zdig1 <= S1(3 downto 0) + 6 when S1 > 9
      else S1(3 downto 0);
  Zdig2 <= "0001" when S1 > 9 else "0000";
end BCDadd;
```

accomplishes the addition of the least significant digits. During the addition of the second digit, the carry digit from the addition of the $XDig0$ and $Ydig0$ is also added. The carry bit $C$ must be converted to the unsigned type before it can be added to $Xdig1 + Ydig1$. The notation `unsigned'(0=>C)` accomplishes this conversion. Thus, the addition of the second digit is accomplished by the statement

```vhdl
S1 <= '0' & Xdig1 + Ydig1 + unsigned'(0=>C);
```

● ● ● ● ● ● ● ● ● ● ● ●
# 4.3 32-Bit Adders

Let us assume that we have to design a 32-bit adder. A simple manner to construct an adder is to build a **ripple-carry adder**, as in Figure 4-7. In this type of adder, 32 copies of a 1-bit full adder are connected in succession to create the 32-bit adder. The carry "ripples" from the least significant bit to the most significant bit. If gate

FIGURE 4-7:
**A 32-Bit Ripple-
Carry Adder**

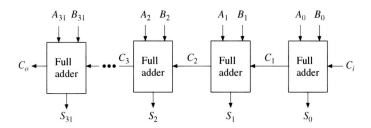

delays are $t_g$, a 1-bit adder delay is $2t_g$ (assuming a sum-of-products expression for sum and carry, and ignoring delay for inverters), and a 32-bit ripple-carry adder will take approximately 64 gate delays. For instance, if gate delays are 1 ns, the maximum frequency at which the 32-bit ripple-carry adder can operate is approximately 16 MHz. This is inadequate for many applications. Hence, designers often resort to faster adders.

### 4.3.1 Carry Look-Ahead Adders

A popular fast-addition technique is carry look-ahead (CLA) addition. In the carry look-ahead adder, the carry signals are calculated in advance, based on the input signals. For any bit position $i$, we can see that a carry will be generated if the corresponding input bits (i.e., $A_i$, $B_i$) are '1' or if there was a carry-in to that bit and at least one of the input bits are '1'. In other words, bit $i$ has carry-out if $A_i$ and $B_i$ are '1' (irrespective of carry-in to bit i); bit $i$ also has a carry-out if $C_i$ = '1' and either $A_i$ or $B_i$ is '1'. Thus, for any stage $i$, the carry-out is

$$C_{i+1} = A_iB_i + (A_i \oplus B_i) \cdot C_i \qquad (4\text{-}1)$$

The "$\oplus$" stands for the exclusive OR operation. Equation (4-1) simply expresses that there is a carry out from a bit position if it **generated** a carry by itself (i.e., $A_iB_i$ = '1') or it simply **propagated** the carry from the lower bit forwarded to it (i.e., $(A_i \oplus B_i) \cdot C_i$).

Since $A_iB_i$ = '1' indicates that a stage generated a carry, a general **generate ($G_i$) function** may be written as

$$G_i = A_iB_i \qquad (4\text{-}2)$$

Similarly, since $(A_i \oplus B_i)$ indicates whether a stage should propagate the carry it receives from the lower stage, a general **propagate ($P_i$) function** may be written as

$$P_i = A_i \oplus B_i \qquad (4\text{-}3)$$

Notice that the propagate and generate functions only depend on the input bits and can be realized with one or two gate delays. Since there will be a carry whether one of $A_i$ or $B_i$ is '1' or both are '1', we can also write the propagate expression as

$$P_i = A_i + B_i \qquad (4\text{-}4)$$

where the OR operation is substituted for the XOR operation. Logically this propagate function also results in the correct carry-out; however, traditionally it has been customary to define the propagate function as the XOR; that is, the bit position simply propagates a carry (without generating a carry by itself). Also, typically, the sum signal is expressed as

$$S_i = A_i \oplus B_i \oplus C_i = P_i \oplus C_i \tag{4-5}$$

The expression $P_i \oplus C_i$ can be used for sum only if $P_i$ is defined as $A_i \oplus B_i$.

The carry-out equation can be rewritten by substituting (4-2) and (4-3) in (4-1) for $G_i$ and $P_i$ as

$$C_{i+1} = G_i + P_iC_i \tag{4-6}$$

In a 4-bit adder, the $C_i$'s can be generated by repeatedly applying Equation (4-6) as follows:

$$C_1 = G_0 + P_0C_0 \tag{4-7}$$
$$C_2 = G_1 + P_1C_1 = G_1 + P_1G_0 + P_1P_0C_0 \tag{4-8}$$
$$C_3 = G_2 + P_2C_2 = G_2 + P_2G_1 + P_2P_1G_0 + P_2P_1P_0C_0 \tag{4-9}$$
$$C_4 = G_3 + P_3C_3 = G_3 + P_3G_2 + P_3P_2G_1 + P_3P_2P_1G_0 + P_3P_2P_1P_0C_0 \tag{4-10}$$

These carry bits are the look-ahead carry bits. They are expressed in terms of $P_i$'s, $G_i$'s, and $C_0$. Thus, the sum and carry from any stage can be calculated without waiting for the carry to ripple through all the previous stages. Since $G_i$'s and $P_i$'s can be generated with one or two gate delays, the $C_i$'s will be available in three or four gate delays. The advantage is that these delays will be the same independent of the number of bits we need to add, in contrast to the ripple counter. Of course, this is achieved with the extra gates to generate the look-ahead carry bits. A 4-bit carry look-ahead adder can now be built, as illustrated in Figure 4-8.

**FIGURE 4-8: Block Diagram of a 4-Bit CLA**

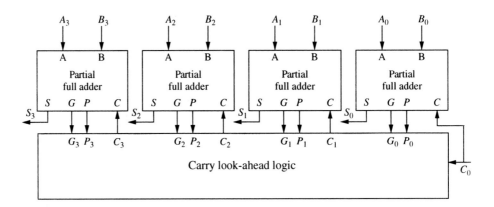

The disadvantage of the carry look-ahead adder is that the look-ahead carry logic, as in Equations (4-7) through (4-10), is not simple. It gets quite complicated for more than 4 bits. For that reason, carry look-ahead adders are usually implemented as 4-bit modules and are used in a hierarchical structure to realize adders that have multiples of 4 bits. Figure 4-9 shows the block diagram for a 16-bit carry look-ahead adder. Four carry look-ahead adders, similar to the one shown in

FIGURE 4-9: **Block Diagram of a 16-Bit CLA**

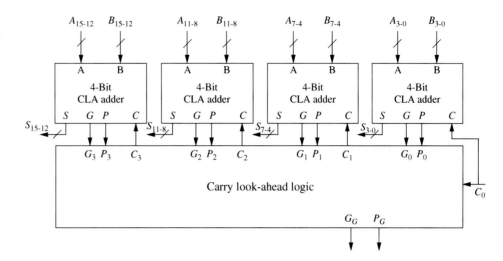

Figure 4-8, are used. Instead of relying on each 4-bit adder to send its carry-out to the next 4-bit adder, the carry look-ahead logic generates input carry bits to be fed to each 4-bit adder. This is accomplished by computing a group propagate $(P_G)$ and group generate $(G_G)$ signal, which is produced by each 4-bit adder. The next level of carry look-ahead logic uses these group propagates/generates and generates the required carry bits in parallel. The propagate for a group is true if all the propagates in that group are true. The generate for a group is true if the MSB generated a carry or if a lower bit generated a carry and every higher bit in the group propagated it. Thus

$$P_G = P_3 P_2 P_1 P_0 \tag{4-11}$$
$$G_G = G_3 + P_3 G_2 + P_3 P_2 G_1 + P_3 P_2 P_1 G_0 \tag{4-12}$$

The group propagate $P_G$ and generate $G_G$ will be available after three and four gate delays, respectively (one or two additional delays than the $P_i$ and $G_i$ signals, respectively). Figure 4-10 illustrates the VHDL description of a 4-bit carry look-ahead adder.

FIGURE 4-10: **VHDL Description of a 4-Bit Carry Look-Ahead Adder**

```
entity CLA4 is
  port(A, B: in bit_vector(3 downto 0); Ci: in bit;        -- Inputs
       S: out bit_vector(3 downto 0); Co, PG, GG: out bit);   -- Outputs
end CLA4;

architecture Structure of CLA4 is
component GPFullAdder
  port(X, Y, Cin: in bit;    -- Inputs
       G, P, Sum: out bit);  -- Outputs
end component;
```

```
component CLALogic is
  port(G, P: in bit_vector(3 downto 0); Ci: in bit;        -- Inputs
       C: out bit_vector(3 downto 1); Co, PG, GG: out bit);  -- Outputs
end component;

signal G, P: bit_vector(3 downto 0); -- carry internal signals
signal C: bit_vector(3 downto 1);
begin         --instantiate four copies of the GPFullAdder
  CarryLogic: CLALogic port map (G, P, Ci, C, Co, PG, GG);
  FA0: GPFullAdder port map (A(0), B(0), Ci, G(0), P(0), S(0));
  FA1: GPFullAdder port map (A(1), B(1), C(1), G(1), P(1), S(1));
  FA2: GPFullAdder port map (A(2), B(2), C(2), G(2), P(2), S(2));
  FA3: GPFullAdder port map (A(3), B(3), C(3), G(3), P(3), S(3));
end Structure;

entity CLALogic is
  port(G, P: in bit_vector(3 downto 0); Ci: in bit;        -- Inputs
       C: out bit_vector(3 downto 1); Co, PG, GG: out bit);  -- Outputs
end CLALogic;

architecture Equations of CLALogic is
signal GG_int, PG_int: bit;
begin         -- concurrent assignment statements
  C(1) <= G(0) or (P(0) and Ci);
  C(2) <= G(1) or (P(1) and G(0)) or (P(1) and P(0) and Ci);
  C(3) <= G(2) or (P(2) and G(1)) or (P(2) and P(1) and G(0)) or
          (P(2) and P(1) and P(0) and Ci);
  PG_int <= P(3) and P(2) and P(1) and P(0);
  GG_int <= G(3) or (P(3) and G(2)) or (P(3) and P(2) and G(1)) or
            (P(3) and P(2) and P(1) and G(0));
  Co <= GG_int or (PG_int and Ci);
  PG <= PG_int;
  GG <= GG_int;
end Equations;

entity GPFullAdder is
  port(X, Y, Cin: in bit;     -- Inputs
       G, P, Sum: out bit);   -- Outputs
end GPFullAdder;

architecture Equations of GPFullAdder is
signal P_int: bit;
begin                          -- concurrent assignment statements
  G <= X and Y;
  P <= P_int;
  P_int <= X xor Y;
  Sum <= P_int xor Cin;
end Equations;
```

VHDL code for a 16-bit carry look-ahead adder can be developed by instantiating four copies of the 4-bit carry look-ahead adder and one additional copy of the carry look-ahead logic. A 64-bit adder can be built by one more level of block carry look-ahead logic. The delay increases only by two gate delays when the adder size increases from 16 bits to 64 bits. Developing VHDL code for 16-bit carry look-ahead logic is left as an exercise.

Figure 4-11 illustrates behavioral VHDL code for a 32-bit adder using the overloaded '+' operator from IEEE numeric_bit library. If this code is synthesized, depending on the tools used and the target technology, an adder with characteristics in between a ripple-carry adder and a fast two-level adder will be obtained. The various topologies result in different area, power, and delay characteristics.

FIGURE 4-11: **Behavioral Model for a 32-Bit Adder**

```
library IEEE;
use IEEE.numeric_bit.all;

entity Adder32 is
  port(A, B: in unsigned(31 downto 0); Ci: in bit;   -- Inputs
        S: out unsigned(31 downto 0); Co: out bit);   -- Outputs
end Adder32;

architecture overload of Adder32 is
signal Sum33: unsigned(32 downto 0);
begin
  Sum33 <= '0' & A + B + unsigned'(0=>Ci);            -- adder
  S <= Sum33(31 downto 0);
  Co <= Sum33(32);
end overload;
```

***Example***

If gate delays are $t_g$, what is the delay of the fastest 32-bit adder? Assume that the amount of hardware consumed is not a constraint. Only speed is important.

**Answer**

We can express each sum bit of a 32-bit adder as a sum of products expression of the input bits. There will be 33 such equations, including one for the carry out bit. These equations will be very long, and some of them could include 60+ variables in the product term. Nevertheless, if gates with any number of inputs are available, theoretically a two-level adder can be made. Although it is not very practical, theoretically, the delay of the fastest adder will be $2t_g$ if gate delays are $t_g$.

**Example**

Is ripple-carry adder the smallest 32-bit adder?

### Answer

A 32-bit ripple-carry adder uses 32 1-bit adders. We could design a 32-bit serial adder using a single 1-bit full adder. The input numbers are shifted into the adder, one bit at a time, and carry output from addition of each pair of bits is saved in a flip-flop and fed back to the next addition. The hardware illustrated in Figure 4-12 accomplishes this. The delay of adder will be $32(2t_g + t_{ff})$, where $2t_g$ is the delay of the 1-bit full adder, and $t_{ff}$ is the delay of the flip-flop (including setup time). If a flip-flop delay is at least two gate delays, the delay of the 32-bit serial adder will be at least $128t_g$. The adder hardware is simple; however, there is also the control circuitry to generate 32 shift signals. The registers storing the operands must have shift capability as well.

**FIGURE 4-12:**
**A 32-Bit Serial Adder Built from a Single 1-Bit Adder**

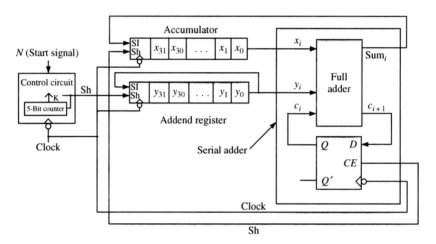

Even if you write VHDL code based on dataflow equations, as in Figure 4-10, that does not guarantee that the synthesizer will produce a carry look-ahead adder with the delay characteristics we discussed. The software might optimize the synthesis output depending on the specific hardware components available in the target technology. For instance, if you are using an FPGA with fast adder support, the software may map some of the functions into the fast adder circuitry. Depending on the number of FPGA logic blocks and interconnects used, the delays will be different from the manual calculations. The delays of a ripple-carry, carry look-ahead, and serial adder for a gate-based implementation are presented in Table 4-1 for various adder sizes. We can see that the carry look-ahead adder is very attractive for large adders.

**TABLE 4-1:**
**Comparison of Ripple-Carry and Carry Look-Ahead Adders**

| Adder size | Ripple-Carry Adder Delay | CLA Delay | Serial Adder Delay |
|---|---|---|---|
| 4 bit | $8t_g$ | $5–6t_g$ | $16t_g$ |
| 16 bit | $32t_g$ | $7–8t_g$ | $64t_g$ |
| 32 bit | $64t_g$ | $9–10t_g$ | $128t_g$ |
| 64 bit | $128t_g$ | $9–10t_g$ | $256t_g$ |

# 4.4 Traffic Light Controller

Let us design a sequential traffic light controller for the intersection of street A and street B. Each street has traffic sensors, which detect the presence of vehicles approaching or stopped at the intersection. $Sa$ = '1' means a vehicle is approaching on street A, and $Sb$ = '1' means a vehicle is approaching on street B. Street A is a main street and has a green light until a car approaches on B. Then the lights change, and B has a green light. At the end of 50 seconds, the lights change back unless there is a car on street B and none on A, in which case the B cycle is extended for 10 additional seconds. If cars continue to arrive on street B and no car appears on street A, B continues to have a green light. When A is green, it remains green at least 60 seconds, and then the lights change only when a car approaches on B. Figure 4-13 shows the external connections to the controller. Three of the outputs ($Ga$, $Ya$, and $Ra$) drive the green, yellow, and red lights on street A. The other three ($Gb$, $Yb$, and $Rb$) drive the corresponding lights on street B.

**FIGURE 4-13: Block Diagram of Traffic Light Controller**

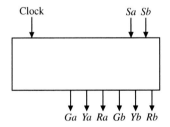

Figure 4-14 shows a Moore state graph for the controller. For timing purposes, the sequential circuit is driven by a clock with a 10-second period. Thus, a state change can occur at most once every 10 seconds. The following notation is used: $GaRb$ in a state means that $Ga = Rb = 1$ and all the other output variables are 0. $Sa'Sb$ on an arc implies that $Sa = 0$ and $Sb = 1$ will cause a transition along that arc. An arc without a label implies that a state transition will occur when the clock

**FIGURE 4-14: State Graph for Traffic Light Controller**

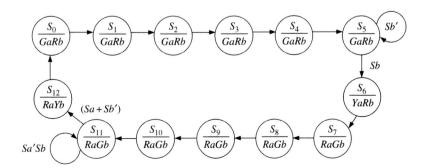

occurs, independent of the input variables. Thus, the green A light will stay on for six clock cycles (60 seconds) and then change to yellow if a car is waiting on B street.

The VHDL code for the traffic light controller (Figure 4-15) represents the state machine with two processes. Whenever the state, *Sa*, or *Sb* changes, the first process updates the outputs and *nextstate*. When the rising edge of the clock occurs, the second process updates the state register. The case statement illustrates use of a **when** clause with a range. Since states $S_0$ through $S_4$ have the same outputs, and the next states are in numeric sequence, we use a **when** clause with a range instead of five separate **when** clauses:

```
when 0 to 4 => Ga <= '1'; Rb <= '1'; nextstate <= state + 1;
```

FIGURE 4-15: **VHDL Code for Traffic Light Controller**

```
entity traffic_light is
  port(clk, Sa, Sb: in bit;
       Ra, Rb, Ga, Gb, Ya, Yb: inout bit);
end traffic_light;

architecture behave of traffic_light is
signal state, nextstate: integer range 0 to 12;
type light is (R, Y, G);
signal lightA, lightB: light;  -- define signals for waveform output
begin
  process(state, Sa, Sb)
  begin
    Ra <= '0'; Rb <= '0'; Ga <= '0'; Gb <= '0'; Ya <= '0'; Yb <= '0';
    case state is
      when 0 to 4 => Ga <= '1'; Rb <= '1'; nextstate <= state+1;
      when 5 => Ga <= '1'; Rb <= '1';
        if Sb = '1' then nextstate <= 6; end if;
      when 6 => Ya <= '1'; Rb <= '1'; nextstate <= 7;
      when 7 to 10 => Ra <= '1'; Gb <= '1'; nextstate <= state+1;
      when 11 => Ra <= '1'; Gb <= '1';
        if (Sa='1' or Sb='0') then nextstate <= 12; end if;
      when 12 => Ra <= '1'; Yb <= '1'; nextstate <= 0;
    end case;
  end process;
  process(clk)
  begin
    if clk'event and clk = '1' then
      state <= nextstate;
    end if;
  end process;
  lightA <= R when Ra='1' else Y when Ya='1' else G when Ga='1';
  lightB <= R when Rb='1' else Y when Yb='1' else G when Gb='1';
end behave;
```

For each state, only the signals that are '1' are listed within the case statement. Since in VHDL a signal will hold its value until it is changed, we should turn off each signal when the next state is reached. In state 6 we should set $Ga$ to '0', in state 7 we should set $Ya$ to '0', and so on. This could be accomplished by inserting appropriate statements in the when clauses. For example, we could insert `Ga <= '0'` in the **when** 6 => clause. An easier way to turn off the outputs is to set them all to '0' before the case statement, as shown in Figure 4-15. At first, it seems that a glitch might occur in the output when we set a signal to '0' that should remain '1'. However, this is not a problem because the sequential statements within a process execute instantaneously. For example, suppose that at time = 20 ns a state change from $S_2$ to $S_3$ occurs. $Ga$ and $Rb$ are '1', but as soon as the process starts executing, the first line of code is executed and $Ga$ and $Rb$ are scheduled to change to '0' at time $20 + \Delta$. The case statement then executes, and $Ga$ and $Rb$ are scheduled to change to '1' at time $20 + \Delta$. Since this is the same time as before, the new value ('1') preempts the previously scheduled value ('0'), and the signals never change to '0'.

Before completing the design of the traffic controller, we will test the VHDL code to see that it meets specifications. As a minimum, our test sequence should cause all of the arcs on the state graph to be traversed at least once. We may want to perform additional tests to check the timing for various traffic conditions, such as heavy traffic on both A and B, light traffic on both, heavy traffic on A only, heavy traffic on B only, and special cases such as a car failing to move when the light is green, a car going through the intersection when the light is red, and so on.

To make it easier to interpret the simulator output, we define a type named light with the values $R$, $Y$, and $G$ and two signals, *lightA* and *lightB*, which can assume these values. Then we add code to set *lightA* to R when the light is red, to Y when the light is yellow, and to G when the light is green. The following simulator command file first tests the case where both self-loops on the graph are traversed and then the case where neither self-loop is traversed:

```
add wave clk SA SB state lightA lightB
force clk 0 0, 1 5 sec -repeat 10 sec
force SA 1 0, 0 40, 1 170, 0 230, 1 250 sec
force SB 0 0, 1 70, 0 100, 1 120, 0 150, 1 210, 0 250, 1 270 sec
```

The test results in Figure 4-16 verify that the traffic lights change at the specified times.

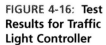

FIGURE 4-16: Test Results for Traffic Light Controller

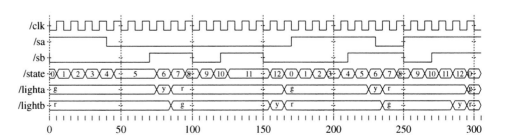

## 4.5 State Graphs for Control Circuits

Before continuing with additional examples, we describe the notation we use on control state graphs, and then state the conditions that must be satisfied to have a proper state graph. We usually label control state graphs using variable names instead of 0's and 1's. This makes the graph easier to read, especially when the number of inputs and outputs is large. If we label an arc on a Mealy state graph $X_iX_j/Z_pZ_q$, this means if inputs $X_i$ and $X_j$ are 1 (we don't care what the other input values are), the outputs $Z_p$ and $Z_q$ are 1 (and the other outputs are 0), and we will traverse this arc to go to the next state. For example, for a circuit with four inputs $(X_1, X_2, X_3, X_4)$ and four outputs $(Z_1, Z_2, Z_3, Z_4)$, the label $X_1X_4'/Z_2Z_3$ is equivalent to 1--0/0110. In general, if we label an arc with an input expression, $I$, we will traverse the arc when $I = 1$. For example, if the input label is $AB + C'$, we will traverse the arc when $AB + C' = 1$.

In order to have a completely specified proper state graph in which the next state is always uniquely defined for every input combination, we must place the following constraints on the input labels for every state $S_k$:

**1.** If $I_i$ and $I_j$ are any pair of input labels on arcs exiting state $S_k$, then $I_iI_j = 0$ if $i \neq j$.
**2.** If $n$ arcs exit state $S_k$ and the $n$ arcs have input labels $I_1, I_2, \ldots, I_n$, respectively, then $I_1 + I_2 + \cdots + I_n = 1$.

Condition 1 assures us that at most one input label can be 1 at any given time, and condition 2 assures us that at least one input label will be 1 at any given time. Therefore, exactly one label will be 1, and the next state will be uniquely defined for every input combination. For example, consider the partial state graph in Figure 4-17, where $I_1 = X_1, I_2 = X_1'X_2'$, and $I_3 = X_1'X_2$:

**FIGURE 4-17:**
**Example Partial**
**State Graph**

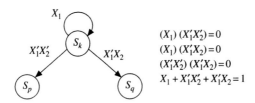

$(X_1)(X_1'X_2') = 0$
$(X_1)(X_1'X_2) = 0$
$(X_1'X_2')(X_1'X_2) = 0$
$X_1 + X_1'X_2' + X_1'X_2 = 1$

Conditions 1 and 2 are satisfied for $S_k$.

An incompletely specified proper state graph must always satisfy condition 2, and it must satisfy condition 1 for all combinations of values of input variables that can occur for each state $S_k$. Thus, the partial state graph in Figure 4-18 represents part of a proper state graph only if input combination $X_1 = X_2 = 1$ cannot occur in state $S_k$.

FIGURE 4-18:
**Example Partial
State Graph**

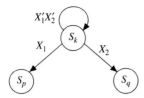

If there are three input variables $(X_1, X_2, X_3)$, the preceding partial state graph represents the following state table row:

| | 000 | 001 | 010 | 011 | 100 | 101 | 110 | 111 |
|---|---|---|---|---|---|---|---|---|
| $S_k$ | $S_k$ | $S_k$ | $S_k$ | $S_q$ | $S_q$ | $S_p$ | $S_p$ | — | — |

• • • • • • • • • • • •

# 4.6 Scoreboard and Controller

In this example, we will design a simple scoreboard, which can display scores from 0 to 99 (decimal). The input to the system should consist of a reset signal and control signals to increment or decrement the score. The two-digit decimal count gets incremented by 1 if increment signal is true and is decremented by 1 if decrement signal is true. If increment and decrement are true simultaneously, no action happens.

The current count is displayed on seven-segment displays. In order to prevent accidental erasure, the reset button must be pressed for five consecutive cycles in order to erase the scoreboard. The scoreboard should allow down counts to correct a mistake (in case of accidentally incrementing more than required).

### 4.6.1 Data Path

At the core of the design will be a two-digit BCD counter to perform the counting. Two seven-segment displays will be needed to display the current score. We will also require BCD to seven-segment decoders to facilitate the display of each BCD digit. Figure 4-19 illustrates a block diagram of the system. Since true reset should happen only after pressing reset for five clock cycles, we will also use a 3-bit reset counter called *rstcnt*.

FIGURE 4-19:
**Overview of
Simple Scoreboard**

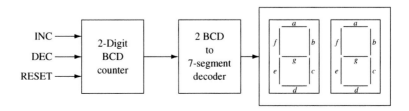

### 4.6.2 **Controller**

The controller for this circuit works as follows. There are two states in this finite state machine (FSM), as indicated in Figure 4-20. In the initial state $(S_0)$, the BCD counter is cleared. The reset counter is also made equal to 0. Essentially, $S_0$ is an initialization state where all the counters are cleared. After the initial start state, the FSM moves to the next state $(S_1)$, which is where counting gets done. In this state, in every clock cycle, incrementing or decrementing is done according to the input signals. If reset signal *rst* arrives, the *rstcnt* is incremented. If reset count has already reached 4, and reset command is still persisting in the fifth clock cycle, a transition to state $S_0$ is made. If the *inc* signal is present and *dec* is not present, the BCD counter is incremented. The notation *add1* on the arc on the top right is used to indicate that the BCD counter is incremented. If the *dec* signal is present and *inc* is not present, the BCD counter is decremented. The notation *sub1* on the arc on the bottom right is used to indicate that the BCD counter is decremented. In any cycle that the reset signal is not present, the *rstcnt* is cleared. If both the *inc* and *dec* signals are true, or neither are true, the reset counter (*rstcnt*) is cleared and the BCD counter is left unchanged.

FIGURE 4-20: **State Graph for Scoreboard**

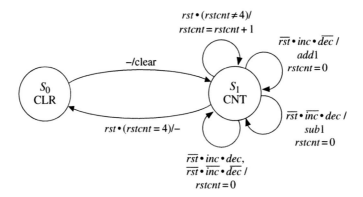

### 4.6.3 **VHDL Model**

The VHDL code for the scoreboard is given in Figure 4-21. The two seven-segment displays, *seg7disp1* and *seg7disp2*, are declared as unsigned 7-bit vectors. The segments of the seven-segment display are labeled *a* through *g*, as in Figure 4-19. The unsigned type is used so that the overloaded '+' operator can be used for incrementing the counter by 1. The decoder for the seven-segment display can be implemented as an array or look-up table. The look-up table consists of ten 7-bit vectors. A new datatype called *sevsegarray* is defined for the array of the seven-segment values corresponding to each BCD digit. It is a two-dimensional array with 10 elements, each of which is a 7-bit unsigned vector. The look-up table must be addressed with an integer data type; hence, the conversion function **to_integer** is used to generate the array index. The expression **to_integer(BCD0)** converts *BCD0* to integer type and the statement

```
seg7disp0 <= seg7rom(to_integer(BCD0));
```

accesses the appropriate element from the array *seg7rom* to convert the decimal digit to the seven-segment form. BCD addition is accomplished with the overloaded '+' operator. If the current count is less than 9, it is incremented. If it is 9, adding 1 results in a 0, but the next digit should be incremented. Similarly, decrementing from 0 is performed by borrowing a 1 from the next higher digit.

FIGURE 4-21: **VHDL Code for Scoreboard**

```vhdl
library IEEE;
use IEEE.numeric_bit.all; -- any package with overloaded add and subtract

entity Scoreboard is
  port(clk, rst, inc, dec: in bit;
       seg7disp1, seg7disp0: out unsigned(6 downto 0));
end Scoreboard;

architecture Behavioral of Scoreboard is
signal State: integer range 0 to 1;
signal BCD1, BCD0: unsigned(3 downto 0) := "0000"; -- unsigned bit vector
signal rstcnt: integer range 0 to 4 := 0;
type sevsegarray is array (0 to 9) of unsigned(6 downto 0);
constant seg7Rom: sevsegarray :=
   ("0111111", "0000110", "1011011", "1001111", "1100110", "1101101", "1111100",
    "0000111", "1111111", "1100111"); -- active high with "gfedcba" order
begin
  process(clk)
  begin
    if clk'event and clk = '1' then
      case State is
        when 0 => -- initial state
          BCD1 <= "0000"; BCD0 <= "0000"; -- clear counter
          rstcnt <= 0; -- reset RESETCOUNT
          State <= 1;
        when 1 => -- state in which the scoreboard waits for inc and dec
          if rst = '1' then
            if rstcnt = 4 then -- checking whether 5th reset cycle
              State <= 0;
            else rstcnt <= rstcnt + 1;
            end if;
          elsif inc = '1' and dec = '0' then
            rstcnt <= 0;
            if BCD0 < "1001" then
              BCD0 <= BCD0 + 1; -- library with overloaded "+" required
            elsif BCD1 < "1001" then
              BCD1 <= BCD1 + 1;
              BCD0 <= "0000";
            end if;
```

```
            elsif dec = '1' and inc = '0' then
               rstcnt <= 0;
               if BCD0 > "0000" then
                  BCD0 <= BCD0 - 1;  -- library with overloaded "-" required
               elsif BCD1 > "0000" then
                  BCD1 <= BCD1 - 1;
                  BCD0 <= "1001";
               end if;
            elsif (inc = '1' and dec = '1') or (inc = '0' and dec = '0') then
               rstcnt <= 0;
            end if;
      end case;
   end if;
 end process;
 seg7disp0 <= seg7rom(to_integer(BCD0));  -- type conversion function from
 seg7disp1 <= seg7rom(to_integer(BCD1));  -- IEEE numeric_bit package used
end Behavioral;
```

• • • • • • • • • • • •
# 4.7 Synchronization and Debouncing

The *inc, dec,* and *rst* signals to the scoreboard in the previous design are external inputs. An issue in systems involving external inputs is synchronization. Outputs from a keypad or push-button switches are not synchronous to the system clock signal. Since they will be used as inputs to a synchronous sequential circuit, they should be synchronized.

Another issue in systems involving external inputs is switch bounce. When a mechanical switch is closed or opened, the switch contact will bounce, causing noise in the switch output, as shown in Figure 4-22(a). The contact may bounce for several milliseconds before it settles down to its final position. After a switch closure has been detected, we must wait for the bounce to settle before reading the key. In any circuit involving mechanical switches, we should debounce the switches. Debouncing means removing the transients in the switch output.

Flip-flops are very useful devices when contacts need to be synchronized and debounced. Figure 4-22(b) shows a proposed debouncing and synchronizing circuit. In this design, the clock period is greater than the bounce time. If the rising edge of the clock occurs during the bounce, either a 0 or 1 will be clocked into the flip-flop at $t_1$. If a 0 was clocked in, a 1 will be clocked in at the next active clock edge ($t_2$). So it appears that $Q_A$ will be a debounced and synchronized version of $K$. However, a possibility of failure exists if *the switch* changes very close to the clock edge such that the setup or hold time is violated. In this case the flip-flop output $Q_A$ may oscillate or otherwise malfunction. Although this situation will occur very infrequently, it is best to guard against it by adding a second flip-flop. We will choose the clock period so that any oscillation at the output of $Q_A$ will have died out before the next active edge of the clock so that the input $D_B$ will always be stable at the active clock edge. The debounced signal, $Q_B$, will always be clean and synchronized with the clock, although it may be delayed up to two clock cycles after the switch is pressed.

FIGURE 4-22:
**Debouncing Mechanical Switches**

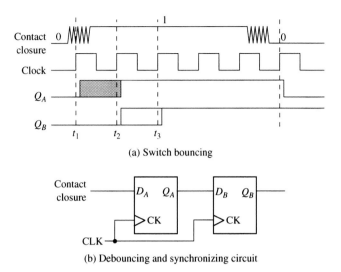

(a) Switch bouncing

(b) Debouncing and synchronizing circuit

### 4.7.1 Single Pulser

One assumption in the scoreboard design is that each time the *inc* and *dec* signals are provided, they last only for one clock cycle. Digital systems generally run at speeds higher than actions by humans, and it is very difficult for humans to produce a signal that only lasts for a clock pulse. If the pressing of the button lasted longer than a clock cycle, the counters will continue to get incremented in the aforementioned design. A solution to the problem is to develop a circuit that generates a single pulse for a human action of pressing a button or switch. Such a circuit can be used in a variety of applications involving humans, push buttons, and switches.

Now, let us design a single pulser circuit that delivers a synchronized pulse that is a single clock cycle long, when a button is pressed. The circuit must sense the pressing of a button and assert an output signal for one clock cycle. Then the output stays inactive until the button is released.

Let us create a state diagram for the single pulser. The single pulser circuit must have two states: one in which it will detect the pressing of the key and one in which it will detect the release of the key. Let us call the first state $S_0$ and the second state $S_1$. Let us use the symbol SYNCPRESS to denote the synchronized key press. When the circuit is in state $S_0$ and the button is pressed, the system produces the single pulse and moves to state $S_1$. The single pulse is a Mealy output as the state changes from $S_0$ to $S_1$. Once the system is in state $S_1$, it waits for the button to be released. As soon as it is released, it moves to the start state $S_0$ waiting for the next button press. The single pulse output is true only during the transition from $S_0$ to $S_1$. The state diagram is illustrated in Figure 4-23.

Since there are only two states for this circuit, it can be implemented using one flip-flop. A single pulser can be implemented as in Figure 4-24. The first block consists of the circuitry in Figure 4-22(b) and generates a synchronized button press, SYNCPRESS. The flip-flop implements the two states of the state machine. Let us assume the state assignments are $S_0 = 0$ and $S_1 = 1$. In such a case, the $Q$ output of

FIGURE 4-23: **State Diagram of Single Pulser**

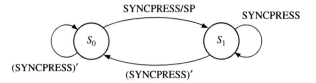

FIGURE 4-24: **Single Pulser and Synchronizer Circuit**

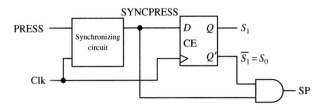

the flip-flop is synonymous with $S_1$, and the $Q'$ output of the flip-flop is synonymous with $S_0$. The equation for the single pulse SP is

$$SP = S0 \cdot SYNCPRESS$$

It may also be noted that $S_0 = S_1'$. Including the two flip-flops inside the synchronizing block, three flip-flops can provide debouncing, synchronization, and single pulsing. If button pushes can be passed through such a circuit, a single pulse that is debounced and synchronized, with respect to the system clock, can be obtained. It is a good practice to feed external push-button signals through such a circuit in order to obtain controlled and predictable operation.

● ● ● ● ● ● ● ● ● ● ● ●

# 4.8  Add-and-Shift Multiplier

In this section, we will design a multiplier for unsigned binary numbers. When we form the product $A \times B$, the first operand ($A$) is called the *multiplicand*, and the second operand ($B$) is called the *multiplier*. As illustrated here, binary multiplication requires only shifting and adding. In the following example, we multiply $13_{10}$ by $11_{10}$ in binary:

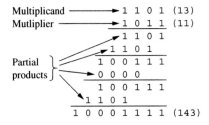

Note that each partial product is either the multiplicand (1101) shifted over by the appropriate number of places or zero. Instead of forming all the partial products first and then adding, each new partial product is added in as soon as it is formed, which eliminates the need for adding more than two binary numbers at a time.

Multiplication of two 4-bit numbers requires a 4-bit multiplicand register, a 4-bit multiplier register, a 4-bit full adder, and an 8-bit register for the product. The product register serves as an accumulator to accumulate the sum of the partial products. If the multiplicand were shifted left each time before it was added to the accumulator, as was done in the previous example, an 8-bit adder would be needed. So it is better to shift the contents of the product register to the right each time, as shown in the block diagram of Figure 4-25. This type of multiplier is sometimes referred to as a serial-parallel multiplier, since the multiplier bits are processed serially, but the addition takes place in parallel. As indicated by the arrows on the diagram, 4 bits from the accumulator (ACC) and 4 bits from the multiplicand register are connected to the adder inputs; the 4 sum bits and the carry output from the adder are connected back to the accumulator. When an add signal ($Ad$) occurs, the adder outputs are transferred to the accumulator by the next clock pulse, thus causing the multiplicand to be added to the accumulator. An extra bit at the left end of the product register temporarily stores any carry that is generated when the multiplicand is added to the accumulator. When a shift signal ($Sh$) occurs, all 9 bits of ACC are shifted right by the next clock pulse.

FIGURE 4-25: **Block Diagram for Binary Multiplier**

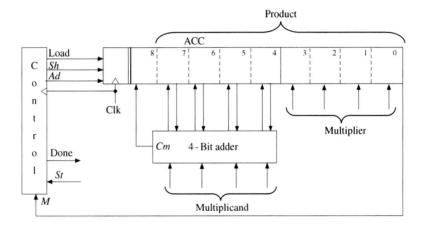

Since the lower 4 bits of the product register are initially unused, we will store the multiplier in this location instead of in a separate register. As each multiplier bit is used, it is shifted out the right end of the register to make room for additional product bits. A shift signal ($Sh$) causes the contents of the product register (including the multiplier) to be shifted right one place when the next clock pulse occurs. The control circuit puts out the proper sequence of add and shift signals after a start signal ($St = 1$) has been received. If the current multiplier bit ($M$) is 1, the multiplicand is added to the accumulator followed by a right shift; if the multiplier bit is 0, the addition is skipped, and only the right shift occurs. The multiplication example

(13 × 11) is reworked below, showing the location of the bits in the registers at each clock time.

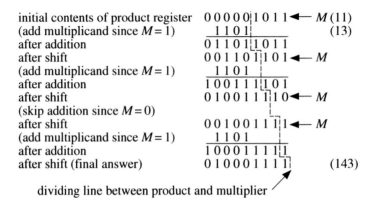

initial contents of product register    0 0 0 0 0|1 0 1 1 ◄— M (11)
(add multiplicand since M = 1)     1 1 0 1!        (13)
after addition            0 1 1 0 1|1 0 1 1
after shift              0 0 1 1 0 1|1 0 1 ◄— M
(add multiplicand since M = 1)     1 1 0 1
after addition            1 0 0 1 1 1|1 0 1
after shift              0 1 0 0 1 1 1|1 0 ◄— M
(skip addition since M = 0)
after shift              0 0 1 0 0 1 1 1|1 ◄— M
(add multiplicand since M = 1)     1 1 0 1
after addition            1 0 0 0 1 1 1 1|1
after shift (final answer)       0 1 0 0 0 1 1 1 1|       (143)

     dividing line between product and multiplier

The control circuit must be designed to output the proper sequence of add and shift signals. Figure 4-26 shows a state graph for the control circuit. In Figure 4-26, $S_0$ is the reset state, and the circuit stays in $S_0$ until a start signal ($St = 1$) is received. This generates a *Load* signal, which causes the multiplier to be loaded into the lower 4 bits of the accumulator (ACC) and the upper 5 bits of the accumulator to be cleared. In state $S_1$, the low-order bit of the multiplier ($M$) is tested. If $M = 1$, an add signal is generated, and if $M = 0$, a shift signal is generated. Similarly, in states $S_3$, $S_5$, and $S_7$, the current multiplier bit ($M$) is tested to determine whether to generate an add or shift signal. A shift signal is always generated at the next clock time following an add signal (states $S_2$, $S_4$, $S_6$, and $S_8$). After four shifts have been generated, the control network goes to $S_9$, and a done signal is generated before returning to $S_0$.

FIGURE 4-26: **State Graph for Binary Multiplier Control**

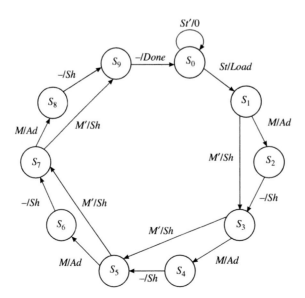

The behavioral VHDL model (Figure 4-27) corresponds directly to the state graph. Since there are 10 states, we have declared an integer ranging from 0 to 9 for the state signal. The signal *ACC* represents the 9-bit accumulator output. The statement

> **alias** M: bit **is** ACC(0);

allows us to use the name *M* in place of *ACC(0)*. The notation **when** 1|3|5|7 => means that when the state is 1 or 3 or 5 or 7, the action that follows occurs. All register operations and state changes take place on the rising edge of the clock. For example, in state 0, if *St* is '1', the multiplier is loaded into the accumulator at the same time the state changes to 1. The expression `'0'` & ACC(7 **downto** 4) + Mcand is used to compute the sum of two 4-bit unsigned vectors to give a 5-bit result. This represents the adder output, which is loaded into *ACC* at the same time the state counter is incremented. The right shift on *ACC* is accomplished by loading *ACC* with '0' concatenated with the upper 8 bits of *ACC*. The expression `'0'` & ACC(8 **downto** 1) could be replaced with ACC **srl** 1.

FIGURE 4-27: **Behavioral Model for 4 × 4 Binary Multiplier**

```
-- This is a behavioral model of a multiplier for unsigned
-- binary numbers. It multiplies a 4-bit multiplicand
-- by a 4-bit multiplier to give an 8-bit product.

-- The maximum number of clock cycles needed for a
-- multiply is 10.

library IEEE;
use IEEE.numeric_bit.all;

entity mult4X4 is
  port(Clk, St: in bit;
       Mplier, Mcand: in unsigned(3 downto 0);
       Done: out bit);
end mult4X4;

architecture behave1 of mult4X4 is
signal State: integer range 0 to 9;
signal ACC: unsigned(8 downto 0); -- accumulator
alias M: bit is ACC(0);        -- M is bit 0 of ACC
begin
  process(Clk)
  begin
    if Clk'event and Clk = '1' then  -- executes on rising edge of clock
      case State is
        when 0 =>                  -- initial State
          if St = '1' then
            ACC(8 downto 4) <= "00000"; -- begin cycle
            ACC(3 downto 0) <= Mplier;  -- load the multiplier
            State <= 1;
          end if;
```

```
          when 1 | 3 | 5 | 7 =>       -- "add/shift" State
             if M = '1' then          -- add multiplicand
                ACC(8 downto 4) <= '0' & ACC(7 downto 4) + Mcand;
                State <= State + 1;
             else
                ACC <= '0' & ACC(8 downto 1);      -- shift accumulator right
                State <= State + 2;
             end if;
          when 2 | 4 | 6 | 8 =>                -- "shift" State
             ACC <= '0' & ACC(8 downto 1);     -- right shift
             State <= State + 1;
          when 9 =>                            -- end of cycle
             State <= 0;
       end case;
    end if;
  end process;
  Done <= '1' when State = 9 else '0';
end behave1;
```

The *Done* signal needs to be turned on only in state 9. If we had used the statement **when** 9 => State <= 0; Done <= '1', *Done* would be turned on at the same time *State* changes to 0. This is too late, since we want *Done* to turn on when *State* becomes 9. Therefore, we used a separate concurrent assignment statement. This statement is placed outside the process so that *Done* will be updated whenever *State* changes.

As the state graph for the multiplier (Figure 4-26) indicates, the control performs two functions—generating add or shift signals as needed and counting the number of shifts. If the number of bits is large, it is convenient to divide the control circuit into a counter and an add-shift control, as shown in Figure 4-28(a). First, we will derive a state graph for the add-shift control that tests $St$ and $M$ and outputs the proper sequence of add and shift signals (Figure 4-28(b)). Then we will add a completion signal ($K$) from the counter that stops the multiplier after the proper number of shifts have been completed. Starting in $S_0$ in Figure 4-28(b), when a start signal $St = 1$ is received, a load signal is generated and the circuit goes to state $S_1$. Then if $M = 1$, an add signal is generated and the circuit goes to state $S_2$; if $M = 0$, a shift signal is generated and the circuit stays in $S_1$. In $S_2$, a shift signal is generated since a shift always follows an add. The graph of Figure 4-28(b) will generate the proper sequence of add and shift signals, but it has no provision for stopping the multiplier.

In order to determine when the multiplication is completed, the counter is incremented each time a shift signal is generated. If the multiplier is $n$ bits, $n$ shifts are required. We will design the counter so that a completion signal ($K$) is generated after $n - 1$ shifts have occurred. When $K = 1$, the circuit should perform one more addition if necessary and then do the final shift. The control operation in Figure 4-28(c) is the same as Figure 4-28(b) as long as $K = 0$. In state $S_1$, if $K = 1$, we test $M$ as usual. If $M = 0$, we output the final shift signal and go to the done state ($S_3$); however, if $M = 1$, we add before shifting and go to state $S_2$. In state $S_2$, if $K = 1$, we output one more shift signal and then go to $S_3$. The last shift signal

FIGURE 4-28:
**Multiplier Control with Counter**

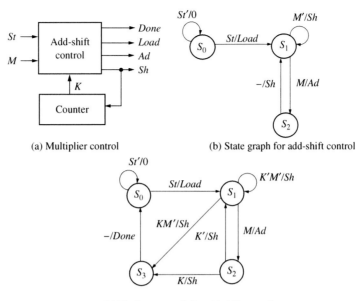

(a) Multiplier control

(b) State graph for add-shift control

(c) Final state graph for add-shift control

will increment the counter to 0 at the same time the add-shift control goes to the done state.

As an example, consider the multiplier of Figure 4-25, but replace the control circuit with Figure 4-28(a). Since $n = 4$, a 2-bit counter is needed to count the four shifts, and $K = 1$ when the counter is in state 3 ($11_2$). Table 4-2 shows the operation of the multiplier when 1101 is multiplied by 1011. $S_0, S_1, S_2$, and $S_3$ represent states of the control circuit (Figure 4-28(c)). The contents of the product register at each step are the same as given on page 212.

At time $t_0$, the control is reset and waiting for a start signal. At time $t_1$, the start signal $St$ is 1, and a *Load* signal is generated. At time $t_2$, $M = 1$, so an $Ad$ signal is generated. When the next clock occurs, the output of the adder is loaded into the accumulator and the control goes to $S_2$. At $t_3$, an $Sh$ signal is generated, so at the next clock shifting occurs and the counter is incremented. At $t_4$, $M = 1$, so $Ad = 1$, and the adder output is loaded into the accumulator at the next clock. At $t_5$ and $t_6$, shifting

TABLE 4-2:
**Operation of Multiplier Using a Counter**

| Time | State | Counter | Product Register | St | M | K | Load | Ad | Sh | Done |
|------|-------|---------|------------------|----|----|----|------|----|----|------|
| $t_0$ | $S_0$ | 00 | 000000000 | 0 | 0 | 0 | 0 | 0 | 0 | 0 |
| $t_1$ | $S_0$ | 00 | 000000000 | 1 | 0 | 0 | 1 | 0 | 0 | 0 |
| $t_2$ | $S_1$ | 00 | 000001011 | 0 | 1 | 0 | 0 | 1 | 0 | 0 |
| $t_3$ | $S_2$ | 00 | 011011011 | 0 | 1 | 0 | 0 | 0 | 1 | 0 |
| $t_4$ | $S_1$ | 01 | 001101101 | 0 | 1 | 0 | 0 | 1 | 0 | 0 |
| $t_5$ | $S_2$ | 01 | 100111101 | 0 | 1 | 0 | 0 | 0 | 1 | 0 |
| $t_6$ | $S_1$ | 10 | 010011110 | 0 | 0 | 0 | 0 | 0 | 1 | 0 |
| $t_7$ | $S_1$ | 11 | 001001111 | 0 | 1 | 1 | 0 | 1 | 0 | 0 |
| $t_8$ | $S_2$ | 11 | 100011111 | 0 | 1 | 1 | 0 | 0 | 1 | 0 |
| $t_9$ | $S_3$ | 00 | 010001111 | 0 | 1 | 0 | 0 | 0 | 0 | 1 |

and counting occur. At $t_7$, three shifts have occurred and the counter state is 11, so $K = 1$. Since $M = 1$, addition occurs and control goes to $S_2$. At $t_8$, $Sh = K = 1$, so at the next clock the final shift occurs and the counter is incremented back to state 00. At $t_9$, a *Done* signal is generated.

The multiplier design given here can easily be expanded to 8, 16, or more bits simply by increasing the register size and the number of bits in the counter. The add-shift control would remain unchanged.

● ● ● ● ● ● ● ● ● ● ● ● ●

## 4.9  Array Multiplier

An array multiplier is a parallel multiplier that generates the partial products in a parallel fashion. The various partial products are added as soon as they are available. Consider the process of multiplication as illustrated in Table 4-3. Two 4-bit unsigned numbers, $X_3X_2X_1X_0$ and $Y_3Y_2Y_1Y_0$, are multiplied to generate a product that is possibly 8 bits. Each of the $X_iY_j$ product bits can be generated by an AND gate. Each partial product can be added to the previous sum of partial products using a row of adders. The sum output of the first row of adders, which adds the first two partial products, is $S_{13}S_{12}S_{11}S_{10}$, and the carry output is $C_{13}C_{12}C_{11}C_{10}$. Similar results occur for the other two rows of adders. (We have used the notation $S_{ij}$ and $C_{ij}$ to represent the sums and carries from the $i$th row of adders.)

**TABLE 4-3: Four-bit Multiplier Partial Products**

| | | | | | $X_3$ | $X_2$ | $X_1$ | $X_0$ | Multiplicand |
|---|---|---|---|---|---|---|---|---|---|
| | | | | | $Y_3$ | $Y_2$ | $Y_1$ | $Y_0$ | Multiplier |
| | | | | | $X_3Y_0$ | $X_2Y_0$ | $X_1Y_0$ | $X_0Y_0$ | Partial product 0 |
| | | | | $X_3Y_1$ | $X_2Y_1$ | $X_1Y_1$ | $X_0Y_1$ | | Partial product 1 |
| | | | | $C_{12}$ | $C_{11}$ | $C_{10}$ | | | First row carries |
| | | | $C_{13}$ | $S_{13}$ | $S_{12}$ | $S_{11}$ | $S_{10}$ | | First row sums |
| | | | | $X_3Y_2$ | $X_2Y_2$ | $X_1Y_2$ | $X_0Y_2$ | | Partial product 2 |
| | | | | $C_{22}$ | $C_{21}$ | $C_{20}$ | | | Second row carries |
| | | $C_{23}$ | $S_{23}$ | $S_{22}$ | $S_{21}$ | $S_{20}$ | | | Second row sums |
| | | | $X_3Y_3$ | $X_2Y_3$ | $X_1Y_3$ | $X_0Y_3$ | | | Partial product 3 |
| | | | $C_{32}$ | $C_{31}$ | $C_{30}$ | | | | Third row carries |
| | $C_{33}$ | $S_{33}$ | $S_{32}$ | $S_{31}$ | $S_{30}$ | | | | Third row sums |
| | $P_7$ | $P_6$ | $P_5$ | $P_4$ | $P_3$ | $P_2$ | $P_1$ | $P_0$ | Final product |

Figure 4-29 shows the array of AND gates and adders to perform this multiplication. If an adder has three inputs, a full adder (FA) is used, but if an adder has only two inputs, a half-adder (HA) is used. A half-adder is the same as a full adder with one of the inputs set to 0. This multiplier requires 16 AND gates, 8 full adders, and 4 half-adders. After the $X$ and $Y$ inputs have been applied, the carry must propagate along each row of cells, and the sum must propagate from row to row. The time required to complete the multiplication depends primarily on the propagation delay in the adders. The longest path from input to output goes through 8 adders. If $t_{ad}$ is

FIGURE 4-29: **Block Diagram of 4 × 4 Array Multiplier**

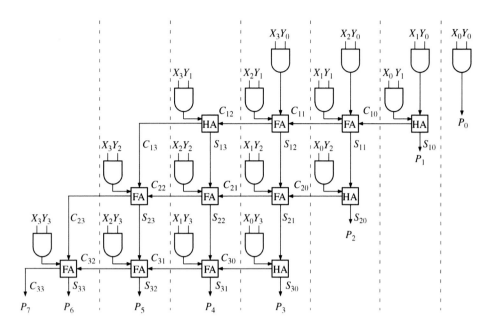

the worst-case (longest possible) delay through an adder, and $t_g$ is the longest AND gate delay, then the worst-case time to complete the multiplication is $8t_{ad} + t_g$.

In general, an $n$-bit-by-$n$-bit array multiplier would require $n^2$ AND gates, $n(n-2)$ full adders, and $n$ half-adders. So the number of components required increases quadratically. For the serial-parallel multiplier previously designed, the amount of hardware required in addition to the control circuit increases linearly with $n$.

For an $n \times n$ array multiplier, the longest path from input to output goes through $n$ adders in the top row, $n - 1$ adders in the bottom row, and $n - 3$ adders in the middle rows. The corresponding worst-case multiply time is $(3n - 4)t_{ad} + t_g$. The longest delay in a circuit is called critical path. The worst-case delay can be improved to $2nt_{ad} + t_g$ by forwarding carry from each adder to the diagonally lower adder rather than the adder on the left side. When $n = 4$, both expressions are the same; however, for larger values of $n$, it is beneficial to pass carry diagonally as opposed to rippling it to the left. Note that this multiplier has no sequential logic or registers.

The shift-and-add multiplier that we previously designed requires $2n$ clocks to complete the multiply in the worst case, although this can be reduced to $n$ clocks using a technique discussed in the next section. The minimum clock period depends on the propagation delay through the $n$-bit adder as well as the propagation delay and setup time for the accumulator flip-flops.

### 4.9.1 VHDL Coding

If the topology has to be exactly what the designer wants, we need to do structural coding as shown in Figure 4-30. If we made a behavioral model of a multiplier without specifying the topology, the topology generated by the synthesizer would depend on the synthesis tool. Here, we present a structural model for an array

FIGURE 4-30: **VHDL Code for 4 × 4 Array Multiplier**

```vhdl
entity Array_Mult is
  port(X, Y: in bit_vector(3 downto 0);
       P: out bit_vector(7 downto 0));
end Array_Mult;

architecture Behavioral of Array_Mult is
signal C1, C2, C3: bit_vector(3 downto 0);
signal S1, S2, S3: bit_vector(3 downto 0);
signal XY0, XY1, XY2, XY3: bit_vector(3 downto 0);
component FullAdder
  port(X, Y, Cin: in bit;
       Cout, Sum: out bit);
end component;
component HalfAdder
  port(X, Y: in bit;
       Cout, Sum: out bit);
end component;
begin
  XY0(0) <= X(0) and Y(0); XY1(0) <= X(0) and Y(1);
  XY0(1) <= X(1) and Y(0); XY1(1) <= X(1) and Y(1);
  XY0(2) <= X(2) and Y(0); XY1(2) <= X(2) and Y(1);
  XY0(3) <= X(3) and Y(0); XY1(3) <= X(3) and Y(1);

  XY2(0) <= X(0) and Y(2); XY3(0) <= X(0) and Y(3);
  XY2(1) <= X(1) and Y(2); XY3(1) <= X(1) and Y(3);
  XY2(2) <= X(2) and Y(2); XY3(2) <= X(2) and Y(3);
  XY2(3) <= X(3) and Y(2); XY3(3) <= X(3) and Y(3);

  FA1: FullAdder port map (XY0(2), XY1(1), C1(0), C1(1), S1(1));
  FA2: FullAdder port map (XY0(3), XY1(2), C1(1), C1(2), S1(2));
  FA3: FullAdder port map (S1(2), XY2(1), C2(0), C2(1), S2(1));
  FA4: FullAdder port map (S1(3), XY2(2), C2(1), C2(2), S2(2));
  FA5: FullAdder port map (C1(3), XY2(3), C2(2), C2(3), S2(3));
  FA6: FullAdder port map (S2(2), XY3(1), C3(0), C3(1), S3(1));
  FA7: FullAdder port map (S2(3), XY3(2), C3(1), C3(2), S3(2));
  FA8: FullAdder port map (C2(3), XY3(3), C3(2), C3(3), S3(3));
  HA1: HalfAdder port map (XY0(1), XY1(0), C1(0), S1(0));
  HA2: HalfAdder port map (XY1(3), C1(2), C1(3), S1(3));
  HA3: HalfAdder port map (S1(1), XY2(0), C2(0), S2(0));
  HA4: HalfAdder port map (S2(1), XY3(0), C3(0), S3(0));

  P(0) <= XY0(0); P(1) <= S1(0); P(2) <= S2(0);
  P(3) <= S3(0); P(4) <= S3(1); P(5) <= S3(2);
  P(6) <= S3(3); P(7) <= C3(3);
end Behavioral;

-- Full Adder and half adder entity and architecture descriptions
-- should be in the project
```

```
entity FullAdder is
  port(X, Y, Cin: in bit;
       Cout, Sum: out bit);
end FullAdder;

architecture equations of FullAdder is
begin
  Sum <= X xor Y xor Cin;
  Cout <= (X and Y) or (X and Cin) or (Y and Cin);
end equations;

entity HalfAdder is
  port(X, Y: in bit;
       Cout, Sum: out bit);
end HalfAdder;

architecture equations of HalfAdder is
begin
  Sum <= X xor Y;
  Cout <= X and Y;
end equations;
```

multiplier. Full-adder and half-adder modules are created and used as components for the array multiplier. The full adders and half adders are interconnected according to the array multiplier topology. Several instantiation (**port map**) statements are used for this purpose.

● ● ● ● ● ● ● ● ● ● ● ●
## 4.10  A Signed Integer/Fraction Multiplier

Several algorithms are available for multiplication of signed binary numbers. The following procedure is a straightforward way to carry out the multiplication:

**1.** Complement the multiplier if negative.
**2.** Complement the multiplicand if negative.
**3.** Multiply the two positive binary numbers.
**4.** Complement the product if it should be negative.

Although this method is conceptually simple, it requires more hardware and computation time than some of the other available methods.

The next method we describe requires only the ability to complement the multiplicand. Complementation of the multiplier or product is not necessary. Although the method works equally well with integers or fractions, we illustrate the method with fractions, since we will later use this multiplier as part of a multiplier for floating-point numbers. Using 2's complement for negative numbers, we will represent signed binary fractions in the following form:

0.101    +5/8    1.011    −5/8

The digit to the left of the binary point is the sign bit, which is 0 for positive fractions and 1 for negative fractions. In general, the 2's complement of a binary fraction $F$ is $F^* = 2 - F$. Thus, $-5/8$ is represented by $10.000 - 0.101 = 1.011$. (This method of defining 2's complement fractions is consistent with the integer case ($N^* = 2^n - N$), since moving the binary point $n - 1$ places to the left is equivalent to dividing by $2^{n-1}$.) The 2's complement of a fraction can be found by starting at the right end and complementing all the digits to the left of the first 1, the same as for the integer case. The 2's complement fraction $1.000\ldots$ is a special case. It actually represents the number $-1$, since the sign bit is negative and the 2's complement of $1.000\ldots$ is $2 - 1 = 1$. We cannot represent $+1$ in this 2's complement fraction system, since $0.111\ldots$ is the largest positive fraction.

---

**Binary Fixed-Point Fractions**

Fixed-point numbers are number formats in which the decimal or binary point is at a fixed location. We can have a fixed-point 8-bit number format where the binary point is assumed to be after 4 bits (i.e., 4 bits for the fractional part and 4 bits for the integer part). If the binary point is assumed to be located two more bits to the right, there will be 6 bits for the integer part and 2 bits for the fraction. The range and precision of the numbers that can be represented in the different formats depend on the location of the binary point. For instance, if there are 4 bits for the fractional part and 4 bits for the integer, the range, assuming unsigned numbers, is 0.00 to 15.9375. If only 2 bits are allowed for the fractional part and 6 bits for the integer, the range increases; however, the precision reduces. Now, the range would be 0.00 to 63.75, but the fractional part can be specified only as a multiple of 0.25.

Let us say we need to represent $-13.45$ in a 2's complement fixed-point number representation with four fractional bits. To convert any decimal fraction into the binary fraction, one technique is to repeatedly multiply the fractional part (only the fractional part in each intermediate step) with 2. So, starting with 0.45, the repeated multiplication results in

$$\textbf{0}.90$$
$$\textbf{1}.80$$
$$\textbf{1}.60$$
$$\textbf{1}.20$$
$$\textbf{0}.40$$
$$\textbf{0}.80$$
$$\textbf{1}.60$$
$$\textbf{1}.20$$

Now, the binary representation can be obtained by considering the digits in bold. An appropriate representation can be obtained depending on the number of bits available (e.g., 0111 if 4 bits are available, 01110011 if 8 bits are available, and so on). The representation for decimal number 13.45 in the fixed-point format with four binary places will be as follows:

$$13.45: \quad 1101.0111$$

Note that the represented number is only an approximation of the actual number. The represented number can be converted back to decimal and seen to be 13.4375 (slightly off from the number we started with). The representation approaches the actual number as more and more binary places are added to the representation.

Negative fractions can be represented in 2's complement form. Let us represent $-13.45$ in 2's complement form. This cannot be done if we have only four places for the integer. We need to have at least 5 bits for the integer in order to handle the sign. Assuming 5 bits are available for the integer, in a 9-bit format,

|  |  |
|---|---|
| 13.45: | 01101.0111 |
| 1's complement | 10010.1000 |
| 2's complement | 10010.1001 |

Hence $-13.45 = 10010.1001$ in this representation.

When multiplying signed binary numbers, we must consider four cases:

| Multiplicand | Multiplier |
|:---:|:---:|
| + | + |
| − | + |
| + | − |
| − | − |

When both the multiplicand and the multiplier are positive, standard binary multiplication is used. For example,

|  |  |  |  |
|---|---|---|---|
| 0.1 1 1 | (+7/8) | ← | Multiplicand |
| × 0.1 0 1 | (+5/8) | ← | Multiplier |
| (0. 0 0)0 1 1 1 | (+7/64) | ← | *Note*: The proper representation |
| (0.)0 1 1 1 | (+7/16) | ← | of the fractional partial products |
| 0. 1 0 0 0 1 1 | (+35/64) |  | requires extension of the sign |

bit past the binary point, as
indicated in parentheses. (Such
extension is not necessary in
the hardware.)

When the multiplicand is negative and the multiplier is positive, the procedure is the same as in the previous case, except that we must extend the sign bit of the multiplicand so that the partial products and final product will have the proper negative sign. For example,

$$
\begin{array}{ll}
1.1\,0\,1 & (-3/8) \\
\underline{\times\ 0.1\,0\,1} & (+5/8) \\
(1.\ 1\ 1)1\ 1\ 0\ 1 & (-3/64) \quad \leftarrow \\
\underline{(1.)1\ 1\ 0\ 1} & (-3/16) \quad \leftarrow \\
1.\ 1\ 1\ 0\ 0\ 0\ 1 & (-15/64)
\end{array}
$$

Note: The extension of the sign bit provides proper representation of the negative products.

When the multiplier is negative and the multiplicand is positive, we must make a slight change in the multiplication procedure. A negative fraction of the form $1.g$ has a numeric value $-1 + 0.g$; for example, $1.011 = -1 + 0.011 = -(1 - 0.011) = -0.101 = -5/8$. Thus, when multiplying by a negative fraction of the form $1.g$, we treat the fraction part $(.g)$ as a positive fraction, but the sign bit is treated as $-1$. Hence, multiplication proceeds in the normal way as we multiply by each bit of the fraction and accumulate the partial products. However, when we reach the negative sign bit, we must add in the 2's complement of the multiplicand instead of the multiplicand itself. The following example illustrates this:

$$
\begin{array}{ll}
0.1\,0\,1 & (+5/8) \\
\underline{\times\ 1.1\,0\,1} & (-3/8) \\
(0.\,0\,0)0\ 1\ 0\ 1 & (+5/64) \\
\underline{(0.)0\ 1\ 0\ 1} & (+5/16) \\
(0.)0\ 1\ 1\ 0\ 0\ 1 & \\
\underline{1.\ 0\ 1\ 1} & (-5/8) \quad \leftarrow \\
1.\ 1\ 1\ 0\ 0\ 0\ 1 & (-15/64)
\end{array}
$$

Note: The 2's complement of the multiplicand is added at this point.

When both the multiplicand and multiplier are negative, the procedure is the same as before. At each step, we must be careful to extend the sign bit of the partial product to preserve the proper negative sign, and at the final step we must add in the 2's complement of the multiplicand, since the sign bit of the multiplier is negative. For example,

$$
\begin{array}{ll}
1.1\,0\,1 & (-3/8) \\
\underline{\times\ 1.1\,0\,1} & (-3/8) \\
(1.\ 1\ 1)1\ 1\ 0\ 1 & (-3/64) \quad \leftarrow \\
\underline{(1.)1\ 1\ 0\ 1} & (-3/16) \\
1.\ 1\ 1\ 0\ 0\ 0\ 1 & \\
\underline{0.\ 0\ 1\ 1} & (+3/8) \quad \leftarrow \\
0.\ 0\ 0\ 1\ 0\ 0\ 1 & (+9/64)
\end{array}
$$

Note: Extend sign bit.

Add the 2's complement of the multiplicand.

In summary, the procedure for multiplying signed 2's complement binary fractions is the same as for multiplying positive binary fractions, except that we must be careful to preserve the sign of the partial product at each step, and if the sign of the multiplier is negative, we must complement the multiplicand before adding it in at the last step. The hardware is almost identical to that used for multiplication of positive numbers, except a complementer must be added for the multiplicand.

Figure 4-31 shows the hardware required to multiply two 4-bit fractions (including the sign bit). A 5-bit adder is used so the sign of the sum is not lost due to a carry into the sign bit position. The $M$ input to the control circuit is the currently active bit of the multiplier. Control signal $Sh$ causes the accumulator to shift right one place with sign extension. $Ad$ causes the ADDER output to be loaded into the left 5 bits of the accumulator. The carry-out from the last bit of the adder is discarded,

**FIGURE 4-31: Block Diagram for 2's Complement Multiplier**

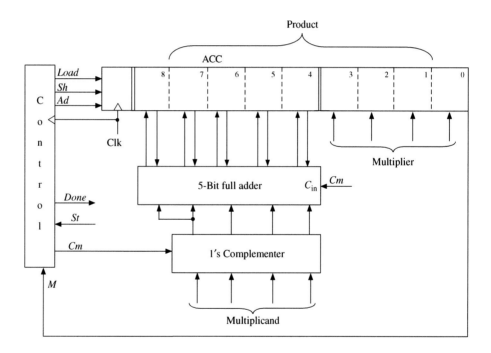

since we are doing 2's complement addition. $Cm$ causes the multiplicand (Mcand) to be complemented (1's complement) before it enters the adder inputs. $Cm$ is also connected to the carry input of the adder so that when $Cm = 1$, the adder adds 1 plus the 1's complement of Mcand to the accumulator, which is equivalent to adding the 2's complement of Mcand. Figure 4-32 shows a state graph for the control circuit. Each multiplier bit ($M$) is tested to determine whether to add and shift or whether to just shift. In state $S_7$, $M$ is the sign bit, and if $M = 1$, the complement of the multiplicand is added to the accumulator.

When the hardware in Figure 4-31 is used, the add and shift operations must be done at two separate clock times. We can speed up operation of the multiplier by

FIGURE 4-32: **State Graph for 2's Complement Multiplier**

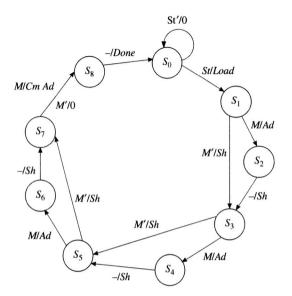

moving the wires from the adder output one position to the right (Figure 4-33) so that the adder output is already shifted over one position when it is loaded into the accumulator. With this arrangement, the add and shift operations can occur at the same clock time, which leads to the control state graph of Figure 4-34. When the multiplication is complete, the product (6 bits plus sign) is in the lower 3 bits of $A$ followed by $B$. The binary point then is in the middle of the $A$ register. If we wanted it between the left 2 bits, we would have to shift $A$ and $B$ left one place.

FIGURE 4-33: **Block Diagram for Faster Multiplier**

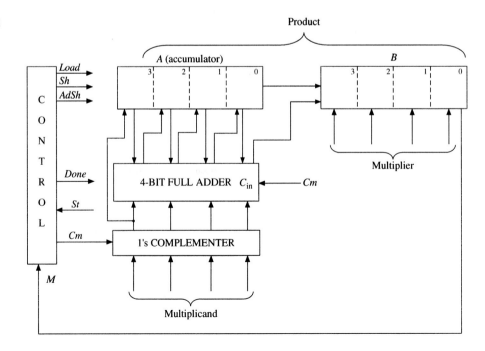

**FIGURE 4-34: State Graph for Faster Multiplier**

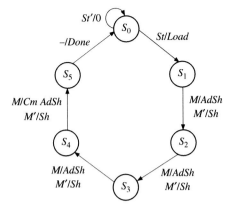

A behavioral VHDL model for this multiplier is shown in Figure 4-35. Shifting the A and B registers together is accomplished by the sequential statements

```
A <= A(3) & A(3 downto 1);
B <= A(0) & B(3 downto 1);
```

Although these statements are executed sequentially, $A$ and $B$ are both scheduled to be updated at the same delta time. Therefore, the old value of $A(0)$ is used when computing the new value of $B$.

**FIGURE 4-35: Behavioral Model for 2's Complement Multiplier**

```
library IEEE;
use IEEE.numeric_bit.all;

entity mult2C is
  port(CLK, St: in bit;
       Mplier, Mcand : in unsigned(3 downto 0);
       Product: out unsigned (6 downto 0);
       Done: out bit);
end mult2C;

architecture behave1 of mult2C is
signal State: integer range 0 to 5;
signal A, B: unsigned(3 downto 0);
alias M: bit is B(0);
begin
  process(CLK)
  variable addout: unsigned(3 downto 0);
  begin
    if CLK'event and CLK = '1' then
      case State is
        when 0 =>                      -- initial State
          if St = '1' then
            A <= "0000";               -- begin cycle
            B <= Mplier;               -- load the multiplier
```

```
               State <= 1;
           end if;
        when 1 | 2 | 3 =>             -- "add/shift" states
          if M = '1' then
            addout := A + Mcand;    -- add multiplicand to A and shift
            A <= Mcand(3) & addout(3 downto 1);
            B <= addout(0) & B(3 downto 1);
          else
            A <= A(3) & A(3 downto 1);   -- arithmetic right shift
            B <= A(0) & B(3 downto 1);
          end if;
          State <= State + 1;
        when 4 =>
          if M = '1' then
            addout := A + not Mcand + 1;
              -- add 2's complement when sign bit of multiplier is 1
            A <= not Mcand(3) & addout(3 downto 1);
            B <= addout(0) & B(3 downto 1);
          else
            A <= A(3) & A(3 downto 1);   -- arithmetic right shift
            B <= A(0) & B(3 downto 1);
          end if;
          State <= 5;
        when 5 =>
          State <= 0;
      end case;
    end if;
  end process;
  Done <= '1' when State = 5 else '0';
  Product <= A(2 downto 0) & B;     -- output product
end behave1;
```

A variable *addout* has been defined to represent the 5-bit output of the adder. In states 1 through 4, if the current multiplier bit $M$ is '1', then the sign bit of the multiplicand followed by 3 bits of *addout* are loaded into $A$. At the same time, the low-order bit of *addout* is loaded into $B$ along with the high-order 3 bits of $B$. The *Done* signal is turned on when control goes to state 5, and then the new value of the product is outputted.

Before continuing with the design, we will test the behavioral level VHDL code to make sure that the algorithm is correct and consistent with the hardware block diagram. At early stages of testing, we will want a step-by-step printout to verify the internal operations of the multiplier and to aid in debugging, if required. When we think that the multiplier is functioning properly, then we will only want to look at the final product output so that we can quickly test a large number of cases.

Figure 4-36 shows the command file and test results for multiplying $+5/8$ by $-3/8$. A clock is defined with a 20-ns period. The *St* signal is turned on at 2 ns and turned off one clock period later. By inspection of the state graph, the multiplication requires six clocks, so the run time is set at 120 ns.

FIGURE 4-36: **Command File and Simulation Results for (+5/8 by –3/8)**

```
-- command file to test signed multiplier
add list CLK St State A B Done Product
force st 1 2, 0 22
force clk 1 0, 0 10 - repeat 20
-- (5/8 * -3/8)
force Mcand 0101
force Mplier 1101
run 120
```

| ns | delta | CLK | St | State | A | B | Done | Product |
|----|-------|-----|----|-------|------|------|------|---------|
| 0 | +1 | 1 | 0 | 0 | 0000 | 0000 | 0 | 0000000 |
| 2 | +0 | 1 | 1 | 0 | 0000 | 0000 | 0 | 0000000 |
| 10 | +0 | 0 | 1 | 0 | 0000 | 0000 | 0 | 0000000 |
| 20 | +1 | 1 | 1 | 1 | 0000 | 1101 | 0 | 0000000 |
| 22 | +0 | 1 | 0 | 1 | 0000 | 1101 | 0 | 0000000 |
| 30 | +0 | 0 | 0 | 1 | 0000 | 1101 | 0 | 0000000 |
| 40 | +1 | 1 | 0 | 2 | 0010 | 1110 | 0 | 0000000 |
| 50 | +0 | 0 | 0 | 2 | 0010 | 1110 | 0 | 0000000 |
| 60 | +1 | 1 | 0 | 3 | 0001 | 0111 | 0 | 0000000 |
| 70 | +0 | 0 | 0 | 3 | 0001 | 0111 | 0 | 0000000 |
| 80 | +1 | 1 | 0 | 4 | 0011 | 0011 | 0 | 0000000 |
| 90 | +0 | 0 | 0 | 4 | 0011 | 0011 | 0 | 0000000 |
| 100 | +2 | 1 | 0 | 5 | 1111 | 0001 | 1 | 1110001 |
| 110 | +0 | 0 | 0 | 5 | 1111 | 0001 | 1 | 1110001 |
| 120 | +1 | 1 | 0 | 0 | 1111 | 0001 | 0 | 1110001 |

To thoroughly test the multiplier, we need to test not only the four standard cases $(+ +, + -, - +,$ and $- -)$ but also special cases and limiting cases. Test values for the multiplicand and multiplier should include 0, the largest positive fraction, the most negative fraction, and all 1's. We will write a VHDL test bench to test the multiplier. The **test bench** will provide a sequence of values for the multiplicand and the multiplier. Thus, it provides stimuli to the system under test, the multiplier. The test bench can also check for the correctness of the multiplier output. The multiplier we are testing will be treated as a component and embedded in the test bench program. The signals generated within the test bench are interfaced to the multiplier as shown in Figure 4-37.

FIGURE 4-37:
**Interface between Multiplier and Its Test Bench**

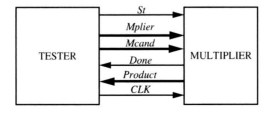

Figure 4-38 shows the VHDL code for the multiplier test bench. The test sequence consists of 11 sets of multiplicands and multipliers, provided in the

FIGURE 4-38: **Test Bench for Signed Multiplier**

```
library IEEE;
use IEEE.numeric_bit.all;

entity testmult is
end testmult;

architecture test1 of testmult is
component mult2C
  port(CLK, St: in bit;
       Mplier, Mcand: in unsigned(3 downto 0);
       Product: out unsigned(6 downto 0);
       Done: out bit);
end component;

constant N: integer := 11;
type arr is array(1 to N) of unsigned(3 downto 0);
type arr2 is array(1 to N) of unsigned(6 downto 0);
constant Mcandarr: arr := ("0111", "1101", "0101", "1101", "0111",
                  "1000", "0111", "1000", "0000", "1111", "1011");
constant Mplierarr: arr := ("0101", "0101", "1101", "1101", "0111",
                  "0111", "1000", "1000", "1101", "1111", "0000");
constant Productarr: arr2 := ("0100011", "1110001", "1110001",
                              "0001001", "0110001", "1001000",
                              "1001000", "1000000", "0000000",
                              "0000001", "0000000");
signal CLK, St, Done: bit;
signal Mplier, Mcand: unsigned(3 downto 0);
signal Product: unsigned(6 downto 0);
begin
  CLK <= not CLK after 10 ns;
  process
  begin
    for i in 1 to N loop
      Mcand <= Mcandarr(i);
      Mplier <= Mplierarr(i);
      St <= '1';
      wait until CLK = '1' and CLK'event;
      St <= '0';
      wait until Done = '0' and Done'event;
      assert Product = Productarr(i)    -- compare with expected answer
        report "Incorrect Product"
        severity error;
    end loop;
    report "TEST COMPLETED";
  end process;
  mult1: mult2c port map(CLK, St, Mplier, Mcand, Product, Done);
end test1;
```

*Mcandarr* and *Mplierarr* arrays. The expected outputs from the multiplier are provided in another array, the *Productarr*, in order to test the correctness of the multiplier outputs. The test values and results are placed in constant arrays in the VHDL code. A component declaration is done for the multiplier. A **port map** statement is used to create an instance of the multiplier. The tester also generates the clock and start signal. The for loop reads values from the *Mcandarr* and *Mplierarr* arrays and then sets the start signal to '1'. After the next clock, the start signal is turned off. Then the test bench waits for the *Done* signal. When the trailing edge of *Done* arrives, the multiplier output is compared against the expected output in the array *Productar*. An error is reported if the answers do not match. Since the *Done* signal is turned off at the same time the multiplier control goes back to $S_0$, the process waits for the falling edge of *Done* before looping back to supply new values of *Mcand* and *Mplier*. Note that the **port map** statement is outside the process that generates the stimulus. The multiplier constantly receives some set of inputs and generates the corresponding set of outputs.

Figure 4-39 shows the command file and simulator output. We have annotated the simulator output to interpret the test results. The –NOtrigger together with

FIGURE 4-39: **Command File and Simulation of Signed Multiplier**

```
-- Command file to test results of signed multiplier
add list -NOtrigger Mplier Mcand product -Trigger done
run 1320
```

| ns | delta | mplier | mcand | product | done | |
|---|---|---|---|---|---|---|
| 0 | +1 | 0101 | 0111 | 0000000 | 0 | |
| 90 | +2 | 0101 | 0111 | 0100011 | 1 | 5/8 * 7/8 = 35/64 |
| 110 | +2 | 0101 | 1101 | 0100011 | 0 | |
| 210 | +2 | 0101 | 1101 | 1110001 | 1 | 5/8 * -3/8 = -15/64 |
| 230 | +2 | 1101 | 0101 | 1110001 | 0 | |
| 330 | +2 | 1101 | 0101 | 1110001 | 1 | -3/8 * 5/8 = -15/64 |
| 350 | +2 | 1101 | 1101 | 1110001 | 0 | |
| 450 | +2 | 1101 | 1101 | 0001001 | 1 | -3/8 * -3/8 = 9/64 |
| 470 | +2 | 0111 | 0111 | 0001001 | 0 | |
| 570 | +2 | 0111 | 0111 | 0110001 | 1 | 7/8 * 7/8 = 49/64 |
| 590 | +2 | 0111 | 1000 | 0110001 | 0 | |
| 690 | +2 | 0111 | 1000 | 1001000 | 1 | 7/8 * -1 = -7/8 |
| 710 | +2 | 1000 | 0111 | 1001000 | 0 | |
| 810 | +2 | 1000 | 0111 | 1001000 | 1 | -1 * 7/8 = -7/8 |
| 830 | +2 | 1000 | 1000 | 1001000 | 0 | |
| 930 | +2 | 1000 | 1000 | 1000000 | 1 | -1 * -1 = -1 (error) |
| 950 | +2 | 1101 | 0000 | 1000000 | 0 | |
| 1050 | +2 | 1101 | 0000 | 0000000 | 1 | -3/8 * 0 = 0 |
| 1070 | +2 | 1111 | 1111 | 0000000 | 0 | |
| 1170 | +2 | 1111 | 1111 | 0000001 | 1 | -1/8 * -1/8 = 1/64 |
| 1190 | +2 | 0000 | 1011 | 0000001 | 0 | |
| 1290 | +2 | 0000 | 1011 | 0000000 | 1 | 0 * -3/8 = 0 |
| 1310 | +2 | 0101 | 0111 | 0000000 | 0 | |

the –Trigger done in the list statement causes the output to be displayed only when the *Done* signal changes. Without the –NOtrigger and –Trigger, the output would be displayed every time any signal on the list changed. All the product outputs are correct, except for the special case of $-1 \times -1$ ($1.000 \times 1.000$), which gives $1.000000$ ($-1$) instead of $+1$. This occurs because no representation of $+1$ is possible without adding another bit.

Next, we refine the VHDL model for the signed multiplier by explicitly defining the control signals and the actions that occur when each control signal is asserted. The VHDL code (Figure 4-40) is organized in a manner similar to the Mealy machine model of Figure 1-17. In the first process, the *Nextstate* and output control signals are defined for each present *State*. In the second process, after waiting for the rising edge of the clock, the appropriate registers are updated and the *State* is updated. We can test the VHDL code of Figure 4-40 using the same test file we used previously and verify that we get the same product outputs.

FIGURE 4-40: **Model for 2's Complement Multiplier with Control Signals**

```
-- This VHDL model explicitly defines control signals.

library IEEE;
use IEEE.numeric_bit.all;

entity mult2C is
  port(CLK, St: in bit;
       Mplier, Mcand: in unsigned(3 downto 0);
       Product: out unsigned (6 downto 0);
       Done: out bit);
end mult2C;

-- This architecture of a 4-bit multiplier for 2's complement numbers
-- uses control signals.

architecture behave2 of mult2C is
signal State, Nextstate: integer range 0 to 5;
signal A, B, compout, addout: unsigned(3 downto 0);
signal AdSh, Sh, Load, Cm: bit;
alias M: bit is B(0);
begin
  process(State, St, M)
  begin
    Load <= '0'; AdSh <= '0'; Sh <= '0'; Cm <= '0'; Done <= '0';
    case State is
      when 0 =>              -- initial state
        if St = '1' then Load <= '1'; Nextstate <= 1; end if;
      when 1 | 2 | 3 = >    -- "add/shift" State
        if M = '1' then AdSh < = '1';
        else Sh <= '1';
        end if;
```

```
         Nextstate <= State + 1;
      when 4 =>              -- add complement if sign
        if M = '1' then      -- bit of multiplier is 1
          Cm <= '1'; AdSh <= '1';
        else Sh <= '1';
        end if;
        Nextstate <= 5;
      when 5 =>              -- output product
        Done <= '1';
        Nextstate <= 0;
    end case;
end process;

compout <= not Mcand when Cm = '1' else Mcand; -- complementer
addout <=  A + compout + unsigned'(0=>Cm);      -- 4-bit adder with carry in

process (CLK)
begin
   if CLK'event and CLK = '1' then   -- executes on rising edge
     if Load = '1' then              -- load the multiplier
       A <= "0000";
       B <= Mplier;
     end if;
     if AdSh = '1' then              -- add multiplicand to A and shift
       A <= compout(3) & addout(3 downto 1);
       B <= addout(0) & B(3 downto 1);
     end if;
     if Sh = '1' then
       A <= A(3) & A(3 downto 1);
       B <= A(0) & B(3 downto 1);
     end if;
     State <= Nextstate;
   end if;
end process;
Product <= A(2 downto 0) & B;
end behave2;
```

● ● ● ● ● ● ● ● ● ● ● ●

# 4.11  Keypad Scanner

In this example, we design a scanner for a keypad with three columns and four rows as in Figure 4-41. The keypad is wired in matrix form with a switch at the intersection of each row and column. Pressing a key establishes a connection between a row and column. The purpose of the scanner is to determine which key has been pressed and output a binary number $N = N_3N_2N_1N_0$, which corresponds to the key number. For example, pressing key 5 must output 0101, pressing the * key must output 1010, and pressing the # key must output 1011. When a valid key has been detected, the scanner should output a signal $V$ for one clock time. Assume that only one key is pressed at a time. The design must include hardware to protect the circuitry from malfunction due to keypad bounces.

FIGURE 4-41:
**Keypad with Three
Columns and Four
Rows**

| 1 | 2 | 3 |
|---|---|---|
| 4 | 5 | 6 |
| 7 | 8 | 9 |
| * | 0 | # |

The overall block diagram of the circuit is presented in Figure 4-42. The keypad contains resistors that are connected to ground. When a switch is pressed, a path is established from the corresponding column line to the ground. If a voltage can be applied on the column lines $C_0$, $C_1$, and $C_2$, then the voltage can be obtained on the row line corresponding to the key that is pressed. One among the rows $R_0$, $R_1$, $R_2$, or $R_3$ will have an active signal.

FIGURE 4-42: **Block
Diagram for
Keypad Scanner**

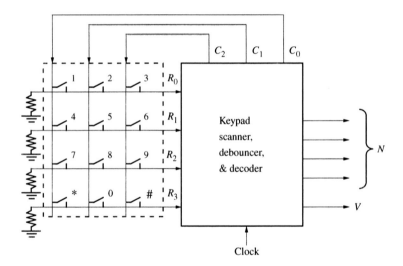

We will divide the design into several modules, as shown in Figure 4-43. The first part of the design will be a scanner that scans the rows and columns of the keypad. The keyscan module generates the column signals to scan the keypad. The debounce module generates a signal $K$ when a key has been pressed and a signal $Kd$ after it has been debounced. When a valid key is detected, the decoder determines the key number from the row and column numbers.

FIGURE 4-43:
**Scanner Modules**

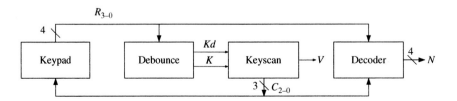

### 4.11.1 **Scanner**

We will use the following procedure to scan the keypad: First apply logic 1's to columns $C_0$, $C_1$, and $C_2$ and wait. If any key is pressed, a 1 will appear on $R_0$, $R_1$, $R_2$, or $R_3$. Then apply a 1 to column $C_0$ only. If any of the $R_i$'s is 1, a valid key is detected. If $R_0$ is received, we know that switch 1 was pressed. If $R_1$, $R_2$, or $R_3$ is received, switch 4, 7, or * was pressed. If so, set $V = 1$ and output the corresponding $N$. If no key is detected in the first column, apply a 1 to $C_1$ and repeat. If no key is detected in the second column, repeat for $C_2$. When a valid key is detected, apply 1's to $C_0$, $C_1$, and $C_2$ and wait until no key is pressed. This last step is necessary so that only one valid signal is generated each time a key is pressed.

### 4.11.2 **Debouncer**

As discussed in the scoreboard example, we need to debounce the keys to avoid malfunctions due to switch bounce. Figure 4-44 shows a proposed debouncing and synchronizing circuit. The four row signals are connected to an OR gate to form signal $K$, which turns on when a key is pressed and a column scan signal is applied. The debounced signal $Kd$ will be fed to the sequential circuit.

**FIGURE 4-44:**
**Debouncing and Synchronizing Circuit**

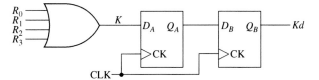

### 4.11.3 **Decoder**

The decoder determines the key number from the row and column numbers using the truth table given in Table 4-4. The truth table has one row for each of the 12 keys. The remaining rows have don't care outputs since we have assumed that only one key is pressed at a time. Since the decoder is a combinational circuit, its output will change

**TABLE 4-4: Truth Table for Decoder**

| $R_3$ | $R_2$ | $R_1$ | $R_0$ | $C_0$ | $C_1$ | $C_2$ | $N_3$ | $N_2$ | $N_1$ | $N_0$ | |
|---|---|---|---|---|---|---|---|---|---|---|---|
| 0 | 0 | 0 | 1 | 1 | 0 | 0 | 0 | 0 | 0 | 1 | |
| 0 | 0 | 0 | 1 | 0 | 1 | 0 | 0 | 0 | 1 | 0 | |
| 0 | 0 | 0 | 1 | 0 | 0 | 1 | 0 | 0 | 1 | 1 | |
| 0 | 0 | 1 | 0 | 1 | 0 | 0 | 0 | 1 | 0 | 0 | |
| 0 | 0 | 1 | 0 | 0 | 1 | 0 | 0 | 1 | 0 | 1 | |
| 0 | 0 | 1 | 0 | 0 | 0 | 1 | 0 | 1 | 1 | 0 | |
| 0 | 1 | 0 | 0 | 1 | 0 | 0 | 0 | 1 | 1 | 1 | |
| 0 | 1 | 0 | 0 | 0 | 1 | 0 | 1 | 0 | 0 | 0 | |
| 0 | 1 | 0 | 0 | 0 | 0 | 1 | 1 | 0 | 0 | 1 | |
| 1 | 0 | 0 | 0 | 1 | 0 | 0 | 1 | 0 | 1 | 0 | (*) |
| 1 | 0 | 0 | 0 | 0 | 1 | 0 | 0 | 0 | 0 | 0 | |
| 1 | 0 | 0 | 0 | 0 | 0 | 1 | 1 | 0 | 1 | 1 | (#) |

Logic Equations for Decoder

$N_3 = R_2 C_0' + R_3 C_1'$

$N_2 = R_1 + R_2 C_0$

$N_1 = R_0 C_0' + R_2' C_2 + R_1' R_0' C_0$

$N_0 = R_1 C_1 + R_1' C_2 + R_3 R_1' C_1'$

as the keypad is scanned. At the time a valid key is detected ($K = 1$ and $V = 1$), its output will have the correct value and this value can be saved in a register at the same time the circuit goes to $S_5$.

### 4.11.4 Controller

Figure 4-45 shows the state diagram of the controller for the keypad scanner. It waits in $S_1$ with outputs $C_0 = C_1 = C_2 = 1$ until a key is pressed. In $S_2$, $C_0 = 1$, so if the key that was pressed is in column 0, $K = 1$, and the circuit outputs a valid signal and goes to $S_5$. Signal $K$ is used instead of $Kd$, since the key press is already debounced. If no key press is found in column 0, column 1 is checked in $S_3$, and if necessary, column 2 is checked in $S_4$. In $S_5$, the circuit waits until all keys are released and $Kd$ goes to 0 before resetting.

FIGURE 4-45: **State Graph for Keypad Scanner**

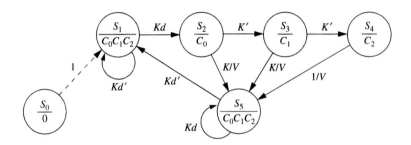

The state diagram in Figure 4-45 works for many cases; however, it does have some timing problems. Let us analyze the following situations.

1. Is $K$ true whenever a button is pressed?
   No. Although $K$ is true if any one of the row signals $R_1$, $R_2$, $R_3$, or $R_4$ is true, if the column scan signals are not active, none of $R_1$–$R_4$ can be true, although the button is pressed.

2. Can $Kd$ be false when a button is continuing to be pressed?
   Yes. Signal $Kd$ is nothing but $K$ delayed by two clock cycles. $K$ can go to 0 during the scan process even when the button is being pressed. For instance, consider the case when a key in the rightmost column is pressed. During scan of the first two columns, $K$ goes to 0. If $K$ goes to 0 at any time, $Kd$ will go to zero two cycles later. Hence, neither $K$ nor $Kd$ is synonymous to pressing the button.

3. Can you go from $S_5$ to $S_1$ when a button is still pressed?
   In the state diagram in Figure 4-45, the $S_4$-to-$S_5$ transition could happen when $Kd$ is false. $Kd$ might have become false while scanning $C_0$ and $C_1$. Hence, it is possible that we reach back to $S_1$ when the key is still being pressed. As an example, let us assume that a button is pressed in column $C_2$. This is to be detected in $S_4$. However, during the scanning process in $S_2$ and $S_3$, $K$ is 0;

hence, two cycles later $Kd$ will be 0 even if the button stays pressed. During the scan in $S_4$, the correct key can be found; however, the system can reach $S_5$ when $Kd$ is still 0 and a malfunction can happen. $S_5$ is intended to sense the release of the key. However, $Kd$ is not synonymous to pressing the button and $Kd'$ does not truly indicate that the button got released. Since $Kd'$ can appear when the button is still pressed, if you reach $S_5$ when $Kd'$ is true due to scanning activity in a previous state, the system can go from $S_5$ to $S_1$ without a key release. In such a case, the same key may be read multiple times.

4. What if a key is pressed for only one or two clock cycles?

If the key is pressed and released very quickly, there would be problems especially if the key is in the third column. By the time the scanner reaches state $S_4$, the key might have been released already. The key should be pressed long enough for the scanner to go through the longest path in the state graph from $S_0$ to $S_5$. This may not be a serious problem because usually the digital system clock is much faster than any mechanical switch.

These problems can be fixed by assuring that we can reach $S_5$ only if $Kd$ is true. A modified state diagram is presented in Figure 4-46. Before transitioning to state $S_5$, this circuit waits in state $S_2$, $S_3$, and $S_4$ until $Kd$ also becomes 1.

**FIGURE 4-46:**
**Modified State Graph for Keypad Scanner**

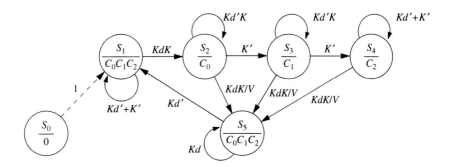

### 4.11.5 VHDL Code

The VHDL code used to implement the design is shown in Figure 4-47. The decoder equations as well as the equations for $K$ and $V$ are implemented by concurrent statements. The process implements the next state equations for the keyscan and debounce flip-flops.

**FIGURE 4-47: VHDL Code for Scanner**

```
entity scanner is
  port(R0, R1, R2, R3, CLK: in bit;
       C0, C1, C2: inout bit;
       N0, N1, N2, N3, V: out bit);
end scanner;
```

```vhdl
architecture behavior of scanner is
signal QA, K,Kd: bit;
signal state, nextstate: integer range 0 to 5;
begin
  K <= R0 or R1 or R2 or R3;    -- this is the decoder section
  N3 <= (R2 and not C0) or (R3 and not C1);
  N2 <= R1 or (R2 and C0);
  N1 <= (R0 and not C0) or (not R2 and C2) or (not R1 and not R0 and C0);
  N0 <= (R1 and C1) or (not R1 and C2) or (not R3 and not R1 and not C1);

  process(state, R0, R1, R2, R3, C0, C1, C2, K, Kd, QA)
  begin
    C0 <= '0'; C1 <= '0'; C2 <= '0'; V <= '0';
    case state is
      when 0 => nextstate < = 1;
      when 1 => C0 <= '1'; C1 <= '1'; C2 <= '1';
        if (Kd and K) = '1' then nextstate <= 2;
        else nextstate <= 1;
        end if;
      when 2 => C0 <= '1';
        if (Kd and K) = '1' then V <= '1'; nextstate <= 5;
        elsif K = '0' then nextstate <= 3;
        else nextstate <= 2;
        end if;
      when 3 => C1 <= '1';
        if (Kd and K) = '1' then V <= '1'; nextstate <= 5;
        elsif K = '0' then nextstate <= 4;
        else nextstate <= 3;
        end if;
      when 4 => C2 <= '1';
        if (Kd and K) = '1' then V <= '1'; nextstate <= 5;
        else nextstate <= 4;
        end if;
      when 5 => C0 <= '1'; C1 <= '1'; C2 <= '1';
        if Kd = '0' then nextstate <= 1;
        else nextstate <= 5;
        end if;
    end case;
  end process;

  process(CLK)
  begin
    if CLK = '1' and CLK'EVENT then
      state <= nextstate;
      QA <= K;
      Kd <= QA;
    end if;
  end process;
end behavior;
```

### 4.11.6 Test Bench for Keypad Scanner

This VHDL code would be very difficult to test by supplying waveforms for the inputs $R_0$, $R_1$, $R_2$, and $R_3$, since these inputs depend on the column outputs ($C_0$, $C_1$, $C_2$). A much better way to test the scanner is by using a test bench in VHDL. The scanner we are testing will be treated as a component and embedded in the test bench program. The signals generated within the test bench are interfaced to the scanner as shown in Figure 4-48. The test bench simulates a key press by supplying the appropriate $R$ signals in response to the $C$ signals from the scanner. When test bench receives $V = 1$ from the scanner, it checks to see if the value of $N$ corresponds to the key that was pressed.

**FIGURE 4-48:**
**Interface for Test Bench**

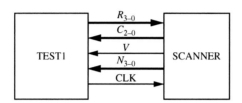

The VHDL code for the keypad test bench is shown in Figure 4-49. A copy of the scanner is instantiated within the *test1* architecture, and connections to the scanner are made by the port map. The sequence of key numbers used for testing is stored in the array *KARRAY*. The tester simulates the keypad operation using

**FIGURE 4-49: VHDL for Scanner Test Bench**

```
library IEEE;
use IEEE.numeric_bit.all;

entity scantest is
end scantest;

architecture test1 of scantest is
component scanner
  port(R0, R1, R2, R3, CLK: in bit;
       C0, C1, C2: inout bit;
       N0, N1, N2, N3, V: out bit);
end component;

type arr is array (0 to 23) of integer;        -- array of keys to test
constant KARRAY: arr := (2,5,8,0,3,6,9,11,1,4,7,10,1,2,3,4,5,6,7,8,9,10,11,0);
signal C0, C1, C2, V, CLK, R0, R1, R2, R3: bit;   -- interface signals
signal N: unsigned(3 downto 0);
signal KN: integer;                              -- key number to test
begin
  CLK <= not CLK after 20 ns;                    -- generate clock signal
```

```
-- this section emulates the keypad
R0 <= '1' when (C0='1' and KN=1) or (C1='1' and KN=2) or (C2='1' and KN=3)
        else '0';
R1 <= '1' when (C0='1' and KN=4) or (C1='1' and KN=5) or (C2='1' and KN=6)
        else '0';
R2 <= '1' when (C0='1' and KN=7) or (C1='1' and KN=8) or (C2='1' and KN=9)
        else '0';
R3 <= '1' when (C0='1' and KN=10) or (C1='1' and KN=0) or (C2='1' and KN=11)
        else '0';

process                                  -- this section tests scanner
begin
  for i in 0 to 23 loop                  -- test every number in key array
    KN <= KARRAY(i);                     -- simulates keypress
    wait until (V = '1' and rising_edge(CLK));
    assert (to_integer(N) = KN)          -- check if output matches
      report "Numbers don't match"
      severity error;
    KN <= 15;                            -- equivalent to no key pressed
    wait until rising_edge(CLK); - wait for scanner to reset
    wait until rising_edge(CLK);
    wait until rising_edge(CLK);
  end loop;
  report "Test Complete.";
end process;
scanner1: scanner port map(R0,R1,R2,R3,CLK,C0,C1,C2,N(0),N(1),N(2),N(3),V);
                                         -- connect test1 to scanner
end test1;
```

concurrent statements for $R_0$, $R_1$, $R_2$, and $R_3$. Whenever $C_0$, $C_1$, $C_2$, or the key number ($KN$) changes, new values for the $R$s are computed. For example, if $KN = 5$ (to simulate pressing key 5), then $R_0R_1R_2R_3 = 0100$ is sent to the scanner when $C_0C_1C_2 = 010$. The test process is as follows:

1. Read a key number from the array to simulate pressing a key.
2. Wait until $V = 1$ and the rising edge of the clock occurs.
3. Verify that the $N$ output from the scanner matches the key number.
4. Set $KN = 15$ to simulate no key pressed. (Since 15 is not a valid key number, all $R$'s will go to 0.)
5. Wait until $Kd = 0$ before selecting a new key.

Key presses in row order and column order are tried using the various numbers in *KARRAY*. The test bench uses **assert** statements to test whether the reported number matches the key pressed. The **report** statement is used to report an error if the scanner generates the wrong key number, and it will report "Test Complete." when all keys have been tested.

# 4.12 Binary Dividers

### 4.12.1 Unsigned Divider

We will consider the design of a parallel divider for positive binary numbers. As an example, we will design a circuit to divide an 8-bit dividend by a 4-bit divisor to obtain a 4-bit quotient. The following example illustrates the division process:

$$
\begin{array}{r}
1010 \\
1101 \overline{)10000111} \\
1101 \\
\hline
0111 \\
0000 \\
\hline
1111 \\
1101 \\
\hline
0101 \\
0000 \\
\hline
0101
\end{array}
$$

Divisor 1101 ⟋ 10000111 — Quotient, Dividend

(135 ÷ 13 = 10 with a remainder of 5)

Remainder

Just as binary multiplication can be carried out as a series of add and shift operations, division can be carried out by a series of subtract and shift operations. To construct the divider, we will use a 9-bit dividend register and a 4-bit divisor register, as shown in Figure 4-50. During the division process, instead of shifting the divisor right before each subtraction, we will shift the dividend to the left. Note that an extra bit is required on the left end of the dividend register so that a bit is not lost when the dividend is shifted left. Instead of using a separate register to store the quotient, we will enter the quotient bit-by-bit into the right end of the dividend register as the dividend is shifted left.

**FIGURE 4-50: Block Diagram for Parallel Binary Divider**

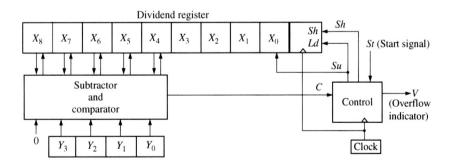

The preceding division example (135 divided by 13) is reworked next, showing the location of the bits in the registers at each clock time. Initially, the dividend and divisor are entered as follows:

| 0 | 1 | 0 | 0 | 0 | 0 | 1 | 1 | 1 |
|---|---|---|---|---|---|---|---|---|

| 1 | 1 | 0 | 1 |
|---|---|---|---|

Subtraction cannot be carried out without a negative result, so we will shift before we subtract. Instead of shifting the divisor one place to the right, we will shift the dividend one place to the left:

```
                                  ┌──── Dividing line between dividend and quotient
    1 0 0 0 0 1 1 1│0
        1 1 0 1        │   ┌── Note that after the shift, the rightmost position
                              in the dividend register is "empty."
```

Subtraction is now carried out and the first quotient digit of 1 is stored in the unused position of the dividend register:

```
    0 0 0 1 1 1 1│1 ◄──────── first quotient digit
```

Next we shift the dividend one place to the left:

```
    0 0 1 1 1 1│1 0
      1 1 0 1   │
```

Since subtraction would yield a negative result, we shift the dividend to the left again, and the second quotient bit remains zero:

```
    0 1 1 1 1│1 0 0
      1 1 0 1 │
```

Subtraction is now carried out, and the third quotient digit of 1 is stored in the unused position of the dividend register:

```
    0 0 0 1 0 1│1 0 1 ◄──────── third quotient digit
```

A final shift is carried out and the fourth quotient bit is set to 0:

```
    0 0 1 0 1│1 0 1 0
    └───┬───┘ └──┬──┘
    remainder   quotient
```

The final result agrees with that obtained in the first example.

If, as a result of a division operation, the quotient contains more bits than are available for storing the quotient, we say that an *overflow* has occurred. For the divider of Figure 4-50, an overflow would occur if the quotient is greater than 15, since only 4 bits are provided to store the quotient. It is not actually necessary to carry out the division to determine if an overflow condition exists, since an initial comparison of the dividend and divisor will tell if the quotient will be too large. For example, if we attempt to divide 135 by 7, the initial contents of the registers are

```
    0 1 0 0 0 0 1 1 1
        0 1 1 1
```

Since subtraction can be carried out with a nonnegative result, we should subtract the divisor from the dividend and enter a quotient bit of 1 in the rightmost place in the dividend register. However, we cannot do this because the rightmost place contains the least significant bit of the dividend, and entering a quotient bit here would destroy that dividend bit. Therefore, the quotient would be too large to store in the 4 bits we have allocated for it, and we have detected an overflow condition. In general, for Figure 4-50, if initially $X_8X_7X_6X_5X_4 \geq Y_3Y_2Y_1Y_0$ (i.e., if the left 5 bits of the dividend register exceed or equal the divisor), the quotient will be greater than 15 and an overflow occurs. Note that if $X_8X_7X_6X_5X_4 \geq Y_3Y_2Y_1Y_0$, the quotient is

$$\frac{X_8X_7X_6X_5X_4X_3X_2X_1X_0}{Y_3Y_2Y_1Y_0} \geq \frac{X_8X_7X_6X_5X_40000}{Y_3Y_2Y_1Y_0} = \frac{X_8X_7X_6X_5X_4 \times 16}{Y_3Y_2Y_1Y_0} \geq 16$$

The operation of the divider can be explained in terms of the block diagram of Figure 4-50. A shift signal ($Sh$) will shift the dividend one place to the left. A subtract signal ($Su$) will subtract the divisor from the five leftmost bits in the dividend register and set the quotient bit (the rightmost bit in the dividend register) to 1. If the divisor is greater than the five leftmost dividend bits, the comparator output is $C = 0$; otherwise, $C = 1$. Whenever $C = 0$, subtraction cannot occur without a negative result, so a shift signal is generated. Whenever $C = 1$, a subtract signal is generated, and the quotient bit is set to 1. The control circuit generates the required sequence of shift and subtract signals.

Figure 4-51 shows the state diagram for the control circuit. When a start signal ($St$) occurs, the 8-bit dividend and 4-bit divisor are loaded into the appropriate registers. If $C$ is 1, the quotient would require five or more bits. Since space is only provided for a 4-bit quotient, this condition constitutes an overflow, so the divider is stopped and the overflow indicator is set by the $V$ output. Normally, the initial value of $C$ is 0, so a shift will occur first, and the control circuit will go to state $S_2$. Then, if $C = 1$, subtraction occurs. After the subtraction is completed, $C$ will always be 0, so the next clock pulse will produce a shift. This process continues until four shifts have occurred and the control is in state $S_5$. Then a final subtraction occurs, if necessary, and the control returns to the stop state. For this example, we will assume that when the start signal ($St$) occurs, it will be 1 for one clock time, and then it will remain 0 until the control circuit is back in state $S_0$. Therefore, $St$ will always be 0 in states $S_1$ through $S_5$.

FIGURE 4-51: **State Diagram for Divider Control Circuit**

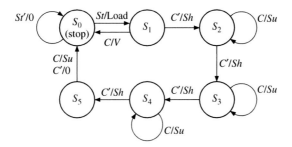

Table 4-5 gives the state table for the control circuit. Since we assumed that $St = 0$ in states $S_1$, $S_2$, $S_3$, and $S_4$, the next states and outputs are "don't cares" for these states when $St = 1$. The entries in the output table indicate which outputs are 1. For example, the entry $Sh$ means $Sh = 1$ and the other outputs are 0.

TABLE 4-5: **State Table for Divider Control Circuit**

| | StC | | | | StC | | | |
|---|---|---|---|---|---|---|---|---|
| State | 00 | 01 | 11 | 10 | 00 | 01 | 11 | 10 |
| $S_0$ | $S_0$ | $S_0$ | $S_1$ | $S_1$ | 0 | 0 | Load | Load |
| $S_1$ | $S_2$ | $S_0$ | — | — | Sh | V | — | — |
| $S_2$ | $S_3$ | $S_2$ | — | — | Sh | Su | — | — |
| $S_3$ | $S_4$ | $S_3$ | — | — | Sh | Su | — | — |
| $S_4$ | $S_5$ | $S_4$ | — | — | Sh | Su | — | — |
| $S_5$ | $S_0$ | $S_0$ | — | — | 0 | Su | — | — |

This example illustrates a general method for designing a divider for unsigned binary numbers, and the design can easily be extended to larger numbers such as 16 bits divided by 8 bits or 32 bits divided by 16 bits. Using a separate counter to count the number of shifts is recommended if more than four shifts are required.

### 4.12.2 Signed Divider

We now design a divider for signed (2's complement) binary numbers that divides a 32-bit dividend by a 16-bit divisor to give a 16-bit quotient. Although algorithms exist to divide the signed numbers directly, such algorithms are rather complex. So we take the easy way out and complement the dividend and divisor if they are negative; when division is complete, we complement the quotient if it should be negative.

Figure 4-52 shows a block diagram for the divider. We use a 16-bit bus to load the registers. Since the dividend is 32 bits, two clocks are required to load the upper and lower halves of the dividend register, and one clock is needed to load the divisor. An extra sign flip-flop is used to store the sign of the dividend. We will use a dividend register with a built-in 2's complementer. The subtracter consists of an adder and a complementer, so subtraction can be accomplished by adding the 2's complement of the divisor to the dividend register. If the divisor is negative, using a separate step to complement it is unnecessary; we can simply disable the complementer and add the negative divisor instead of subtracting its complement. The control circuit is divided into two parts—a main control, which determines the sequence of shifts and subtracts, and a counter, which counts the number of shifts. The counter outputs a signal $K = 1$ when 15 shifts have occurred. Control signals are defined as follows:

*LdU*   Load upper half of dividend from bus.
*LdL*   Load lower half of dividend from bus.
*Lds*   Load sign of dividend into sign flip-flop.
*S*       Sign of dividend.
*Cm1*   Complement dividend register (2's complement).
*Ldd*   Load divisor from bus.
*Su*     Enable adder output onto bus (*Ena*) and load upper half of dividend from bus.

*Cm2*    Enable complementer. (*Cm2* equals the complement of the sign bit of the divisor, so a positive divisor is complemented and a negative divisor is not.)

*Sh*    Shift the dividend register left one place and increment the counter.

*C*    Carry output from adder. (If $C = 1$, the divisor can be subtracted from the upper dividend.)

*St*    Start.

*V*    Overflow.

*Qneg*    Quotient will be negative. (*Qneg* = 1 when the sign of the dividend and divisor are different.)

FIGURE 4-52: **Block Diagram for Signed Divider**

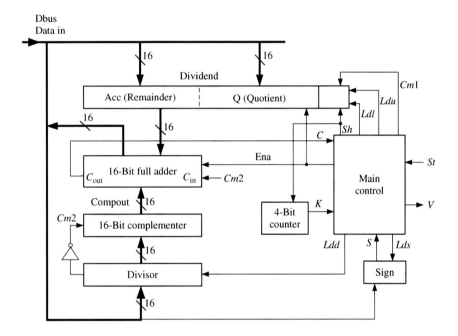

The procedure for carrying out the signed division is as follows:

1. Load the upper half of the dividend from the bus, and copy the sign of the dividend into the sign flip-flop.
2. Load the lower half of the dividend from the bus.
3. Load the divisor from the bus.
4. Complement the dividend if it is negative.
5. If an overflow condition is present, go to the done state.
6. Else carry out the division by a series of shifts and subtracts.
7. When division is complete, complement the quotient if necessary, and go to the done state.

Testing for overflow is slightly more complicated than for the case of unsigned division. First, consider the case of all positive numbers. Since the divisor and quotient are each 15 bits plus sign, their maximum value is 7FFFh. Since the remainder must be less than the divisor, its maximum value is 7FFEh. Therefore, the maximum dividend for no overflow is

$$\text{divisor} \times \text{quotient} + \text{remainder} = 7FFFh \times 7FFFh + 7FFEh = 3FFF7FFFh$$

If the dividend is 1 larger (3FFF8000h), division by 7FFFh (or anything smaller) will give an overflow. We can test for the overflow condition by shifting the dividend left one place and then comparing the upper half of the dividend (divu) with the divisor. If divu ≥ divisor, the quotient would be greater than the maximum value, which is an overflow condition. For the preceding example, shifting 3FFF8000h left once gives 7FFF0000h. Since 7FFFh equals the divisor, there is an overflow. On the other hand, shifting 3FFF7FFFh left gives 7FFEFFFEh, and since 7FFEh < 7FFFh, no overflow occurs when dividing by 7FFFh.

Another way of verifying that we must shift the dividend left before testing for overflow is as follows. If we shift the dividend left one place and then divu ≥ divisor, we could subtract and generate a quotient bit of 1. However, this bit would have to go in the sign bit position of the quotient. This would make the quotient negative, which is incorrect. After testing for overflow, we must shift the dividend left again, which gives a place to store the first quotient bit after the sign bit. Since we work with the complement of a negative dividend or a negative divisor, this method for detecting overflow will work for negative numbers, except for the special case where the dividend is 80000000h (the largest negative value). Modifying the design to detect overflow in this case is left as an exercise.

Figure 4-53 shows the state graph for the control circuit. When $St = 1$, the registers are loaded. In $S_2$, if the sign of the dividend ($S$) is 1, the dividend is complemented. In $S_3$, we shift the dividend left one place and then we test for overflow in $S_4$. If $C = 1$, subtraction is possible, which implies an overflow, and the circuit goes to the done state. Otherwise, the dividend is shifted left. In $S_5$, $C$ is tested. If $C = 1$, then $Su = 1$, which implies $Ldu$ and $Ena$, so the adder output is enabled onto the bus and loaded into the upper dividend register to accomplish the subtraction. Otherwise, $Sh = 1$ and the dividend register is shifted. This continues until $K = 1$, at which time the last shift occurs if $C = 0$, and the circuit goes to $S_6$. Then if the sign of the divisor and the saved sign of the dividend are different, the dividend register is complemented so that the quotient will have the correct sign.

FIGURE 4-53: **State Graph for Signed Divider Control Circuit**

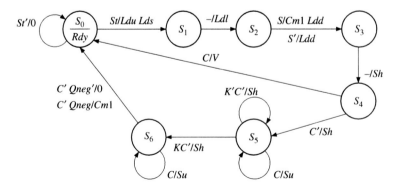

The VHDL code for the signed divider is shown in Figure 4-54. Since the 1's complementer and adder are combinational circuits, we have represented their operation by concurrent statements. All the signals that represent register outputs are updated on the rising edge of the clock, so these signals are updated in the process after waiting for *CLK* to change to '1'. The counter is simulated by a signal, *count*. For convenience in listing the simulator output, we have added a ready signal (*Rdy*), which is turned on in $S_0$ to indicate that the division is completed.

FIGURE 4-54: **VHDL Model of 32-Bit Signed Divider**

```
library IEEE;
use IEEE.numeric_bit.all;

entity sdiv is
  port(CLK, St: in bit;
       Dbus: in unsigned(15 downto 0);
       Quotient: out unsigned(15 downto 0);
       V, Rdy: out bit);
end sdiv;

architecture Signdiv of Sdiv is
signal State: integer range 0 to 6;
signal Count: unsigned(3 downto 0); -- integer range 0 to 15
signal Sign, C, Cm2: bit;
signal Divisor, Sum, Compout: unsigned(15 downto 0);
signal Dividend: unsigned(31 downto 0);
alias Acc: unsigned(15 downto 0) is Dividend(31 downto 16);
begin                                    -- concurrent statements
  Cm2 <= not divisor(15);
  compout <= divisor when Cm2 = '0'          -- 1's complementer
            else not divisor;
  Sum <= Acc + compout + unsigned'(0=>Cm2);  -- adder output
  C <= not Sum(15);
  Quotient <= Dividend(15 downto 0);
  Rdy <= '1' when State = 0 else '0';
  process(CLK)
  begin
    if CLK'event and CLK = '1' then  -- wait for rising edge of clock
      case State is
        when 0 =>
          if St = '1' then
            Acc <= Dbus;                     -- load upper dividend
            Sign <= Dbus(15);
            State <= 1;
            V <= '0';                        -- initialize overflow
            Count <= "0000";                 -- initialize counter
          end if;
```

```
        when 1 =>
          Dividend (15 downto 0) <= Dbus;                -- load lower dividend
          State <= 2;
        when 2 =>
          Divisor <= Dbus;
          if Sign = '1' then  -- two's complement Dividend if necessary
            dividend <= not dividend + 1;
          end if;
          State <= 3;
        when 3 =>
          Dividend <= Dividend(30 downto 0) & '0';       -- left shift
          Count <= Count+1; State <= 4;
        when 4 =>
          if C = '1' then                                -- C
            v <= '1'; State <= 0;
          else                                           -- C'
            Dividend <= Dividend(30 downto 0) & '0';     -- left shift
            Count <= Count+1; State <= 5;
          end if;
        when 5 =>
          if C = '1' then                                -- C
            ACC <= Sum;                                  -- subtract
            dividend(0) <= '1';
          else
            Dividend <= Dividend(30 downto 0) & '0';     -- left shift
            if Count = 15 then State <= 6; end if;       -- KC'
            Count <= Count+1;
          end if;
        when 6 =>
          state <= 0;
          if C = '1' then                                -- C
            Acc <= Sum;                                  -- subtract
            dividend(0) <= '1'; State <= 6;
          elsif (Sign xor Divisor(15)) = '1' then        -- C'Qneg
            Dividend <= not Dividend + 1;
          end if;                                        -- 2's complement Dividend
      end case;
    end if;
  end process;
end signdiv;
```

We are now ready to test the divider design by using the VHDL simulator. We will need a comprehensive set of test examples that will test all the different special cases that can arise in the division process. To start with, we need to test the basic operation of the divider for all the different combinations of signs for the divisor and dividend ($++$, $+-$, $-+$, and $--$). We also need to test the overflow detection for these four cases. Limiting cases must also be tested, including largest quotient, zero quotient, and so on. Use of a VHDL test bench is convenient because the test data must be supplied in sequence at certain times, and the length of time to complete the division is dependent on the test data. Figure 4-55 shows a test bench for the divisor. The test bench contains a dividend array and a divisor array for the test data. The notation X"07FF00BB" is the hexadecimal representation of a bit string. The process in testsdiv first puts the upper dividend on *Dbus* and supplies a start signal. After waiting for the clock, it puts the lower dividend on *Dbus*. After the next clock, it puts the divisor on *Dbus*. It then waits until the *Rdy* signal indicates that division is complete before continuing. *Count* is set equal to the loop-index, so that the change in *Count* can be used to trigger the listing output.

FIGURE 4-55: **Test Bench for Signed Divider**

```
library IEEE;
use IEEE.numeric_bit.all;

entity testsdiv is
end testsdiv;

architecture test1 of testsdiv is
component sdiv
  port(CLK, St: in bit;
       Dbus: in unsigned(15 downto 0);
       Quotient: out unsigned(15 downto 0);
       V, Rdy: out bit);
end component;

constant N: integer : = 12;                    -- test sdiv1 N times
type arr1 is array(1 to N) of unsigned(31 downto 0);
type arr2 is array(1 to N) of unsigned(15 downto 0);
constant dividendarr: arr1 := (X"0000006F", X"07FF00BB", X"FFFFFE08",
     X"FF80030A", X"3FFF8000", X"3FFF7FFF", X"C0008000", X"C0008000",
     X"C0008001", X"00000000", X"FFFFFFFF", X"FFFFFFFF");
constant divisorarr: arr2 := (X"0007", X"E005", X"001E", X"EFFA", X"7FFF",
   X"7FFF", X"7FFF", X"8000", X"7FFF", X"0001", X"7FFF", X"0000");
signal CLK, St, V, Rdy: bit;
signal Dbus, Quotient, divisor: unsigned(15 downto 0);
signal Dividend: unsigned(31 downto 0);
signal Count: integer range 0 to N;

begin
  CLK <= not CLK after 10 ns;
  process
  begin
    for i in 1 to N loop
      St <= '1';
      Dbus <= dividendarr(i) (31 downto 16);
      wait until (CLK'event and CLK = '1');
```

```
        Dbus <= dividendarr(i) (15 downto 0);
        wait until (CLK'event and CLK = '1');
        Dbus <= divisorarr(i);
        St <= '0';
        dividend <= dividendarr(i) (31 downto 0);    -- save dividend for listing
        divisor <= divisorarr(i);                    -- save divisor for listing
        wait until (Rdy = '1');
        count <= i;                                  -- save index for triggering
      end loop;
  end process;
  sdiv1: sdiv port map(CLK, St, Dbus, Quotient, V, Rdy);
end test1;
```

Figure 4-56 shows the simulator command file and output. The –NOtrigger, together with the –Trigger count in the list statement, causes the output to be displayed only when the *count* signal changes. Examination of the simulator output shows that the divider operation is correct for all of the test cases, except for the following case:

$$C0008000h \div 7FFFh = -3FFF8000 \div 7FFFh = -8000h = 8000h$$

In this case, the overflow is turned on, and division never occurs. In general, the divider will indicate an overflow whenever the quotient should be 8000h (the most negative value). This occurs because the divider basically divides positive numbers, and the largest positive quotient is 7FFFh. If it is important to be able to generate the quotient 8000h, the overflow detection can be modified so it does not generate an overflow in this special case.

FIGURE 4-56: **Simulation Test Results for Signed Divider**

```
-- Command file to test results of signed divider
add list -hex -NOtrigger dividend divisor Quotient V -Trigger count
run 5300
```

| ns | delta | dividend | divisor | quotient | v | count |
|---|---|---|---|---|---|---|
| 0 | +0 | 00000000 | 0000 | 0000 | 0 | 0 |
| 470 | +3 | 0000006F | 0007 | 000F | 0 | 1 |
| 910 | +3 | 07FF00BB | E005 | BFFE | 0 | 2 |
| 1330 | +3 | FFFFFE08 | 001E | FFF0 | 0 | 3 |
| 1910 | +3 | FF80030A | EFFA | 07FC | 0 | 4 |
| 2010 | +3 | 3FFF8000 | 7FFF | 0000 | 1 | 5 |
| 2710 | +3 | 3FFF7FFF | 7FFF | 7FFF | 0 | 6 |
| 2810 | +3 | C0008000 | 7FFF | 0000 | 1 | 7 |
| 3510 | +3 | C0008000 | 8000 | 7FFF | 0 | 8 |
| 4210 | +3 | C0008001 | 7FFF | 8001 | 0 | 9 |
| 4610 | +3 | 00000000 | 0001 | 0000 | 0 | A |
| 5010 | +3 | FFFFFFFF | 7FFF | 0000 | 0 | B |
| 5110 | +3 | FFFFFFFF | 0000 | 0002 | 1 | C |

In this chapter, we presented several design examples. The examples included several arithmetic and nonarithmetic circuits. A seven-segment display, a BCD adder, a traffic light controller, a scoreboard, and a keypad scanner are examples of non-arithmetic circuits presented in the chapter. We also described algorithms for addition, multiplication, and division of unsigned and signed binary numbers. Specific designs such as the carry look-ahead adder and the array multiplier were presented. We designed digital systems to implement these algorithms. After developing a block diagram for such a system and defining the required control signals, we used state graphs to define a sequential machine that generates control signals in the proper sequence. We used VHDL to describe the systems at several different levels so that we can simulate and test for correct operation of the systems we have designed.

● ● ● ● ● ● ● ● ● ● ● ● ●

# Problems

4.1 Design the correction circuit for a BCD adder that computes $Z$ digit 0 and $C$ for $S_0$ (see Figures 4-5 and 4-6). This correction circuit adds "0110" to $S_0$ if $S_0 > 9$. This is the same as adding "$0AA0$" to $S_0$, where $A = $ '1' if $S_0 > 9$. Draw a block diagram for the correction circuit using one full adder, three half-adders, and a logic circuit to compute $A$. Design a circuit for $A$ using a minimum number of gates. Note that the maximum possible value of $S_0$ is 10010.

4.2 **(a)** If gate delays are 5 ns, what is the delay of the fastest 4-bit ripple carry adder? Explain your calculation.

**(b)** If gate delays are 5 ns, what is the delay of the fastest 4-bit adder? What kind of an adder will it be? Explain your calculation.

4.3 Develop a VHDL model for a 16-bit carry look-ahead adder utilizing the 4-bit adder from Figure 4-10 as a component.

4.4 Derive generates, propagates, group generates, group propagates, and the final sum and carry out for the 16-bit carry look ahead adder of Figure 4-9, while adding 0101 1010 1111 1000 and 0011 1100 1100 0011.

4.5 **(a)** Write a VHDL module that describes one bit of a full adder with accumulator. The module should have two control inputs, $Ad$ and $L$. If $Ad = 1$, the $Y$ input (and carry input) are added to the accumulator. If $L = 1$, the $Y$ input is loaded into the accumulator.

**(b)** Using the module defined in (a), write a VHDL description of a 4-bit subtracter with accumulator. Assume negative numbers are represented in 1's complement. The subtracter should have control inputs $Su$ (subtract) and $Ld$ (load).

4.6 **(a)** Implement the traffic-light controller of Figure 4-14 using a modulo 13 counter with added logic. The counter should increment every clock, with two exceptions. Use a ROM to generate the outputs.

**(b)** Write a VHDL description of your answer to (a).

**(c)** Write a test bench for part (b) and verify that your controller works correctly. Use concurrent statements to generate test inputs for *Sa* and *Sb*.

4.7 Make the necessary additions to the following state graph so that it is a proper, completely specified state graph. Demonstrate that your answer is correct. Convert the graph to a state table using 0's and 1's for inputs and outputs.

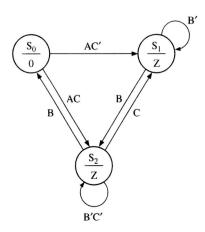

4.8 Write synthesizable VHDL code that will generate the given waveform ($W$). Use a single process. Assume that a clock with a 1 μs period is available as an input.

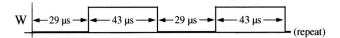

4.9 A BCD adder adds two BCD numbers (each of range 0 to 9) and produces the sum in BCD form. For example, if it adds 9 (1001) and 8 (1000) the result would be 17 (1 0111). Implement such a BCD adder using a 4-bit binary adder and appropriate control circuitry. Assume that the two BCD numbers are already loaded into two 4-bit registers ($A$ and $B$), and there is a 5-bit sum register ($S$) available. You need some kind of correction to get the sum in the BCD form, because the binary adder produces results in the range 0000 to 1111 (plus a carry in some cases). If any addition is required for this correction, use the same adder (i.e., you can use only one adder). Use multiplexers at the adder inputs to steer the appropriate numbers to the adder in each cycle. Assume a start signal to initiate the addition and a done signal to indicate completion.

**(a)** Draw a block diagram of the system. Label each component appropriately to indicate its functionality and size.

**(b)** Describe step-by-step the algorithm that you would use to perform the addition. Explain and illustrate the correction step.

**(c)** Draw a state graph for the controller.

4.10 Write VHDL code for a shift register module that includes a 16-bit shift register, a controller, and a 4-bit down counter. The shifter can shift a variable number of bits depending on a count provided to the shifter module. Inputs to the module are a number $N$ (indicating shift count) in the range 1 to 15, a 16-bit vector *par_in*, a clock, and a start signal, *St*. When $St = \text{'1'}$, $N$ is loaded into the down counter, and *par_in* is loaded into the shift register. Then the shift register does a cycle left shift $N$ times, and the controller returns to the start state. Assume that *St* is only '1' for one clock time. All operations are synchronous on the falling edge of the clock.

   **(a)** Draw a block diagram of the system and define any necessary control signals.
   **(b)** Draw a state graph for the controller (two states).
   **(c)** Write VHDL code for the shift-register module. Use two processes (one for the combinational part of the circuit, and one for updating the registers).

4.11 **(a)** Figure 4-12 shows the block diagram for a 32-bit serial adder with accumulator. The control circuit uses a 5-bit counter, which outputs a signal $K = 1$ when it is in state 11111. When a start signal ($St$) is received, the registers should be loaded. Assume that $St$ will remain 1 until the addition is complete. When the addition is complete, the control circuit should go to a stop state and remain there until $St$ is changed back to 0. Draw a state diagram for the control circuit (excluding the counter).
   **(b)** Write the VHDL for the complete system, and verify its correct operation.

4.12 A block diagram for a 16-bit 2's complement serial subtracter is given here. When $St = 1$, the registers are loaded and then subtraction occurs. The shift counter, $C$, produces a signal $C15 = 1$ after 15 shifts. $V$ should be set to 1 if an overflow occurs. Set the carry flip-flop to 1 during load in order to form the 2's complement. Assume that $St$ remains 1 for one clock time.

   **(a)** Draw a state diagram for the control (two states).
   **(b)** Write VHDL code for the system. Use two processes. The first process should determine the next state and control signals; the second process should update the registers on the rising edge of the clock.

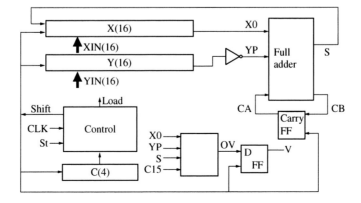

4.13  This problem involves the design of a BCD to binary converter. Initially a three-digit BCD number is placed in the $A$ register. When a $St$ signal is received, conversion to binary takes place, and the resulting binary number is stored in the $B$ register. At each step of the conversion, the entire BCD number (along with the binary number) is shifted one place to the right. If the result in a given decade is greater than or equal 1000, the correction circuit subtracts 0011 from that decade. (If the result is less than 1000, the correction circuit leaves the contents of the decade unchanged.) A shift counter is provided to count the number of shifts. When conversion is complete, the maximum value of $B$ will be 999 (in binary). *Note: B is 10 bits.*

(a) Illustrate the algorithm starting with the BCD number 857, showing $A$ and $B$ at each step.

(b) Draw the block diagram of the BCD-to-binary converter.

(c) Draw a state diagram of the control circuit (three states). Use the following control signals: $St$: start conversion; $Sh$: shift right; $Co$: subtract correction if necessary; and $C9$: counter is in state 9, or $C10$: counter is in state 10. (Use either $C9$ or $C10$ but not both.)

(d) Write a VHDL description of the system.

4.14  This problem involves the design of a circuit that finds the square root of an 8-bit unsigned binary number $N$ using the method of subtracting out odd integers. To find the square root of $N$, we subtract 1, then 3, then 5, and so on, until we can no longer subtract without the result going negative. The number of times we subtract is equal to the square root of $N$. For example, to find $\sqrt{27}$: $27 - 1 = 26$; $26 - 3 = 23$; $23 - 5 = 18$; $18 - 7 = 11$; $11 - 9 = 2$; $2 - 11$ (can't subtract). Since we subtracted five times, $\sqrt{27} = 5$. Note that the final odd integer is $11_{10} = 1011_2$, and this consists of the square root ($101_2 = 5_{10}$) followed by a 1.

(a) Draw a block diagram of the square rooter that includes a register to hold $N$, a subtracter, a register to hold the odd integers, and a control circuit. Indicate where to read the final square root. Define the control signals used on the diagram.

(b) Draw a state graph for the control circuit using a minimum number of states. The $N$ register should be loaded when $St = 1$. When the square root is complete, the control circuit should output a done signal and wait until $St = 0$ before resetting.

4.15  This problem concerns the design of a multiplier for unsigned binary numbers that multiplies a 4-bit number by a 16-bit number to give a 20-bit product. To speed up the multiplication, a 4-by-4 array multiplier is used so that we can multiply by 4 bits in one clock time instead of only by 1 bit at each clock time. The hardware includes a 24-bit accumulator register that can be shifted right 4 bits at a time using a control signal $Sh4$. The array multiplier multiplies 4 bits by 4 bits to give an 8-bit product. This product is added to the accumulator using an $Ad$ control signal. When a $St$ signal occurs, the 16-bit multiplier is loaded into the lower part of the $A$ register. A done signal should be turned on when the multiplication is complete. Since both the array multiplier and adder are combinational circuits, the 4-bit multiply and the 8-bit add can both be completed in the same clock cycle. Do NOT include the array multiplier logic in your code, just use the overloaded "*" operator. If $D$ and $E$ are 4-bit unsigned numbers, $D * E$ will compute an 8-bit product.

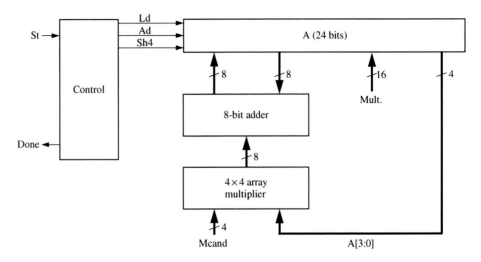

**(a)** Draw a state graph for the controller (10 states)
**(b)** Write VHDL code for the multiplier. Use two processes (a combinational process and a clocked process). All signals should be of type unsigned or bit.

4.16 **(a)** Estimate how many AND gates and adders will be required for a 16-bit × 16-bit array multiplier.
**(b)** What is the longest delay in a 16 × 16 array multiplier, assuming an AND gate delay is $t_g$, and adder delay (full adder and half adder) is $t_{ad}$?

4.17 **(a)** Draw the organization of an 8 × 8 array multiplier and calculate how many full adders, half-adders, and AND gates are required.
**(b)** Highlight the critical path in your answer to (a) (If there are many equivalent ones, highlight any one of them.)
**(c)** What is the longest delay in an 8 × 8 array multiplier, assuming an AND gate delay is $t_g$ = 1 ns, and adder delay (full adder and half adder) is $t_{ad}$ = 2 ns?
**(d)** For an 8-bit × 8-bit add-and-shift multiplier (similar to Figure 4-25), how fast must the clock be in order to complete the multiplication in the same time as in part (c)?

4.18 An $n \times n$ array multiplier, as in Figure 4-29, takes $3n - 4$ adder delays + 1 gate delay to calculate a product. Design an array multiplier which is faster than this for $n > 4$. (*Hint*: Instead of passing carry output to the left adder, pass it to the diagonally lower one, speeding up the critical path. This topology is called "multiplier using carry-save adder.")

4.19 The block diagram for a multiplier for signed (2's complement) binary numbers is shown in Figure 4-33. Give the contents of the $A$ and $B$ registers after each clock pulse when multiplicand = $-1/8$ and multiplier = $-3/8$.

4.20 In Section 4.10 we developed an algorithm for multiplying signed binary fractions, with negative fractions represented in 2's complement.

**(a)** Illustrate this algorithm by multiplying 1.0111 by 1.101.

**(b)** Draw a block diagram of the hardware necessary to implement this algorithm for the case where the multiplier is 4 bits, including sign, and the multiplicand is 5 bits, including sign.

4.21 The objective of this problem is to use VHDL to describe and simulate a multiplier for signed binary numbers using Booth's algorithm. Negative numbers should be represented by their 2's complement. Booth's algorithm works as follows, assuming each number is $n$ bits including sign: Use an $(n + 1)$-bit register for the accumulator $(A)$ so the sign bit will not be lost if an overflow occurs. Also, use an $(n + 1)$-bit register $(B)$ to hold the multiplier and an $n$-bit register $(C)$ to hold the multiplicand.

**1.** Clear $A$ (the accumulator), load the multiplier into the upper $n$ bits of $B$, clear $B_0$, and load the multiplicand into $C$.

**2.** Test the lower two bits of $B$ $(B_1 B_0)$.
 If $B_1 B_0 = 01$, then add $C$ to $A$ ($C$ should be sign-extended to $n + 1$ bits and added to $A$ using an $(n + 1)$-bit adder).
 If $B_1 B_0 = 10$, then add the 2's complement of $C$ to $A$.
 If $B_1 B_0 = 00$ or 11, skip this step.

**3.** Shift $A$ and $B$ together right one place with sign extended.

**4.** Repeat steps 2 and 3, $n - 1$ more times.

**5.** The product will be in $A$ and $B$, except ignore $B_0$.

Example for $n = 5$: Multiply $-9$ by $-13$.

|  | $A$ | $B$ | $B_1 B_0$ |  |
|---|---|---|---|---|
| **1.** Load registers. | 000000 | 100110 | 10 | $C = 10111$ |
| **2.** Add 2's comp. of $C$ to $A$. | <u>001001</u> | | | |
| | 001001 | 100110 | | |
| **3.** Shift $A\&B$. | 000100 | 110011 | 11 | |
| **3.** Shift $A\&B$. | 000010 | 011001 | 01 | |
| **2.** Add $C$ to $A$. | <u>110111</u> | | | |
| | 111001 | 011001 | | |
| **3.** Shift $A\&B$. | 111100 | 101100 | 00 | |
| **3.** Shift $A\&B$. | 111110 | 010110 | 10 | |
| **2.** Add 2's comp. of $C$ to $A$. | <u>001001</u> | | | |
| | 000111 | 010110 | | |
| **3.** Shift $A\&B$. | 000011 | 101011 | | |

Final result: 0001110101 = +117

**(a)** Draw a block diagram of the system for $n = 8$. Use 9-bit registers for $A$ and $B$, a 9-bit full adder, an 8-bit complementer, a 3-bit counter, and a control circuit. Use the counter to count the number of shifts.

**(b)** Draw a state graph for the control circuit. When the counter is in state 111, return to the start state at the time the last shift occurs (three states should be sufficient).

(c) Write behavioral VHDL code for the multiplier.

(d) Simulate your VHDL design using the following test cases (in each pair, the second number is the multiplier):

$$01100110 \times 00110011$$
$$10100110 \times 01100110$$
$$01101011 \times 10001110$$
$$11001100 \times 10011001$$

Verify that your results are correct.

4.22 Design a multiplier that will multiply two 16-bit signed binary integers to give a 32-bit product. Negative numbers should be represented in 2's complement form. Use the following method: First complement the multiplier and multiplicand if they are negative, multiply the positive numbers, and then complement the product if necessary. Design the multiplier so that after the registers are loaded, the multiplication can be completed in 16 clocks.

(a) Draw a block diagram of the multiplier. Use a 4-bit counter to count the number of shifts. (The counter will output a signal $K = 1$ when it is in state 15.) Define all condition and control signals used on your diagram.

(b) Draw a state diagram for the multiplier control using a minimum number of states (five states). When the multiplication is complete, the control circuit should output a done signal and then wait for $ST = 0$ before returning to state $S_0$.

(c) Write a VHDL behavioral description of the multiplier without using control signals (for example, see Figure 4-35) and test it.

(d) Write a VHDL behavioral description using control signals (for example, see Figure 4-40) and test it.

4.23 This problem involves the design of a parallel adder-subtracter for 8-bit numbers expressed in sign and magnitude notation. The inputs $X$ and $Y$ are in sign and magnitude, and the output $Z$ must be in sign and magnitude. Internal computation may be done in either 2's complement or 1's complement (specify which you use), but no credit will be given if you assume the inputs $X$ and $Y$ are in 1's or 2's complement. If the input signal $Sub = 1$, then $Z = X - Y$, else $Z = X + Y$. Your circuit must work for all combinations of positive and negative inputs for both add and subtract. You may use only the following components: an 8-bit adder, a 1's complementer (for the input $Y$), a second complementer (which may be either 1's complement or 2's complement—specify which you use), and a combinational logic circuit to generate control signals. (*Hint*: $-X + Y = -(X - Y)$.) Also generate an overflow signal that is 1 if the result cannot be represented in 8-bit sign and magnitude.)

(a) Draw the block diagram. No registers, multiplexers, or tristate busses are allowed.

(b) Give a truth table for the logic circuit that generates the necessary control signals. Inputs for the table should be $Sub$, $Xs$, and $Ys$ in that order, where $Xs$ is the sign of $X$ and $Ys$ is the sign of $Y$.

(c) Explain how you would determine the overflow and give an appropriate equation.

4.24 Four push buttons ($B_0$, $B_1$, $B_2$, and $B_3$) are used as inputs to a logic circuit. Whenever a button is pushed, it is debounced and then the circuit loads the button number in binary into a 2-bit register ($N$). For example, if $B_2$ is pushed, the register output becomes $N = 10_2$. The register holds this value until another button is pushed. Use a total of two flip-flops for debouncing. Use a 10-bit counter as a clock divider to provide a slow clock for debouncing. $Kd$ is a signal which is 1 when any button has been pushed and debounced.

**(a)** Draw a state graph (two states) to generate the signal that loads the register when $Kd = 1$.
**(b)** Draw a logic circuit diagram showing the 10-bit counter, the 2-bit register $N$, and all necessary gates and flip-flops.

4.25 Design a $4 \times 4$ keypad scanner for the following keypad layout.

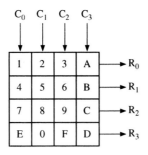

**(a)** Assuming only one key can be pressed at a time, find the equations for a number decoder given $R_{3-0}$ and $C_{3-0}$, whose output corresponds to the binary value of the key. For example, the F key will return $N_{3-0} = 1111$ in binary, or 15.
**(b)** Design a debouncing circuit that detects when a key has been pressed or depressed. Assume switch bounce will die out in one or two clock cycles. When a key has been pressed, $K = 1$ and $Kd$ is the debounced signal.
**(c)** Design and draw a state graph that performs the keyscan and issues a valid pulse when a valid key has been pressed using inputs from part (b).
**(d)** Write a VHDL description of your keypad scanner and include the decoder, debouncing circuit, and scanner.

4.26 This problem concerns the design of a divider for unsigned binary numbers that will divide a 16-bit dividend by an 8-bit divisor to give an 8-bit quotient. Assume that the start signal ($ST = 1$) is 1 for exactly one clock time. If the quotient would require more than 8 bits, the divider should stop immediately and output $V = 1$ to indicate an overflow. Use a 17-bit dividend register and store the quotient in the lower 8 bits of this register. Use a 4-bit counter to count the number of shifts, together with a subtract-shift controller.

**(a)** Draw a block diagram of the divider.
**(b)** Draw a state graph for the subtract-shift controller (three states).
**(c)** Write a VHDL description of the divider. Use two processes, similar to Figure 4-40.
**(d)** Write a test bench for your divider (similar to Figure 4-55).

4.27 A block diagram and state graph for a divider for unsigned binary numbers is shown below. This divider divides a 16-bit dividend by a 16-bit divisor to give a 16-bit quotient. The divisor can be any number in the range 1 to $2^{16} - 1$. The only case where an overflow can occur is when the divisor is 0. Control signals are defined as follows: $Ld1$: load the divisor from the input bus; $Ld2$: load the dividend from the input bus and clear ACC; $Sh$: left shift ACC & Dividend; $Su$: load the subtractor output into ACC and set the lower quotient bit to 1; $K = 1$ when 15 shifts have been made. Write complete VHDL code for the divider. All signals must be of type unsigned or bit. Use two processes.

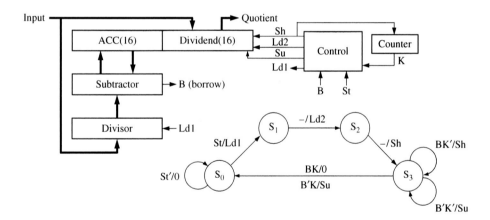

4.28 A block diagram for a divider that divides an 8-bit unsigned number by a 4-bit unsigned number to give a 4-bit quotient is given below. Note that the $X_i$ inputs to the subtractors are shifted over one position to the left. This means that the shift-and-subtract operation can be completed in one clock time instead of two. Depending on the borrow from the subtractor, a shift or shift-and-subtract operation occurs at each clock time, and the division can always be completed in four clock times after the registers are loaded. Ignore overflow. When the start signal ($St$) is 1, the $X$ and $Y$ registers are loaded. Assume that the start signal ($St$) is 1 for only one clock time. $Sh$ causes $X$ to shift left with 0 fill. $SubSh$ causes the subtractor output to be loaded into the left part of $X$, and at the same time the rest of $X$ is shifted left.

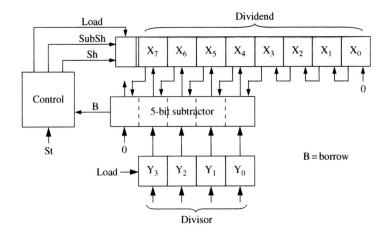

**(a)** Draw a state graph for the controller (5 states).
**(b)** Complete the VHDL code given below. Registers and signals should be of type unsigned so that overloaded operators may be used. Write behavioral code that uses a single process.

```vhdl
library IEEE;
use IEEE.numeric_bit.all;

entity divu is
  port(dividend: in unsigned(7 downto 0);
       divisor: in unsigned(3 downto 0);
       St, clk: in bit;
       quotient: out unsigned(3 downto 0));
end entity divu;

architecture div of divu is
```

4.29 An older model Thunderbird car has three left (LA, LB, LC) and three right (RA, RB, RC) tail lights which flash in unique patterns to indicate left and right turns.

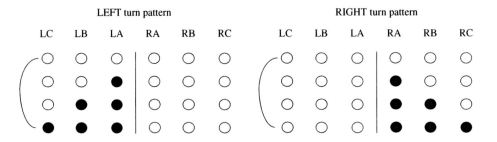

Design a Moore sequential circuit to control these lights. The circuit has three inputs *LEFT*, *RIGHT*, and *HAZ*. *LEFT* and *RIGHT* come from the driver's turn signal switch and cannot be 1 at the same time. As indicated above, when *LEFT* = 1

the lights flash in a pattern $LA$ on; $LA$ and $LB$ on; $LA, LB,$ and $LC$ on; all off; and then the sequence repeats. When $RIGHT = 1,$ a similar sequence appears on lights RA, RB, and RC, as indicated on the right side of the picture. If a switch from $LEFT$ to $RIGHT$ (or vice versa) occurs in the middle of a flashing sequence, the circuit should immediately go to the IDLE (lights off) state and then start the new sequence. $HAZ$ comes from the hazard switch, and when $HAZ = 1,$ all six lights flash on and off in unison. $HAZ$ takes precedence if $LEFT$ or $RIGHT$ is also on.

Assume that a clock signal is available with a frequency equal to the desired flashing rate.

**(a)** Draw the state graph (eight states).
**(b)** Realize the circuit using six D flip-flops, and make a one-hot state assignment such that each flip-flop output drives one of the six lights directly. (You may use *LogicAid*.)
**(c)** Realize the circuit using three D flip-flops, using the guidelines from Section 1.7 to determine a suitable encoded state assignment. Note the tradeoff between more flip-flops and more gates in (b) and (c).

4.30 Design a sequential circuit to control the motor of a tape player. The logic circuit will have five inputs and three outputs. Four of the inputs are the control buttons on the tape player. The input $PL$ is 1 if the play button is pressed, the input $RE$ is 1 if the rewind button is pressed, the input $FF$ is 1 if the fast forward button is pressed, and the input $ST$ is 1 if the stop button is pressed. The fifth input to the control circuit is $M,$ which is 1 if the special "music sensor" detects music at the current tape position. The three outputs of the control circuit are $P,$ $R,$ and $F,$ which make the tape play, rewind, and fast forward, respectively, when 1. No more than one output should ever be on at a time; all outputs off causes the motor to stop. The buttons control the tape as follows: If the play button is pressed, the tape player will start playing the tape (output $P = 1$). If the play button is held down and the rewind button is pressed and released, the tape player will rewind to the beginning of the current song (output $R = 1$ until $M = 0$) and then start playing. If the play button is held down and the fast forward button is pressed and released, the tape player will fast forward to the end of the current song (output $F = 1$ until $M = 0$) and then start playing. If rewind or fast forward is pressed while play is released, the tape player will rewind or fast forward the tape. Pressing the stop button at any time should stop the tape player motor.

**(a)** Construct a state graph chart for the tape player controller. You may assume that only one of the four buttons can be pressed at any given time.
**(b)** Write VHDL code for the controller.

# SM Charts and Microprogramming

## CHAPTER 5

A state machine is often used to control a digital system that carries out a step-by-step procedure or algorithm. State diagrams or state graphs with circles representing states and arcs representing transitions have traditionally been used to specify the operation of the controller state machine. As an alternative to using state graphs, a special type of flow chart, called a *state machine chart*, or **SM chart**, may be used to describe the behavior of a state machine. These charts are also called *algorithmic state machine charts*, or **ASM charts**. SM charts are often used to design control units for digital systems.

In this chapter, we first describe the properties of SM charts and how they are used in the design of state machines. Then we show examples of SM charts for a multiplier and a dice game controller. We construct VHDL descriptions of these systems from the SM charts, and we simulate the VHDL code to verify correct operation. We then proceed with the design and show how the SM chart can be realized with hardware. We then introduce **microprogramming** as a technique to implement the SM chart.

• • • • • • • • • • • •

## 5.1 State Machine Charts

SM charts resemble software flow charts. Flow charts have been very useful in software design for decades, and in a similar fashion, SM charts have been useful in hardware design. This is especially true in behavioral-level design entry.

SM charts offer several advantages over state graphs. It is often easier to understand the operation of a digital system by inspection of the SM chart instead of the equivalent state graph. A proper state graph has to obey some conditions: (1) One and exactly one transition from a state must be true at any time, and (2) the next state must be uniquely defined for every input combination. These conditions are automatically satisfied for an SM chart. An SM chart also directly leads to a hardware realization. A given SM chart can be converted into several equivalent forms, and different forms might naturally result in different implementations. Hence, a designer may optimize and transform SM charts to suit the implementation style/technology that he or she is looking for.

An SM chart differs from an ordinary flow chart in that certain specific rules must be followed in constructing the SM chart. When these rules are followed, the SM chart is equivalent to a state graph, and it leads directly to a hardware realization.

Figure 5-1 shows the three principal components of an SM chart. The state of the system is represented by a *state box*. The state box contains a *state name*, followed by a slash (/) and an optional *output list*. After a state assignment has been made, a *state code* may be placed outside the box at the top. A *decision box* is represented by a diamond-shaped symbol with true and false branches. The *condition* placed in the box is a Boolean expression that is evaluated to determine which branch to take. The *conditional output box*, which has curved ends, contains a *conditional output list*. The conditional outputs depend on both the state of the system and the inputs.

**FIGURE 5-1:**
**Principal Components of an SM Chart**

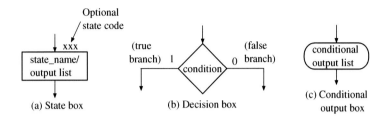

(a) State box  (b) Decision box  (c) Conditional output box

An SM chart is constructed from *SM blocks*. Each SM block (Figure 5-2) contains exactly one state box, together with the decision boxes and conditional output boxes associated with that state. An SM block has one *entrance path* and one or more *exit paths*. Each SM block describes the machine operation during the time that the machine is in one state. When a digital system enters the state associated with a given SM block, the outputs on the output list in the state box become true. The conditions in the decision boxes are evaluated to determine which paths are followed through the SM block. When a conditional output box is encountered along such a path, the corresponding conditional outputs become true. If an output is not encountered along a path, that output is false by default. A path through an SM block from entrance to exit is referred to as a *link path*.

**FIGURE 5-2:**
**Example of an SM Block**

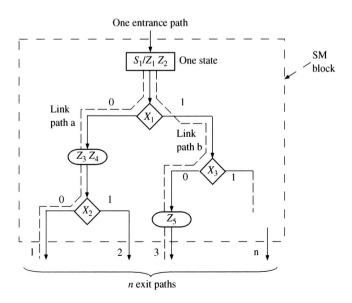

For the example of Figure 5-2, when state $S_1$ is entered, outputs $Z_1$ and $Z_2$ become 1. If input $X_1 = 0$, $Z_3$ and $Z_4$ also become 1. If $X_1 = X_2 = 0$, at the end of the state time, the machine goes to the next state via exit path 1. On the other hand, if $X_1 = 1$ and $X_3 = 0$, the output $Z_5$ is l, and exiting to the next state will occur via exit path 3. Since $Z_3$ and $Z_4$ are not encountered along this link path, $Z_3 = Z_4 = 0$ by default.

A given SM block can generally be drawn in several different forms. Figure 5-3 shows two equivalent SM blocks. In both (a) and (b), the output $Z_2 = 1$ if $X_1 = 0$; the next state is $S_2$ if $X_2 = 0$ and $S_3$ if $X_2 = 1$. As illustrated in this example, the order in which the inputs are tested may affect the complexity of the SM chart.

**FIGURE 5-3:**
**Equivalent SM**
**Blocks**

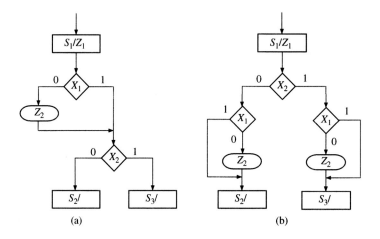

(a)                    (b)

The SM charts of Figures 5-4(a) and (b) each represent a combinational circuit, since there is only one state and no state change occurs. The output is $Z_1 = 1$ if $A + BC = 1$; otherwise $Z_1 = 0$. Figure 5-4(b) shows an equivalent SM chart in which the input variables are tested individually. The output is $Z_1 = 1$ if $A = 1$ or if $A = 0$, $B = 1$, and $C = 1$. Hence

$$Z_1 = A + A'BC = A + BC$$

which is the same output function realized by the SM chart of Figure 5-4(a).

**FIGURE 5-4:**
**Equivalent SM**
**Charts for a**
**Combinational**
**Circuit**

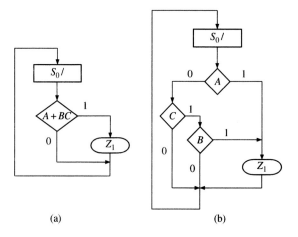

(a)                    (b)

Certain rules must be followed when constructing an SM block. First, for every valid combination of input variables, there must be exactly one exit path defined. This is necessary since each allowable input combination must lead to a single next state. Second, no internal feedback within an SM block is allowed. Figure 5-5 shows incorrect and correct ways of drawing an SM block with feedback.

**FIGURE 5-5:**
**SM Block with**
**Feedback**

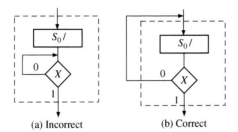

(a) Incorrect      (b) Correct

As shown in Figure 5-6(a), an SM block can have several parallel paths that lead to the same exit path, and more than one of these paths can be active at the same time. For example, if $X_1 = X_2 = 1$ and $X_3 = 0$, the link paths marked with dashed lines are active, and the outputs $Z_1$, $Z_2$, and $Z_3$ are 1. Although Figure 5-6(a) would not be a valid flow chart for a program for a serial computer, it presents no problems for a state machine implementation. The state machine can have a multiple-output circuit that generates $Z_1$, $Z_2$, and $Z_3$ at the same time. Figure 5-6(b) shows a serial SM block, which is equivalent to Figure 5-6(a). In the serial block, only one active link path between entrance and exit is possible. For any combination of input values, the outputs will be the same as in the equivalent parallel form. The link path for $X_1 = X_2 = 1$ and $X_3 = 0$

**FIGURE 5-6:**
**Equivalent SM**
**Blocks**

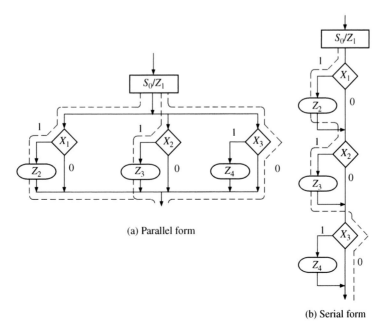

(a) Parallel form

(b) Serial form

is shown with a dashed line, and the outputs encountered on this path are $Z_1$, $Z_2$, and $Z_3$. Regardless of whether the SM block is drawn in serial or parallel form, all the tests take place within one clock time. In the rest of this text, we use only the serial form for SM charts.

It is easy to convert a state graph for a sequential machine to an equivalent SM chart. The state graph of Figure 5-7(a) has both Moore and Mealy outputs. The equivalent SM chart has three blocks—one for each state. The Moore outputs ($Z_a$, $Z_b$, $Z_c$) are placed in the state boxes, since they do not depend on the input. The Mealy outputs ($Z_1$, $Z_2$) appear in conditional output boxes, since they depend on both the state and input. In this example, each SM block has only one decision box, since only one input variable must be tested. For both the state graph and SM chart, $Z_c$ is always 1 in state $S_2$. If $X = 0$ in state $S_2$, $Z_1 = 1$ and the next state is $S_0$. If $X = 1$, $Z_2 = 1$ and the next state is $S_2$. We have added a state assignment ($S_0 = 00$, $S_1 = 01$, $S_2 = 11$) next to the state boxes.

**FIGURE 5-7:**
**Conversion of a State Graph to an SM Chart**

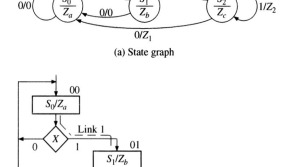

(a) State graph

(b) Equivalent SM chart

Figure 5-8 shows a timing chart for the SM chart of Figure 5-7 with an input sequence $X = 1, 1, 1, 0, 0, 0$. In this example, all state changes occur immediately after the rising edge of the clock. Since the Moore outputs ($Z_a$, $Z_b$, $Z_c$) depend on the state, they can change only immediately following a state change. The Mealy outputs ($Z_1$, $Z_2$) can change immediately after a state change or an input change. In any case, all outputs will have their correct values at the time of the active clock edge.

**FIGURE 5-8: Timing Chart for Figure 5-7**

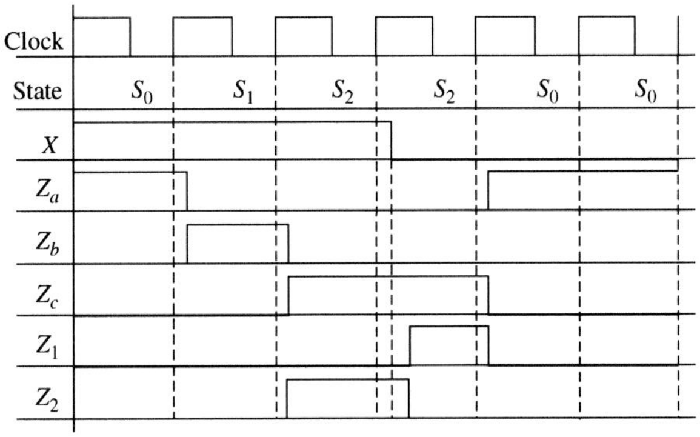

• • • • • • • • • • •

## 5.2 Derivation of SM Charts

The method used to derive an SM chart for a sequential control circuit is similar to that used to derive the state graph. First, we should draw a block diagram of the system we are controlling. Next, we should define the required input and output signals to the control circuit. Then we can construct an SM chart that tests the input signals and generates the proper sequence of output signals. In this section, we give two examples of derivation of SM charts.

### 5.2.1 Binary Multiplier

The first example is an SM chart for control of the binary multiplier shown in Figures 4-25 and 4-28(a). The add-shift control generates the required sequence of add and shift signals. The counter counts the number of shifts and outputs $K = 1$ just before the last shift occurs. The SM chart for the multiplier control (Figure 5-9) corresponds closely to the state graph of Figure 4-28(c). In state $S_0$, when the start signal $St$ is 1, the registers are loaded. In $S_1$, the multiplier bit $M$ is tested. If $M = 1$, an add signal is generated and the next state is $S_2$. If $M = 0$, a shift signal is generated and $K$ is tested. If $K = 1$, this will be the last shift and the next state is $S_3$. In $S_2$, a shift signal is generated, since a shift must always follow an add. If $K = 1$, the circuit goes to S3 at the time of the last shift; otherwise, the next state is $S_1$. In $S_3$, the done signal is turned on.

Conversion of an SM chart to a VHDL process is straightforward. A **case** statement can be used to specify what happens in each state. Each condition box corresponds directly to an **if** statement (or an **elsif**). Figure 5-10 shows the VHDL code for the SM chart in Figure 5-9. Two processes are used. The first process represents the combinational part of the circuit, and the second process updates the state register on the rising edge of the clock. The signals *Load*, *Sh*, and *Ad* are turned on in the appropriate states, and they must be turned off when the state changes. A convenient way to do this is to set them all to 0 at the start of the process. This VHDL code only models the controller. It assumes the presence of adders and shifters (shift registers) in the architecture and generates the appropriate signals to load the registers, to add and/or to shift.

FIGURE 5-9: **SM Chart for Binary Multiplier**

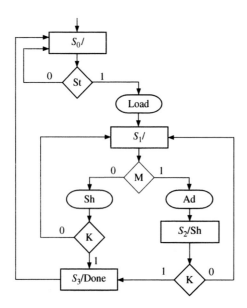

FIGURE 5-10: **Behavioral VHDL for Multiplier Controller (SM Chart of Figure 5-9)**

```vhdl
entity Mult is
  port(CLK, St, K, M: in bit;
      Load, Sh, Ad, Done: out bit);
end Mult;

architecture SMbehave of Mult is
signal State, Nextstate: integer range 0 to 3;
begin
  process(St, K, M, State)                    -- start if state or inputs change
  begin
    Load <= '0'; Sh <= '0'; Ad <= '0'; Done <= '0';
    case State is
      when 0 =>
        if St = '1' then               -- St (state 0)
          Load <= '1';
          Nextstate <= 1;
        else Nextstate <= 0;           -- St'
        end if;
      when 1 =>
        if M = '1' then                -- M (state 1)
          Ad <= '1';
          Nextstate <= 2;
        else                           -- M'
          Sh <= '1';
          if K = '1' then Nextstate <= 3;    -- K
          else Nextstate <= 1;               -- K'
          end if;
        end if;
```

```
      when 2 =>
        Sh <= '1';                         -- (state 2)
        if K = '1' then Nextstate <= 3;    -- K
        else Nextstate <= 1;               -- K'
        end if;
      when 3 =>
        Done <= '1';                       -- (state 3)
        Nextstate <= 0;
    end case;
  end process;
  process(CLK)
  begin
    if CLK = '1' and CLK'event then
      State <= Nextstate;                  -- update state on rising edge
    end if;
  end process;
end SMbehave;
```

### 5.2.2 A Dice Game

As a second example of SM chart construction, we will design an electronic dice game. This game is popularly known as craps in the United States. The game involves two dice, each of which can have a value between 1 and 6. Two counters are used to simulate the roll of the dice. Each counter counts in the sequence 1, 2, 3, 4, 5, 6, 1, 2, . . . . Thus, after the "roll" of the dice, the sum of the values in the two counters will be in the range 2 through 12. The rules of the game are as follows:

**1.** After the first roll of the dice, the player wins if the sum is 7 or 11. The player loses if the sum is 2, 3, or 12. Otherwise, the sum the player obtained on the first roll is referred to as a point, and he or she must roll the dice again.
**2.** On the second or subsequent roll of the dice, the player wins if the sum equals the point, and he or she loses if the sum is 7. Otherwise, the player must roll again until he or she finally wins or loses.

Figure 5-11 shows the block diagram for the dice game. The inputs to the dice game come from two push buttons, *Rb* (roll button) and *Reset*. *Reset* is used to

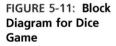
FIGURE 5-11: **Block Diagram for Dice Game**

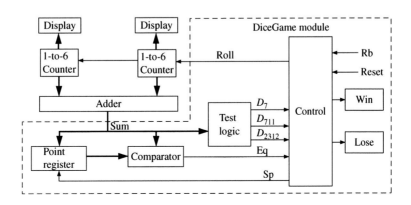

initiate a new game. When the roll button is pushed, the dice counters count at a high speed, so the values cannot be read on the display. When the roll button is released, the values in the two counters are displayed.

Figure 5-12 shows a flow chart for the dice game. After rolling the dice, the sum is tested. If it is 7 or 11, the player wins; if it is 2, 3, or 12, he or she loses. Otherwise the sum is saved in the point register, and the player rolls again. If the new sum equals the point, the player wins; if it is 7, he or she loses. Otherwise, the player rolls again. If the *Win* light or *Lose* light is not on, the player must push the roll button again. After winning or losing, he or she must push *Reset* to begin a new game. We will assume at this point that the push buttons are properly debounced and that changes in *Rb* are properly synchronized with the clock. A method for debouncing and synchronization was discussed in Chapter 4.

**FIGURE 5-12: Flow Chart for Dice Game**

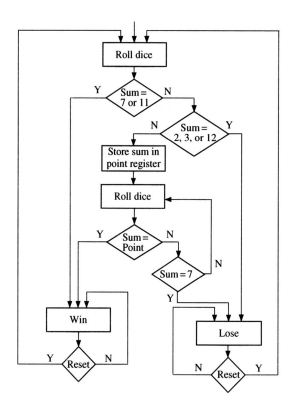

The components for the dice game shown in the block diagram (Figure 5-11) include an adder, which adds the two counter outputs, a register to store the point, test logic to determine conditions for win or lose, and a control circuit. Input signals to the control circuit are defined as follows:

$D_7$ = 1 if the sum of the dice is 7

$D_{711}$ = 1 if the sum of the dice is 7 or 11

$D_{2312}$ = 1 if the sum of the dice is 2, 3, or 12

$Eq$ = 1 if the sum of the dice equals the number stored in the point register

$Rb$ = 1 when the roll button is pressed

$Reset$ = 1 when the reset button is pressed

Outputs from the control circuit are defined as follows:

$Roll$ = 1 enables the dice counters

$Sp$ = 1 causes the sum to be stored in the point register

$Win$ = 1 turns on the win light

$Lose$ = 1 turns on the lose light

The $Rb$ and $Roll$ signals may look synonymous; however, they are different. We are using electronic dice counters, and $Roll$ is the signal to let the counters continue to count. $Rb$ is a push-button signal requesting that the dice be rolled. Thus, $Rb$ is an input to the control circuit, while $Roll$ is an output from the control circuit. When the control circuit is in a state looking for a new roll of the dice, whenever the push button is pressed (i.e., $Rb$ is activated), the control circuit will generate the $Roll$ signal to the electronic dice.

We now convert the flow chart for the dice game to an SM chart for the control circuit using the control signals defined above. Figure 5-13 shows the resulting SM chart.

The control circuit waits in state $S_0$ until the roll button is pressed ($Rb = 1$). Then, it goes to state $S_1$, and the roll counters are enabled as long as $Rb = 1$. As soon as the roll button is released ($Rb = 0$), $D_{711}$ is tested. If the sum is 7 or 11, the circuit goes to state $S_2$ and turns on the $Win$ light; otherwise, $D_{2312}$ is tested. If the sum is 2, 3, or 12, the circuit goes to state $S_3$ and turns on the $Lose$ light; otherwise, the signal $Sp$ becomes 1 and the sum is stored in the point register. It then enters $S_4$ and waits for the player to "roll the dice" again. In $S_5$, after the roll button is released, if $Eq = 1$, the sum equals the point and state $S_2$ is entered to indicate a win. If $D_7 = 1$, the sum is 7 and $S_3$ is entered to indicate a loss. Otherwise, control returns to $S_4$ so that the player can roll again. When in $S_2$ or $S_3$, the game is reset to $S_0$ when the $Reset$ button is pressed.

Instead of using an SM chart, we could construct an equivalent state graph from the flow chart. Figure 5-14 shows a state graph for the dice game controller. The state graph has the same states, inputs, and outputs as the SM chart. The arcs have been labeled consistently with the rules for proper state graphs given in Section 4.5. Thus, the arcs leaving state S1 are labeled $Rb$, $Rb'D_{711}$, $Rb'D'_{711}D_{2312}$, and $Rb'D'_{711}D'_{2312}$.

Before proceeding with the design, it is important to verify that the SM chart (or state graph) is correct. We will write a behavioral VHDL description based on the SM chart and then write a test bench to simulate the roll of the dice. Initially, we will write a dice game module that contains the control circuit, point register, and comparator (see Figure 5-11). Later, we will add the counters and adder so that we can simulate the complete dice game.

FIGURE 5-13: **SM Chart for Dice Game**

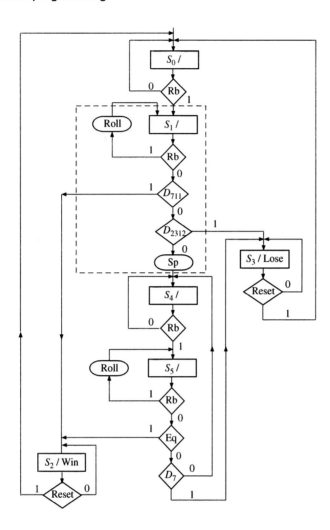

The VHDL code for the dice game in Figure 5-15 corresponds directly to the SM chart of Figure 5-13. The **case** statement in the first **process** tests the state, and in each state nested **if-then-else** (or **elsif**) statements are used to implement the conditional tests. In State 1 the *Roll* signal is turned on when *Rb* is 1. If all conditions test false, *Sp* is set to 1 and the next state is 4. In the second **process**, the state is updated after the rising edge of the clock, and if *Sp* is 1, the sum is stored in the point register.

We are now ready to test the behavioral model of the dice game. It is not convenient to include the counters that generate random numbers in the initial test, since we want to specify a sequence of dice rolls that will test all paths on the SM chart. We could prepare a simulator command file that would generate a sequence of data for *Rb, Sum,* and *Reset*. This would require careful analysis of the timing to make sure that the input signals change at the proper time. A better approach for testing the dice game is to design a VHDL test bench module to monitor the output signals from the dice game module and supply a sequence of inputs in response.

FIGURE 5-14: **State Graph for Dice Game Controller**

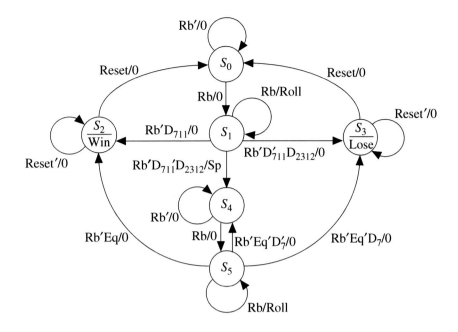

FIGURE 5-15: **Behavioral Model for Dice Game Controller**

```
entity DiceGame is
  port(Rb, Reset, CLK: in bit;
       Sum: in integer range 2 to 12;
       Roll, Win, Lose: out bit);
end DiceGame;

architecture DiceBehave of DiceGame is
signal State, Nextstate: integer range 0 to 5;
signal Point: integer range 2 to 12;
signal Sp: bit;
begin
  process(Rb, Reset, Sum, State)
  begin
    Sp <= '0'; Roll <= '0'; Win <= '0'; Lose <= '0';
    case State is
      when 0 => if Rb = '1' then Nextstate <= 1; end if;
      when 1 =>
        if Rb = '1' then Roll <= '1';
        elsif Sum = 7 or Sum = 11 then Nextstate <= 2;
        elsif Sum = 2 or Sum = 3 or Sum = 12 then Nextstate <= 3;
        else Sp <= '1'; Nextstate <= 4;
        end if;
      when 2 => Win <= '1';
        if Reset = '1' then Nextstate <= 0; end if;
      when 3 => Lose <= '1';
        if Reset = '1' then Nextstate <= 0; end if;
```

```
      when 4 => if Rb = '1' then Nextstate <= 5; end if;
      when 5 =>
        if Rb = '1' then Roll <= '1';
        elsif Sum = Point then Nextstate <= 2;
        elsif Sum = 7 then Nextstate <= 3;
        else Nextstate <= 4;
        end if;
    end case;
  end process;

  process(CLK)
  begin
    if CLK'event and CLK = '1' then
      State <= Nextstate;
      if Sp = '1' then Point <= Sum; end if;
    end if;
  end process;
end DiceBehave;
```

Figure 5-16 shows the DiceGame connected to a module called GameTest. GameTest needs to perform the following functions:

1. Initially supply the *Rb* signal.
2. When the DiceGame responds with a *Roll* signal, supply a *Sum* signal, which represents the sum of the two dice.
3. If no *Win* or *Lose* signal is generated by the DiceGame, repeat steps 1 and 2 to roll again.
4. When a *Win* or *Lose* signal is detected, generate a *Reset* signal and start again.

**FIGURE 5-16: Dice Game with Test Bench**

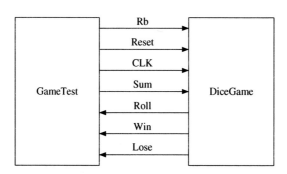

Figure 5-17 shows an SM chart for the GameTest module. *Rb* is generated in state $T_0$. When DiceGame detects *Rb*, it goes to $S_1$ and generates *Roll*. When GameTest detects *Roll*, the *Sum* that represents the next roll of the dice is read from *Sumarray(i)* and *i* is incremented. When the state goes to $T_1$, *Rb* goes to 0. The DiceGame goes to $S_2$, $S_3$, or $S_4$ and GameTest goes to $T_2$. The *Win* and *Lose* outputs are tested in state $T_2$. If *Win* or *Lose* is detected, a *Reset* signal is generated before the next roll of the dice. After *N* rolls of the dice, GameTest goes to state $T_3$, and no further action occurs.

**FIGURE 5-17: SM
Chart for Dice
Game Test**

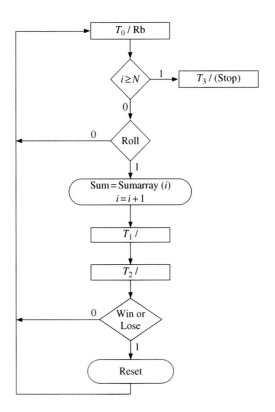

GameTest (Figure 5-18) implements the SM chart for the GameTest module. It contains an array of test data, a concurrent statement that generates the clock, and two processes. The first **process** generates *Rb, Reset*, and *Tnext* (the next state) whenever *Roll, Win, Lose*, or *Tstate* changes. The second **process** updates *Tstate* (the state of GameTest). When running the simulator, we want to display only one line of output for each roll of the dice. To facilitate this, we have added a signal *Trig1*, which changes every time state $T_2$ is entered.

Tester (Figure 5-19) connects the DiceGame and GameTest components so that the game can be tested. Figure 5-20 shows the simulator command file and output. The listing is triggered by *Trig1* once for every roll of the dice. The run 2000 command runs for more than enough time to process all the test data.

**FIGURE 5-18: Dice Game Test Module**

```
entity GameTest is
  port(Rb, Reset: out bit;
       Sum: out integer range 2 to 12;
       CLK: inout bit;
       Roll, Win, Lose: in bit);
end GameTest;
```

```vhdl
architecture dicetest of GameTest is
signal Tstate, Tnext: integer range 0 to 3;
signal Trig1: bit;
type arr is array(0 to 11) of integer;
constant Sumarray:arr := (7, 11, 2, 4, 7, 5, 6, 7, 6, 8, 9, 6);
begin
  CLK <= not CLK after 20 ns;
  process(Roll, Win, Lose, Tstate)
  variable i: natural;                      -- i is initialized to 0
  begin
    case Tstate is
      when 0 => Rb <= '1';                  -- wait for Roll
        Reset <= '0';
        if i >= 12 then Tnext <= 3;
        elsif Roll = '1' then
          Sum <= Sumarray(i);
          i := i + 1;
          Tnext <= 1;
        end if;
      when 1 => Rb <= '0'; Tnext <= 2;
      when 2 => Tnext <= 0;
        Trig1 <= not Trig1;                 -- toggle Trig1
        if (Win or Lose) = '1' then
          Reset <= '1';
        end if;
      when 3 => null;                        -- Stop state
    end case;
  end process;

  process(CLK)
  begin
    if CLK = '1' and CLK'event then
      Tstate <= Tnext;
    end if;
  end process;
end dicetest;
```

FIGURE 5-19: **Tester for DiceGame**

```vhdl
entity tester is
end tester;

architecture test of tester is
component GameTest
  port(Rb, Reset: out bit;
       Sum: out integer range 2 to 12;
       CLK: inout bit;
       Roll, Win, Lose: in bit);
end component;
```

```
component DiceGame
  port(Rb, Reset, CLK: in bit;
       Sum: in integer range 2 to 12;
       Roll, Win, Lose: out bit);
end component;

signal rb1, reset1, clk1, roll1, win1, lose1: bit;
signal sum1: integer range 2 to 12;
begin
  Dice: Dicegame port map (rb1, reset1, clk1, sum1, roll1, win1, lose1);
  Dicetest: GameTest port map (rb1, reset1, sum1, clk1, roll1, win1, lose1);
end test;
```

FIGURE 5-20: **Simulation and Command File for Dice Game Tester**

```
add list /dicetest/trig1 -NOTrigger sum1 win1 lose1 /dice/point
run 2000
```

| ns | delta | trig1 | sum1 | win1 | lose1 | point |
|----|-------|-------|------|------|-------|-------|
| 0 | +0 | 0 | 2 | 0 | 0 | 2 |
| 100 | +3 | 0 | 7 | 1 | 0 | 2 |
| 260 | +3 | 0 | 11 | 1 | 0 | 2 |
| 420 | +3 | 0 | 2 | 0 | 1 | 2 |
| 580 | +2 | 1 | 4 | 0 | 0 | 4 |
| 740 | +3 | 1 | 7 | 0 | 1 | 4 |
| 900 | +2 | 0 | 5 | 0 | 0 | 5 |
| 1060 | +2 | 1 | 6 | 0 | 0 | 5 |
| 1220 | +3 | 1 | 7 | 0 | 1 | 5 |
| 1380 | +2 | 0 | 6 | 0 | 0 | 6 |
| 1540 | +2 | 1 | 8 | 0 | 0 | 6 |
| 1700 | +2 | 0 | 9 | 0 | 0 | 6 |
| 1860 | +3 | 0 | 6 | 1 | 0 | 6 |

● ● ● ● ● ● ● ● ● ● ●
# 5.3 Realization of SM Charts

Methods used to realize SM charts are similar to the methods used to realize state graphs. As with any sequential circuit, the realization will consist of a combinational subcircuit, together with flip-flops for storing the state of the circuit. In some cases, it may be possible to identify equivalent states in an SM chart and eliminate redundant states using the same method as was used for reducing state tables. However, an SM chart is usually incompletely specified in the sense that all inputs are not tested in every state, which makes the reduction procedure more difficult. Even if the number of states in an SM chart can be reduced, it is not always desirable to do so, since combining states may make the SM chart more difficult to interpret.

Before deriving next state and output equations from an SM chart, a state assignment must be made. The best way of making the assignment depends on how the SM chart is realized. If gates and flip-flops (or the equivalent PLD realization) are used, the guidelines for state assignment given in Section 1.7 may be useful. If programmable gate arrays are used, a one-hot assignment may be best, as explained in Section 6.9.

As an example of realizing an SM chart, consider the SM chart in Figure 5-21.

FIGURE 5-21:
**Example SM Chart for Implementation**

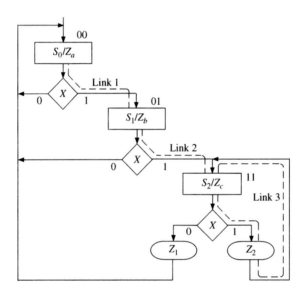

We have made the state assignment $AB = 00$ for S0, $AB = 01$ for $S_1$, and $AB = 11$ for $S_2$. After a state assignment has been made, output and next-state equations can be read directly from the SM chart. Since the Moore output $Z_a$ is 1 only in state 00, $Z_a = A'B'$. Similarly, $Z_b = A'B$ and $Z_c = AB$. The conditional output $Z_1 = ABX'$, since the only link path through $Z_1$ starts with $AB = 11$ and takes the $X = 0$ branch. Similarly, $Z_2 = ABX$. There are three link paths (labeled link 1, link 2, and link 3 in Figure 5-21), which terminate in a state that has $B = 1$. Link 1 starts with a present state $AB = 00$, takes the $X = 1$ branch, and terminates on a state in which $B = 1$. Therefore, the next state of $B$ ($B^+$) equals 1 when $A'B'X = 1$. Link 2 starts in state 01, takes the $X = 1$ branch, and ends in state 11, so $B^+$ has a term $A'BX$. Similarly, $B^+$ has a term $ABX$ from link 3. The next state equation for $B$ thus has three terms corresponding to the three link paths:

$$B^+ = A'B'X + A'BX + ABX$$

$$\text{link 1} \quad \text{link 2} \quad \text{link 3}$$

Similarly, two link paths terminate in a state with $A = 1$, so

$$A^+ = A'BX + ABX$$

These output and next state equations can be simplified with Karnaugh maps using the unused state assignment ($AB = 10$) as a "don't care" condition.

As illustrated above for flip-flops $A$ and $B$, the procedure for deriving the next state equation for a flip-flop $Q$ from the SM chart is as follows:

1. Identify all of the states in which $Q = 1$.
2. For each of these states, find all the link paths that lead *into* the state.
3. For each of these link paths, find a term that is 1 when the link path is followed. That is, for a link path from $S_i$ to $S_j$, the term will be 1 if the machine is in state $S_i$ and the conditions for exiting to $S_j$ are satisfied.
4. The expression for $Q^+$ (the next state of $Q$) is formed by OR'ing together the terms found in step 3.

### 5.3.1 Implementation of Binary Multiplier Controller

Next, consider the SM chart for the multiplier control repeated here, in Figure 5-22.

**FIGURE 5-22: SM Chart for Multiplier Controller**

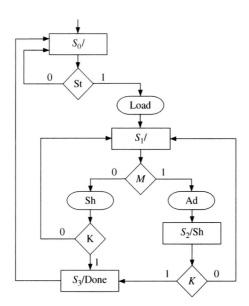

We can realize this SM chart with two $D$ flip-flops and a combinational circuit. Let us assume that the state assignments are $AB = 00$ for $S_0$, $AB = 01$ for $S_1$, $AB = 10$ for $S_2$, and $AB = 11$ for $S_3$.

The logic equations for the multiplier control and the next state equations can be derived by tracing link paths on the SM chart and then simplifying the resulting equations. First, let us consider the control signals. *Load* is true only in $S_0$ and only if *St* is true. Hence, $Load = S_0 St = A'B' St$. Similarly, $Ad$ is true only in $S_1$ and only if $M$ is true. Hence, $Ad = A'BM$. *Done* is a Moore output in $S_3$, and hence *Done* = $S_3 = AB$. In summary, the logic equations for the multiplier control are

$$Load = A'B'St$$
$$Sh = A'BM'(K' + K) + AB'(K' + K) = A'BM' + AB'$$
$$Ad = A'BM$$
$$Done = AB$$

The next state equations can be derived by inspection of the SM chart and considering the state assignments. $A$ is true in states $S_2$ and $S_3$. State $S_2$ is the next state when current state is $S_1$ and $M$ is true ($A'BM$). State $S_3$ is the next state when current state is $S_1$, $M$ is false, and $K$ is true ($A'BM'K$) and when current state is $S_2$ and $K$ is true ($AB'K$). Hence, we can write that

$$A^+ = A'BM'K + A'BM + AB'K = A'B(M + K) + AB'K$$

Similarly, we can derive the next state equation for $B$ by inspection of the ASM diagram:

$$B^+ = A'B'St + A'BM'(K' + K) + AB'(K' + K) = A'B'St + A'BM' + AB'$$

The multiplier controller can be implemented in a hardwired fashion by two flip-flops and a few logic gates. The logic gates implement the next state equations and control signal equations. The circuit can be implemented with discrete gates or in a PLA, CPLD, or FPGA.

Table 5-1 illustrates a state transition table for the multiplier control. Each row in the table corresponds to one of the link paths in the SM chart. Since $S_0$ has two exit paths, the table has two rows for present state $S_0$. The first row corresponds to the $St = 0$ exit path, so the next state and outputs are 0. In the second row, $St = 1$, so the next state is 01 and the other outputs are 1000. Since $St$ is not tested in states $S_1$, $S_2$, and $S_3$, $St$ is a "don't care" in the corresponding rows. The outputs for each row can be filled in by tracing the corresponding link paths on the SM chart. For example, the link path from $S_1$ to $S_2$ passes through conditional output $Ad$, so $Ad = 1$ in this row. Since $S_2$ has a Moore output $Sh$, $Sh = 1$ in both of the rows for which $AB = 10$.

TABLE 5-1: State Transition Table for Multiplier Control

| | A | B | St | M | K | A$^+$ | B$^+$ | Load | Sh | Ad | Done |
|---|---|---|---|---|---|---|---|---|---|---|---|
| $S_0$ | 0 | 0 | 0 | — | — | 0 | 0 | 0 | 0 | 0 | 0 |
| | 0 | 0 | 1 | — | — | 0 | 1 | 1 | 0 | 0 | 0 |
| $S_1$ | 0 | 1 | — | 0 | 0 | 0 | 1 | 0 | 1 | 0 | 0 |
| | 0 | 1 | — | 0 | 1 | 1 | 1 | 0 | 1 | 0 | 0 |
| | 0 | 1 | — | 1 | — | 1 | 0 | 0 | 0 | 1 | 0 |
| $S_2$ | 1 | 0 | — | — | 0 | 0 | 1 | 0 | 1 | 0 | 0 |
| | 1 | 0 | — | — | 1 | 1 | 1 | 0 | 1 | 0 | 0 |
| $S_3$ | 1 | 1 | — | — | — | 0 | 0 | 0 | 0 | 0 | 1 |

The design may also be implemented with ROM. If it has to be implemented using the ROM method, we can calculate the size of the ROM as follows. There are five different inputs to the combinational circuit here (*A, B, St, M,* and *K*). Hence, the ROM will have 32 entries. The combinational circuit should generate six signals (four control signals plus two next states). Hence, each entry has to be 6 bits wide. Thus, this design can be implemented using a $32 \times 6$ ROM and two *D* flip-flops. If the combinational logic is implemented with a PLA instead of a ROM, the PLA table is the same as the state transition table. The PLA would have 5 inputs, 6 outputs, and 8 product terms.

If a ROM is used, the table must be expanded to $2^5 = 32$ rows since there are five inputs. To expand the table, the dashes in each row must be replaced with all possible combinations of 0's and 1's. If a row has *n* dashes, it must be replaced with $2^n$ rows. For example, the fifth row in Table 5-1 would be replaced with the following 4 rows:

| 0 | 1 | **0** | 1 | **0** | 1 | 0 | 0 | 0 | 1 | 0 |
|---|---|---|---|---|---|---|---|---|---|---|
| 0 | 1 | **0** | 1 | **1** | 1 | 0 | 0 | 0 | 1 | 0 |
| 0 | 1 | **1** | 1 | **0** | 1 | 0 | 0 | 0 | 1 | 0 |
| 0 | 1 | **1** | 1 | **1** | 1 | 0 | 0 | 0 | 1 | 0 |

The added entries are printed in boldface.

● ● ● ● ● ● ● ● ● ● ● ●

## 5.4 **Implementation of the Dice Game**

We can realize the SM chart for the dice game (Figure 5-13) using combinational circuitry and three *D* flip-flops, as shown in Figure 5-23. We use a straight binary state assignment. The combinational circuit has nine inputs and seven outputs. Three of the inputs correspond to current state, and three of the outputs provide the next state information. All inputs and outputs are listed at the top of Table 5-2. The state

**FIGURE 5-23:**
**Realization of Dice Game Controller**

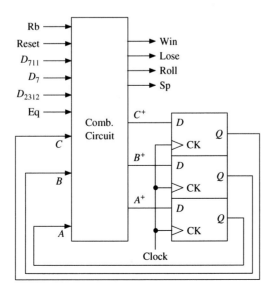

transition table has one row for each link path on the SM chart. In state $ABC = 000$, the next state is $A^+B^+C^+ = 000$ or $001$, depending on the value of $Rb$. Since state $001$ has four exit paths, the table has four corresponding rows. When $Rb$ is 1, $Roll$ is 1 and there is no state change. When $Rb = 0$ and $D_{711}$ is 1, the next state is $010$. When $Rb = 0$ and $D_{2312} = 1$, the next state is $011$. For the link path from state $001$ to $100$, $Rb, D_{711}$, and $D_{2312}$ are all 0, and $Sp$ is a conditional output. This path corresponds to row 4 of the state transition table, which has $Sp = 1$ and $A^+B^+C^+ = 100$. In state $010$, the $Win$ signal is always on, and the next state is $010$ or $000$, depending on the value of $Reset$. Similarly, $Lose$ is always on in state $011$. In state $101$, $A^+B^+C^+ = 010$ if $Eq = 1$; otherwise, $A^+B^+C^+ = 011$ or $100$, depending on the value of $D_7$. Since states $110$ and $111$ are not used, the next states and outputs are don't cares when $ABC = 110$ or $111$.

We can use Table 5-2 and derive equations for the control signals and the next state equations. The required equations can be derived from Table 5-2 using the method of map-entered variables (see Chapter 1) or using a CAD program such as *LogicAid*. These equations can also be derived by tracing link paths on the SM chart and then simplifying the resulting equations using the "don't care" next states.

Figure 5-24 shows K-maps for $A^+$, $B^+$, and $Win$, which were plotted directly from the table. Since $A$, $B$, $C$, and $Rb$ have assigned values in most of the rows of the table, these four variables are used on the map edges, and the remaining variables are entered within the map. (Chapter 1 described the K-map technique that uses map-entered variables.) $E_1$, $E_2$, $E_3$, and $E_4$ on the maps represent the expressions given below the maps. From the $A^+$ column in the table, $A^+$ is 1 in row 4, so we should enter $D'_{711}D'_{2312}$ in the $ABCRb = 0010$ square of the map. To save space, we define $E_1 = D'_{711}D'_{2312}$ and place $E_1$ in the square. Since $A^+$ is 1 in rows 11, 12, and 16, 1's are placed on the map squares $ABCRb = 1000, 1001$, and $1011$. From row 13, we place

| | ABC | Rb | Reset | $D_7$ | $D_{711}$ | $D_{2312}$ | Eq | $A^+$ | $B^+$ | $C^+$ | Win | Lose | Roll | Sp |
|---|---|---|---|---|---|---|---|---|---|---|---|---|---|---|
| 1 | 000 | 0 | — | — | — | — | — | 0 | 0 | 0 | 0 | 0 | 0 | 0 |
| 2 | 000 | 1 | — | — | — | — | — | 0 | 0 | 1 | 0 | 0 | 0 | 0 |
| 3 | 001 | 1 | — | — | — | — | — | 0 | 0 | 1 | 0 | 0 | 1 | 0 |
| 4 | 001 | 0 | — | — | 0 | 0 | — | 1 | 0 | 0 | 0 | 0 | 0 | 1 |
| 5 | 001 | 0 | — | — | 0 | 1 | — | 0 | 1 | 1 | 0 | 0 | 0 | 0 |
| 6 | 001 | 0 | — | — | 1 | — | — | 0 | 1 | 0 | 0 | 0 | 0 | 0 |
| 7 | 010 | — | 0 | — | — | — | — | 0 | 1 | 0 | 1 | 0 | 0 | 0 |
| 8 | 010 | — | 1 | — | — | — | — | 0 | 0 | 0 | 1 | 0 | 0 | 0 |
| 9 | 011 | — | 1 | — | — | — | — | 0 | 0 | 0 | 0 | 1 | 0 | 0 |
| 10 | 011 | — | 0 | — | — | — | — | 0 | 1 | 1 | 0 | 1 | 0 | 0 |
| 11 | 100 | 0 | — | — | — | — | — | 1 | 0 | 0 | 0 | 0 | 0 | 0 |
| 12 | 100 | 1 | — | — | — | — | — | 1 | 0 | 1 | 0 | 0 | 0 | 0 |
| 13 | 101 | 0 | — | 0 | — | — | 0 | 1 | 0 | 0 | 0 | 0 | 0 | 0 |
| 14 | 101 | 0 | — | 1 | — | — | 0 | 0 | 1 | 1 | 0 | 0 | 0 | 0 |
| 15 | 101 | 0 | — | — | — | — | 1 | 0 | 1 | 0 | 0 | 0 | 0 | 0 |
| 16 | 101 | 1 | — | — | — | — | — | 1 | 0 | 1 | 0 | 0 | 1 | 0 |
| 17 | 110 | — | — | — | — | — | — | — | — | — | — | — | — | — |
| 18 | 111 | — | — | — | — | — | — | — | — | — | — | — | — | — |

TABLE 5-2: State Transition Table (PLA Table) for Dice Game

FIGURE 5-24: **Maps Derived from Table 5-2**

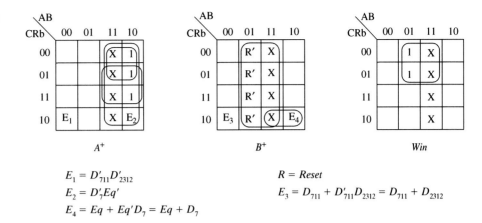

$$E_1 = D'_{711}D'_{2312}$$
$$E_2 = D'_7Eq'$$
$$E_4 = Eq + Eq'D_7 = Eq + D_7$$

$$R = Reset$$
$$E_3 = D_{711} + D'_{711}D_{2312} = D_{711} + D_{2312}$$

$E_2 = D'_7Eq'$ in the 1010 square. In rows 7 and 8, *Win* is always 1 when $ABC = 010$, so 1's are plotted in the corresponding squares of the *Win* map.

The resulting equations are

$$A^+ = A'B'C\,Rb'D'_{711}D'_{2312} + AC' + ARb + AD'_7Eq' \qquad (5\text{-}1)$$
$$B^+ = A'B'C\,Rb'(D_{711} + D_{2312}) + BReset' + AC\,Rb'(Eq + D_7)$$
$$C^+ = B'Rb + A'B'C\,D'_{711}D_{2312} + BC\,Reset' + AC\,D_7Eq'$$
$$Win = BC'$$
$$Lose = BC$$
$$Roll = B'CRb$$
$$Sp = A'B'C\,Rb'D'_{711}D'_{2312}$$

These equations can be implemented in any standard technology (using discrete gates, PALs, GALs, CPLDs, or FPGAs).

The dice game controller can also be realized using a ROM. A ROM (LUT) implementation of the game controller will need 512 entries (since there are 9 inputs). Each entry must be 7 bits wide (3 bits for next states and 4 bits for outputs). The ROM is very large because of the large number of inputs involved. The ROM method is hence not very desirable for state machines with a large number of inputs.

We now write a dataflow VHDL model for the dice game controller based on the block diagram of Figure 5-11 and Equations (5-1). The corresponding VHDL architecture is shown in Figure 5-25. The process updates the flip-flop states and the point register when the rising edge of the clock occurs. Generation of the control signals and *D* flip-flop input equations is done using concurrent statements. In particular, $D_7$, $D_{711}$, $D_{2312}$, and *Eq* are implemented using conditional signal assignments. As an alternative, all the signals and *D* input equations could have been implemented in a process with a sensitivity list containing *A, B, C, Sum, Point, Rb,* $D_7$, $D_{711}$, $D_{2312}$, *Eq,* and *Reset*. If the architecture of Figure 5-25 is used with the test bench of Figure 5-19, the results are identical to those obtained with the behavioral architecture in Figure 5-15.

FIGURE 5-25: **Dataflow Model for Dice Game (Based on Equations (5-1))**

```
architecture Dice_Eq of DiceGame is
signal Sp,Eq,D7,D711,D2312: bit:= '0';
signal DA,DB,DC,A,B,C: bit:='0';
signal Point: integer range 2 to 12;
begin
  process(CLK)
  begin
    if CLK = '1' and CLK'event then
      A <= DA; B <= DB; C <= DC;
      if Sp = '1' then Point <= Sum; end if;
    end if;
  end process;
  Win <= B and not C;
  Lose <= B and C;
  Roll <= not B and C and Rb;
  Sp <= not A and not B and C and not Rb and not D711 and not D2312;
  D7 <= '1' when Sum = 7 else '0';
  D711 <= '1' when (Sum = 11) or (Sum = 7) else '0';
  D2312 <= '1' when (Sum = 2) or (Sum = 3) or (Sum = 12) else '0';
  Eq <= '1' when Point = Sum else '0';
  DA <= (not A and not B and C and not Rb and not D711 and not D2312) or
        (A and not C) or (A and Rb) or (A and not D7 and not Eq);
  DB <= ((not A and not B and C and not Rb) and (D711 or D2312)) or
        (B and not Reset) or ((A and C and not Rb) and (Eq or D7));
  DC <= (not B and Rb) or (not A and not B and C and not D711 and D2312) or
        (B and C and not Reset) or (A and C and D7 and not Eq);
end Dice_Eq;
```

To complete the VHDL implementation of the dice game, we add two modulo-6 counters as shown in Figures 5-26 and 5-27. The counters are initialized to 1, so the sum of the two dice will always be in the range 2 through 12. When *Cnt1* is in state 6, the next clock sets it to state 1, and *Cnt2* is incremented (or *Cnt2* is set to 1 if it is in state 6).

FIGURE 5-26: **Counter for Dice Game**

```
entity Counter is
  port(Clk, Roll: in bit;
       Sum: out integer range 2 to 12);
end Counter;

architecture Count of Counter is
signal Cnt1, Cnt2: integer range 1 to 6:= 1;
begin
  process(Clk)
  begin
    if Clk = '1' then
      if Roll = '1' then
```

```
                    if Cnt1 = 6 then Cnt1 <= 1; else Cnt1 <= Cnt1 + 1; end if;
                    if Cnt1 = 6 then
                      if Cnt2 = 6 then Cnt2 <= 1; else Cnt2 <= Cnt2 + 1; end if;
                    end if;
                  end if;
                end if;
              end process;
            Sum <= Cnt1 + Cnt2;
          end Count;
```

FIGURE 5-27: **Complete Dice Game**

```
entity Game is
  port(Rb, Reset, Clk: in bit;
       Win, Lose: out bit);
end Game;

architecture Play1 of Game is
component Counter
  port(Clk, Roll: in bit;
       Sum: out integer range 2 to 12);
end component;

component DiceGame
  port(Rb, Reset, CLK: in bit;
       Sum: in integer range 2 to 12;
       Roll, Win, Lose: out bit);
end component;

signal roll1: bit;
signal sum1: integer range 2 to 12;
begin
  Dice: Dicegame port map (Rb, Reset, Clk, sum1, roll1, Win, Lose);
  Count: Counter port map (Clk, roll1, sum1);
end Play1;
```

This section has illustrated one way of realizing an SM chart. The implementation can use discrete gates, a PLA, a ROM, or a PAL. Alternative procedures are available that make it possible to reduce the size of the PLA or ROM by adding some components to the circuit. These methods are generally based on transformation of the SM chart to different forms and techniques, such as microprogramming.

• • • • • • • • • • • •

## 5.5  Microprogramming

Microprogramming is a technique to implement the control unit of a digital system. In order to realize a control unit, we can inspect the state diagram or SM chart, write the logic equations for the control outputs and the next states, and implement the state machine using gates and flip-flops. Sections 5.3 and 5.4 demonstrated this process for the binary multiplier and the dice game, respectively. This method of implementation is called **hardwiring**, to indicate that the control signals are generated using fixed (hardwired) logic circuitry.

In contrast, an alternative approach called **microprogramming** has been developed for designing control units for complex digital systems. Proposed by Maurice Wilkes in 1951, microprogramming is building a special computer for executing the algorithmic flow chart describing the controller of a system. This development stemmed from the separation of architecture and controller, which we described at the beginning of Chapter 4. Once the architecture and controller are clearly delineated, the controller flow chart systematically specifies all the controller signals that should be generated at each time during the flow of control from the reset state through each of the other states. By inspection of the SM chart for the shift and add binary multiplier in Figure 5-28(a), we can write pseudocode for the multiplier controller operation, as illustrated in Figure 5-28(b). This multiplier was presented in detail in Chapter 4.

Such a description of the controller easily makes us see the correspondence of the controller activity to a normal computer program. Microprogramming developed from exactly this realization.

If a memory can store all control signals and the next state information corresponding to each state for each input condition, we should be able to realize the controller by just **"sequencing"** through the memory. For this reason, microprogrammed controllers are also often called sequencers. The memory that stores the control words is called the **control store** or **microprogram memory**.

Microprogramming seemed extremely attractive in an era where the complexity of digital systems was growing prohibitively. Since debugging was done manually in those days, it was very hard to identify and correct errors. The systematic nature of microprogramming made debugging systems easier. Changes to systems can be implemented relatively easily. Errors can be identified and corrected easily. This made microprogramming very popular.

The disadvantage of microprogramming is that it is slow. A memory access is required to access the control word from the control store. Hardwiring results in faster systems because hardwired control signals are generated by logic gates, and they are typically faster than memory.

Early microprocessors such as Intel 8086 and Motorola 68000 were microprogrammed. These microprocessors supported a variety of memory addressing modes with base registers and index registers. They allowed operands to be accessed directly from memory and results be written directly to memory. Many complex instructions that performed a series of fundamental operations were available on these processors. Microprogramming was convenient when the control signals for the several operations needed for a complex instruction could be systematically specified in the microprogram word. It would have been extremely hard to implement these microprocessors with hardwiring.

Many things have changed since then. In the late 1970s, it was observed that in many microprocessors, more than half of the chip area was spent in the controller (i.e., the data path of the processor occupied less than half the chip area). The complexity

FIGURE 5-28: **SM Chart and Operation Flow of the Multiplier**

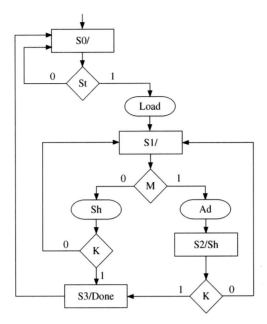

(a) SM chart for Multiplier

```
S0:  if St is true, produce Load Signal and go to S1,
            else return to S0
S1:  if M is true, produce Ad and go to S2,
            else produce Sh, check whether K is 1;
            if K is 1 go to S3;
            if K is 0, go to S1;
S2:  produce Sh;
            if K = 0, go to S1;
            else go to S3;
S3:  produce Done and go to S0
```

(b) Pseudo code representing the operation of the
multiplier controller

of the microprocessors led researchers and designers to the RISC (Reduced Instruction Set Computing) era. RISC microprocessors are simpler, have fewer memory addressing modes, and need simpler control units. Computer-aided design (CAD) tools have improved, and the designers' capability to debug has improved. Today microprogramming may be used only for microprocessors with complex instruction set architectures (ISAs); however, it is a powerful concept and a very elegant one.

Microprogramming can be implemented in a variety of ways. The general idea is to store a control word corresponding to each state. The control word is also called a **microinstruction**. The microinstruction specifies the outputs to be generated. It also specifies where the next microinstruction can be found. This corresponds to the state transitions in the state diagram or SM chart.

### 5.5.1 Two-Address Microcode

Figure 5-29 illustrates a suitable hardware arrangement for a typical microprogram implementation. Each ROM location stores a control word or microinstruction. The only inputs to the ROM come from the state register. A multiplexer with each of the inputs can be used to selectively test at most one variable in each state. This multiplexer is used to indicate whether the selected control signal (as indicated by TEST) is true or false. Another multiplexer is used to select which next state should control branch to. This technique is called two-address microcoding because the next states corresponding to both true and false conditions of the test signal are explicitly specified in the microinstruction.

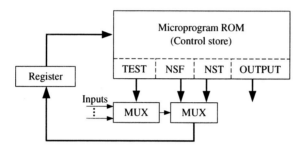

The ROM output has four fields: TEST, NSF, NST, and OUTPUT. TEST controls the input MUX, which selects one of the inputs to be tested in each state. If this input is 0 (false), then the second MUX selects the NSF field as the next state. If the input is 1 (true), it selects the NST field as the next state. The OUTPUT bits correspond to the control signals. Note that in order to use this hardware arrangement, the SM chart must have only Moore outputs, since the outputs can be a function only of the state.

### SM Chart Transformations for Microprogramming

Transformations are performed on the SM chart to facilitate easy and efficient microprogramming. We do not want a naïve look-up table method where all combinations of inputs and present states are directly specified. We transform the SM chart in such a way that only one entry is required per state. Some of the transformations do increase the number of states; however, the achieved microprogram size is still significantly smaller than the ROM size in a naïve LUT method.

### Eliminate Conditional Outputs

It is desirable to construct the controller as a Moore machine so that there will be no conditional control signals. If control signals are conditional on some inputs, we should store control signals corresponding to different combinations of inputs. Hence, the first step in transforming a state diagram or SM chart for easy microprogramming is to convert it into a Moore state machine. Any Mealy machine can be converted into a Moore machine by adding an appropriate number of additional states.

## Allow Only Single Qualifier per State

The inputs that are tested in each state of the state machine are called **qualifiers** in the microprogram literature. For example, in Figure 5-28, $St$, $M$, and $K$ are qualifiers. States $S_0$ and $S_2$ contain only one qualifier, but state $S_1$ tests qualifiers $M$ and $K$. The multiple qualifiers in $S_1$ led to nested **if** statements in the pseudo code in Figure 5-28. Microprogramming can be done with multiple qualifiers per state; however, it is simpler to implement microprogramming when only one variable is tested in each state.

Thus, microprogramming becomes easy if the following two transformations are done on SM charts:

**1.** Eliminate all conditional outputs by transforming to a Moore machine
**2.** Test only one input (qualifier) in each state

Let us transform the SM chart of the multiplier for microprogramming. First, we will convert it to a Moore machine by adding a state for each conditional output (i.e., each oval in the SM chart). That results in additional states $S_{01}$ in state $S_0$ for the conditional output *Load*, $S_{11}$ in the original state $S_1$ for the conditional output *Ad*, and $S_{12}$ in $S_1$ for the conditional output *Sh*. Fortunately, no more than one qualifier is tested in any state. The modified SM chart is shown in Figure 5-30.

FIGURE 5-30:
**Multiplier SM Chart with No Conditional Outputs (Derived from Figure 5-28)**

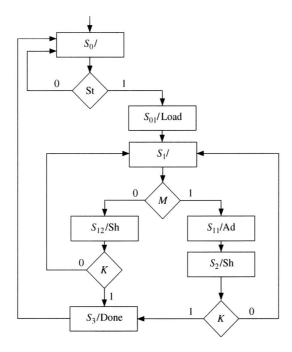

The corresponding actions can be described by the following pseudocode:

```
S0:  if St is true, go to S01,
          else go to S0;
S01: produce Load; Go to S1;
```

```
S1:  if M is true, go to S11, else go to S12;
S11: produce Ad; go to S2;
S12: produce Sh; if K = 0, go to S1; else go to S3;
S2:  produce Sh;
             if K=0, go to S1;
             else go to S3;
S3:  produce Done; go to S0;
```

At this stage, the transformed SM chart can be inspected for eliminating redundant states. Can states $S_{11}$ and $S_2$ be combined? Since the add operation has to be performed before shift, the $Ad$ control signal should appear ahead of the $Sh$ control signal. Hence, $S_{11}$ and $S_2$ cannot be combined.

Now, let us inspect states $S_{12}$ and $S_2$. States $S_{12}$ and $S_2$ perform exactly the same tasks and have the same next states. Hence, they can be combined. This is an example of potential state minimizations after the transformation. Let us denote the new combined state as $S_2$. The improved SM chart is shown in Figure 5-31.

**FIGURE 5-31:**
**Modified Multiplier SM Chart After State Minimization is Applied to Figure 5-30**

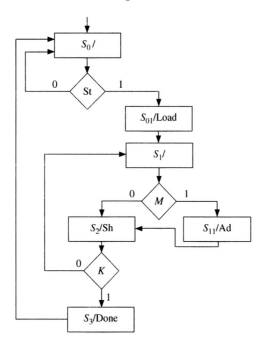

The microprogram will look as in Table 5-3, assuming a straight binary state assignment in the sequence $S_0, S_{01}, S_1, S_{11}, S_2,$ and $S_3$. Since there are three inputs, $St$, $M$, and $K$, a 4-to-1 MUX will be sufficient to select the appropriate qualifier. The multiplexer connections are assumed to be as in Figure 5-32.

Let us look at the first row in Table 5-3. It corresponds to state $S_0$, which is encoded as 000. The input tested is $St$. Since $St$ is connected to input 0 of the multiplexer, the TEST field for this row is 00. If $St$ is false, the next state is $S_0$, leading to 000 in the NSF field. If $St$ is true, the next state is $S_{01}$, leading to the 001 bits in the NST field. The control signals $Load, Ad, Sh,$ and $Done$ are 0 in state $S_0$.

TABLE 5-3: **Two Address Microprogram for Multiplier. Both NST and NSF Specified (Corresponds to Figure 5-29)**

| State | ABC | TEST | NSF | NST | Load | Ad | Sh | Done |
|---|---|---|---|---|---|---|---|---|
| $S_0$ | 000 | 00 | 000 | 001 | 0 | 0 | 0 | 0 |
| $S_{01}$ | 001 | 11 | 010 | 010 | 1 | 0 | 0 | 0 |
| $S_1$ | 010 | 01 | 100 | 011 | 0 | 0 | 0 | 0 |
| $S_{11}$ | 011 | 11 | 100 | 100 | 0 | 1 | 0 | 0 |
| $S_2$ | 100 | 10 | 010 | 101 | 0 | 0 | 1 | 0 |
| $S_3$ | 101 | 11 | 000 | 000 | 0 | 0 | 0 | 1 |

FIGURE 5-32: **4-to-1 MUX for Microprogramming the Multiplier (Two Address Microcode)**

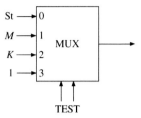

The microcode for state $S_{01}$ is shown in the second row. State $S_{01}$ generates the *Load* signal and the controller transitions to state $S_1$. No input signals are tested. In the multiplexer in Figure 5-32, we provide a value of '1' to the last unused muliplexer input. So we can mark the TEST field as 11, corresponding to the last input of the multiplexer. In state $S_1$, input signal $M$ is tested. Since $M$ is connected to input 1 of the multiplexer, the TEST field for the third row is 01. In a similar fashion, all rows of Table 5-3 are filled.

Since there are six states, three flip-flops will be required. The ROM that stores this microprogram will need six entries, one for each state. Each entry will need 12 bits, including 2 bits for TEST, 3 bits for NSF, 3 for NST, and 4 bits for control signals *Load, Ad, Sh*, and *Done. ABC* represents the address at which the microinstruction is stored.

The hardware arrangement in Figure 5-29 is for microprogramming with two next state addresses and single qualifier per state. Single qualifier microprogramming means that only one input can be tested in a state. Two address microcoding means that next states for both possible input values (i.e., next state if the input is true (NST) and next state if the input is false (NSF)), are explicitly specified in the control word. (Figure 5-29 could be modified to allow Mealy outputs by replacing the OUTPUT field with OUTPUTF and OUTPUTT, and adding a MUX to select one of the two output fields.)

### 5.5.2 Single-Qualifier, Single-Address Microcode

In the microprogram of Table 5-3, each microinstruction can specify two potential next states, the next state if the input is true and the next state if the input is false. The microcode for the different states can be located in any sequence because the next microinstruction for each state is specified without assuming any default flow of control.

The aforementioned microprogram resembles software, but in conventional programs, control flows in sequence except when branch and jump instructions alter the control flow. If a branch is not taken, control simply flows to the next instruction. If we could take advantage of a similar structure, each microprogram entry will need to specify only one next state address.

Let us consider what we should do in order to make the default next state be the state located in the next row. In that case, the state assignments should be such that, if the qualifier (input) is false, the next state should be the current state incremented by 1. The next state when the qualifier is true will be the only next state explicitly specified in the microcode. If the qualifier is false, control simply goes to the next row to get the succeeding microinstruction.

This type of microprogram can be implemented using the hardware arrangement shown in Figure 5-33. Since control normally just advances to the next location, a counter can be effectively used. This counter is analogous to a program counter (PC) in a microprocessor. The counter points to the current state of the controller, analogous to a PC pointing to the next instruction to be fetched. Each ROM location stores a control word or microinstruction. The OUTPUT bits correspond to the control signals. The TEST bits specify the qualifier being tested and the NST bits indicate the target microinstruction if the qualifier is true. A multiplexer is used to indicate whether the selected control signal (as indicated by TEST) is true or false. If the qualifier is false, the counter increments to point to the next microinstruction. This corresponds to the default next state. If the qualifier is true, the counter should load the NST bits as the location of the next microinstruction. This is the explicitly specified next state. A counter with parallel load capability is the ideal building block for this module. The multiplexer selects the relevant qualifier and its output is used to decide whether the counter should count sequentially or load the next state indicated by NST.

**FIGURE 5-33:**
**Microprogrammed System with Single Address Microcode**

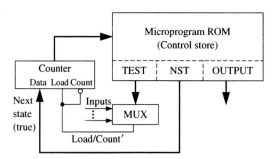

The state assignment for the single-address microcoding has to be done carefully. (In contrast, in the two-address microcoding that was discussed earlier, any state assignment was acceptable.) In the current technique, the assignments should meet the condition that for every state, one of the next states should be current state's assignment incremented by one (the default next state). For each condition box, for the false branch, the next state must be assigned in sequence, if possible. If this is not possible, extra states (called $X$-states) must be added. The required number of

$X$-states can be reduced by assigning long strings of states in sequence. To facilitate this, it may be necessary to complement some of the variables that are tested.

Figure 5-34 illustrates the modified SM chart for a binary multiplier with a serial state assignment for single address microcoding. For state $S_0$, input $St$ is complemented, so that $S_{01}$ can be the default next state, as in Figure 5-34(a). If input $St$ is not complemented, an extra state will be required as in Figure 5-34(b). State $S_2$ is the default successor for state $S_1$. In state $S_2$, we use $K'$, so that $S_3$ can be the default successor to $S_2$. Thus, in Figure 5-34(a), states $S_0$, $S_{01}$, $S_1$, $S_2$, and $S_3$ can be assigned sequential values from 0 to 4. The explicit next state, corresponding to the qualifier being true, can have any assignment. We assign 5 to state $S_{11}$. If variable $K$ was used instead of $K'$, an extra state would be required on the path from $S_2$ to $S_1$, when $K$ equals 0. As Figure 5-34(b) illustrates, two extra states will be required if input variables cannot be complemented.

**FIGURE 5-34:**
**Modified SM Chart for Binary Multiplier with Serial State Assignment for Single-Address Microcoding**

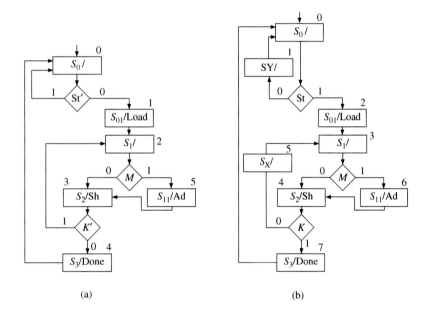

(a)                                        (b)

Table 5-4 illustrates the single-address microprogram for the multiplier. The modified SM chart, with the minimum number of states (Figure 5-34(a)), is used. Since there are three inputs, $St'$, $M$, and $K'$, a 4-to-1 MUX will be sufficient to select the appropriate qualifier. The multiplexer connections are assumed to be as in Figure 5-35.

**FIGURE 5-35:**
**Multiplexer for Microprogramming the Multiplier (Single-Address Microcode)**

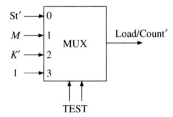

TABLE 5-4:
Single-Address
Microprogram for
Multiplier (Only
NST Specified)

| State | ABC | TEST | NST | Load | Ad | Sh | Done |
|-------|-----|------|-----|------|-----|-----|------|
| $S_0$ | 000 | 00 | 000 | 0 | 0 | 0 | 0 |
| $S_{01}$ | 001 | 11 | 010 | 1 | 0 | 0 | 0 |
| $S_1$ | 010 | 01 | 101 | 0 | 0 | 0 | 0 |
| $S_2$ | 011 | 10 | 010 | 0 | 0 | 1 | 0 |
| $S_3$ | 100 | 11 | 000 | 0 | 0 | 0 | 1 |
| $S_{11}$ | 101 | 11 | 011 | 0 | 1 | 0 | 0 |

The single-address microprogram in Table 5-4 consists of six entries of 9 bits each in contrast to the two-address microprogram in Table 5-3, which needs six entries of 12 bits each.

If the multiplier controller is implemented by a standard ROM (LUT) method, the ROM size must be 32 × 6. There are four states, necessitating two flip-flops and two next state equations. There are three inputs *St, M*, and *K*. Hence, the state table for this state machine will have 32 rows. There will be two next state equations and four outputs, necessitating 6 bits in each entry. A comparison of the ROM (LUT) method with the microcoded implementations is shown in Table 5-5. If the state machine had a large number of inputs, the size of the ROM in naïve LUT method will be prohibitively large.

TABLE 5-5:
Comparison of
Different
Implementations of
the Multiplier
Control

| Method | Size of ROM | |
|--------|------------------------|----------|
| | # entries × width | # bits |
| ROM method with original SM chart | 32 × 6 | 192 bits |
| Two-address microcode | 6 × 12 | 72 bits |
| Single-address microcode | 6 × 9 | 54 bits |

### 5.5.3 Microprogramming the Dice Controller

Let us realize the dice controller that we described earlier by microprogramming. It can be microprogrammed using two-address microcoding or single-address microcoding.

#### Two-Address Microcode Implementation for the Dice Controller

We first discuss the two-address microcoding of the dice controller using the hardware arrangement in Figure 5-29. In order to perform microcoding, we need to modify the SM chart. First, all the outputs must be converted to Moore outputs. Second, only one input variable must be tested in each state. This corresponds directly to the block diagram of Figure 5-29, since the TEST field can select only one input to test in each state and the output depends only on the state. Figure 5-36 shows a modified version of the dice game SM chart.

Next, we derive the microprogram (Table 5-6) using a straight binary state assignment. The variables $Rb$, $D_{711}$, $D_{2312}$, $Eq$, $D_7$, and *Reset* must be tested, so we will

**FIGURE 5-36: SM Chart with Moore Outputs and One Qualifier per State**

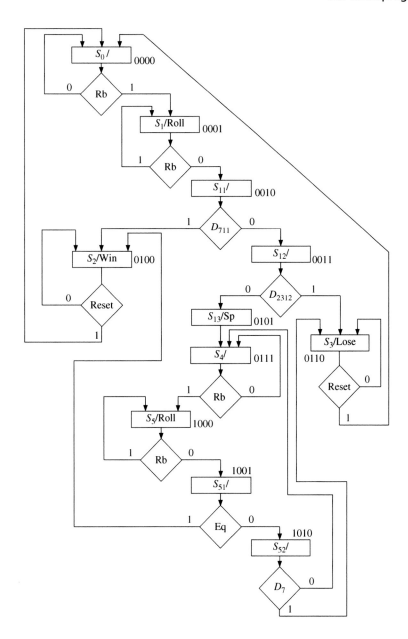

use an 8-to-1 MUX (Figure 5-37). When $TEST = 001$, $Rb$ is selected, and so on. In state $S_{13}$ the next state is always 0111, so $NSF = NST = 0111$ and the TEST field is a "don't care." Each row in the ROM table corresponds to a link path on the SM chart. For example, in $S_2$, the test field 110 selects *Reset*. If $Reset = 0$, $NSF = 0100$ is selected, and if $Reset = 1$, $NST = 0000$ is selected. In $S_2$, the output *Win* = 1 and the other outputs are 0.

TABLE 5-6:
Two-Address
Microprogram for
Dice Game

| State | ABCD | TEST | NSF | NST | Roll | Sp | Win | Lose |
|-------|------|------|------|------|------|------|------|------|
| $S_0$ | 0000 | 001 | 0000 | 0001 | 0 | 0 | 0 | 0 |
| $S_1$ | 0001 | 001 | 0010 | 0001 | 1 | 0 | 0 | 0 |
| $S_{11}$ | 0010 | 010 | 0011 | 0100 | 0 | 0 | 0 | 0 |
| $S_{12}$ | 0011 | 011 | 0101 | 0110 | 0 | 0 | 0 | 0 |
| $S_2$ | 0100 | 110 | 0100 | 0000 | 0 | 0 | 1 | 0 |
| $S_{13}$ | 0101 | xxx | 0111 | 0111 | 0 | 1 | 0 | 0 |
| $S_3$ | 0110 | 110 | 0110 | 0000 | 0 | 0 | 0 | 1 |
| $S_4$ | 0111 | 001 | 0111 | 1000 | 0 | 0 | 0 | 0 |
| $S_5$ | 1000 | 001 | 1001 | 1000 | 1 | 0 | 0 | 0 |
| $S_{51}$ | 1001 | 100 | 1010 | 0100 | 0 | 0 | 0 | 0 |
| $S_{52}$ | 1010 | 101 | 0111 | 0110 | 0 | 0 | 0 | 0 |

FIGURE 5-37: MUX
for Two-Address
Microcoding of
Dice Game

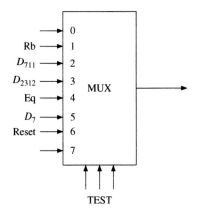

## Single-Address Microcode for the Dice Controller

Single-address microcode will use the hardware as in the block diagram of Figure 5-33. This circuit uses a counter instead of the state register. Only one target, the NST field, is specified. The TEST field selects one of the inputs to be tested in each state. If the selected input is 1 (true), the NST field is loaded into the counter. If the selected input is 0, the counter is incremented.

This method requires that the SM chart be modified, as shown in Figure 5-38, and that the state assignment be made in a serial fashion. If serial state assignment is not possible, extra states are added. The required number of $X$-states can be reduced by assigning long strings of states in sequence. To facilitate this, it may be necessary to complement some of the variables that are tested. In Figure 5-38, *Rb* and *Reset* have each been complemented in two places, and the 0 and 1 branches have been interchanged accordingly. With this change, states 0000, 0001, ..., 1000 are in sequence. $S_3$ has been assigned 1001, and before adding an $X$-state, *NSF* was 0000 and *NST* was 1001, so neither next state was in sequence. Therefore, $X$-state $S_X$ was added with a sequential assignment 1010; the next state of $S_X$ is always 0000. If we assign 1011 to $S_2$, the next states would be 1011 and 0000, and neither next state would be in sequence. We could solve the problem by adding an $X$-state. A better approach is to

**FIGURE 5-38:** **SM Chart with Serial State Assignment and Added *X*-State**

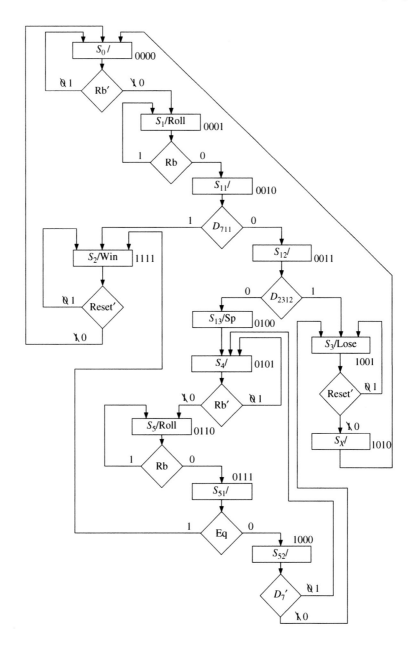

assign 1111 to $S_2$, as shown. Since incrementing 1111 goes to 0000, one of the next states is in sequence, and no *X*-state is required.

The inputs tested by the MUX in Figure 5-39 are similar to Figure 5-37, except $D_7$ and *Reset* have been complemented, and both *Rb* and *Rb'* are needed. Since NST is always 0000 in state $S_x$, a 1 input to multiplexer is needed. The corresponding microprogram ROM table is given in Table 5-7.

FIGURE 5-39: **MUX for Single-Address Microcoding of Dice Game**

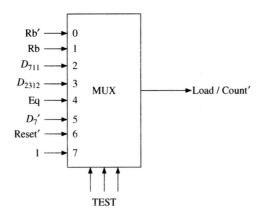

TABLE 5-7: **Microprogram for Dice Game with Single-Address Microcoding**

| State | ABCD | TEST | NST | Roll | Sp | Win | Lose |
|-------|------|------|------|------|-----|------|------|
| $S_0$ | 0000 | 000 | 0000 | 0 | 0 | 0 | 0 |
| $S_1$ | 0001 | 001 | 0001 | 1 | 0 | 0 | 0 |
| $S_{11}$ | 0010 | 010 | 1111 | 0 | 0 | 0 | 0 |
| $S_{12}$ | 0011 | 011 | 1001 | 0 | 0 | 0 | 0 |
| $S_{13}$ | 0100 | 111 | 0101 | 0 | 1 | 0 | 0 |
| $S_4$ | 0101 | 000 | 0101 | 0 | 0 | 0 | 0 |
| $S_5$ | 0110 | 001 | 0110 | 1 | 0 | 0 | 0 |
| $S_{51}$ | 0111 | 100 | 1111 | 0 | 0 | 0 | 0 |
| $S_{52}$ | 1000 | 101 | 0101 | 0 | 0 | 0 | 0 |
| $S_3$ | 1001 | 110 | 1001 | 0 | 0 | 0 | 1 |
| $S_x$ | 1010 | 111 | 0000 | 0 | 0 | 0 | 0 |
| $S_2$ | 1111 | 110 | 1111 | 0 | 0 | 1 | 0 |

A comparison of the naïve LUT (ROM) method implementation with the microprogrammed implementations is given in Table 5-8. The ROM method with original SM chart (Figure 5-13) needs $2^9$ entries because it needs three state variables and six inputs. The three next state variables and four outputs necessitate 7 bits in each entry. The two-address microcode entry is based on Table 5-6 and the single-address microcode entry is based on Table 5-7.

TABLE 5-8: **Comparison of Different Implementations of Dice Controller**

| Method | Size of ROM | |
|--------|-------------------------|-----------|
| | # entries × width | # bits |
| ROM method with original SM chart | 512 × 7 | 3584 bits |
| Two-address microcode | 11 × 15 | 165 bits |
| Single-address microcode | 12 × 11 | 132 bits |

The methods we have just studied for implementing SM charts are examples of microprogramming. The counter in Figure 5-33 is analogous to the program counter in a computer, which provides the address of the next instruction to be executed. The ROM output is a **microinstruction**, which is executed by the remaining hardware. Each microinstruction is like a conditional branch instruction that tests an input and branches to a different address if the test is true; otherwise, the next instruction in sequence is executed. The output field in the microinstruction has bits that control the operation of the hardware.

● ● ● ● ● ● ● ● ● ● ● ●

## 5.6 Linked State Machines

When a sequential machine becomes large and complex, it is desirable to divide the machine up into several smaller machines that are linked together. Each of the smaller machines is easier to design and implement. Also, one of the submachines may be "called" in several different places by the main machine. This is analogous to dividing a large software program into procedures that are called by the main program.

Figure 5-40 shows the SM charts for two serially linked state machines. The main machine (machine A) executes a sequence of "some states" until it is ready to call the submachine (machine B). When state $S_A$ is reached, the output signal ZA activates machine B. Machine B then leaves its idle state and executes a sequence of "other states." When it is finished, it outputs ZB before returning to the idle state. When machine A receives ZB, it continues to execute "other states." Figure 5-40 assumes that the two machines have a common clock.

**FIGURE 5-40: SM Charts for Serially Linked State Machines**

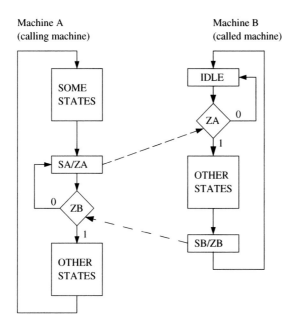

As an example of using linked state machines, we split the SM chart of Figure 5-13 into two linked SM charts. In Figure 5-13, $Rb$ is used to control the roll of the dice in states $S_0$ and $S_1$ and in an identical way in states $S_4$ and $S_5$. Since this function is repeated in two places, it is logical to use a separate machine for the roll control (Figure 5-41(b)). Use of the separate roll control allows the main dice control

(Figure 5-41(a)) to be reduced from six states to four states. The main control generates an *En_roll* (enable roll) signal in $T_0$ and then waits for a *Dn_roll* (done rolling) signal before continuing. Similar action occurs in $T_1$. The roll control machine waits in state $S_0$ until it gets an *En_roll* signal from the main dice game control. Then, when the roll button is pressed ($Rb = 1$), the machine goes to $S_1$ and generates a *Roll* signal. It remains in $S_1$ until $Rb = 0$, in which case the *Dn_roll* signal is generated, and the machine goes back to state $S_0$.

**FIGURE 5-41:**
**Linked SM Charts**
**for Dice Game**

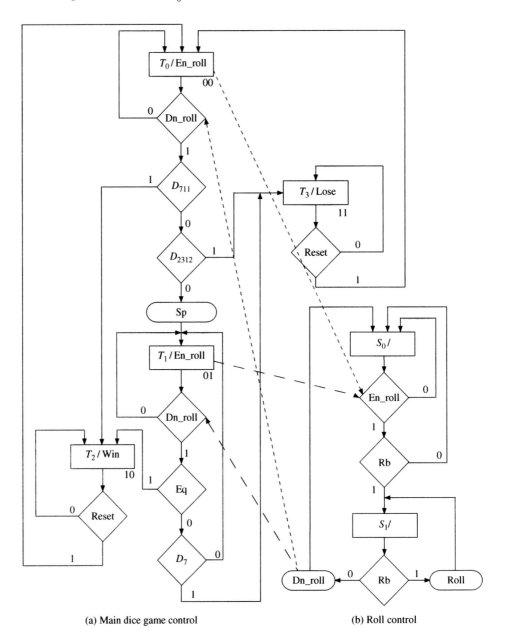

(a) Main dice game control  (b) Roll control

In this chapter we described a procedure for digital system design based on SM charts. An SM chart is equivalent to a state graph, but it is usually easier to understand the system operation by inspection of the SM chart. After we have drawn a block diagram for a digital system, we can represent the control unit by an SM chart. Next we can write a behavioral VHDL description of the system based on this chart. Using a test bench written in VHDL, we can simulate the VHDL code to verify that the system functions according to specifications. After making any necessary corrections to the VHDL code and SM chart, we can proceed with the detailed logic design of the system. Rewriting the VHDL architecture to describe the system operation, in terms of control signals and logic equations, allows us to verify that our design is correct.

We also presented techniques for implementing control units: hardwiring and microprogramming. We showed how logic equations can easily be derived by tracing link paths on an SM chart. Hardwired control units can easily be implemented from these equations. Then we presented microprogramming. In this technique, control words are stored in the microprogram memory. The size of the microprogram is reduced by transforming the SM chart into a form in which only one input is tested in each state. For complex systems, we can split the control unit into several sections by using linked state machines.

● ● ● ● ● ● ● ● ● ● ● ●
# Problems

5.1 **(a)** Construct an SM chart equivalent to the following state table. Test only one variable in each decision box. Try to minimize the number of decision boxes.
**(b)** Write a VHDL description of the state machine based on the SM chart.

| Present State | $X_1X_2 =$ | Next State 00 | 01 | 10 | 11 | $X_1X_2 =$ | Output ($Z_1Z_2$) 00 | 01 | 10 | 11 |
|---|---|---|---|---|---|---|---|---|---|---|
| $S_0$ | | $S_3$ | $S_2$ | $S_1$ | $S_0$ | | 00 | 10 | 11 | 01 |
| $S_1$ | | $S_0$ | $S_1$ | $S_2$ | $S_3$ | | 10 | 10 | 11 | 11 |
| $S_2$ | | $S_3$ | $S_0$ | $S_1$ | $S_1$ | | 00 | 10 | 11 | 01 |
| $S_3$ | | $S_2$ | $S_2$ | $S_1$ | $S_0$ | | 00 | 00 | 01 | 01 |

5.2 Construct an SM chart that is equivalent to the following state table. Test only one variable in each decision box. Try to minimize the number of decision boxes. Show Mealy and Moore outputs on the SM chart.

| Present State | $X_1X_2 =$ | Next State 00 | 01 | 10 | 11 | $X_1X_2 =$ | Output ($Z_1Z_2Z_3$) 00 | 01 | 10 | 11 |
|---|---|---|---|---|---|---|---|---|---|---|
| $S_0$ | | $S_1$ | $S_1$ | $S_1$ | $S_1$ | | 000 | 100 | 110 | 010 |
| $S_1$ | | $S_1$ | $S_1$ | $S_0$ | $S_0$ | | 001 | 001 | 001 | 001 |

5.3 An association has 15 voting members. Executive meetings of this association can be held only if more than half (i.e., at least 8) the members are present (i.e., 8 is the minimum quorum required to hold meetings). Classified matters can be discussed and voted on only if two-thirds the members are present. The chairman can cast two votes if the quorum is met, but an even number of members (including the chairman) are present. Above the room door there are three lights, GREEN, BLUE, and RED, to indicate the quorum status. Derive an SM chart for a system that will indicate whether minimum quorum is met (GREEN), classified matters can be discussed (BLUE), or quorum met, but even members (RED). GREEN and RED lights may be present at the same time or GREEN, BLUE, and RED lights may be present simultaneously.

Assume that there is a single door to the meeting room and that it is fitted with two photocells. One photocell (PHOTO1) is on the inner side of the door and the other (PHOTO2) is on the outer side. Light beams shine on each photocell, producing a false output from the cell; a true output from a photocell arises when the light beam is interrupted. Assume that once a person starts through a door, the process is completed before another one can enter or leave (i.e., only one person enters or leaves at a time). If PHOTO1 is followed by PHOTO2, a sequencer generates a *LEAVE* signal and if PHOTO2 is followed by PHOTO1, the sequencer generates an *ENTER* signal. At most one *ENTER* or *LEAVE* will be true at any time. Assume that these signals will be true until you read them. Basically you read the signal and provide a signal to the door controller indicating that the door is *READY* to let the next person in or out.

**(a)** Draw a block diagram for the data section of this circuit. Assume that *ENTER* and *LEAVE* signals are available for you (i.e., you do not need to generate them for this part of the question).

**(b)** Draw an SM chart for the controller. Write the steps required to accomplish the design. Define all control signals used.

**(c)** Draw an SM chart for a circuit that generates *ENTER* and *LEAVE*.

5.4 **(a)** Draw the block diagram for a divider that divides an 8-bit dividend by a 5-bit divisor to give a 3-bit quotient. The dividend register should be loaded when $St = 1$.

**(b)** Draw an SM chart for the control circuit.

**(c)** Write a VHDL description of the divider based on your SM chart. Your VHDL should explicitly generate the control signals.

**(d)** Give a sequence of simulator commands that would test the divider for the case 93 divided by 17.

5.5 Draw an SM chart for the BCD to binary converter of Problem 4.13.

5.6 Draw an SM chart for the square root circuit of Problem 4.14.

5.7 Draw an SM chart for the binary multiplier of Problem 4.22.

5.8 Design a binary-to-BCD converter that converts a 10-bit binary number to a 3-digit BCD number. Assume that the binary number is $\leq 999$. Initially the binary number is placed in register B. When a $St$ signal is received, conversion to BCD takes place, and the resulting BCD number is stored in the A register (12 bits). Initially A contains

0000 0000 0000. The conversion algorithm is as follows: If the digit in any decade of A is $\geq 0101$, add 0011 to that decade. Then shift the A register together with the B register one place to the left. Repeat until 10 shifts have occurred. At each step, as the left shift occurs, this effectively multiplies the BCD number by 2 and adds in the next bit of the binary number.

**(a)** Illustrate the algorithm by converting 100011101 to BCD.
**(b)** Draw the block diagram of the binary-to-BCD converter. Use a counter to count the number of shifts. The counter should output a signal $C_{10}$ after 10 shifts have occurred.
**(c)** Draw an SM chart for the converter (three states).
**(d)** Write a VHDL description of the converter.

5.9 Design a multiplier for 16-bit binary integers. Use a design similar to Figures 4-33 and 4-34.

**(a)** Draw the block diagram. Add a counter to the control circuit to count the number of shifts.
**(b)** Draw the SM chart for the controller (three states). Assume that the counter outputs $K = 1$ after 15 shifts have occurred.
**(c)** Write VHDL code for your design.

5.10 The block diagram for an elevator controller for a building with two floors is shown below. The inputs $FB_1$ and $FB_2$ are floor buttons in the elevator. The inputs $CALL_1$ and $CALL_2$ are call buttons in the hall. The inputs $FS_1$ and $FS_2$ are floor switches that output a 1 when the elevator is at the first or second floor landing. Outputs $UP$ and $DOWN$ control the motor, and the elevator is stopped when $UP = DOWN = 0$. $N_1$ and $N_2$ are flip-flops that indicate when the elevator is needed on the first or second floor. $R_1$ and $R_2$ are signals that reset these flip-flops. $DO = 1$ causes the door to open, and $DC = 1$ indicates that the door is closed. Draw an SM chart for the elevator controller (four states).

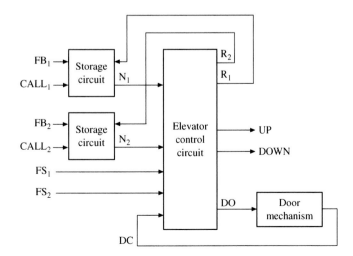

5.11 Write a test bench for the elevator controller of Problem 5.10. The test bench has two functions: to simulate the operation of the elevator (including the door operation) and to provide a sequence of button pushes to test the operation of the controller.

To simulate the elevator: If the elevator is on the first floor ($FS_1 = 1$) and an *UP* signal is received, wait 1 second and turn off $FS_1$; then wait 10 seconds and turn on $FS_2$; this simulates the elevator moving from the first floor to the second. Similar action should occur if the elevator is on the second floor ($FS_2 = 1$) and a *DOWN* signal is received. When a door open signal is received ($DO = 1$), set door closed ($DC$) to 0, wait 5 seconds, and then set $DC = 1$.

Test sequence: CALL1, 2, FB2, 4, FB1, 1, CALL2, 10, FB2.

Assume each button is held down for 1 s and then released. The numbers between buttons are the delays in seconds between button pushes; this delay is in addition to the 1 s the button is held down.

Complete the following test bench:

```
entity test_el is
end test_el;

architecture eltest of test_el is
   component elev_control
      port(CALL1, CALL2, FB1, FB2, FS1, FS2, DC, CLK: in bit;
           UP, DOWN, DO: out bit);
   end component;
```

5.12 For the following SM chart:

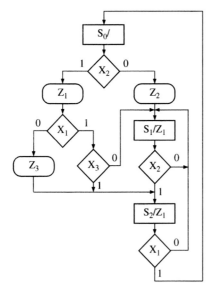

**(a)** Draw a timing chart that shows the clock, the state ($S_0$, $S_1$, or $S_2$), the inputs ($X_1$, $X_2$, and $X_3$), and the outputs. The input sequence is $X_1 X_2 X_3 = 011, 101, 111,$

010, 110, 101, 001. Assume that all state changes occur on the rising edge of the clock, and the inputs change between clock pulses.

**(b)** Use the state assignment $S_0: AB = 00; S_1: AB = 01; S_2: AB = 10$. Derive the next state and output equations by tracing link paths. Simplify these equations using the don't care state $(AB = 11)$.

**(c)** Realize the chart using a PLA and D flip-flops. Give the PLA table (state transition table).

**(d)** If a ROM is used instead of a PLA, what size ROM is required? Give the first five rows of the ROM table. Assume a naïve ROM method is used (i.e., a full look-up table).

5.13 For the given SM chart:

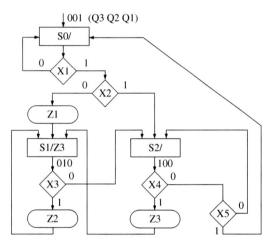

**(a)** Complete the following timing diagram (assume that $X_1 = 1$, $X_2 = 0$, $X_3 = 0$, $X_5 = 1$, and $X_4$ is as shown). Flip-flops change state on falling edge of clock.

Clock

X4

Q2

Q3

Z3

**(b)** Using the given one-hot state assignment, derive the minimum next state and output equations by inspection of the SM chart.

**(c)** Write a VHDL description of the digital system.

5.14 **(a)** Draw an SM chart that is equivalent to the state graph of Figure 4-46.

**(b)** If the SM chart is implemented using a PLA and three flip-flops $(A, B, C)$, give the PLA table (state transition table). Use a straight binary state assignment.

**(c)** Give the equation for $A^+$ determined by inspection of the PLA table.

**(d)** If a one-hot state assignment is used, give the next-state and output equations.

5.15 **(a)** Write VHDL code that describes the following SM chart. Assume that state changes occur on the falling edge of the clock. Use two processes.

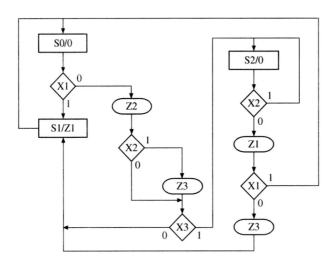

**(b)** The SM chart is to be implemented using a PLA and two flip-flops $(A$ and $B)$. Complete the following state transition table (PLA table) by tracing link paths. Find the equation for $A^+$ by inspection of the PLA table.

$$A \quad B \quad X_1 \quad X_2 \quad X_3 \mid A^+ \quad B^+ \quad Z_1 \quad Z_2 \quad Z_3$$

**(c)** Complete the following timing diagram.

| | |
|---|---|
| Clock | |
| X1 | |
| X2 | |
| X3 | |
| State S0 | |
| Z1 | |
| Z2 | |
| Z3 | |

5.16 Realize the following SM chart using a ROM with a minimum number of inputs, a multiplexer, and a loadable counter (like the 74163). The ROM should generate NST. The multiplexer inputs are selected as shown in the table beside the SM chart.

**(a)** Draw the block diagram.

**(b)** Convert the SM chart to the proper format. Add a minimum number of extra states.

(c) Make a suitable state assignment and give the first five rows of the ROM table.
(d) Write a VHDL description of the system using a ROM.

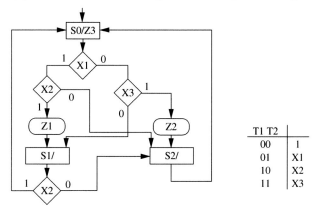

| T1 T2 | |
|---|---|
| 00 | 1 |
| 01 | X1 |
| 10 | X2 |
| 11 | X3 |

5.17 Realize the SM chart of Problem 5.16 using the two-address microprogramming structure shown in Figure 5-29.

(a) Convert the SM chart to the proper form by adding a minimum number of states to the given chart.
(b) Write the microprogram required to implement the circuit.
(c) What is the size of the ROM required for microprogramming?
(d) What is the size of the ROM if no microprogram is used, but the traditional ROM method is used to implement the original SM chart?

5.18 The following SM chart is to be realized using the two-address microprogramming structure shown in Figure 5-29.

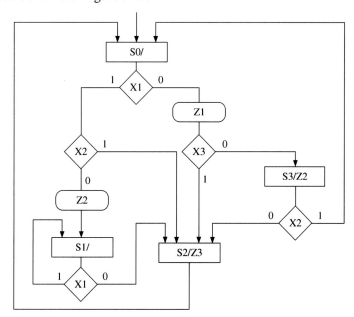

| $Q_c$ | $Q_b$ | $Q_a$ | T1 | T0 | CF | BF | AF | CT | BT | AT | Z1 | Z2 | Z3 |
|---|---|---|---|---|---|---|---|---|---|---|---|---|---|
| 0 | 0 | 0 | | | | | | | | | | | |

**(a)** Convert the SM chart to the proper form by adding a minimum number of states to the given diagram. Make a suitable state assignment.

**(b)** Write the microprogram required to implement this SM chart.

**(c)** Draw a block diagram showing how the SM chart can be realized using a ROM, multiplexers, and flip-flops.

5.19 **(a)** What are the conditions an SM chart must satisfy in order to realize it using single-address microprogramming with a counter, ROM, and multiplexer as in Figure 5-33?

**(b)** Give the modified SM chart and the required state assignment if the SM chart of Problem 5.16 is realized with this kind of microprogramming.

5.20 **(a)** What are the conditions an SM chart must satisfy in order to realize it using single-address microprogramming with a counter, ROM, and multiplexer as in Figure 5-33?

**(b)** Give the modified SM chart and the required state assignment if the SM chart of Problem 5.18 is realized with this kind of microprogramming.

5.21 Realize the SM chart given here using a ROM, a counter, and a 4-to-1 multiplexer.

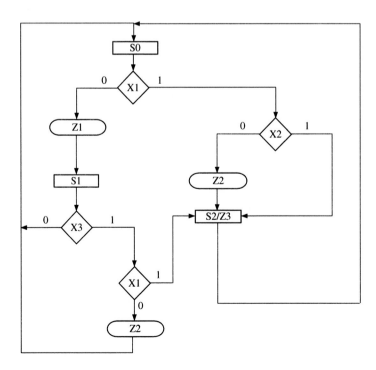

(a) Draw a block diagram. Show the MUX inputs.

(b) Change the SM chart to the proper form. Mark required changes on the given chart.

(c) Make a suitable state assignment. Give the first six rows of the ROM table.

5.22 Realize the SM chart of Problem 5.21 using the two-address microprogramming hardware structure shown in Figure 5-29.

(a) Convert the SM chart to the proper form by adding a minimum number of states to the given diagram. What are the changes needed?

(b) Write the microcode for implementing this state machine using the indicated hardware. You may indicate states in the microcode using the state names $S_0, S_1$, and so on instead of using a bit assignment. Indicate the MUX connections (inputs) necessary to understand your microcode.

(c) What is the size of the microcode ROM? Explain your calculation.

(d) If the given (original) SM chart is implemented using a traditional ROM method, how big a ROM is needed? Explain your calculation.

5.23 The following SM chart is to be realized using single-address microprogramming.

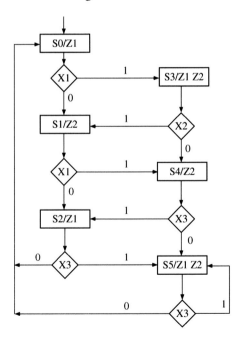

(a) Show the new SM chart and show the state assignments. The MUX inputs are 1, $X1$, $X2$, and $X3$. Do not invert inputs. Add extra states if necessary.

(b) Write the microcode for implementing this state machine using single-address microprogramming.

(c) If the given (original) SM chart is implemented using a traditional ROM method, how big a ROM is needed? Explain your calculation.

5.24 Given the following SM chart,

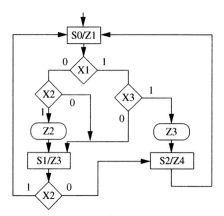

**(a)** Derive the next state and output equations, assuming the following state assignment: $S_0 = 00, S_1 = 01, S_2 = 10$.

**(b)** Convert the SM chart to a form where it can be implemented by single- address microprogramming, with only next state true (NST) specified in the microprogram. Show the new SM chart and show the new state assignments.

**(c)** Write the single-address microprogram required to implement this circuit.

**(d)** What is the size of the microprogram ROM for single-address microprogramming of the modified SM chart?

5.25 The SM charts for three linked machines are given below. All state changes occur during the falling edge of a common clock. Complete a timing chart including *ST*, *Wa*, *A*, *B*, *C*, and *D*. All state machines start in the state with an asterisk (*).

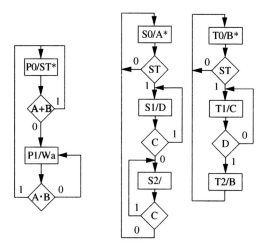

5.26 SM charts for two linked state machines are shown below. Machine *T* starts in state $T_0$, and machine *S* starts in $S_0$. Draw a timing chart that shows CLK, the states of

*T* and *S*, and signals *P, R*, and *D* for 10 clocks. All state changes occur on the rising edge of the clock.

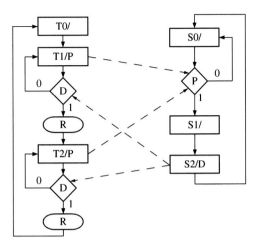

5.27 The SM charts for two linked state machines are given below.

   **(a)** Complete the timing diagram given below.
   **(b)** For the SM chart on the left, make a one-hot state assignment, and derive D flip-flop input equations and output equations by inspection.

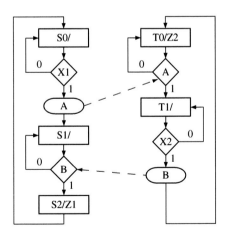

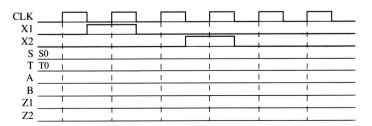

# Designing with Field Programmable Gate Arrays

This chapter describes various issues related to implementing designs in FPGAs. A few simple designs are hand-mapped into FPGA building blocks to illustrate tradeoffs arising from the structure of the basic FPGA building block. Shannon's expansion for decomposition of large functions into smaller functions is presented. Issues of the one-hot method of state assignment, which is particularly suitable for FPGA-like technology, are discussed. The design flow is described, and synthesis, mapping, and placement issues are discussed briefly. Features of several commercial FPGAs appear in discussions and examples, but we avoid presenting the entire architecture of any commercial FPGA family. Instead, the basic principles are presented in a general fashion. Once you understand the fundamentals, you will be able to refer to manufacturers' data books and Web pages for more detailed descriptions of the particular devices you want to use/understand in more detail.

## 6.1 Implementing Functions in FPGAs

Typically behavioral, RTL, or structural models of designs are created in a language such as VHDL or Verilog, and automatic CAD software is used to synthesize, map, partition, place, and route the design into an FPGA. To understand issues associated with partitioning a design into an FPGA, let us design some small components using FPGAs.

Let us assume that we want to design a 4-to-1 multiplexer using an FPGA whose logic block is represented by Figure 6-1(a). This building block contains two 4-variable function generators, X and Y, and two flip-flops. The X function generator can generate any functions of $X_1$, $X_2$, $X_3$, and $X_4$. Similarly, the Y function generator can create any function of $Y_1$, $Y_2$, $Y_3$, and $Y_4$. Latched or unlatched forms of the generated functions can be brought to the output of the logic block. The latched outputs are $QX$ and $QY$; the combinational outputs are $X$ and $Y$. Assuming that the multiplexer inputs are $I_0$, $I_1$, $I_2$, and $I_3$, and that the multiplexer selects are $S_1$ and $S_0$, the output equation for the multiplexer can be written as follows:

$$M = S_1'S_0'I_0 + S_1'S_0I_1 + S_1S_0'I_2 + S_1S_0I_3 \tag{6-1}$$

A 4-to-1 multiplexer can be decomposed into three 2-to-1 multiplexers as illustrated in Figure 6-1(b):

$$M_1 = S_0'I_0 + S_0I_1$$
$$M_2 = S_0'I_2 + S_0I_3$$

A third 2-to-1 multiplexer must now be used to create the output of the 4-to-1 multiplexer:

$$M = S_1'M_1 + S_1M_2$$

FIGURE 6-1:
(a) Example
Building Block for
an FPGA; (b) 4-to-1
Multiplexer Using
2-to-1 Multiplexers

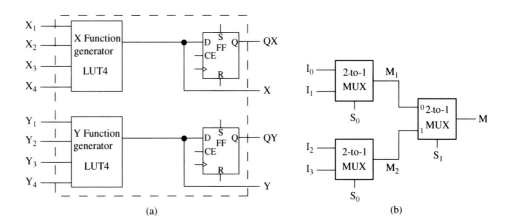

(a)                                           (b)

The output is the same as the expected output of the 4-to-1 multiplexer $(M)$. Two of the 2-to-1 multiplexers $(M_1$ and $M_2)$ can be implemented in one logic block, and a second logic block can be used to implement the third multiplexer $(M)$. Thus, two logic blocks will be required to implement a 4-to-1 multiplexer using this type of logic block. The functions generated by the first logic block are

$$X = M_1 = S_0'I_0 + S_0I_1$$
$$Y = M_2 = S_0'I_2 + S_0I_3$$

Only half of the second logic block is used. The X function generator creates the function

$$M = S_1'M_1 + S_1M_2$$

The path used by $M_1$ and $M_2$ is highlighted in Figure 6-2. The flip-flops are unused in this design.

Many modern FPGAs use a four-input look-up table (**LUT**) as a basic building block. Many designers refer to this building block as **LUT4**. It can implement a function (1-bit) of any four variables. It takes 16 bits of SRAM in order to realize the four-input LUT using the SRAM technology.

FIGURE 6-2:
**Highlighting Paths
for a 4-to-1 Mux**

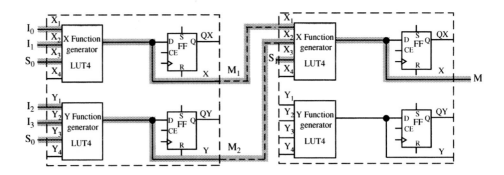

**Example**

What are the contents of the look-up tables implementing the multiplexers in Figure 6-2?

**Answer**

As illustrated in the figure, three look-up tables are used to implement functions M1, M2, and M. All of them are essentially 2-to-1 multiplexers. Assuming $X_1$ and $Y_1$ are the LSBs and $X_4$ and $Y_4$ are the MSBs of the LUT addresses, one can create the truth tables for each LUT as shown. When $S_0$ is 0, the output ($X$) equals $I_0$, and when $S_0$ is 1, the output equals $I_1$. Let us denote the three LUTs as LUT-M1, LUT-M2, and LUT-M.

| | Inputs | | | Output |
|---|---|---|---|---|
| $X_4$ | $X_3(S_0)$ | $X_2(I_1)$ | $X_1(I_0)$ | $X$ |
| x | 0 | 0 | 0 | 0 |
| x | 0 | 0 | 1 | 1 |
| x | 0 | 1 | 0 | 0 |
| x | 0 | 1 | 1 | 1 |
| x | 1 | 0 | 0 | 0 |
| x | 1 | 0 | 1 | 0 |
| x | 1 | 1 | 0 | 1 |
| x | 1 | 1 | 1 | 1 |

The MSB of each LUT is unused. The contents of the first 8 locations of the LUT should be duplicated for the next 8 locations, since irrespective of the value of $X_4$, we expect it to behave like a 2-to-1 multiplexer. Hence, the contents of LUT-M1 are the following:

LUT-M1 – 0, 1, 0, 1, 0, 0, 1, 1, 0, 1, 0, 1, 0, 0, 1, 1

Since all three LUTs in Figure 6-2 are implementing 2-to-1 multiplexers, they have identical contents for the input connections shown. The contents of the second and third LUTs are the following:

LUT-M2 – 0, 1, 0, 1, 0, 0, 1, 1, 0, 1, 0, 1, 0, 0, 1, 1

LUT-M  – 0, 1, 0, 1, 0, 0, 1, 1, 0, 1, 0, 1, 0, 0, 1, 1

Some FPGAs provide two 4-variable function generators and a method to combine the output of the two function generators. Consider the logic block in Figure 6-3. This programmable logic block has nine logic inputs ($X_1, X_2, X_3, X_4,$ $Y_1, Y_2, Y_3, Y_4,$ and $C$). It can generate two independent functions of four variables:

$$f_1(X_1, X_2, X_3, X_4) \text{ and } f_2(Y_1, Y_2, Y_3, Y_4)$$

The logic block can also generate a function $Z$, which depends on $f_1, f_2,$ and $C$. Several programmable multiplexers are used to select what is brought out at the combinational outputs ($X$out, $Y$out) and the sequential outputs ($QX, QY$). The block can generate any function of five variables in the form $Z = f_1(F_1, F_2, F_3, F_4) \cdot C' + f_2(F_1, F_2, F_3, F_4) \cdot C$. It can also generate some functions of six, seven, eight, and nine variables. A Xilinx FPGA from the past, the XC4000, uses a similar structure for its logic blocks.

**FIGURE 6-3:**
**Example Programmable Logic Block with Three Look-Up Tables**

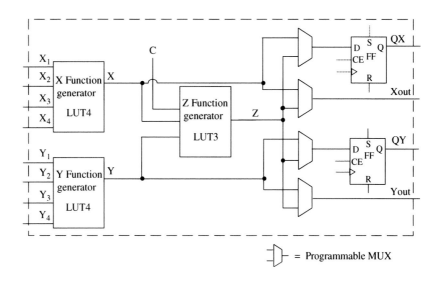

= Programmable MUX

Now consider the implementation of a 4-to-1 multiplexer using this FPGA building block. A 4-to-1 multiplexer can be implemented using a single logic block of this FPGA, as highlighted in Figure 6-4. The X function generator (LUT4) implements the function $M_1 = S_0'I_0 + S_0I_1$, the Y function generator (LUT4) implements the function $M_2 = S_0'I_2 + S_0I_3$, and the Z function generator implements the function $M = S_1'M_1 + S_1M_2$. The input $C$ is used to feed in select signal $S_1$ for use in the Z function generator. This design needs no flip-flops or latches.

Often, there are many ways to map the same design. The 4-to-1 multiplexer (shown in Figure 6-4) was generated using the $C$ input of the block. The multiplexer can be created even without using the $C$ input. The first two terms of the multiplexer's equation (Equation (6-1)) have four variables $S_0, S_1, I_0$ and $I_1$.

FIGURE 6-4:
**A 4-to-1 Multiplexer in a Programmable Logic Block with Three Function Generators**

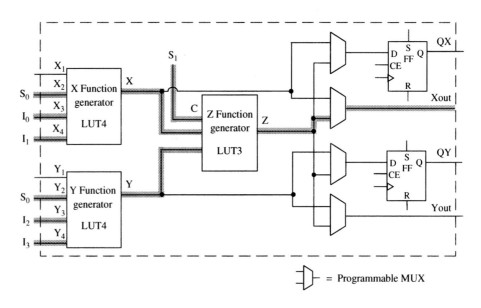

The third and fourth terms of the equation have four variables $S_0$, $S_1$, $I_2$, and $I_3$. Thus, a four-variable function generator can implement the first two terms, and another four-variable function generator can implement the third and fourth terms. However, now the outputs of the two four-variable function generators need to be combined. The Z function generator can be used for this purpose. In this case, the X function generator (LUT4) generates the function

$$F_1 = S_1'S_0'I_0 + S_1'S_0I_1 \tag{6-1a}$$

which is the first half of the function in Equation (6-1). The Y function generator (LUT4) generates the function

$$F_2 = S_1S_0'I_2 + S_1S_0I_3 \tag{6-1b}$$

which is the second half of the function in Equation (6-1). The Z function generator (LUT3) performs an OR function of the $F_1$ and $F_2$ functions

$$Z = F_1 + F_2 \tag{6-2}$$

In this case, the $C$ input is not required. This is an example of how mapping software has choices in the mapping of circuitry into resources available in the target technology.

The preceding example illustrated that it is very expensive to create multiplexers using LUTs. Three 4-input function generators (LUTs) are required to create a 4-to-1 multiplexer. Since 16 SRAM cells are required to create a four-variable function generator, 48 memory cells are required to create a 4-to-1 multiplexer using the FPGA building block in Figure 6-2.

Eight memory cells are required to create a three-variable function generator (LUT3). Hence, the multiplexer in Figure 6-3 needs 40 memory cells (16 cells for X, 16 cells for Y, and 8 cells for Z). The contents of these memory cells are part of what we need to download into the FPGA in order to program it.

When the programmable logic block of an FPGA is a large unit with the ability to realize a fairly complex multivariable function, it is possible that a large part of each logic block may go unused. Let us consider an example. Assume that we must design a 4-bit circular shift register in an FPGA, whose building block is similar to the one in Figure 6-1(a). In a circular shift register, the output of the rightmost flip-flop is fed back to the input of the leftmost flip-flop. Such a shift register is also called a **ring counter**. Since four flip-flops are required for a 4-bit shift register, two such basic building blocks will be required to realize this circuit. The four next state equations are $D_1 = Q_4$, $D_2 = Q_1$, $D_3 = Q_2$, and $D_4 = Q_3$. Two next state equations can be realized using the combinatorial function generators in one logic block. Figure 6-5b highlights the active paths for the shift register. The X function generator is used to generate $D_1 = Q_4$ and the Y function generator is used to generate $D_2 = Q_1$.

**FIGURE 6-5:**
**(a) Circular Shift Register;**
**(b) Implementation Using Simple FPGA Building Block**

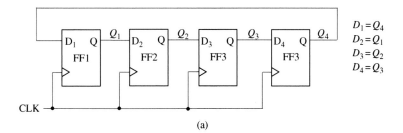

(a)

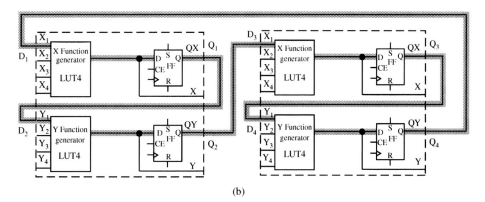

(b)

Notice that the four-variable function generators are largely unused in this example, because the next state equations for the flip-flops are rather simple; they depend only on the current state of the preceding flip-flop (i.e., a single-variable function). However, even if a function generator is used for a single-variable function, the rest of the function generator cannot be used for anything else.

*Example*

How many programmable logic blocks similar to the one in Figure 6-1(a) will be required to create a 3-to-8 decoder?

**Answer**

4. A 3-to-8 decoder has three inputs and eight outputs. Each output will need a three-variable function generator. Since what is available in the logic block in Figure 6-1(a) is a four-variable function generator, we will have to use one such function generator to create one output. Thus, eight function generators (i.e., eight 4-input LUTs) will be required to create a 3-to-8 decoder. One logic block shown in Figure 6-1(a) can generate two outputs. So four such programmable logic blocks will be required to create a 3-to-8 decoder.

If the LUTs are SRAM based, 128 SRAM cells are required to implement the 3-to-8 decoder using the LUT-based FPGA. This decoder will only need eight 3-input AND gates and three inverters, if implemented using logic gates. Thus, LUTs are very expensive for implementing certain functions.

Some FPGAs use multiplexers and gates as a basic building block. Some FPGAs (e.g., the Xilinx Spartan) provide LUTs and multiplexers. The mapping software looks at the resources available in the target technology (i.e., the specific FPGA that is used) and translates the design into the available building blocks.

## 6.2 Implementing Functions Using Shannon's Decomposition

Shannon's expansion theorem can be used to decompose functions of large numbers of variables into functions of fewer variables. In the previous section, we decomposed a 4-to-1 multiplexer into 2-to-1 multiplexers in order to implement it in a logic block with four-variable function generators. Shannon's expansion offers a general decomposition technique for any function.

Let us illustrate Shannon's decomposition for realizing any six-variable function $Z(a, b, c, d, e, f)$. First, expand the function as follows:

$$Z(a, b, c, d, e, f) = a' \cdot Z(0, b, c, d, e, f) + a \cdot Z(1, b, c, d, e, f) = a'Z_0 + aZ_1 \quad (6\text{-}3)$$

We can verify that Equation (6-3) is correct by first setting $a$ to 0 on both sides and then setting $a$ to 1 on both sides. Since the equation is true for both $a = 0$ and $a = 1$, it is always true. Equation (6-3) leads directly to the circuit of Figure 6-6(a), which uses two cells to realize $Z_0$ and $Z_1$. Half of a third cell is used to realize the three-variable function, $Z = a'Z_0 + aZ_1$.

As an example, consider the following function:

$$Z = abcd'ef' + a'b'c'def' + b'cde'f$$

Setting $a = 0$ gives

$$Z_0 = 0 \cdot bcd'ef' + 1 \cdot b'c'def' + b'cde'f = b'c'def' + b'cde'f$$

and setting $a = 1$ gives

$$Z_1 = 1 \cdot bcd'ef' + 0 \cdot b'c'def' + b'cde'f = bcd'ef' + b'cde'f.$$

Since $Z_0$ and $Z_1$ are five-variable functions, each of them needs a five-input LUT. Irrespective of the number of terms in a function, as long as there are only five variables, it can be realized by one five-input LUT. Then a 2-to-1 multiplexer or another LUT5 will be required to generate $Z$ from $Z_0$ and $Z_1$.

If only four-input LUTs are available, the five-variable functions should be further decomposed into four-variable functions. This can be done by applying Shannon's expansion theorem twice, first expanding about $a$ and then expanding about $b$. Or it can be done in one step by decomposing into four component functions as follows:

$$\begin{aligned} Z(a, b, c, d, e, f) &= a'b' \cdot Z(0, 0, c, d, e, f) + a'b \cdot Z(0, 1, c, d, e, f) \\ &\quad + ab' \cdot Z(1, 0, c, d, e, f) + ab \cdot Z(1, 1, c, d, e, f) \\ &= a'b' \cdot Y_0 + a'b \cdot Y_1 + ab' \cdot Y_2 + ab \cdot Y_3 \end{aligned} \qquad (6\text{-}4)$$

Figure 6-6(b) illustrates the realization of a general six-variable function using four-variable functions.

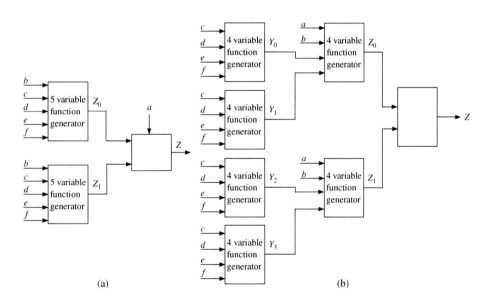

FIGURE 6-6:
**Realization of Six-Variable Functions Using (a) Five-Variable and (b) Four-Variable Function Generators**

(a)

(b)

Now let us consider the decomposition of function

$$Z = abcd'ef' + a'b'c'def' + b'cde'f$$

into four-variable functions. Let us apply Shannon's expansion around $a$ and $b$.

- Substituting    $a = b = 0$      gives    $Y_0 = c'def' + cde'f$
- Substituting    $a = 0, b = 1$    gives    $Y_1 = 0$
- Substituting    $a = 1, b = 0$    gives    $Y_2 = cde'f,$
- Substituting    $a = b = 1$      gives    $Y_3 = cd'ef'$

In a general implementation, seven 4-variable function generators will be required to implement a six-variable function as in Figure 6-6(b). However, in this example, one of the four-variable functions obtained by decomposing is the null function, which results in a simpler function:

$$Z = a'b' \cdot Y_0 + ab' \cdot Y_2 + ab \cdot Y_3$$

Five 4-variable function generators will be sufficient to implement this function, one each for $Y_0$, $Y_2$, and $Y_3$, one for generating $Z_1 = ab' \cdot Y_2 + ab \cdot Y_3$, and another one for generating $a'b' \cdot Y_0 + Z_1$. Figure 6-7 illustrates the implementation of the function $Z = abcd'ef' + a'b'c'def' + b'cde'f$, using only four-variable function generators.

**FIGURE 6-7:**
**Example Function Implementation Using Four-Variable Function Generators**

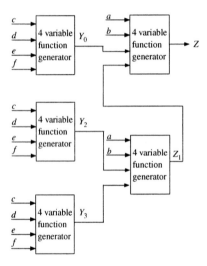

Any seven-variable function can be realized with six or fewer LUT5s. The expansion for a general seven-variable function is

$$Z(a, b, c, d, e, f, g) = a'b' \cdot Z(0, 0, c, d, e, f, g) + a'b \cdot Z(0, 1, c, d, e, f, g)$$
$$+ ab' \cdot Z(1, 0, c, d, e, f, g) + ab \cdot Z(1, 1, c, d, e, f, g)$$
$$= a'b' \cdot Y_0 + a'b \cdot Y_1 + ab' \cdot Y_2 + ab \cdot Y_3 \qquad (6\text{-}5)$$

Here $Y_0$, $Y_1$, $Y_2$, and $Y_3$ are five-variable functions of $c, d, e, f,$ and $g$. Equation (6-5) can be obtained by applying the expansion theorem twice, first expanding about $a$ and then expanding about $b$. As an example, consider the seven-variable function:

$$Z = c'de'fg + bcd'e'fg' + a'c'def'g + a'b'd'ef'g' + ab'defg'$$

- Substituting    $a = b = 0$      gives    $Y_0 = c'de'fg + c'def'g + d'ef'g'$
- Substituting    $a = 0, b = 1$    gives    $Y_1 = c'de'fg + cd'e'fg' + c'def'g$

- Substituting $a = 1, b = 0$ gives $Y_2 = c'de'fg + defg'$
- Substituting $a = b = 1$ gives $Y_3 = c'de'fg + cd'e'fg'$

This function can be implemented using six 5-variable function generators. Four of the function generators will implement the functions, $Y_0$, $Y_1$, $Y_2$, and $Y_3$. A fifth function generator implements the four-variable function, $Z_0 = a'b' \cdot Y_0 + a'b \cdot Y_1$, and the remaining function generator implements a five-variable function, $Z = Z_0 + ab' \cdot Y_2 + ab \cdot Y_3$.

Shannon's decomposition allows us to decompose an $n$-variable function into two $n - 1$ variable functions and multiplexers. As we saw in the earlier part of this chapter, it is very inefficient to realize multiplexers using LUTs. As the number of variables ($n$) increases, the number of look-up tables required to realize an $n$-variable function increases rapidly. Availability of multiplexers can greatly reduce the number of LUTs needed. For this reason, some FPGAs provide multiplexers in addition to LUT4s.

*Example*

Implement a seven-variable function using four-input LUTs and 2-to-1 multiplexers.

**Answer**

Shannon's expansion can be used to obtain the following decompositions:

7-variable function generator = two 6-variable function generators + a 2-to-1 mux . . .   (i)

6-variable function generator = two 5-variable function generators + a 2-to-1 mux . . .   (ii)

5 variable function generator = two 4-variable function generators + a 2-to-1 mux . . .   (iii)

Substituting (iii) into (ii), we obtain

6-variable function generator = four 4-variable function generators
+ three 2-to-1 muxes . . .   (iv)

Substituting (iv) into (i), we obtain

7-variable function generator = eight 4-variable function generators + seven 2-to-1 muxes

Thus a seven-variable function can be implemented as in Figure 6-8.

If only four-variable LUTs are available, a seven-variable function needs fifteen 4-variable LUTs. A 2-to-1 multiplexer is cheaper than a four-input LUT, and hence it is implemented using eight 4-input LUTs and seven 2-to-1 multiplexers in Figure 6-8.

The Xilinx Spartan FPGA is an example of an FPGA that provides multiplexers in addition to the general four-variable LUTs. A logic unit in these FPGAs is called a **slice**, and a slice may be represented in a simple fashion as in Figure 6-9. It contains two 4-input LUTs and three 2-to-1 multiplexers (plus other logic not shown here). A seven-variable function can be realized using four such slices, as in Figure 6-10. Dotted lines are used to indicate each slice.

FIGURE 6-8: **A Seven-Variable Function Using Four-Input LUTs and 2-to-1 Muxes**

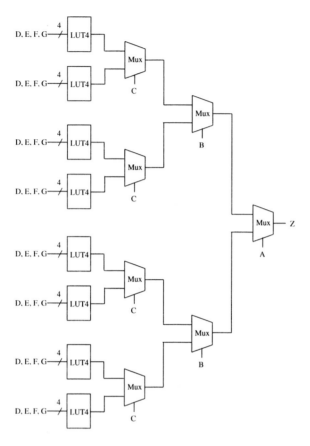

FIGURE 6-9: **Simplified View of a Xilinx Spartan Slice**

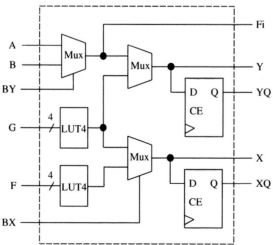

As another example, let us generate a parity function using four-variable function generators. The parity function is defined as

$$F = A \oplus B \oplus C \oplus D \oplus E$$

FIGURE 6-10:
**Implementing a
Seven-Variable
Function Using
Four Xilinx Spartan
Slices**

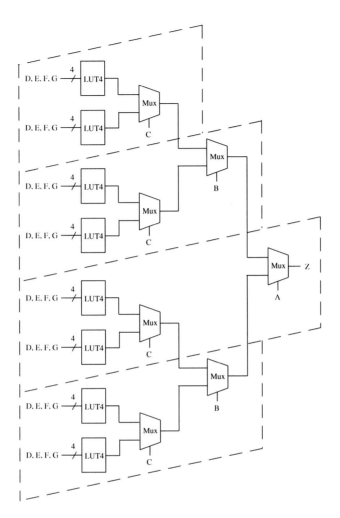

which has 16 terms when expanded to a sum of products, but it is a five-variable function. Any five-variable function can be decomposed into two 4-variable functions using Shannon's expansion and can be realized using two 4-input LUTs and a 2-to-1 multiplexer. Two four-variable function generators are sufficient for this specific function because it can be broken down into a 4-variable parity function and an XOR with the fifth variable.

● ● ● ● ● ● ● ● ● ● ● ●

# 6.3 **Carry Chains in FPGAs**

The most naïve method for creating an adder with FPGAs would be to use FPGA logic blocks to generate the sum and carry for each bit. A four-variable look-up table (which is the standard building block nowadays) can generate the sum, and another LUT4 will typically be required to realize the carry equation. The carry output from each bit has to be forwarded to the next bit using interconnect resources.

But since addition is a fundamental and commonplace operation, many FPGAs provide dedicated circuitry for generating and propagating carry bits to subsequent higher bits. Typically, a dedicated carry chain is implemented. As an example, consider the carry chain illustrated in Figure 6-11. Each LUT generates the sum bit of the corresponding input bits ($a$, $b$, and *Carry-in*). The carry chain generates the carry in parallel and feeds it using the dedicated interconnect to the LUT implementing the sum of the next bit.

FIGURE 6-11: **Carry Chains for Fast Addition**

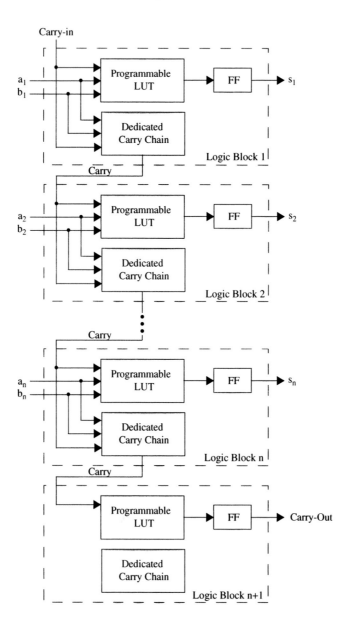

Without such a carry chain, an $n$-bit adder typically will take $2n$ logic blocks (if a logic block is an LUT4), whereas with the carry chain, $n$ logic blocks (albeit with additional dedicated circuitry) are sufficient. Dedicated circuitry generates the carry and routes it directly to the next LUT4. The hardware for the carry generation will be unused in many circuits, but because addition is a common operation, it is generally worthwhile to include such circuitry in the FPGA logic block.

# 6.4 Cascade Chains in FPGAs

Some FPGAs contain support for cascading outputs from FPGA blocks in series. The common types of cascading are the AND configuration and the OR configuration. Instead of using separate function generators to perform AND or OR functions of logic block outputs, the output from one logic block can be directly fed to the cascade circuitry to create AND or OR functions of the logic block outputs. Figure 6-12 illustrates the cascade chains in an example FPGA that uses four-input LUTs for function generation. So if an OR operation of 32 variables is desired, we can accomplish this using eight logic blocks. Each logic block will generate a four-variable OR, and the cascading OR gate can be used to OR the output from the previous logic block. Cascading AND and exclusive OR gates are also provided in some FPGAs. In look-up table–based FPGAs, these types of cascade chains may be called LUT chains.

FIGURE 6-12:
**Cascade Chain Operations**

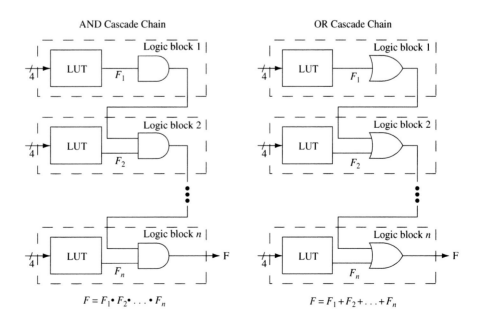

● ● ● ● ● ● ● ● ● ● ● ●

# 6.5 Examples of Logic Blocks in Commercial FPGAs

We provide three examples of commercial FPGA logic blocks. They are from Xilinx, Altera, and Actel. The Xilinx and Altera architectures both use four-variable look-up tables as their basic building block. The Actel basic block uses multiplexers and gates.

### 6.5.1 The Xilinx Configurable Logic Block

Xilinx Spartan and Virtex family FPGAs use two or four copies of a basic block called a **slice**, illustrated in Figure 6-13, to form a configurable logic block (CLB). CLB is the Xilinx terminology for the programmable logic block in Xilinx's FPGAs. Each slice contains two function generators, the G function generator and the F function generator. Additionally, there are two multiplexers, F5 and FX, for function implementation. In order to implement a four-variable LUT, 16 SRAM bits are required, so a slice contains 32 bits of SRAM in order to generate the combinational function. The F5 multiplexer can be used to combine the outputs of two 4-variable function generators to form a five-variable function generator. The select input of the

FIGURE 6-13:
**Simplified View of the Xilinx Spartan and Virtex Slice**

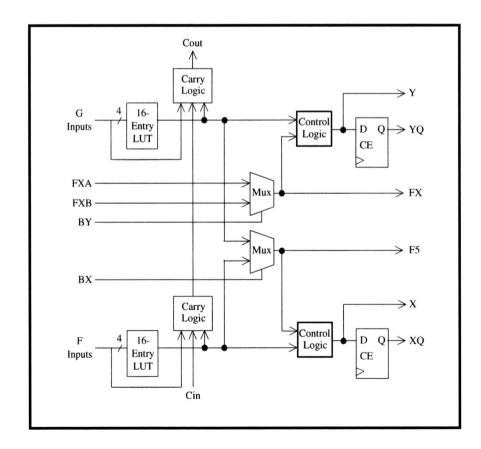

multiplexer is available to feed in the fifth input variable. All inputs of the FX multiplexer are accessible, allowing the creation of several two-variable functions. This multiplexer can be used to combine the F5 outputs from two slices to form a six-input function. Each slice also contains two flip-flops that can be configured as edge-sensitive D flip-flops or as level-sensitive latches. There is support for fast carry generation for addition. There is also additional logic to generate a few specific logic functions in addition to the general four-variable LUT.

### 6.5.2 The Altera Logic Element

Altera's name for its basic logic block is the logic element (LE). Figure 6-14 illustrates a simplified view of the logic element of the Altera Stratix FPGA. Each LE contains a four-variable LUT and a flip-flop. It can implement any function of four variables. The output can come out directly from the combinational logic or from the flip-flop. A cascade chain provides connections to adjacent LEs so that functions of more than four variables can be implemented. There is also a fast carry chain in order to allow high-speed addition. The flip-flop can be cleared or set asynchronously. Since it is a simplified view, many details are left out. The flip-flop output can be fed back as an input to the LUT. There are also additional logic gates to manipulate some of the LUT inputs.

**FIGURE 6-14:**
**Simplified View of the Altera Stratix Logic Element**

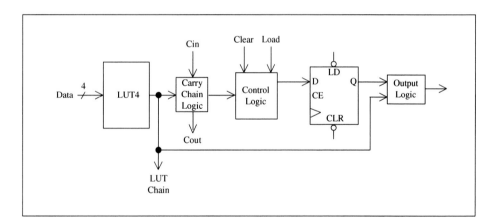

### 6.5.3 The Actel Fusion VersaTile

The building block in the Actel Fusion architecture, illustrated in Figure 6-15, consists of multiplexers and gates. Actel calls their basic block the **VersaTile**. The VeraTile block has four inputs, $X_1$, $X_2$, $X_3$, and $X_c$ as illustrated in Figure 6-15. Each VersaTile can be configured to be any of the following:

- a three-input logic function
- a latch with a clear or set
- a D-flip-flop with clear or set
- a D flip-flop with enable, clear, or set

When used as a three-input logic function, the inputs are $X_1$, $X_2$, and $X_3$. When used for the latch/flip-flop, input $X_2$ is typically used for the clock. Inputs $X_1$ and $X_c$

are used for flip-flop enable and clear signals. The logic block provides duplicate outputs tailored for fast local connections or efficient long-line connections, but for simplicity we only show one output in Figure 6-15. The VersaTile is of significantly finer grain than the four-input LUTs in many other FPGAs. The granularity of this building block is comparable to that of standard gate arrays (i.e., traditional gate arrays that are mask programmable).

**FIGURE 6-15: Simplified View of the Actel Fusion and ProASIC Logic Block (© 2006 Actel Corporation)**

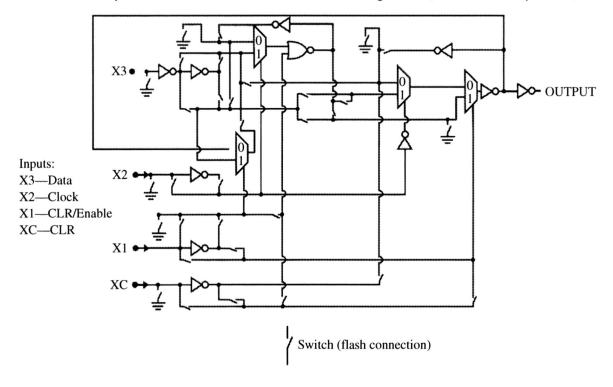

Inputs:
X3—Data
X2—Clock
X1—CLR/Enable
XC—CLR

Switch (flash connection)

## 6.6 Dedicated Memory in FPGAs

Many applications need memory. It could be for storing a table of constants to be used as coefficients during processing, or it could be for implementing instruction and data memories for an embedded processor that you are designing using the FPGA. Early FPGAs did not contain any dedicated memory. Designers typically interfaced the FPGAs to external memory chips when memory was desired. As chip densities have increased, FPGA designers started to incorporate dedicated memory on FPGA chips, eliminating the need to interface them with external memory chips.

Modern FPGAs include 16K to 10M bits of dedicated memory. Table 6-1 presents the amount of dedicated RAM in some FPGAs. As an example, the Xilinx Virtex-5 contains 1 to 10M bits of dedicated memory. Similarly, the Altera Stratix II contains

409K to 9M bits of memory. The Actel Fusion contains 27 to 270K bits of memory. The dedicated memory is typically implemented using a few (4–1000) large blocks of dedicated SRAM located in the FPGA. Figure 6-16 indicates a typical organization for the dedicated RAM blocks. In many FPGAs, they are situated outside the region of the logic block arrays (e.g., Xilinx Virtex/Spartan and Actel Fusion). In some FPGAs (e.g., Altera Stratix), there are columns of memory in a few different locations in the FPGA. In many FPGAs, the SRAM blocks are of one size (e.g., 18Kb in Xilinx Virtex). In some FPGAs, there are blocks of different sizes. For example, the Altera Stratix II has 512b, 4Kb, and 512Kb blocks). The dedicated memory on the Xilinx FPGAs is called **block RAM**. The dedicated memory on the Altera FPGAs is called **TriMatrix** memory. Some FPGAs provide parity bits in the SRAM. The parity bits are included when calculating the dedicated RAM size in the literature from some vendors; other vendors exclude the parity bits and count only the usable dedicated RAM.

**FIGURE 6-16:**
**Embedded RAMs**
**in FPGAs**

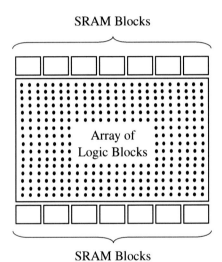

A key feature of the dedicated RAM on modern FPGAs is the ability to adjust the width of the RAM. As shown in Table 6-1, there are several tiles or blocks of memory. They can be placed in various ways to achieve different aspect ratios. Let us assume that there are 32K bits of SRAM provided as blocks of RAM. This RAM can be used as 32K × 1, 16K × 2, 8K × 4, or 4K × 8. Thus, the width of the RAM can be adjusted depending on the needs of the application. One application may need byte-wide memories; another application may need 64-bit-wide memories.

LUT-based FPGAs offer another alternative for memory. If only small amounts of memory are required, it is possible to create that memory using the bits in the LUTs (i.e., without using the dedicated memory). As you know, a four-variable LUT contains 16 bits of storage. We can create small amounts of memory by combining the storage cells from the LUTs. Two 4-input LUTs (as in Figure 6-17) can be used to create a 32 × 1 memory or a 16 × 2 memory. When used as a 32 × 1 memory, there must be five address lines and one data line (i.e., $D_1$ and $D_2$ must be connected). The top LUT must be enabled when the MSB of the address is 0, and the bottom LUT

TABLE 6-1:
**Size of Dedicated RAM in Example FPGAs**

| FPGA Family | Dedicated RAM Size (Kb) | Organization |
|---|---|---|
| Xilinx Virtex 5 | 1152–10368 | 64–576 18Kb blocks |
| Xilinx Virtex 4 | 864–9936 | 48–552 18Kb blocks |
| Xilinx Virtex-II | 72–3024 | 4–168 18Kb blocks |
| Xilinx Spartan 3E | 72–648 | 4–36 18Kb blocks |
| Altera Stratix II | 409–9163 | 104–930 512b blocks<br>78–768 4Kb blocks<br>0–9 512Kb blocks |
| Altera Cyclone II | 117–1125 | 26–250 4Kb blocks |
| Lattice SC | 1054–7987 | 56–424 18Kb blocks |
| Actel Fusion | 27–270 | 6–60 4Kb blocks |

FIGURE 6-17:
**Creating Memory from LUTs**

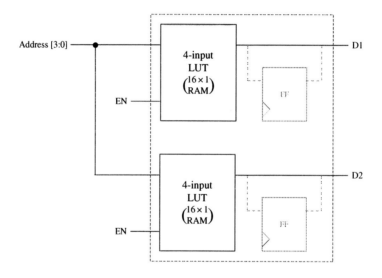

must be enabled when the MSB of the address is 1. This can be done using the highest bit of the address and an inverter. When used as a $16 \times 2$ memory, both LUTs are enabled, and data lines $D_1$ and $D_2$ are brought out in parallel. Memory created from LUT cells is called **distributed memory** (in Xilinx terminology). As the term indicates, this memory is distributed throughout the chip inside the logic blocks. A disadvantage of distributed memory is that once the LUT memory is used in this fashion, the logic block is generally unusable. The LUT memory can be used as asynchronous memory; it can also be combined with the logic block flip-flops to create synchronous memory. Table 6-2 presents the amount of LUT-based memory available in some FPGAs.

### 6.6.1 VHDL Models for Inferring Memory in FPGAs

Embedded memory on FPGAs can be instantiated using behavioral VHDL models. Memories can be synchronous or asynchronous. An asynchronous read operation means that the data from the addressed location is available on the output bus after

TABLE 6-2:
**LUT-Based RAM
in Some FPGAs**

| FPGA Family | LUT-Based RAM (Kb) | No. of LUTs |
|---|---|---|
| Xilinx Virtex 5 | 320–3420 | 19200–207,360 |
| Xilinx Virtex 4 | 96–987 | 12288–126,336 |
| Xilinx Virtex-II | 8–1456 | 512–93,184 |
| Xilinx Spartan 3E | 15–231* | 1920–29,504 |
| Altera Stratix II | 195–2242** | 12480–143,520 |
| Altera Cyclone II | 72–1069** | 4608–68,416 |
| Lattice SC | 245–1884 | 15200–115,200 |
| Lattice ECP2 | 12–136 | 6000–68,000 |

\* does not use all of the LUTs as distributed RAM

\*\* calculated from LUT counts

the access time, irrespective of the clock. In contrast, in synchronous memory, read and write control lines will have an impact only if the clock is active. In some memories, write is synchronous and read is asynchronous.

Modern synthesis tools are capable of inferring embedded memory from high-level constructs. Figure 6-18 illustrates VHDL code that creates a synchronous-write, asynchronous-read memory. The memory array is represented by an array of unsigned vectors. Since *Address* is typed as an unsigned vector, it must be converted to an integer in order to index the memory array; hence we use the IEEE.numeric_bit library and its conversion functions. A data type called *RAM* is defined as an array of 128 elements, each of which is 32 bits. The signal *DATAMEM* is of type *RAM*. The memory array is not initialized here; however, it may be initialized to any desired values. The write operation is performed inside the process, and only at the positive edge of the clock. The read operation is outside the process; hence it occurs irrespective of the clock. Synthesis using current Xilinx tools results in distributed memory for this code. Distributed memory is ideal for asynchronous memory, since the LUT generates its output asynchronously. In contrast, the code in Figure 6-19 infers block RAM. In this code sequence, the read statement appears inside the process, and read also happens only at the clock edge.

**FIGURE 6-18: Behavioral VHDL Code That Typically Infers LUT-Based Memory**

```
library IEEE;
use IEEE.numeric_bit.all;

entity Memory is
  port(Address: in unsigned(6 downto 0);
       CLK, MemWrite: in bit;
       Data_In: in unsigned(31 downto 0);
       Data_Out: out unsigned(31 downto 0));
end Memory;

architecture Behavioral of Memory is
type RAM is array (0 to 127) of unsigned(31 downto 0);
```

```
signal DataMEM: RAM;  -- no initial values
begin
  process(CLK)
  begin
    if CLK'event and CLK = '1' then
      if MemWrite = '1' then
        DataMEM(to_integer(Address)) <= Data_In;  -- Synchronous Write
      end if;
    end if;
  end process;

  Data_Out <= DataMEM(to_integer(Address));  -- Asynchronous Read
end Behavioral;
```

FIGURE 6-19: **Behavioral VHDL Code That Typically Infers Dedicated Memory**

```
library IEEE;
use IEEE.numeric_bit.all;

entity Memory is
  port(Address: in unsigned(6 downto 0);
       CLK, MemWrite: in bit;
       Data_In: in unsigned(31 downto 0);
       Data_Out: out unsigned(31 downto 0));
end Memory;

architecture Behavioral of Memory is
type RAM is array (0 to 127) of unsigned(31 downto 0);
signal DataMEM: RAM;  -- no initial values
begin
  process(CLK)
  begin
    if CLK'event and CLK = '1' then
      if MemWrite = '1' then
        DataMEM(to_integer(Address)) <= Data_In;  -- Synchronous Write
      end if;
      Data_Out <= DataMEM(to_integer(Address));  -- Synchronous Read
    end if;
  end process;
end Behavioral;
```

If the ROM method is used for implementing circuits, the synthesis tools may infer RAM in order to implement the look-up tables. As an example, consider the creation of a $4 \times 4$ multiplier using a look-up table method, as illustrated by the VHDL code in Figure 6-20. Since it uses the look-up table method, the product values for each of the input combinations are stored in a look-up table. Since the multiplicand and multiplier are 4 bits each, there are 256 possible combinations of inputs. A constant array is used to store the product array. The multiplicand is 0000 for the first 16 entries; hence the product is 0 for the first 16 entries. The multiplicand is 0001 for the next 16 entries; hence the product ranges from 0 to 15 (decimal) as the multiplier changes from 0 to 15. VHDL code for this multiplier is presented in Figure 6-20. If this code is synthesized, current Xilinx tools infer distributed RAM to store the product values. Distributed RAM is inferred to implement asynchronous reads since the LUTs in the logic blocks can continuously update the outputs as the inputs change. No clock is required. However, it might be desirable to store the arrays in the dedicated block RAM, especially if we do not want to waste LUTs for realizing memory. If the read operation is made synchronous, as in

```
process(CLK)
begin
        if CLK'event and CLK = '1' then
        Product <= PROD_ROM(to_integer(Mplier & Mcand));
                        -- read Product LUT (Synchronously)
        end if;
end process;
```

current synthesis tools from Xilinx infer dedicated block RAM to store the 256 product values.

FIGURE 6-20: **Look-Up Table–Based 4 × 4 Multiplier**

```
library IEEE;
use IEEE.numeric_bit.all;

entity LUTmult is
  port(Mplier, Mcand: in unsigned(3 downto 0);
       Product: out unsigned(7 downto 0));
end LUTmult;

architecture ROM1 of LUTmult is
type ROM is array (0 to 255) of unsigned(7 downto 0);
constant PROD_ROM: ROM: =
      (x"00", x"00", x"00", x"00", x"00", x"00", x"00", x"00", x"00", x"00", x"00", x"00", x"00", x"00", x"00", x"00",
      x"00", x"01", x"02", x"03", x"04", x"05", x"06", x"07", x"08", x"09", x"0A", x"0B", x"0C", x"0D", x"0E", x"0F",
      x"00", x"02", x"04", x"06", x"08", x"0A", x"0C", x"0E", x"10", x"12", x"14", x"16", x"18", x"1A", x"1C", x"1E",
      x"00", x"03", x"06", x"09", x"0C", x"0F", x"12", x"15", x"18", x"1B", x"1E", x"21", x"24", x"27", x"2A", x"2D",
      x"00", x"04", x"08", x"0C", x"10", x"14", x"18", x"1C", x"20", x"24", x"28", x"2C", x"30", x"34", x"38", x"3C",
      x"00", x"05", x"0A", x"0F", x"14", x"19", x"1E", x"23", x"28", x"2D", x"32", x"37", x"3C", x"41", x"46", x"4B",
      x"00", x"06", x"0C", x"12", x"18", x"1E", x"24", x"2A", x"30", x"36", x"3C", x"42", x"48", x"4E", x"54", x"5A",
      x"00", x"07", x"0E", x"15", x"1C", x"23", x"2A", x"31", x"38", x"3F", x"46", x"4D", x"54", x"5B", x"62", x"69",
      x"00", x"08", x"10", x"18", x"20", x"28", x"30", x"38", x"40", x"48", x"50", x"58", x"60", x"68", x"70", x"78",
      x"00", x"09", x"12", x"1B", x"24", x"2D", x"36", x"3F", x"48", x"51", x"5A", x"63", x"6C", x"75", x"7E", x"87",
      x"00", x"0A", x"14", x"1E", x"28", x"32", x"3C", x"46", x"50", x"5A", x"64", x"6E", x"78", x"82", x"8C", x"96",
```

```
        x"00", x"0B", x"16", x"21", x"2C", x"37", x"42", x"4D", x"58", x"63", x"6E", x"79", x"84", x"8F", x"9A", x"A5",
        x"00", x"0C", x"18", x"24", x"30", x"3C", x"48", x"54", x"60", x"6C", x"78", x"84", x"90", x"9C", x"A8", x"B4",
        x"00", x"0D", x"1A", x"27", x"34", x"41", x"4E", x"5B", x"68", x"75", x"82", x"8F", x"9C", x"A9", x"B6", x"C3",
        x"00", x"0E", x"1C", x"2A", x"38", x"46", x"54", x"62", x"70", x"7E", x"8C", x"9A", x"A8", x"B6", x"C4", x"D2",
        x"00", x"0F", x"1E", x"2D", x"3C", x"4B", x"5A", x"69", x"78", x"87", x"96", x"A5", x"B4", x"C3", x"D2", x"E1");

begin
Product <= PROD_ROM(to_integer(Mplier&Mcand));   -- read Product LUT
end ROM1;
```

## 6.7 Dedicated Multipliers in FPGAs

Many modern FPGAs provide dedicated multipliers. Suppose that a designer wants a 16 × 16 multiplier. If dedicated multipliers are not provided, several programmable logic blocks will be used to create the 16 × 16 multiplier. Such a multiplier will be expensive in terms of the number of blocks and interconnect resources used; it will also be slow due to the switches involved in interconnecting the parts of the multiplier. Dedicated multipliers will be more area-efficient and will be faster than multipliers realized using logic blocks. Since multiplication is an important operation in many applications involving FPGAs, many commercial FPGAs provide dedicated multipliers. For instance, Xilinx Virtex-4/Spartan-3, and Altera Stratix/Cyclone FPGAs contain 18 × 18 multipliers. These multipliers take two 18-bit operands and produce a 36-bit product, as illustrated in Figure 6-21. It is possible to load the multiplicand and multiplier into optional registers and load the product into an optional product register. The inputs to the multipliers can come from external pins or they can come from other logic in the FPGA.

When multiplication of numbers larger than 18 bits is required, several of the dedicated built-in multipliers can be put together. If A and B are 32 bits, and C, D, E, and F are the 16-bit components of A and B such that

$$A = C \times 2^{16} + D$$
$$B = E \times 2^{16} + F$$

then $AB = CE \times 2^{32} + (DE + CF) \times 2^{16} + DF$. This means that four multipliers to generate the partial products CE, DE, CF, and DF, and several adders to add the partial products are required.

**FIGURE 6-21:**
**Dedicated**
**Multipliers**

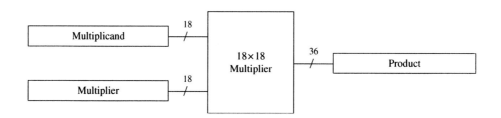

Synthesis tools are capable of inferring dedicated multipliers on FPGAs that provide them. For instance, if the VHDL code in Figure 6-22 is synthesized for Xilinx Spartan devices using Xilinx ISE tools, the synthesis tool infers four dedicated $18 \times 18$ multipliers. When the code in Figure 6-22 is synthesized, several logic blocks in the FPGA are used in addition to the four multipliers. The logic blocks are used to realize the adders for the partial products. Sixty-four I/O pins are used to provide the multiplicand and multiplier, and 64 I/O pins are used for the output. Here external pins are used to provide inputs to the multipliers, but the inputs to the multipliers may also come from the embedded memory in the FPGAs or the optional registers.

FIGURE 6-22: **VHDL Code That Infers Dedicated Multipliers**

```
library IEEE;
use IEEE.numeric_bit.all;

entity multiplier is
  port(A, B: in unsigned (31 downto 0);
       C: out unsigned (63 downto 0));
end multiplier;

architecture mult of multiplier is
begin
  C <= A * B;
end mult;
```

● ● ● ● ● ● ● ● ● ● ●
## 6.8  Cost of Programmability

The programmability in an FPGA comes with a significant amount of hardware cost. In a SRAM-based FPGA, such as the Xilinx XC4000, Virtex, and Spartan families, SRAM is used for creating the logic blocks, the programmable interconnects, and the programmable I/O blocks. The logic blocks in many modern FPGAs contain four-variable function generators. A four-variable function generator takes 16 bits of SRAM. Logic functions are realized by loading appropriate bits into the LUTs. Additionally, several multiplexers are used to select among various generated functions, to choose between latched and unlatched outputs, or to generate functions of more variables. One bit of SRAM are required to implement the select input of the 2-to-1 multiplexers, and 2 bits of SRAM are required for select lines of the programmable 4-to-1 multiplexers. Consider the logic block in Figure 6-23. The small boxes with M marked in them indicate memory cells required to program the multiplexers. A memory cell is used to select an external clock-enable signal. Another memory cell is used to invert the clock. A total of 46 memory cells

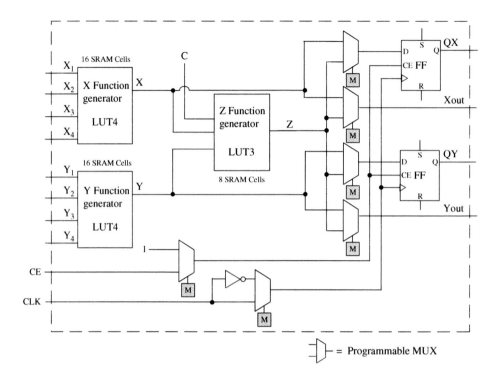

FIGURE 6-23: **Logic Block with Several Programmable SRAM Cells**

= Programmable MUX

are required to configure this logic block. The 40 memory cells in the three function generators (LUTs) might be implementing a simple one-variable function or a complex five-variable function.

We will use one more example to illustrate the overhead of programmability. Figure 6-15 illustrated a logic block of the Actel Fusion FPGA. Each switch shown in the figure needs a flash memory cell. The various flash memory cells required to program this logic block constitute the overhead of programmability of this logic block.

The I/O blocks also contain several programmable points. Consider the I/O block in Figure 6-24. Memory bits for controlling the configuration are indicated by the boxes marked with M. They are used to enable tristate output, to invert outputs, to enable the latching of output, to control the slew rate of the signal, to enable pull-up resistors, and so on.

Each SRAM cell typically takes six transistors. A flash memory cell consumes approximately 25% of an SRAM cell's area. The various programmable points add flexibility to the FPGA; however, the flexibility comes with the cost associated with the SRAM/flash memory cells. Table 6-3 shows the number of configuration bits in a few Xilinx Spartan and Virtex FPGAs. A Virtex-II FPGA, the XC2V40, which has 512 four-variable LUTs, needs 338,976 configuration bits. Another Virtex-II FPGA, the XC2V8000, has 93,184 four-variable LUTs and needs more than 26 million configuration bits. Thus, it is clear that the flexibility and programmability of the FPGA comes at a high cost.

FIGURE 6-24:
**Programmable Points in FPGA I/O Block (Indicated by Boxes with 'M')**

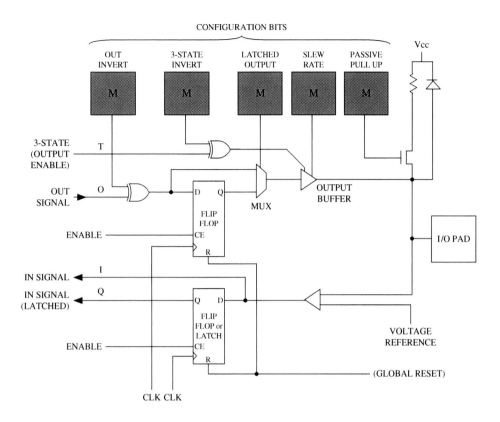

TABLE 6-3:  **Number of Configuration Bits in Example FPGAs**

| Vendor | Device Family | Device | # of Configuration Bits | # of Logic Blocks | # of LUTs | # Usable I/O Pins |
|---|---|---|---|---|---|---|
| Xilinx | Virtex-5 | XC5VLX30 | 8.4M | 4,800 | 19,200 | 400 |
| | | XC5VLX330 | 79.7M | 51,840 | 207,360 | 1200 |
| Xilinx | Virtex-II | XC2V40 | 0.3M | 256 | 512 | 88 |
| | | XC2V8000 | 26.2M | 46,592 | 93,184 | 1108 |
| Xilinx | Spartan 3E | XC3S100E | 0.6M | 960 | 1,920 | 108 |
| | | XC3S1600E | 6.0M | 14,752 | 29,504 | 376 |
| Altera | Stratix II | EP2S15 | 4.7M | 6,240 | 12,480 | 366 |
| | | EP2S180 | 49.8M | 71,760 | 143,520 | 1170 |
| Altera | Stratix | EP1S10 | 3.5M | 10,570 | 10,570 | 426 |
| | | EP1S80 | 23.8M | 79,040 | 79,040 | 1238 |
| Altera | Cyclone II | EP2C5 | 1.3M | 4,608 | 4,608 | 158 |
| | | EP2C70 | 14.3M | 68,416 | 68,416 | 622 |

● ● ● ● ● ● ● ● ● ● ● ● ●

# 6.9 FPGAs and One-Hot State Assignment

When designing with FPGAs, it may not be important to minimize the number of flip-flops used in the design. Instead, we should try to reduce the total number of logic cells used and try to reduce the interconnections between cells. In order to design faster logic, we should try to reduce the number of cells required to realize each equation. Using a *one-hot state assignment* will often help to accomplish this. One-hot assignment takes more flip-flops than encoded assignment; however, the next state equations for flip-flops are often simpler in the one-hot method than the equations in the encoded method.

The one-hot assignment uses one flip-flop for each state, so a state machine with $N$ states requires $N$ flip-flops. Exactly one flip-flop is set to 1 in each state. For example, a system with four states ($T_0$, $T_1$, $T_2$, and $T_3$) could use four flip-flops ($Q_0$, $Q_1$, $Q_2$, and $Q_3$) with the following state assignment:

$$T_0 : Q_0 Q_1 Q_2 Q_3 = 1000, \quad T_1 : 0100, \quad T_2 : 0010, \quad T_3 : 0001 \qquad (6\text{-}6)$$

The other 12 combinations are not used.

We can write next state and output equations by inspection of the state graph or by tracing link paths on an SM chart. Consider the partial state graph given in Figure 6-25. The next state equation for flip-flop $Q_3$ could be written as

$$Q_3{}^+ = X_1 Q_0 Q_1{}' Q_2{}' Q_3{}' + X_2 Q_0{}' Q_1 Q_2{}' Q_3{}' + X_3 Q_0{}' Q_1{}' Q_2 Q_3{}' + X_4 Q_0{}' Q_1{}' Q_2{}' Q_3$$

However, since $Q_0 = 1$ implies $Q_1 = Q_2 = Q_3 = 0$, the $Q_1{}' Q_2{}' Q_3{}'$ term is redundant and can be eliminated. Similarly, all the primed state variables can be eliminated from the other terms, so the next state equation reduces to

$$Q_3{}^+ = X_1 Q_0 + X_2 Q_1 + X_3 Q_2 + X_4 Q_3$$

Note that each term contains exactly one state variable. Similarly, each term in each output equation contains exactly one state variable:

$$Z_1 = X_1 Q_0 + X_3 Q_2, \qquad Z_2 = X_2 Q_1 + X_4 Q_3$$

When a one-hot assignment is used, the next state equation for each flip-flop will contain one term for each arc leading into the corresponding state (or for each link

**FIGURE 6-25:**
**Partial State Graph**

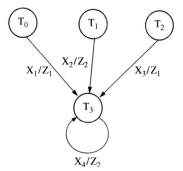

path leading into the state). In general, each term in every next state equation and in every output equation will contain exactly one state variable.

When a one-hot assignment is used, resetting the system requires that one flip-flop be set to 1 instead of resetting all flip-flops to 0. If the flip-flops used do not have a preset input (as is the case for the Xilinx 3000 series), then we can modify the one-hot assignment by replacing $Q_0$ with $Q_0{'}$ throughout. For the preceding assignment, the modification is

$$T_0: Q_0Q_1Q_2Q_3 = 0000, \qquad T_1: 1100, \qquad T_2: 1010, \qquad T_3: 1001 \qquad (6\text{-}7)$$

and the modified equations are

$$Q_3{}^+ = X_1Q_0{'} + X_2Q_1 + X_3Q_2 + X_4Q_3$$
$$Z_1 = X_1Q_0{'} + X_3Q_2, \qquad Z_2 = X_2Q_1 + X_4Q_3$$

Another way to solve the reset problem without modifying the one-hot assignment is to add an extra term to the equation for the flip-flop, which should be 1 in the starting state. If the system is reset to state 0000 after power-up, we can add the term $Q_0{'}Q_1{'}Q_2{'}Q_3{'}$ to the equation for $Q_0{}^+$. Then, after the first clock, the state will change from 0000 to 1000 ($T_0$), which is the correct starting state. In general, both an assignment with a minimum number of state variables and a one-hot assignment should be tried to see which one leads to a design with the smallest number of logic cells. Alternatively, if speed of operation is important, the design that leads to the fastest logic should be chosen. When a one-hot assignment is used, more next state equations are required, but in general both the next state and output equations will contain fewer variables. An equation with fewer variables generally requires fewer logic cells to realize. The more cells cascaded, the longer the propagation delay and the slower the operation.

• • • • • • • • • • • •

## 6.10 FPGA Capacity: Maximum Gates versus Usable Gates

You often come across gate counts of FPGAs. As you know, many FPGAs are not structured as arrays of gates. Some are simply arrays of look-up tables rather than arrays of gates. So what does the gate count of an FPGA mean?

The number of raw gates that have gone into building an FPGA is not an interesting or useful metric to an FPGA user. What is useful to the user is a count of the circuitry that can fit into a particular FPGA. This is called the **equivalent gate count**. But, as you might know, this type of achievable gate count will depend on the type of circuitry, the type of interconnections between different parts of the circuitry, and so on.

Gate counts are estimated in many different ways. An approximate equivalent gate count can be established for a logic block by considering circuits that can be implemented in a logic block, and the total gate count can be estimated by multiplying it with the number of logic blocks in the FPGA. This gate count is likely to be higher than the gate count of practical circuitry that can be realized in the FPGA. A better gate count estimate can be derived using benchmark circuits. PREP is an organization that facilitates standard benchmark circuits for ASIC and FPGA benchmarking. Assume that a particular circuitry typically takes 2000 gates in ASIC, and if an FPGA device can fit 20 copies of that circuitry, an FPGA vendor may estimate the maximum gate count of its FPGA as 40K. Since the circuit is simply replicated and no actual interconnection exists between the copies, this count is also likely to be higher than the gate count of practical circuitry that can be realized in the FPGA. Some FPGA vendors provide a typical gate count by adjusting the maximum gate count with some weighting schemes. The benchmarks gathered and distributed by PREP can be useful in the benchmarking of FPGAs.

It is very difficult to estimate gate counts of FPGAs in which logic is implemented with LUTs. A four-input LUT may be used to implement a four-variable logic function with one or more product terms, or it can be used to store 16 bits of information. When the LUTs are used as RAM, higher gate counts may be obtained. Hence, depending on the portion of LUTs used as RAM, we can estimate different gate counts for the same FPGA. Vendors often compute their "system gates" count by considering a fraction of CLBs (say, 20–30%) as RAM.

Altera provides two types of gate counts for its APEX family: maximum gates and usable gates. The APEX II devices range from 1.9 million to 5.25 million maximum gates, but the typical gate count is published as 600K to 3 million. Some FPGA vendors provide their chip capacities with a count of the logic blocks (logic elements) rather than a gate count.

---

**PREP Benchmarks**
The **Programmable Electronics Performance Company (PREP)** was a non-profit organization that gathered and distributed a series of benchmarks for programmable ASICs. The nine PREP benchmark circuits in the PREP 1.3 suite were as follows:

1. An 8-bit datapath consisting of a 4-to-1 MUX, a register, and a shift register
2. An 8-bit timer-counter consisting of two registers, a 4-to-1 MUX, a counter, and a comparator
3. A small state machine (8 states, 8 inputs, and 8 outputs)
4. A larger state machine (16 states, 8 inputs, and 8 outputs)
5. An ALU consisting of a $4 \times 4$ multiplier, an 8-bit adder, and an 8-bit register
6. A 16-bit accumulator
7. A 16-bit counter with synchronous load and enable
8. A 16-bit prescaled counter with load and enable
9. A 16-bit address decoder

PREP's online information included Verilog and VHDL source code and test benches (provided by Synplicity). PREP also made additional synthesis benchmarks available, including a bit-slice processor, multiplier, and R4000 MIPS RISC microprocessor.

• • • • • • • • • • • •
# 6.11 Design Translation (Synthesis)

In the early sections of this chapter, we hand-mapped some designs into FPGA logic blocks. This process is analogous to writing assembly language programs for microprocessors. It is tedious. Productivity of designers will be very low if they can only enter designs at that level. Just as the majority of the programs in the modern-day world are written in high-level languages like C and translated by a compiler, modern-day digital designs are done at behavioral or RTL level and translated to target devices. This applies not only for FPGAs, but also for ASIC design.

A number of CAD tools are now available that take a VHDL description of a digital system and automatically generate a circuit description that implements the digital system. The term **synthesis** refers to the translation of an abstract high-level design to a circuit description, typically in the form of a logic schematic. The input to the CAD tool is a behavioral or structural VHDL/Verilog model. The output from the synthesis tools may be a logic schematic together with an associated wirelist, which implements the digital system as an interconnection of gates, flip-flops, registers, counters, multiplexers, adders, and other basic logic blocks. This representation is called a **netlist**. The circuit can now be targeted for an FPGA, a CPLD, or an ASIC.

Typical computer-aided design flow involves the following steps:
Design translation (synthesis) and optimization
Mapping
Placement
Routing

These steps are illustrated in Figure 6-26. In this section, we describe design translation and optimization techniques. The mapping, placement, and routing of designs are described in the next section.

Even if VHDL code compiles and simulates correctly, it may not necessarily synthesize correctly. And even if the VHDL code does synthesize correctly, the resulting implementation may not be very efficient. In general, synthesis tools will accept only a subset of VHDL as input. Other changes must be made in the VHDL code so the synthesis tool "understands" the intent of the designer. Further changes in the VHDL code may be required in order to produce an efficient implementation.

In VHDL, a signal may represent the output of a flip-flop or register, or it may represent the output of a combinational logic block. The synthesis tool will attempt to determine what is intended from the context. For example, the concurrent statement

```
A <= B and C;
```

implies that $A$ should be implemented using combinational logic. On the other hand, if the sequential statements

```
wait until clock'event and clock = '1';
A <= B and C;
```

appear in a process, this implies that $A$ represents a register (or flip-flop) that changes state on the rising edge of the clock.

FIGURE 6-26: **CAD Design Flow**

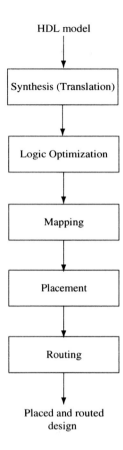

When integer signals are used, specifying the integer range is important. If no range is specified, the VHDL synthesizer may interpret an integer signal to represent a 32-bit register, since the maximum size of a VHDL integer is 32 bits. When the integer range is specified, most synthesizers will implement integer addition and subtraction using binary adders with the appropriate number of bits.

Most VHDL synthesizers do a line-by-line translation of VHDL into gates, registers, multiplexers, and other general components with very little optimization up front. Then the resulting design is optimized. Synthesizers associate particular VHDL constructs with particular hardware structures. For instance, **case** statements typically result in multiplexers. Use of '+,' '−,' and comparison results in the use of an adder, use of shift operators results in the use of a shift register, and so on.

During the initial translation of the VHDL code and during the optimization phase, the synthesis tool will select components from those available in its library. Several different component libraries may be provided to allow implementation with different technologies.

### 6.11.1 Synthesis of a Case Statement

The example of Figure 6-27 shows how the Synopsis Design Compiler implements a **case** statement using multiplexers and gates. Figure 6-27(a) shows the code. The integers *a* and *b* are each implemented with 2-bit binary numbers. Two 4-to-1 multiplexers

FIGURE 6-27: **Synthesis of a Case Statement**

```
entity case_example is
  port(a: in integer range 0 to 3;
       b: out integer range 0 to 3);
end case_example;

architecture test1 of case_example is
begin
  process(a)
  begin
    case a is
      when 0 => b <= 1;
      when 1 => b <= 3;
      when 2 => b <= 0;
      when 3 => b <= 1;
    end case;
  end process;
end test1;
```

(a) VHDL code for case example

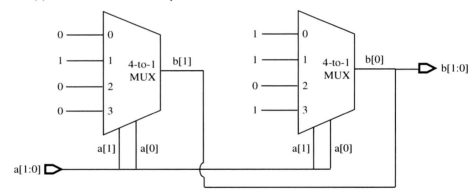

(b) Synthesized circuit before optimization

| $a_1$ | $a_0$ | $b_1$ | $b_0$ |
|---|---|---|---|
| 0 | 0 | 0 | 1 |
| 0 | 1 | 1 | 1 |
| 1 | 0 | 0 | 0 |
| 1 | 1 | 0 | 1 |

$$b_1 = a_1' \cdot a_0$$
$$= (a_1 + a_0')'$$
$$b_0 = (a_1 \cdot a_0')'$$

(c) Logic optimization

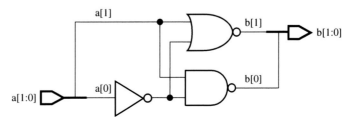

(d) Synthesized circuit after optimization

are required. The 2 bits of $a$ are used as control inputs to the multiplexer. The multiplexer inputs are hardwired to a logic 1 or a logic 0. Figure 6-27(b) shows the hardware that will be generated by a typical synthesizer.

Most modern synthesizers will also perform optimizations to reduce the logic that is generated. Because the MUX inputs are constants, elimination of the mux and several gates are possible by inspection of the truth table in Figure 6-27(c). The optimized output equations are $b_1 = a_1'a_0 = (a_1 + a_0')'$ and $b_0 = (a_1a_0)'$. An optimized circuit for the code in Figure 6-27(a) consists only of a NOR, a NAND, and a NOT gate. Figure 6-27(d) shows the resulting circuit after optimization.

## Unintentional Latch Creation

In general, when a VHDL signal is assigned a value, it will hold that value until it is assigned a new value. Because of this property, some VHDL synthesizers will infer a latch when none is intended by the designer. Figure 6-28(a) shows an example of a **case** statement that creates an unintended latch. The **case** statement results in a 4-to-1 multiplexer whose data inputs are set to the values in each case. The select lines are controlled by the value of $a$. Since the value of $b$ is not specified if $a$ is not equal to 0, 1, or 2, the synthesizer assumes that the value of $b$ should be held in a latch if $a = 3$.

When $a = 3$, the previous value of $b$ should be used as the output. This necessitates a latch whose D input $= b_0$. In order to hold the value in the latch, the latch gate control signal $G$ should be 0 when $a = 3$. Thus $G = (a_1a_0)'$. A naïve synthesizer might generate a 4-to-1 multiplexer and a latch as in Figure 6-28(c). The latch can be eliminated by replacing the word **null** in the VHDL code with b <= '0' as in Figure 6-28(b). If this change is made, most synthesizers will generate only a multiplexer and no latch.

FIGURE 6-28: **Example of Unintentional Latch Creation**

```
entity latch_example is
  port(a: in integer range 0 to 3;
       b: out bit);
end latch_example;

architecture test1 of latch_example is
begin
  process(a)
  begin
    case a is
      when 0 => b <= '1';
      when 1 => b <= '0';
      when 2 => b <= '1';
      when others => null;
    end case;
  end process;
end test1;
```
                    (a) VHDL code that infers a latch

```
entity latch_example is
  port(a: in integer range 0 to 3;
       b: out bit);
end latch_example;

architecture test1 of latch_example is
begin
  process(a)
  begin
    case a is
      when 0 => b <= '1';
      when 1 => b <= '0';
      when 2 => b <= '1';
      when 3 => b <= '0';
    end case;
  end process;
end test1;
```

(b) Modified code not resulting in latch

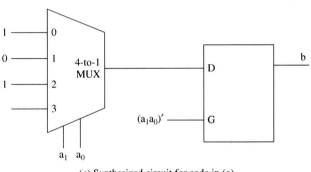

(c) Synthesized circuit for code in (a)

| $a_1$ | $a_0$ | b |
|---|---|---|
| 0 | 0 | 1 |
| 0 | 1 | 0 |
| 1 | 0 | 1 |
| 1 | 1 | previous b |

| $a_1$ | $a_0$ | b |
|---|---|---|
| 0 | 0 | 1 |
| 0 | 1 | 0 |
| 1 | 0 | 1 |
| 1 | 1 | 0 |

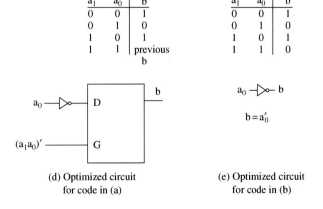

(d) Optimized circuit
for code in (a)

(e) Optimized circuit
for code in (b)

$b = a_0'$

Most modern synthesizers also perform optimizations to reduce the logic that is generated. For example, a 4-to-1 multiplexer is not required for this circuit. As easy way to derive the optimized circuit is by inspection of the truth table in Figure 6-28(d). We may easily observe that when $a$ equals 0, 1, or 2, $b = a_0'$. An optimizing synthesizer might generate a single NOT gate for the code, as in Figure 6-28(e). If the **null** statement was not removed, this optimizing synthesizer would generate a latch also as in Figure 6-28(d) (i.e., with the unintended latch).

### 6.11.2 Synthesis of if Statements

When **if** statements are used, care should be taken to specify a value for each branch. For example, if a designer writes

```
if A = '1' then Nextstate <= 3; Z<= 1;
end if;
```

he or she may intend for *Nextstate* to retain its previous value if $A \neq$ '1', and the code will simulate correctly. However, the synthesizer might interpret this code to mean if $A \neq$ '1', then *Nextstate* is unknown ('X'), and the result of the synthesis may be incorrect. Also, it will result in latches for $Z$. For this reason, it is always best to include an **else** clause in every **if** statement. For example,

```
if A = '1' then Nextstate <= 3; Z<=1;
   else Nextstate <= 2; Z<= 0;
end if;
```

is unambiguous.

The example of Figure 6-29 shows how a typical synthesizer implements an **if-then-elsif-else** statement using a multiplexer and gates. Figure 6-29(b) represents the truth table corresponding to the various input combinations. $C$ is selected if $A = 1$; $D$ is selected if $A = 0$ and $B = 0$; and $E$ is selected if $A = 0$ and $B = 1$. Figure 6-29(c) indicates the synthesized hardware. $A$ and $B$ are used as select signals of the multiplexer.

FIGURE 6-29: Synthesis of an if Statement

```
entity if_example is
  port(A, B: in bit;
       C, D, E: in bit_vector(2 downto 0);
       Z: out bit_vector(2 downto 0));
end if_example;

architecture test1 of if_example is
begin
  process(A, B)
  begin
    if A = '1' then Z <= C;
    elsif B = '0' then Z <= D;
    else Z <= E;
    end if;
  end process;
end test1;
```

(a) VHDL code for **if** example

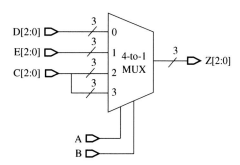

(b) Equivalent truth table

(c) Synthesized hardware for code in (a)

**Example**

What hardware does the statement

    LE <= (A <= B);

result in? Assume that A and B are 4-bit vectors.

### Answer

The result is a 4-bit comparator. Only one of the <= symbols indicates an assignment. The <= symbol between *A* and *B* is a relational operator. The right side of the assignment symbol returns a TRUE or '1' if *A* is less than or equal to *B*. Hence, if *A* is less than or equal to *B, LE* is set to '1.' Otherwise, *LE* will be '0.'

Most standard comparators come with EQUAL_TO (EQ), GREATER_THAN (GT), and LESS_THAN (LT) outputs. In this case, *LE* should be '1' if EQUAL_TO or LESS_THAN is true. Figure 6-30 illustrates the hardware.

**FIGURE 6-30: Hardware for Less Than or Equal To Checker**

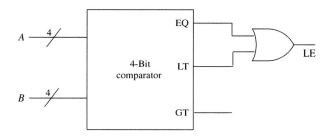

## 6.11.3 Synthesis of Arithmetic Components

CAD tools for synthesis have design libraries that include components to implement the operations defined in the numeric packages. The example of Figure 6-31 uses IEEE numeric_std library. When this code is synthesized, the result includes library components that implement a 4-bit comparator, a 4-bit binary adder with a 4-bit accumulator register, and a 4-bit counter. Some synthesis tools will implement the counter with a 4-bit adder with a "0001" input and then optimize the result to eliminate unneeded gates. The resulting hardware is shown in Figure 6-31(b).

FIGURE 6-31: **VHDL Code Example for Synthesis and Corresponding Hardware**

```
library IEEE;
use IEEE.numeric_bit.all;

entity examples is
  port(signal clock: in bit;
       signal A, B: in signed(3 downto 0);
       signal ge: out boolean;
       signal acc: inout signed(3 downto 0) := "0000";
       signal count: inout unsigned(3 downto 0) := "0000");
end examples;

architecture x1 of examples is
begin
  ge <= (A >= B);  -- 4-bit comparator
  process
  begin
    wait until clock'event and clock = '1';
    acc <= acc + B;  -- 4-bit register and 4-bit adder
    count <= count + 1;  -- 4-bit counter
  end process;
end x1;
```

(a) VHDL code

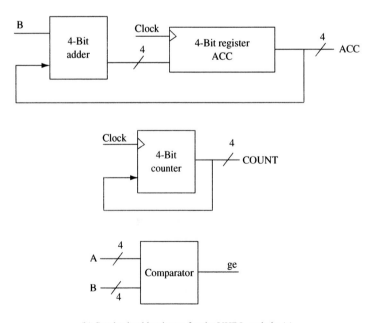

(b) Synthesized hardware for the VHDL code in (a)

*Example*

Generate optimized hardware for the following statement, assuming $A$ is a 4-bit vector:

> EQ3 <= (A = 3);

**Answer**

A 4-bit comparator can be used to realize this statement. One input to the comparator will be $A$, and the other input will be the number 3 (i.e., 0011) (binary).

But since we know that one input is constantly 3, we could optimize it further to result in an AND gate and two inverters as in Figure 6-32.

FIGURE 6-32:
**Optimized
Hardware for
Equality Checker**

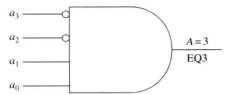

Some synthesizers will not automatically provide this optimized hardware. Under such circumstances, we can alter the VHDL source code to

> EQ3 <= **not** A(3) **and not** A(2) **and** A(1) **and** A(0);

This statement will result in the four-input AND gate of Figure 6-32.

Different kinds of optimizations are required for different target technologies. For instance, reduction in absolute number of gates is important for a gate-based target technology, but if an FPGA with LUTs is the target technology, optimization does not need to consider absolute number of gates in the design. It only needs to optimize the number of LUTs.

## 6.11.4 Area, Power, and Delay Optimizations

Most VHDL synthesizers allow the design to be optimized for maximum speed or for minimum chip area. Power consumption has also recently become a major design constraint along with area and delay. Typically, optimizing for one constraint will worsen the performance of another. For example, when improving speed, area might worsen. Improving speed often means that some operation that is being performed serially, reusing some gates, may have to be performed in parallel. Hence, often improving the speed results in increasing the number of components. Consider a serial adder, which is used to perform 4-bit addition, versus a fully parallel combinational 4-bit adder that uses a lot more hardware to achieve much better speed. When optimizing for area, an effort is made to decrease the number of components, which in turn often increases the critical path. **Critical path** means the longest delay in the circuit.

CAD tools incorporate gate libraries. The libraries provide various options for achieving requirements on area, speed, and power. Gates and building blocks that are optimized individually for area, speed, or power or collectively for two or more of these can be obtained, and depending on the designer's specifications, appropriate elements from the libraries can be used.

Area and delay of a circuit are often inversely related to each other. Energy and delay are also inversely related. The **Area-Time** (AT) product and **Energy-Delay** (ED) product are popularly used metrics to describe the quality of a circuit. **Area-Time$^2$** (AT$^2$) and **Energy-Delay$^2$** (ED$^2$) are also used as metrics to measure the quality of circuits and systems.

In spite of the inverse relationships between area and delay or energy and delay, there are optimizations that simultaneously improve area, delay, and power. For example, consider the optimizations in Figure 6-27(b) to (d) and the optimization in Figure 6-28(c) to (e). These optimizations at the logic level perform the required task in an effective way resulting in less hardware, less area, less power, and surprisingly smaller critical path, too.

When designing with FPGAs, we should keep in mind that optimizations for discrete gates are not necessarily the best optimizations for FPGAs. As an example, consider function minimization. In a SRAM FPGA, the important issue is to minimize the number of variables in an expression. Minimizing the number of terms in an equation is not required because the entire truth table is stored in LUT form.

---

**Major Vendors of CAD Tools**

Cadence
Synopsis
Mentor Graphics

**Major Vendors of FPGA CAD tools**

Xilinx
Altera
Actel

---

● ● ● ● ● ● ● ● ● ● ● ●
# 6.12  Mapping, Placement, and Routing

Once the design is translated by synthesis and the netlist is generated, the resulting design must be mapped into a specific implementation technology. Implementation technologies include gate arrays, FPGAs, CPLDs, and ASIC standard cell designs. Mapping, placement, and routing are the three major steps that happen in order to transform the design in netlist form to the appropriate target technology.

### 6.12.1 **Mapping**

Mapping is the process of binding technology-dependent circuits of the target technology to the technology-independent circuits in the design. As you know, a design can be implemented in multiple ways: using multiplexers, using ROM or LUTs, using NAND gates, using NOR gates, or using AND-OR gates. Designs can also be implemented as a combination of several of these technologies.

If we are using a gate-array based on standard cells, the netlist needs to be "mapped" into the standard cells. If we are using a field programmable gate array with LUTs, the design needs to be transferred or "mapped" into the LUTs. If we are using a field programmable gate array with only 4-to-1 multiplexers, the design needs to be mapped into a structure which only needs multiplexers. If a target technology contains only two-input NAND gates, the design needs to be mapped to a form that uses only two-input NAND gates. We did this process manually for a shift register and multiplexer at the beginning of this chapter. CAD tools use mapping software to accomplish this task.

---

**Standard Cell Approach**   Standard cell design is a common technique for integrated circuit design. The design is mapped into a library of standard logic gates. Typically NOT, AND, NAND, OR, NOR, XOR, XNOR, and so on are available. CAD tools that support standard cell design methodology will also usually contain a library of complex functions and standard building blocks such as multiplexers, decoders, encoders, comparators, and counters. The design is mapped into a form that contains only cells available in the library. The cells are placed in rows that are separated by routing channels as in Figure 6-33. Some cells may be used only for routing between rows of cells. Such cells are called feedthrough cells. For the standard cell methodology to be effective, the height of cells should be the same. But it is possible to include memory modules, specialized arithmetic modules, and so on.

---

### 6.12.2 **Place and Route**

**Placement** is the process of taking defined logic and input/output (I/O) blocks (modules) from the technology mapper and assigning them to physical locations of the target implementation. It involves determining the positions of the sub-blocks in the design area. Placement choices matter because they impact subsequent routing. A good placement algorithm will try to reduce area and delay. Area and delay are partly determined by wiring. Algorithms typically estimate wire length and decide on appropriate placement choices. Complicated placement algorithms are not desirable because they consume too much run time.

**Routing** is the process of interconnecting the sub-blocks in a design. The choices for routing are greatly dependent on placement; hence place and route

FIGURE 6-33:
**Overview of a
Standard Cell
Design**

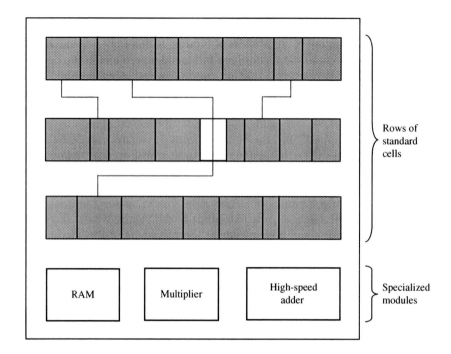

Rows of
standard
cells

Specialized
modules

RAM

Multiplier

High-speed
adder

are often done in tandem. Routing may be done in multiple steps. Global routing decisions can be made to minimize routing wire length, and then detailed routing of sections can be done. When only a part of a circuit is changed, incremental routing is useful.

Usually heuristics are used to perform placement. Most placement techniques start with an initial solution and then try to improve it with alternate placements. For instance, two blocks in one placement can be swapped to get an alternate placement, and wire length is evaluated for both the choices. The process is repeated until no further improvements are possible.

**Simulated annealing** techniques are used in the place and route process. *Annealing* is a term from metallurgy. Simulated annealing algorithms quickly and effectively optimize solutions over large state spaces. Simulated annealing does not guarantee the optimal solution, but it can produce a solution close to the global minimum in much less time than an exhaustive search. The simulated annealing process starts with a feasible solution (i.e., legal but not necessarily optimal) and searches for better solutions by making random modifications (permutations). An **iterative improvement** algorithm accepts only better solutions in each step. Algorithms that accept only better moves are considered **greedy algorithms**. But if we only accept better placements, we could be caught in a local minimum. It has been shown that it is beneficial occasionally to accept "bad moves." Often, these "bad moves" will let the algorithm reach a global minimum.

Accepting a bad move is certainly a risk. We can take more risks in the beginning of the simulated annealing process, but we need to be more conservative toward the later stages because there might not be sufficient time left to refine the solution to an acceptable level. In simulated annealing algorithms, the algorithms have a concept of a temperature, as in physical annealing in metallurgy. The temperature is high in the beginning and keeps reducing. Simulated annealing algorithms allow risky moves depending on the temperature. As the temperature is reduced, the probability of accepting bad moves decreases. Eventually, the algorithm defaults to a greedy algorithm that only accepts positive moves. Figure 6-34 illustrates the difference between simulated annealing and iterative improvement algorithms. The *y*-axis is the cost (or figure of merit) of the solution. The *x*-axis indicates the steps during the process.

**FIGURE 6-34:**
**Simulated**
**Annealing versus**
**Iterative**
**Improvement**
**Algorithms**

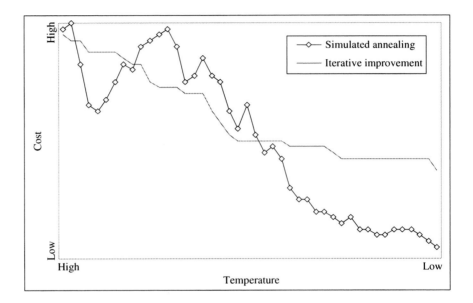

In simulated-annealing place-and-route algorithms, an initial placement is assumed and cost of alternate placement is estimated. Typically, the cost of a placement indicates the amount of routing that is needed. A move is considered better if it produces a better cost figure (for instance, wire length).

The ability of the tools to map and route designs depends on the algorithms in the tools and the granularity of the resources. Figure 6-35 shows a routed FPGA implementing an example design. (It is actually the dice game of Chapter 5, implemented in an early Xilinx FPGA, the XC3000.) The boxes on the periphery are the I/O blocks. Obviously, only a few of them on the top left corner and on the bottom side are used. The logic blocks in the middle are utilized, while several logic blocks are unused. Synthesis tools will provide a synthesis report giving the number and percentage of logic blocks used, number and percentage of flip-flops used, and so on.

The utilization of an FPGA depends on the nature of the logic blocks, the efficiency of the mapping tools, the routing resources, the efficiency of the routing tools, and so on. If logic blocks are of large granularity, it is very likely that parts

FIGURE 6-35: **A Routed FPGA**

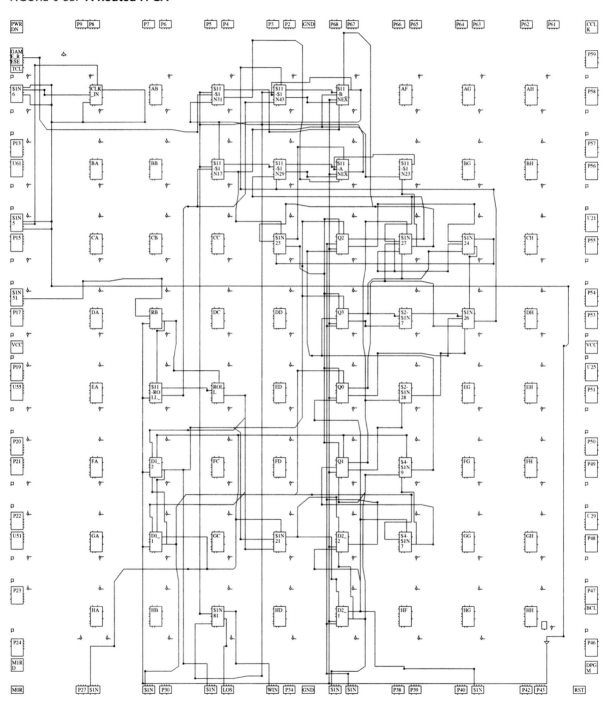

of logic blocks are unused. For instance, we saw that the shift register design in Figure 6-5 did not utilize a large part of the function generator. Similarly, the multiplexer designs in Figures 6-2 and 6-4 did not utilize the flip-flops on the logic block.

In this chapter we described several types of FPGAs and procedures for designing with these devices. Nowadays, sophisticated CAD tools are available to assist with the design of systems using programmable gate arrays. However, in this chapter, several hand designs were presented first to illustrate the underlying steps in CAD tools. Techniques to decompose functions of several variables into functions with fewer variables were illustrated. Features of modern FPGAs, such as embedded memory, embedded multipliers, and carry and cascade chains, were described. A brief overview of the synthesis, mapping, placement, and routing process was presented.

● ● ● ● ● ● ● ● ● ● ● ● ●

# Problems

6.1  An 8-bit right shift register with parallel load is to be implemented using an FPGA with logic blocks as in Figure 6-1(a). The flip-flops are labeled $X_7X_6X_5X_4X_3X_2X_1X_0$. The control signals $N$ and $S$ operate as follows: $N = 0$, do nothing; $NS = 11$, right shift; $NS = 10$, load. The serial input for right shift is $SI$.

   **(a)** How many logic blocks are required?
   **(b)** Show the required connections for the rightmost block on a copy of Figure 6-1(a). Connect $N$ to $CE$.
   **(c)** Give the function generator outputs for this block.

6.2  Implement a 2-bit binary counter using one logic block as in Figure 6-1(a). $A0$ is the least significant bit, and $A1$ is the most significant bit of the counter. The counter has a synchronous load ($Ld$). The counter operates as follows:

   $En = 0$              No change.
   $En = 1,\ Ld = 1$     Load $A0$ and $A1$ with external inputs $U$ and $V$ on rising edge of clock.
   $En = 1,\ Ld = 0$     Increment counter on rising edge of clock.

   **(a)** Give the next-state equations for $A0$ and $A1$.
   **(b)** Show all required inputs and connections on a copy of Figure 6-1(a). Show the connection paths with heavy lines. Use the $CE$ input. Give the function realized by each four-input LUT.

6.3  Design a 4-bit right-shift register using an FPGA with logic blocks as in Figure 6-1(a). When the register is clocked, the register loads if $Ld = 1$ and $En = 1$, it shifts right when $Ld = 0$ and $En = 1$, and nothing happens when $En = 0$. $Si$ and $So$ are the shift input and output of the register. $D_{3-0}$ and $Q_{3-0}$ are the parallel inputs and outputs, respectively. The next-state equation for the leftmost flip-flop is $Q_3{}^+ = En'Q_3 + En$ $(Ld\ D_3 + Ld'\ Si)$.

**(a)** Give the next-state equations for the other three flip-flops.
**(b)** Determine the minimum number of Figure 6-1(a) logic blocks required to implement the shift register.
**(c)** For the left block, give the input connections and the internal paths on a copy of Figure 6-1(a). Also, give the $X$ and $Y$ functions.

6.4 The next-state equations for a sequential circuit with two flip-flops ($Q_1$ and $Q_2$), input signals $R$, $S$, $T$, and an output $P$ are

$$D_1 = Q_1{}^+ = Q_2 R + Q_1 S$$
$$D_2 = Q_2{}^+ = Q_1 + Q_2' T$$

The output equation is $P = Q_2 RT + Q_1 ST$.

**(a)** Explain how this sequential circuit can be implemented using a single Figure 6-3 logic block. Write the equation that each function generator in the block will implement.
**(b)** Mark (highlight) the input signals, state and output variables, and the activated paths on a copy of Figure 6-3.

6.5 **(a)** Implement an 8-to-1 multiplexer using a minimum number of logic blocks of the type shown in Figure 6-1(a). Give the $X$ and $Y$ functions for each block and show the connections between blocks.
**(b)** Repeat (a) using logic blocks of Figure 6-3. Give $X$, $Y$, and $Z$ for each block.
**(c)** What are the LUT contents for the design in part (a)?
**(d)** What are the LUT contents for the design in part (b)?

6.6 **(a)** Write VHDL code that describes the logic block of Figure 6-1(a). Use the following entity:

```
entity Figure6_1a is
  port(X_in, Y_in: in unsigned(1 to 4);
       clk, CE: in bit;
       Qx, Qy: out bit;
       X, Y: inout bit;
       XLUT, YLUT: in unsigned(0 to 15));
  end Figure6_1a;
```

**(b)** Write structural VHDL code that instantiates two `Figure6_1a` block components to implement the 4-to-1 MUX of Figure 6-2. When you instantiate a block, use the actual bit patterns stored in *XLUT* and *YLUT* to specify the function generated by each of the LUTs.

6.7 **(a)** Write VHDL code that describes the logic block of Figure 6-3. Use an entity similar to Problem 6.6(a), except add *ZLUT* and *SA, SB, SC*, and *SD. SA, SB, SC*, and *SD* represent the programmable select bits that control the four

MUXes. These bits should be assigned values of '0' or '1' when the block component is instantiated.

**(b)** Write structural VHDL code that instantiates two `Figure6_3` block components to implement the code converter of Figure 1-26. When you instantiate a block component, use the actual bit patterns stored in *XLUT, YLUT,* and *ZLUT* to specify the function generated by each of the LUTs.

6.8 **(a)** How many logic blocks as in Figure 6-1(a) are required to create a 4-to-16 decoder?
   **(b)** Give the contents of the LUTs in the first logic block.

6.9 **(a)** How many logic blocks as in Figure 6-3 are required to create an 8-to-3 priority encoder?
   **(b)** Give the contents of the LUTs in the first logic block.

6.10 Show how to realize the following combinational function using two Figure 6-1(a) logic blocks. Show the connections on a copy of Figure 6-1(a) and give the functions *X* and *Y* for both blocks.

$$F = X_1'X_2X_3'X_6 + X_2'X_3'X_4X_6' + X_2X_3'X_4' + X_2X_3X_4'X_6 + X_3'X_4X_5X_6' + X_7$$

6.11 Realize the following next-state equation using a minimum number of Figure 6-1(a) logic blocks. Draw a diagram that shows the connections to the logic blocks and give the functions *X* and *Y* for each cell. (The equation is already in minimum form.)

$$Q^+ = UQV'W + U'Q'VX'Y' + UQX'Y + U'Q'V'Y + U'Q'XY + UQVW' \\ + U'Q'V'X$$

6.12 What is the minimum number of Figure 6-3 logic blocks required to realize the following function?

$$X = X_1'X_2'X_3'X_4'X_5 + X_1X_2X_3X_4X_5 + X_5'X_6X_7'X_8'X_9 + X_5'X_6'X_7X_8X_9'$$

If your answer is 1, show the required input connections on a copy of Figure 6-3, and mark the internal connection paths with heavy lines. If your answer is greater than 1, draw a block diagram showing the cell inputs and interconnections between cells. In any case, give the functions to be realized by each *X*, *Y*, and *Z* function generator.

6.13 Given $Z(T, U, V, W, X, Y) = VW'X + U'V'WY + TV'WY'$,

   **(a)** Show how *Z* can be realized using a single Figure 6-3 logic block. Show the cell inputs on a copy of Figure 6-3, indicate the internal connections in the cell, and specify the functions *X*, *Y*, and *Z*.
   **(b)** Show how *Z* can be realized using two Figure 6-1(a) logic blocks. Draw a diagram showing the inputs to each cell, the interconnections between cells, and the *X* and *Y* functions for each cell.

6.14 Use Shannon's expansion theorem around *a* and *b* for the function

$$Y = abcde + cde'f + a'b'c'def + bcdef' + ab'cd'ef' + a'bc'de'f + abcd'e'f$$

so that it can be implemented using only four-variable function generators. Draw a block diagram to indicate how $Y$ can be implemented using only four-variable function generators. Indicate the function realized by each four-variable function generator.

6.15  Use Shannon's expansion theorem around $e$ and $f$ for the function

$$Y = ab'cdef + a'bc'd'e + b'c'ef' + abcde'f$$

so that it can be implemented using a minimum number of four-variable functions. Rewrite $Y$ to indicate how it will be implemented using four-variable function generators and draw a block diagram. Indicate the function generated by each function generator.

6.16  **(a)** Use Shannon's expansion theorem around $a$ for the function

$$Y = ab'cd'e + a'bc'd'e + b'c'e + abcde$$

so that it can be implemented using four-variable functions.

**(b)** Use the expanded function to show how $Y$ can be implemented using one Figure 6-3 logic block. Mark (highlight) the input signals and the activated paths on a copy of Figure 6-3.

**(c)** Give the contents of the three LUTs.

6.17  **(a)** If logic blocks of Figure 6-1(a) are used, how many LUTs are required to build a 4-bit adder with accumulator?

**(b)** If an FPGA with built-in carry chain logic as in Figure 6-11 is used, how many four-input LUTs are required?

**(c)** Design a 4-bit adder-subtractor with accumulator using an FPGA with carry chain logic and four-input LUTs. Assume a control signal $Su$ which is 0 for addition and 1 for subtraction. Show the required connections on a diagram similar to Figure 6-11 and give the function realized by each LUT.

6.18  A $4 \times 4$ array multiplier (Figure 4-29) is to be implemented using an FPGA.

**(a)** Partition the logic so that it fits in a minimum number of Figure 6-1(a) logic blocks. Draw loops around each set of components that will fit in a single logic block. Determine the total number of four-input LUTs required.

**(b)** Repeat part (a), except assume that carry chain logic is available.

6.19  **(a)** Use Shannon's expansion theorem to expand the following function around $A$ and then expand each sub-function around $D$:

$$Z = AB'CD'E'F + A'BC'D'EF' + B'C'E'F + A'BC'E'F' + ABCDE$$

**(b)** Explain how the expanded function could be implemented using two Xilinx Virtex FPGA slices (Figure 6-13). On the slice diagrams, label the inputs to the LUTs (function generators) and draw the connection paths within the slice. Give the function implemented by each LUT.

6.20 **(a)** Indicate the connections of the switches in Figure 6-15 to realize the function

$$Z = AB'C + A'BC' + BC$$

**(b)** Indicate the connections of the switches in Figure 6-15 to realize the function

$$F = AB + A'C$$

**(c)** Indicate the connections of the switches in Figure 6-15 to realize a latch as in Figure 2-17.

**(d)** Indicate the connections of the switches in Figure 6-15 to realize a D flip-flop.

6.21 The logic equations for a sequential network with five inputs, two flip-flops, and two outputs are

$$Q_1^+ = Q_1(Q_2ABC) + Q_1'(Q_2'CDE)$$
$$Q_2^+ = Q_1'$$
$$Z_1 = Q_1'Q_2'AB + Q_1'Q_2'A'B' + Q_1Q_2'AB' + Q_1Q_2(A' + B + C)$$
$$Z_2 = Q_1A' + Q_1B + Q_2'$$

How many Virtex slices (Figure 6-13) are required to implement the logic equations, including the flip-flops? Specify the inputs to each slice and the functions realized by each LUT.

6.22 Perform a survey of FPGA chips now on the market.

**(a)** Generate a table like Table 6-1 for current FPGAs.
**(b)** Generate a table like Table 6-2 for current FPGAs.

6.23 Show how $32 \times 32$-bit unsigned multiplication can be accomplished using four $16 \times 16$-bit multipliers and several adders. Draw a block diagram showing the required connections.

6.24 Fast shifting can be accomplished by using dedicated multipliers. Shifting left $N$ places is equivalent to multiplying by $2^N$.

**(a)** Given that $A$ is a 16-bit unsigned number and $0 \le N \le 15$, show how to construct a left shifter using a multiplier and a decoder.
**(b)** Write VHDL code that infers this type of shifter.
**(c)** Repeat (a) and (b) for a right shifter. *Hint:* Multiply by $2^{15-N}$ and select the appropriate 16 bits of the 32-bit product.

6.25 Make a one-hot state assignment for Figure 4-28(c). Derive the next state and output equations by inspection.

6.26 Make a one-hot state assignment for Figure 4-53 and write the next state and output equations by inspection. Then change the state assignment so that $S_0$ is assigned 0000000, $S_1$ is assigned 1100000, $S_2$ is 1010000, and so on and rewrite the equations for this assignment.

6.27 Assume that a sequential system with four states is to be implemented using a one-hot state assignment, but the flip-flops do not have preset input. The flip-flops do have a reset input; hence, it is beneficial to have 0000 as the starting state. What should be the state assignments for the other states if one wants to take advantage of the one-hot assignment scheme? Explain.

6.28 For the given state graph,

   **(a)** Derive the simplified next-state and output equations by inspection. Use the following one-hot state assignment for flip-flops $Q_0Q_1Q_2Q_3$: $S_0$, 1000; $S_1$, 0100; $S_2$, 0010; $S_3$, 0001.

   **(b)** How many Virtex slices (Figure 6-13) are required to implement these equations?

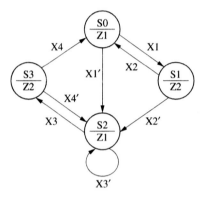

6.29 Make any necessary changes in the VHDL code for the traffic light controller (Figure 4-15) so that it can be synthesized without latches using whatever synthesis tool you have available. Synthesize the code using a suitable FPGA or CPLD as a target.

6.30 Synthesize the behavioral model of the 2's complement multiplier (Figure 4-35) using whatever synthesis tool you have available. Then synthesize the model with control signals (Figure 4-40) and compare the results (number of flip-flops, number of LUTs, number of slices, etc.). Try different synthesis options such as optimizing for area or speed, and different finite-state machine encoding algorithms such as one-hot, compact, and so on and compare the results. Which combination of options uses the least resources?

6.31 Consider the VHDL code

```
entity example is
   port(a: in integer range 0 to 3;
        b: out integer range 0 to 3);
end example;
```

```vhdl
architecture test2 of example is
begin
  process(a)
  begin
    case a is
      when 0 => b <= 3;
      when 1 => b <= 2;
      when 2 => b <= 1;
      when 3 => b <= 1;
    end case;
  end process;
end test2;
```

**(a)** Show the hardware you would obtain if you synthesize the preceding VHDL code without any optimizations. Explain your reasoning.

**(b)** Show optimized hardware emphasizing minimum area. Show the steps/reasoning by which you obtained the optimized hardware.

6.32 Draw the hardware structures that will be inferred by typical synthesizers from the code excerpts that follow. *A*, *B*, and *E* are 4-bit vectors. *C* and *D* are 2-bit numbers. *clock* is a 1-bit signal. Draw the structure and mark the inputs and outputs.

**(a)**
```vhdl
process(clock)
begin
  A <= A(3) & A(3 downto 1);
  B <= A(0) & B(3 downto 1);
end process;
```

**(b)**
```vhdl
architecture test2 of example is
begin
  process(C)
  begin
    case C is
      when 0 => D <= 3;
      when 1 => D <= 2;
      when 2 => D <= 0;
      when others => null;   -- preserve value
    end case;
  end process;
end test2;
```

**(c)**
```vhdl
architecture test2 of example is
begin
  process(C)
  begin
    case C is
      when 0 => E <= A + B;
      when 1 => E <= A sra 2;
      when 2 => E <= A - B;
```

```
          when 3 => E <= A;
        end case;
      end process;
    end test2;
```

6.33 **(a)** Draw a logic diagram (use gates, adders, muxes, D flip-flops, etc.) that shows the result of synthesizing the following VHDL code. *A*, *B*, and *C* are unsigned vectors dimensioned 2 downto 0.

```
process(CLK)
if CLK'event and CLK = 0 then
  if C0 = '1' then C <= not A; end if;
  if Ad = '1' then C <= A + B; end if;
  if Sh = '1' then C <= C sra 1; end if;
end if;
```

**(b)** Describe in one or two sentences what this circuit does.

6.34 Draw the hardware structures that will be inferred by typical synthesizers from the code excerpts below. If any ambiguities exist in the code, mention what you are assuming. Show optimized and unoptimized hardware.

**(a)**
```
architecture test2 of example is
begin
  process(a)
  begin
    case a is
      when 0 => b <= 2;
      when 1 => b <= 0;
      when 2 => b <= 3;
      when 3 => b <= 1;
    end case;
  end process;
end test2;
```

**(b)**
```
if arg1 > arg2 and arg1 > arg3 then
  result <= arg1;
else
  result <= '0';
end if;
```

6.35 What hardware does the statement

```
F <= (A >= B);
```

result in? Assume that *A* and *B* are 8-bit vectors.

6.36 Generate optimized hardware for the following statement, assuming *A* is a 4-bit vector:

```
F <= (A = 9);
```

# Floating-Point Arithmetic

Floating-point numbers are frequently used for numerical calculations in computing systems. Arithmetic units for floating-point numbers are considerably more complex than those for fixed-point numbers. Floating-point numbers allow very large or very small numbers to be specified. This chapter first describes a simple representation for floating-point numbers. Then it describes the IEEE floating-point standard. Next, an algorithm for floating-point multiplication is developed and tested using VHDL. Then the design of the floating-point multiplier is completed and implemented using an FPGA. Floating-point addition, subtraction, and division are also briefly described.

## 7.1 Representation of Floating-Point Numbers

A simple representation of a floating-point (or real) number ($N$) uses a fraction ($F$), base ($B$), and exponent ($E$), where $N = F \times B^E$. The base can be 2, 10, 16, or any other number. The fraction and the exponent can be represented in many formats. For example, they can be represented by 2's complement formats, sign-magnitude form, or another number representation. There are a variety of floating-point formats depending on how many bits are available for $F$ and $E$, what the base is, and how negative numbers are represented for $F$ and $E$. The base can be implied or explicit. Depending on all these choices, a wide variety of floating-point formats have existed in the past.

### 7.1.1 A Simple Floating-Point Format Using 2's Complement

In this section, we describe a floating-point format where negative exponents and fractions are represented using the 2's complement form. The base for the exponent is 2. Hence, the value of the number is $N = F \times 2^E$. In a typical floating-point number system, $F$ is 16 to 64 bits long and $E$ is 8 to 15 bits long. In order to keep the examples in this section simple and easy to follow, we will use a 4-bit fraction and a 4-bit exponent, but the concepts presented here can easily be extended to more bits.

The fraction and the exponent in this system will use 2's complement. (Refer to Section 4.10 for a discussion of 2's complement fractions.) We will use 4 bits for the fraction and 4 bits for the exponent. The fractional part will have a leading sign bit and three actual fraction bits. The implied binary point is after the first bit. The sign bit is 0 for positive numbers and 1 for negative numbers.

**361**

As an example, let us represent decimal 2.5 in this 8-bit 2's complement floating-point format.

$$2.5 = 0010.1000$$
$$= 1.010 \times 2^1 \quad \text{(standardized normal representation)}$$
$$= 0.101 \times 2^2 \quad \text{(4-bit 2's complement fraction)}$$

Therefore,

$$F = 0.101 \qquad E = 0010 \qquad N = 5/8 \times 2^2$$

If the number was $-2.5$, the same exponent can be used, but the fraction must have a negative sign. The 2's complement representation for the fraction is 1.011. Therefore,

$$F = 1.011 \qquad E = 0010 \qquad N = -5/8 \times 2^2$$

Other examples of floating-point numbers using a 4-bit fraction and a 4-bit exponent are

$$F = 0.101 \qquad E = 0101 \qquad N = 5/8 \times 2^5$$
$$F = 1.011 \qquad E = 1011 \qquad N = -5/8 \times 2^{-5}$$
$$F = 1.000 \qquad E = 1000 \qquad N = -1 \times 2^{-8}$$

In order to utilize all the bits in $F$ and have the maximum number of significant figures, $F$ should be normalized so that its magnitude is as large as possible. If $F$ is not normalized, we can normalize $F$ by shifting it left until the sign bit and the next bit are different. Shifting $F$ left is equivalent to multiplying by 2, so every time we shift we must decrement $E$ by 1 to keep $N$ the same. After normalization, the magnitude of $F$ will be as large as possible, since any further shifting would change the sign bit. In the following examples, $F$ is unnormalized to start with and then it is normalized by shifting left.

Unnormalized: $\quad F = 0.0101 \quad E = 0011 \quad N = 5/16 \times 2^3 = 5/2$
Normalized: $\quad F = 0.101 \quad E = 0010 \quad N = 5/8 \times 2^2 = 5/2$
Unnormalized: $\quad F = 1.11011 \quad E = 1100 \quad N = -5/32 \times 2^{-4} = -5 \times 2^{-9}$
(shift $F$ left) $\quad F = 1.1011 \quad E = 1011 \quad N = -5/16 \times 2^{-5} = -5 \times 2^{-9}$
Normalized: $\quad F = 1.011 \quad E = 1010 \quad N = -5/8 \times 2^{-6} = -5 \times 2^{-9}$

The exponent can be any number between $-8$ and $+7$. The fraction can be any number between $-1$ and $+0.875$.

Zero cannot be normalized, so $F = 0.000$ when $N = 0$. Any exponent could then be used; however, it is best to have a uniform representation of 0. In this format, we will associate the negative exponent with the largest magnitude with the fraction 0. In a 4-bit 2's complement integer number system, the most negative number is 1000, which represents $-8$. Thus when $F$ and $E$ are 4 bits, 0 is represented by

$$F = 0.000 \qquad E = 1000 \qquad N = 0.000 \times 2^{-8}$$

Some floating-point systems use a biased exponent, so $E = 0$ is associated with $F = 0$.

### 7.1.2 **The IEEE 754 Floating-Point Formats**

The IEEE 754 is a floating-point standard established by IEEE in 1985. It contains two representations for floating-point numbers, the IEEE single precision format and the IEEE double precision format. The IEEE 754 single precision representation uses 32 bits and the double precision system uses 64 bits.

Although 2's complement representations are very common for negative numbers, the IEEE floating-point representations do not use 2's complement for either the fraction or the exponent. The designers of IEEE 754 desired a format that was easy to sort and hence adopted **a sign-magnitude system** for the **fractional part** and a **biased notation** for the **exponent**.

The IEEE 754 floating-point formats need three subfields: sign, fraction, and exponent. The fractional part of the number is represented using a sign-magnitude representation in the IEEE floating-point formats (i.e., there is an explicit sign bit ($S$) for the fraction). The sign is 0 for positive numbers and 1 for negative numbers. In a binary normalized scientific notation, the leading bit before the binary point is always 1 and hence the designers of the IEEE format decided to make it implied, representing only the bits after the binary point. In general, the number is of the form

$$N = (-1)^S \times (1 + F) \times 2^E$$

where $S$ is the sign bit, $F$ is the fractional part, and $E$ is the exponent. The base of the exponent is 2. The base is implied (i.e., it is not stored anywhere in the representation). The magnitude of the number is $1 + F$ because of the omitted leading 1. The terms significand means the magnitude of the fraction and is $1 + F$ in the IEEE format. But often the terms *significand* and *fraction* are used interchangeably by many, including in this book.

The exponent in the IEEE floating-point formats uses what is known as a biased notation. A biased representation is one in which every number is represented by the number plus a certain bias. In the IEEE single precision format, the bias is 127. Hence, if the exponent is $+1$, it will be represented by $+1 + 127 = 128$. If the exponent is $-2$, it will be represented by $-2 + 127 = 125$. Thus, exponents less than 127 indicate actual negative exponents and exponents greater than 127 indicate actual positive exponents. The bias is 1023 in the double precision format.

If a positive exponent becomes too large to fit in the exponent field, the situation is called **overflow**, and if a negative exponent is too large to fit in the exponent field, that situation is called **underflow**.

### The IEEE Single Precision Format

The IEEE single precision format uses 32 bits for representing a floating-point number, divided into three subfields, as illustrated in Figure 7-1. The first field is the sign bit for the fractional part. The next field consists of 8 bits which are used for the exponent. The third field consists of the remaining 23 bits and is used for the fractional part.

The sign bit reflects the sign of the fraction. It is 0 for positive numbers and 1 for negative numbers. In order to represent a number in the IEEE single precision

**FIGURE 7-1: IEEE Single Precision Floating-Point Format**

| S | Exponent | Fraction |
|---|----------|----------|
| 1 bit | 8 bits | 23 bits |

format, first it should be converted to a normalized scientific notation with exactly one bit before the binary point, simultaneously adjusting the exponent value.

The exponent representation that goes into the second field of the IEEE 754 representation is obtained by adding 127 to the actual exponent of the number when represented in the normalized form. Exponents in the range 1–254 are used for representing normalized floating-point numbers. Exponent values 0 and 255 are reserved for special cases, which will be discussed later.

The representation for the 23-bit fraction is obtained from the normalized scientific notation by dropping the leading 1. Zero cannot be represented in this fashion; hence it is treated as a special case (explained later). Since every number in the normalized scientific notation will have a leading 1, this leading 1 can be dropped so that one more bit can be packed into the significand (fraction). Thus, a 24-bit fraction can be represented using the 23 bits in the representation. The designers of the IEEE formats wanted to make highest use of all the bits in the exponent and fraction fields.

In order to understand the IEEE format, let us represent 13.45 in the IEEE floating-point format. We can see that 0.45 is a recurring binary fraction and hence

$$13.45 = 1101.01\ 1100\ 1100\ 1100 \ldots\ldots\ldots \text{ with the bits } 1100 \text{ continuing to recur}$$

Normalized scientific representation yields

$$13.45 = 1.10101\ 1100\ 1100 \ldots \times 2^3$$

Since the number is positive, the sign bit for the IEEE 754 representation is 0.

The exponent in the biased notation will be $127 + 3 = 130$, which in binary format is 10000010.

The fraction is $1.10101\ 1100\ 1100 \ldots \ldots \ldots$ (with 1100 recurring). Omitting the leading 1, the 23 bits for the fractional part are

$$10101\ 1100\ 1100\ 1100\ 1100\ 11$$

Thus, the 32 bits are

$$0\ 10000010\ 10101\ 1100\ 1100\ 1100\ 1100\ 11$$

as illustrated in Figure 7-2.

**FIGURE 7-2: IEEE Single Precision Floating-Point Representation for 13.45**

| S | Exponent | Fraction |
|---|----------|----------|
| 0 | 1 0 0 0 0 0 1 0 | 1 0 1 0 1 1 1 0 0 1 1 0 0 1 1 0 0 1 1 0 0 1 1 |

The 32 bits can be expressed more conveniently in a hexadecimal (hex) format as

4157 3333

The number $-13.45$ can be represented by changing only the sign bit (i.e., the first bit must be 1 instead of 0). Hence, the hex number C157 3333 represents $-13.45$ in IEEE 754 single precision format.

### The IEEE Double Precision Format

The IEEE double precision format uses 64 bits for representing a floating-point number, as illustrated in Figure 7-3. The first bit is the sign bit for the fractional part. The next 11 bits are used for the exponent, and the remaining 52 bits are used for the fractional part.

FIGURE 7-3: IEEE
Double Precision
Floating-Point
Format

| S | Exponent | Fraction |
|---|---|---|
| 1 bit | 11 bits | 52 bits |

As in the single precision format, the sign bit is 0 for positive numbers and 1 for negative numbers.

The exponent representation used in the second field is obtained by adding the bias value of 1023 to the actual exponent of the number in the normalized form. Exponents in the range 1–2046 are used for representing normalized floating-point numbers. Exponent values 0 and 2047 are reserved for special cases.

The representation for the 52-bit fraction is obtained from the normalized scientific notation by dropping the leading 1 and considering only the next 52 bits.

As an example, let us represent 13.45 in IEEE double precision floating-point format. Converting 13.45 to a binary representation,

$13.45 = 1101.01\ 1100\ 1100\ 1100 \ldots \ldots \ldots$ with the bits 1100 continuing to recur

In normalized scientific representation,

$$13.45 = 1.10101\ 1100\ 1100 \ldots \times 2^3$$

The exponent in biased notation will be $1023 + 3 = 1026$, which in binary representation is

$$10000000010$$

The fraction is $1.10101\ 1100\ 1100 \ldots \ldots \ldots$ (with 1100 recurring). Omitting the leading 1, the 52 bits of the fractional part are

$$10101\ 1100\ 1100\ 1100\ 1100\ 1100\ 1100\ 1100\ 1100\ 1100\ 1100\ 1100\ 110$$

Thus, the 64 bits are

$$0\ 10000000010\ 10101\ 1100\ 1100\ 1100\ 1100\ 1100\ 1100\ 1100\ 1100\ 1100\ 1100\ 1100\ 110$$

as illustrated in Figure 7-4. The 64 bits can be expressed more conveniently in a hexadecimal format as

$$402A\ E666\ 6666\ 6666$$

**FIGURE 7-4: IEEE Double Precision Floating-Point Representation for 13.45**

| S | Exponent | Fraction |
|---|----------|----------|
| 0 | 1 0 0 0 0 0 0 0 0 1 0 | 1 0 1 0 1 1 1 0 0 1 1 0 0 1 1 0 0 1 1 0 |

| Fraction (cont'd) |
|-------------------|
| 0 1 1 0 0 1 1 0 0 1 1 0 0 1 1 0 0 1 1 0 0 1 1 0 0 1 1 0 0 1 1 0 |

The number $-13.45$ can be represented by changing only the sign bit (i.e., the first bit must be 1 instead of 0). Hence, the hex number C02A E666 6666 6666 represents $-13.45$ in IEEE 754 double precision format.

### Special Cases in the IEEE 754 Standard

The IEEE 754 standard has several special cases, which are illustrated in Figure 7-5. These include 0, infinity, denormalized numbers, and NaN (Not a Number) representations. The smallest and the highest exponents are used to denote these special cases.

**FIGURE 7-5: Special Cases in the IEEE 754 Floating-Point Formats**

| Single Precision | | Double Precision | | Object Represented |
|---|---|---|---|---|
| Exponent | Fraction | Exponent | Fraction | |
| 0 | 0 | 0 | 0 | 0 |
| 0 | nonzero | 0 | nonzero | $\pm$ denormalized number |
| 255 | 0 | 2047 | 0 | $\pm$ infinity |
| 255 | nonzero | 2047 | nonzero | NaN (not a number) |

**Zero**  The IEEE format specifies 0 to be the representation with 0's in all bits (i.e., all exponent and fraction bits are 0). Zero is specified as a special case in the format due to the difficulty in representing 0 in a normalized format. When using the usual convention for IEEE format normalized numbers, we would add a leading 1 to the fractional part, but that would make it impossible to represent 0.

**Denormalized Numbers**  The smallest normalized number that the single precision format can represent is

$$1.0 \times 2^{-126}$$

Numbers between this number and 0 cannot be expressed in the normalized format. If normalization is not made a requirement of the format, we could represent numbers smaller than $1.0 \times 2^{-126}$. Hence, the IEEE floating-point format allows denormalized numbers as a special case. If the exponent is 0 and the fraction is nonzero, the number is considered denormalized. Now, the smallest number that can be represented is

$$0.00000000000000000000001 \times 2^{-126}, \text{ which is } 1.0 \times 2^{-149}.$$

Thus, denormalization allows numbers between $1.0 \times 2^{-126}$ and $1.0 \times 2^{-149}$ to be represented.

For double precision, the denormalized range allows numbers between $1.0 \times 2^{-1022}$ and $1.0 \times 2^{-1074}$.

**Infinity**    Infinity is represented by the highest exponent value together with a fraction of 0. In the case of single precision representation, the exponent is 255, and for double precision, it is 2047.

**Not a Number (NaN)**    The IEEE 754 standard has a special representation to represent the result of invalid operations, such as 0/0. This special representation is called NaN or Not a Number. If the exponent is 255 and the fraction is any nonzero number, it is considered to be NaN or Not a Number.

**Rounding**    When the number of bits available is fewer than the number of bits required to represent a number, rounding is employed. It is desirable to round to the nearest value. We can round up if the number is higher than halfway between and round down if the number is less than halfway between. Another option is to truncate, ignoring the bits beyond the allowable number of bits. We must keep more bits in intermediate representations to achieve higher accuracy. The IEEE standard requires two extra bits in intermediate representations in order to facilitate better rounding. The two bits are called **guard and round**. Sometimes, a third intermediate bit is used in rounding in addition to the guard and round bits. It is called **sticky bit**. The sticky bit is set whenever there are nonzero bits to the right of the round bit.

The biggest challenge is when the number falls halfway in between. The IEEE standard has four different rounding modes:

- **Round up**        Round toward positive infinity. Round up to the next higher number.
- **Round down**      Round toward negative infinity. Round down to the nearest smaller number.
- **Truncate**        Round toward zero. Ignore bits beyond the allowable number of bits. Same as truncation in sign magnitude.
- **Unbiased**        If the number falls halfway between, round up half the time and round down half the time. In order to achieve rounding up half the time, add 1 if the lowest bit retained is 1, and truncate if it is 0. This is based on the assumption that a 0 or 1 appears in the lowest retained bit with an equal probability. One consequence of this rounding scheme is that the rounded number always has a 0 in the lowest place.

## 7.2 Floating-Point Multiplication

Given two floating-point numbers, $(F_1 \times 2^{E_1})$ and $(F_2 \times 2^{E_2})$, the product is

$$(F_1 \times 2^{E_1}) \times (F_2 \times 2^{E_2}) = (F_1 \times F_2) \times 2^{(E_1 + E_2)} = F \times 2^E$$

The fraction part of the product is the product of the fractions, and the exponent part of the product is the sum of the exponents. Hence, a floating-point multiplier consists of two major components: a fraction multiplier, and an exponent adder. The details of floating-point multiplication will depend on the precise formats in which the fraction multiplication and exponent addition are performed.

Fraction multiplication can be done in many ways. If the IEEE format is used, multiplication of the magnitude can be done and then the signs can be adjusted. If 2's complement fractions are used, we can use a fraction multiplier that handles signed 2's complement numbers directly. We discussed such a fraction multiplier in Chapter 4.

Addition of the exponents can be done with a binary adder. If the IEEE formats are directly used, the representations must be carefully adjusted in order to obtain the correct result. For instance, if exponents of two floating-point numbers in the biased format are added, the sum contains twice the bias value. To get the correct exponent, the bias value must be subtracted from the sum.

The 2's complement system has several interesting properties for performing arithmetic. Hence, many floating-point arithmetic units convert the IEEE notation to 2's complement and then use the 2's complement internally for carrying out the floating-point operations. Then the final result is converted back to IEEE standard notation.

The general procedure for performing floating-point multiplication is the following:

1. Add the two exponents.
2. Multiply the two fractions (significands).
3. If the product is 0, adjust the representation to the proper representation for 0.
4. a. If the product fraction is too big, normalize by shifting it right and incrementing the exponent.
   b. If the product fraction is too small, normalize by shifting left and decrementing the exponent.
5. If an exponent underflow or overflow occurs, generate an exception or error indicator.
6. Round to the appropriate number of bits. If rounding resulted in loss of normalization, go to step 4 again.

Note that, in addition to adding the exponents and multiplying the fractions, several steps—such as normalizing the product, handling overflow and underflow, and rounding to the appropriate number of bits—also need to be done. We assume that the two numbers are properly normalized to start with, and we want the final result to be normalized.

Now, we discuss the design of a floating-point multiplier. We use 4-bit fractions and 4-bit exponents, with negative numbers represented in 2's complement.

The fundamental steps are to add the exponents (step 1) and multiply the fractions (step 2). However, several special cases must be considered. If $F$ is 0, we must set the exponent $E$ to the largest negative value (1000) (step 3). A special situation occurs if we multiply $-1$ by $-1$ ($1.000 \times 1.000$). The result should be $+1$. Since we cannot represent $+1$ as a 2's complement fraction with a 4-bit fraction, this special case necessitates right shifting as in step 4. To correct this situation, we right shift the significand (fraction) and increment the exponent. Essentially, we set $F = 1/2$ (0.100) and add 1 to $E$. This results in the correct answer, since $1 \times 2^E = 1/2 \times 2^{E+1}$.

When we multiply the fractions, the result could be unnormalized. For example,

$$(0.1 \times 2^{E_1}) \times (0.1 \times 2^{E_2}) = 0.01 \times 2^{E_1 + E_2} = 0.1 \times 2^{E_1 + E_2 - 1}$$

This is situation 4.b in the preceding list. In this case, we normalize the result by shifting the fraction left one place and subtracting 1 from the exponent to compensate. Finally, if the resulting exponent is too large in magnitude to represent in our number system, we have an exponent overflow. (An overflow in the negative direction is referred to as an *underflow*.) Since we are using 4-bit exponents, if the exponent is not in the range 1000 to 0111 ($-8$ to $+7$), an overflow has occurred. Since an exponent overflow cannot be corrected, an overflow indicator should be turned on (step 5).

A flow chart for this floating-point multiplier is shown in Figure 7-6. After multiplying the fraction, all the special cases are tested for. Since $F_1$ and $F_2$ are normalized, the smallest possible magnitude for the product is 0.01, as indicated in the preceding example. Therefore, only one left shift is required to normalize $F$.

FIGURE 7-6: **Flow Chart for Floating-Point Multiplication with 2's Complement Fractions/ Exponents**

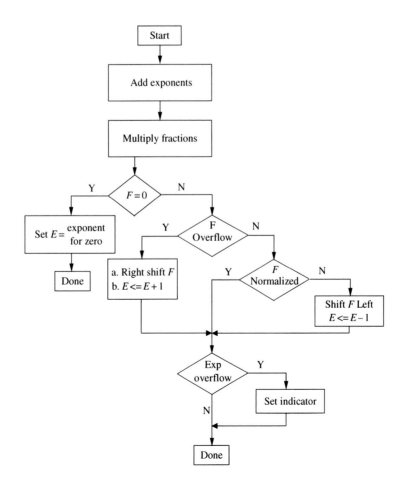

The hardware required to implement the multiplier (Figure 7-7) consists of an exponent adder, a fraction multiplier, and a control unit that provides the signals to perform the appropriate operations of right shifting, left shifting, exponent incrementing/decrementing, and so on.

*Exponent Adder:* Since 2's complement addition results with the sum in the proper format, the design of the exponent adder is straightforward. A 5-bit full adder is used as the exponent adder as demonstrated in Figure 7-7. When the fraction is normalized, the exponent will have to be correspondingly incremented or decremented. Also, in

FIGURE 7-7: **Major Components of a Floating-Point Multiplier**

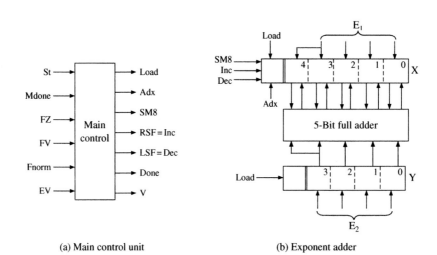

(a) Main control unit             (b) Exponent adder

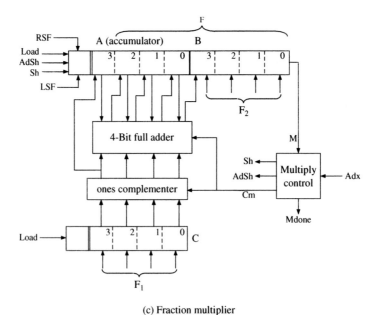

(c) Fraction multiplier

the special case when product is 0, the register should be set to the value 1000. The register has control signals for incrementing, decrementing, and setting to the most negative value ($SM8$).

The register which holds the sum is made into a 5-bit register to handle special situations. When the exponents are added, an overflow can occur. If $E_1$ and $E_2$ are positive and the sum ($E$) is negative, or if $E_1$ and $E_2$ are negative and the sum is positive, the result is a 2's complement overflow. However, this overflow might be corrected when 1 is added to or subtracted from $E$ during normalization or correction of fraction overflow. To allow for this case, we have made the $X$ register 5 bits long. When $E_1$ is loaded into $X$, the sign bit must be extended so that we have a correct 2's complement representation. Since there are two sign bits, if the addition of $E_1$ and $E_2$ produces an overflow, the lower sign bit will get changed, but the high-order sign bit will be unchanged. Each of the following examples has an overflow, since the lower sign bit has the wrong value:

$$7 + 6 = 00111 + 00110 = 01101 = 13 \qquad \text{(maximum allowable value is 7)}$$
$$-7 + (-6) = 11001 + 11010 = 10011 = -13 \qquad \text{(maximum allowable negative value is } -8\text{)}$$

The following example illustrates the special case where an initial fraction overflow and exponent overflow occurs, but the exponent overflow is corrected when the fraction overflow is corrected:

$$(1.000 \times 2^{-3}) \times (1.000 \times 2^{-6}) = 01.000000 \times 2^{-9} = 00.100000 \times 2^{-8}$$

*Fraction Multiplier:*   The fraction multiplier that we designed in Section 4.10 handles 2's complement fractions in a straightforward manner. Hence, we adapt that design for the floating-point multiplier. It implements a shift and add multiplier algorithm. Since we are multiplying 3 bits plus sign by 3 bits plus sign, the result will be 6 bits plus sign. After the fraction multiplication, the 7-bit result ($F$) will be the lower 3 bits of $A$ concatenated with $B$. The multiplier has its own control unit that generates appropriate shift and add signals depending on the multiplier bits.

*Main Control Unit:*   The SM chart for the main controller (Figure 7-8) of the floating-point multiplier is based on the flow chart. This controller is called main controller to distinguish it from the controller for the multiplier, which is a separate state machine that is linked into the main controller.

The SM chart uses the following inputs and control signals:

| | |
|---|---|
| St | Start the floating-point multiplication. |
| Mdone | Fraction multiply is done. |
| FZ | Fraction is zero. |
| FV | Fraction overflow (fraction is too big). |
| Fnorm | $F$ is normalized. |
| EV | Exponent overflow. |
| Load | Load $F_1$, $E_1$, $F_2$, $E_2$ into the appropriate registers (also clear $A$ in preparation for multiplication). |
| Adx | Add exponents; this signal also starts the fraction multiplier. |

SM8    Set exponent to minus 8 (to handle special case of 0).
RSF    Shift fraction right; also increment *E*.
LSF    Shift fraction left; also decrement *E*.
V      Overflow indicator.
Done   Floating-point multiplication is complete.

The SM chart for the main controller has four states. In $S_0$, the registers are loaded when the start signal is 1. In $S_1$, the exponents are added, and the fraction multiply is started. In $S_2$, we wait until the fraction multiply is done and then test for special cases and take appropriate action. It may seem surprising that the tests on *FZ*, *FV*, and *Fnorm* can all be done in the same state since they are done in sequence on the flow chart. However, *FZ*, *FV*, and *Fnorm* are generated by combinational circuits that operate in parallel and hence can be tested in the same state. However, we must wait until the exponent has been incremented or decremented at the next clock

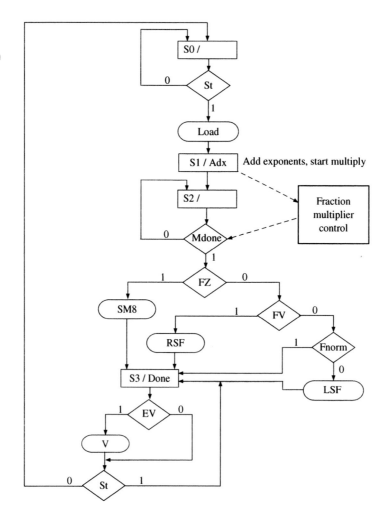

**FIGURE 7-8: SM Chart for Floating-Point Multiplication**

before we can check for exponent overflow in $S_3$. In $S_3$, the *Done* signal is turned on and the controller waits for $St = 0$ before returning to $S_0$.

The state graph for the multiplier control (Figure 7-9) is similar to Figure 4-34, except that the load state is not needed because the registers are loaded by the main controller. Add and shift operations are performed in one state because as seen in Figure 7-7(c), the sum wires from the adder are shifted by 1 before loading into the accumulator register. When $Adx = 1$, the multiplier is started, and *Mdone* is turned on when the multiplication is completed.

FIGURE 7-9:
**State Graph for Multiplier Control**

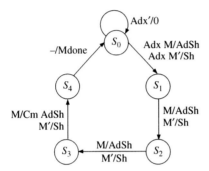

The VHDL behavioral description (Figure 7-10) uses three processes. The main process generates control signals based on the SM chart. A second process generates the control signals for the fraction multiplier. The third process tests the control signals and updates the appropriate registers on the rising edge of the clock. In state $S_2$ of the main process, $A = $ "0000" implies that $F = 0$ ($FZ = 1$ on the SM chart). If we multiply $1.000 \times 1.000$, the result is $A$ & $B = $ "01000000", and a fraction overflow has occurred ($FV = 1$). If $A(2) = A(1)$, the sign bit of $F$ and the following bit are the same and $F$ is unnormalized ($Fnorm = 0$). In state $S_3$, if the two high-order bits of $X$ are different, an exponent overflow has occurred ($EV = 1$).

The registers are updated in the third process. The variable *addout* represents the output of the 4-bit full adder, which is part of the fraction multiplier. This adder adds the 2's complement of $C$ to $A$ when $Cm = 1$. When $Load = 1$, the sign-extended exponents are loaded into $X$ and $Y$. When $Adx = 1$, vectors $X$ and $Y$ are

FIGURE 7-10: **VHDL Code for Floating-Point Multiplier**

```vhdl
library IEEE;
use IEEE.numeric_bit.all;

entity FMUL is
  port(CLK, St: in bit;
       F1, E1, F2, E2: in unsigned(3 downto 0);
       F: out unsigned(6 downto 0);
       V, done: out bit);
end FMUL;
```

```
architecture FMULB of FMUL is
signal A, B, C: unsigned(3 downto 0);          -- fraction registers
signal X, Y: unsigned(4 downto 0);             -- exponent registers
signal Load, Adx, SM8, RSF, LSF: bit;
signal AdSh, Sh, Cm, Mdone: bit;
signal PS1, NS1: integer range 0 to 3;         -- present and next state
signal State, Nextstate: integer range 0 to 4; -- multiplier control state
begin
  main_control: process(PS1, St, Mdone, X, A, B)
  begin
    Load <= '0'; Adx <= '0'; NS1 <= 0;         -- clear control signals
    SM8 <= '0'; RSF <= '0'; LSF <= '0'; V <= '0'; F <= "0000000";
    done <= '0';
    case PS1 is
      when 0 => F <= "0000000";                -- clear outputs
        done <= '0'; V <= '0';
        if St = '1' then Load <= '1'; NS1 <= 1; end if;
      when 1 => Adx <= '1'; NS1 <= 2;
      when 2 =>
        if Mdone = '1' then                    -- wait for multiply
          if A = 0 then                        -- zero fraction
            SM8 <= '1';
          elsif A = 4 and B = 0 then
            RSF <= '1';                         -- shift AB right
          elsif A(2) = A(1) then                -- test for unnormalized
            LSF <= '1';                         -- shift AB left
          end if;
          NS1 <= 3;
        else
          NS1 <= 2;
        end if;
      when 3 =>                                 -- test for exp overflow
        if X(4) /= X(3) then V <= '1'; else V <= '0'; end if;
        done <= '1';
        F <= A(2 downto 0) & B;                 -- output fraction
        if ST = '0' then NS1 <= 0; end if;
    end case;
  end process main_control;

  mul2c: process(State, Adx, B)                 -- 2's complement multiply
  begin
    AdSh <= '0'; Sh <= '0'; Cm <= '0'; Mdone <= '0';  -- clear control signals
    Nextstate <= 0;
    case State is
      when 0 =>                                 -- start multiply
        if Adx = '1' then
          if B(0) = '1' then AdSh <= '1'; else Sh <= '1'; end if;
          Nextstate <= 1;
        end if;
```

```
      when 1 | 2 =>                                     -- add/shift state
        if B(0) = '1' then AdSh <= '1'; else Sh <= '1'; end if;
        Nextstate <= State + 1;
      when 3 =>
        if B(0) = '1' then Cm <= '1'; AdSh <= '1'; else Sh <= '1'; end if;
        Nextstate <= 4;
      when 4 =>
        Mdone <= '1'; Nextstate <= 0;
    end case;
  end process mul2c;

  update: process                                       -- update registers
  variable addout: unsigned(3 downto 0);
  begin
    wait until CLK = '1' and CLK'event;
    PS1 <= NS1;
    State <= Nextstate;
    if Cm = '0' then addout := A + C;
    else addout := A - C;
    end if;                                             -- add 2's comp. of C
    if Load = '1' then
      X <= E1(3) & E1; Y <= E2(3) & E2;
      A <= "0000"; B <= F1; C <= F2;
    end if;
    if ADX = '1' then X <= X + Y; end if;
    if SM8 = '1' then X <= "11000"; end if;
    if RSF = '1' then A <= '0' & A(3 downto 1);
      B <= A(0) & B(3 downto 1);
      X <= X + 1;
    end if;                                             -- increment X
    if LSF = '1' then
      A <= A(2 downto 0) & B(3); B <= B(2 downto 0) & '0';
      X <= X + 31;
    end if;                                             -- decrement X
    if AdSh = '1' then
      A <= (C(3) xor Cm) & addout(3 downto 1);          -- load shifted adder
      B <= addout(0) & B(3 downto 1);                   -- output into A & B
    end if;
    if Sh = '1' then
      A <= A(3) & A(3 downto 1);                        -- right shift A & B
      B <= A(0) & B(3 downto 1);                        -- with sign extend
    end if;
  end process update;
end FMULB;
```

added. When $SM8 = 1$, $-8$ is loaded into $X$. When $AdSh = 1$, $A$ is loaded with the sign bit of $C$ (or the complement of the sign bit if $Cm = 1$), concatenated with bits **3 downto** 1 of the adder output, and the remaining bit of *addout* is shifted into the $B$ register.

Testing the VHDL code for the floating-point multiplier must be done carefully to account for all the special cases in combination with positive and negative fractions, as well as positive and negative exponents. Figure 7-11 shows a command file and some test results. This is not a complete test.

When the VHDL code was synthesized for the Xilinx Spartan-3/Virtex-4 architectures using the Xilinx ISE tools, the result was 38 slices, 29 flip-flops, 72 four-input LUTs, 27 I/O blocks, and one global clock circuitry. The output signals *V*, *Done*, and

**FIGURE 7-11:** Test Data and Simulation Results for Floating-Point Multiplier

```
add list f x f1 e1 f2 e2 v done
force f1 0111 0, 1001 200, 1000 400, 0000 600, 0111 800
force e1 0001 0, 1001 200, 0111 400, 1000 600, 0111 800
force f2 0111 0, 1001 200, 1000 400, 0000 600, 1001 800
force e2 1000 0, 0001 200, 1001 400, 1000 600, 0001 800
force st 1 0, 0 20, 1 200, 0 220, 1 400, 0 420, 1 600, 0 620, 1 800, 0 820
force clk 0 0, 1 10 -repeat 20
run 1000
```

| ns | delta | f | x | f1 | e1 | f2 | e2 | v | done | |
|----|-------|---|---|----|----|----|----|---|------|---|
| 0 | +0 | 0000000 | 00000 | 0000 | 0000 | 0000 | 0000 | 0 | 0 | |
| 0 | +1 | 0000000 | 00000 | 0111 | 0001 | 0111 | 1000 | 0 | 0 | $(0.111 \times 2^1) \times (0.111 \times 2^{-8})$ |
| 10 | +1 | 0000000 | 00001 | 0111 | 0001 | 0111 | 1000 | 0 | 0 | |
| 30 | +1 | 0000000 | 11001 | 0111 | 0001 | 0111 | 1000 | 0 | 0 | |
| 150 | +2 | 0110001 | 11001 | 0111 | 0001 | 0111 | 1000 | 0 | 1 | $= 0.110001 \times 2^{-7}$ |
| 170 | +2 | 0000000 | 11001 | 0111 | 0001 | 0111 | 1000 | 0 | 0 | |
| 200 | +0 | 0000000 | 11001 | 1001 | 1001 | 1001 | 0001 | 0 | 0 | $(1.001 \times 2^{-7}) \times (1.001 \times 2^1)$ |
| 250 | +1 | 0000000 | 11010 | 1001 | 1001 | 1001 | 0001 | 0 | 0 | |
| 370 | +2 | 0110001 | 11010 | 1001 | 1001 | 1001 | 0001 | 0 | 1 | $= 0.110001 \times 2^{-6}$ |
| 390 | +2 | 0000000 | 11010 | 1001 | 1001 | 1001 | 0001 | 0 | 0 | |
| 400 | +0 | 0000000 | 11010 | 1000 | 0111 | 1000 | 1001 | 0 | 0 | $(1.000 \times 2^7) \times (1.000 \times 2^{-7})$ |
| 430 | +1 | 0000000 | 00111 | 1000 | 0111 | 1000 | 1001 | 0 | 0 | |
| 450 | +1 | 0000000 | 00000 | 1000 | 0111 | 1000 | 1001 | 0 | 0 | |
| 570 | +1 | 0000000 | 00001 | 1000 | 0111 | 1000 | 1001 | 0 | 0 | |
| 570 | +2 | 0100000 | 00001 | 1000 | 0111 | 1000 | 1001 | 0 | 1 | $= 0.100000 \times 2^1$ |
| 590 | +2 | 0000000 | 00001 | 1000 | 0111 | 1000 | 1001 | 0 | 0 | |
| 600 | +0 | 0000000 | 00001 | 0000 | 1000 | 0000 | 1000 | 0 | 0 | $(0.000 \times 2^{-8}) \times (0.000 \times 2^{-8})$ |
| 630 | +1 | 0000000 | 11000 | 0000 | 1000 | 0000 | 1000 | 0 | 0 | |
| 650 | +1 | 0000000 | 10000 | 0000 | 1000 | 0000 | 1000 | 0 | 0 | |
| 770 | +1 | 0000000 | 11000 | 0000 | 1000 | 0000 | 1000 | 0 | 0 | |
| 770 | +2 | 0000000 | 11000 | 0000 | 1000 | 0000 | 1000 | 0 | 1 | $= 0.0000000 \times 2^{-8}$ |
| 790 | +2 | 0000000 | 11000 | 0000 | 1000 | 0000 | 1000 | 0 | 0 | |
| 800 | +0 | 0000000 | 11000 | 0111 | 0111 | 1001 | 0001 | 0 | 0 | $(0.111 \times 2^7) \times (1.001 \times 2^1)$ |
| 830 | +1 | 0000000 | 00111 | 0111 | 0111 | 1001 | 0001 | 0 | 0 | |
| 850 | +1 | 0000000 | 01000 | 0111 | 0111 | 1001 | 0001 | 0 | 0 | |
| 970 | +2 | 1001111 | 01000 | 0111 | 0111 | 1001 | 0001 | 1 | 1 | $= 1.001111 \times 2^8$ (overflow) |
| 990 | +2 | 0000000 | 01000 | 0111 | 0111 | 1001 | 0001 | 0 | 0 | |

*F* were set to zero at the beginning of the process to eliminate unwanted latches. An RTL-level design was also attempted, but the RTL design was not superior to the synthesized behavioral design.

Now that the basic design has been completed, we need to determine how fast the floating-point multiplier will operate and determine the maximum clock frequency. Most CAD tools provide a way of simulating the final circuit taking into account both the delays within the logic blocks and the interconnection delays. If this timing analysis indicates that the design does not operate fast enough to meet specifications, several options are possible. Most FPGAs come in several different speed grades, so one option is to select a faster part. Another approach is to determine the longest delay path in the circuit and attempt to reroute the connections or redesign that part of the circuit to reduce the delays.

● ● ● ● ● ● ● ● ● ● ● ● ●
## 7.3 Floating-Point Addition

Next, we consider the design of an adder for floating-point numbers. Two floating-point numbers will be added to form a floating-point sum:

$$(F_1 \times 2^{E_1}) + (F_2 \times 2^{E_2}) = F \times 2^E$$

Again, we will assume that the numbers to be added are properly normalized and that the answer should be put in normalized form. In order to add two fractions, the associated exponents must be equal. Thus, if the exponents $E_1$ and $E_2$ are different, we must unnormalize one of the fractions and adjust the exponent accordingly. The smaller number is the one that should be adjusted so that if significant digits are lost, the effect is not significant. To illustrate the process, we add

$$F_1 \times 2^{E_1} = 0.111 \times 2^5 \text{ and } F_2 \times 2^{E_2} = 0.101 \times 2^3$$

Since $E_2 \neq E_1$, we unnormalize the smaller number $F_2$ by shifting right two times and adding 2 to the exponent:

$$0.101 \times 2^3 = 0.0101 \times 2^4 = 0.00101 \times 2^5$$

Note that shifting right one place is equivalent to dividing by 2, so each time we shift we must add 1 to the exponent to compensate. When the exponents are equal, we add the fractions:

$$(0.111 \times 2^5) + (0.00101 \times 2^5) = 01.00001 \times 2^5$$

This addition caused an overflow into the sign bit position, so we shift right and add 1 to the exponent to correct the fraction overflow. The final result is

$$F \times 2^E = 0.100001 \times 2^6$$

When one of the fractions is negative, the result of adding fractions may be unnormalized, as illustrated in the following example:

$(1.100 \times 2^{-2}) + (0.100 \times 2^{-1})$

$= (1.110 \times 2^{-1}) + (0.100 \times 2^{-1})$ (after shifting $F_1$)

$= 0.010 \times 2^{-1}$ (result of adding fractions is unnormalized)

$= 0.100 \times 2^{-2}$ (normalized by shifting left and subtracting 1 from exponent)

In summary, the steps required to carry out floating-point addition are as follows:

1. Compare exponents. If the exponents are not equal, shift the fraction with the smaller exponent right and add 1 to its exponent; repeat until the exponents are equal.
2. Add the fractions (significands).
3. If the result is 0, set the exponent to the appropriate representation for 0 and exit.
4. If fraction overflow occurs, shift right and add 1 to the exponent to correct the overflow.
5. If the fraction is unnormalized, shift left and subtract 1 from the exponent until the fraction is normalized.
6. Check for exponent overflow. Set overflow indicator, if necessary.
7. Round to the appropriate number of bits.
   Still normalized? If not, go back to step 4.

Figure 7-12 illustrates this procedure graphically. An optimization can be added to step 1. We can identify cases where the two numbers are vastly different. If $E_1 \gg E_2$ and $F_2$ is positive, $F_2$ will become all 0's as we right shift $F_2$ to equalize the exponents. In this case, the result is $F = F_1$ and $E = E_1$, so it is a waste of time to do the shifting. If $E_1 \gg E_2$ and $F_2$ is negative, $F_2$ will become all 1's (instead of all 0's) as we right shift $F_2$ to equalize the exponents. When we add the fractions, we will get the wrong answer. To avoid this problem, we can skip the shifting when $E_1 \gg E_2$ and set $F = F_1$ and $E = E_1$. Similarly, if $E_2 \gg E_1$, we can skip the shifting and set $F = F_2$ and $E = E_2$.

For the 4-bit fractions in our example, if $|E_1 - E_2| > 3$, we can skip the shifting. For IEEE single precision numbers, there are 23 bits after the binary point; hence if the exponent difference is greater than 23, the smaller number will become 0 before the exponents are equal. In general, if the exponent difference is greater than the number of available fractional bits, the sum should be set to the larger number. If $E_1 \gg E_2$, set $F = F_1$ and $E = E_1$. If $E_2 \gg E_1$, set $F = F_2$ and $E = E_2$.

Inspection of this procedure illustrates that the following hardware units are required to implement a floating-point adder:

- Adder (subtractor) to compare exponents (step 1a)
- Shift register to shift the smaller number to the right (step 1b)
- ALU (adder) to add fractions (step 2)
- Bidirectional shifter, incrementer/decrementer (steps 4, 5)
- Overflow detector (step 6)
- Rounding hardware (step 7)

FIGURE 7-12: **Flow Chart for Floating-Point Addition**

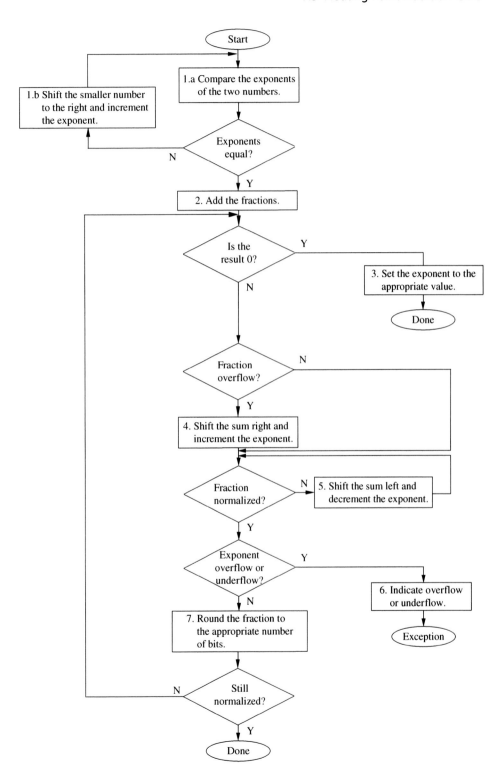

Many of these components can be combined. For instance, the register that stores the fractions can be made a shift register in order to perform the shifts. The register that stores the exponent can be a counter with increment/decrement capability. Figure 7-13 shows a hardware arrangement for the floating-point adder. The major components are the exponent comparator and the fraction adder. Fraction addition can be done using 2's complement addition. It is assumed that the operands are delivered on an I/O bus. If the numbers are in a sign-magnitude form as in the IEEE format, they can be converted to 2's complement numbers and then added. Special cases should be handled according to the requirements of the format. The sum is written back into the Addend register in Figure 7-13.

FIGURE 7-13:
**Overview of a Floating-Point Adder**

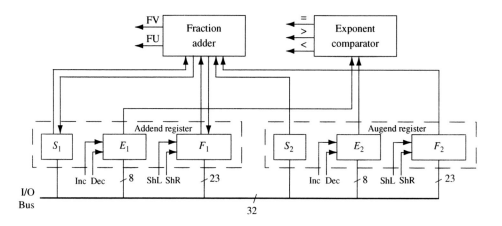

Figure 7-14 shows VHDL code for a floating-point adder based on the IEEE single precision floating-point format. This code is not a complete implementation of the standard. It handles the special case of 0, but it does not deal with infinity, unnormalized, and not-a-number formats. The final result is truncated instead of rounded. Sign and magnitude format and biased exponents are used throughout, except 2's complement is used for the fraction addition.

FPinput is an input bus, and we assume that the input numbers represent normalized floating-point numbers in IEEE standard format. In state 0, the first number is split and loaded into $S_1$, $F_1$, and $E_1$. These represent the sign of the fraction, the magnitude of the fraction, and the biased exponent. When $F_1$ is loaded, the 23-bit fraction is prefixed by a 1 except in the special case of 0, in which case the leading bit is a 0. Two 0's are appended at the end of the fraction to conform to the IEEE standard requirements (guard and round bits). In state 1, the second number to be added is loaded into $S_2$, $F_2$, and $E_2$. In state 2, the fraction with the smallest exponent is unnormalized by shifting right and incrementing the exponent. When this operation is complete, the exponents are equal, except in the special case when $F_1$ or $F_2$ equals 0.

FIGURE 7-14: **VHDL Code for a Floating-Point Adder**

```vhdl
library IEEE;
use IEEE.numeric_bit.all;

entity FPADD is
  port(CLK, St: in bit; done, ovf, unf: out bit;
       FPinput: in unsigned(31 downto 0);   -- IEEE single precision FP format
       FPsum: out unsigned(31 downto 0));   -- IEEE single precision FP format
end FPADD;

architecture FPADDER of FPADD is
 -- F1 and F2 store significand with leading 1 and trailing 0's added
signal F1, F2: unsigned(25 downto 0);
signal E1, E2: unsigned(7 downto 0);   -- exponents
signal S1, S2, FV, FU: bit;
 -- intermediate results for 2's complement addition
signal F1comp, F2comp, Addout, Fsum: unsigned(27 downto 0);
signal State: integer range 0 to 6;
begin  -- convert fractions to 2's comp and add
  F1comp <= not ("00" & F1) + 1 when S1 = '1' else "00" & F1;
  F2comp <= not ("00" & F2) + 1 when S2 = '1' else "00" & F2;
  Addout <= F1comp + F2comp;
   -- find magnitude of sum
  Fsum <= Addout when Addout(27) = '0' else not Addout + 1;
  FV <= Fsum(27) xor Fsum(26);        -- fraction overflow
  FU <= not F1(25);                   -- fraction underflow
  FPsum <= S1 & E1 & F1(24 downto 2); -- pack output word
  process(CLK)
  begin
    if CLK'event and CLK = '1' then
      case State is
        when 0 =>
          if St = '1' then -- load E1 and F1
            E1 <= FPinput(30 downto 23); S1 <= FPinput(31);
            F1(24 downto 0) <= FPinput(22 downto 0) & "00";
            -- insert 1 in significand (or 0 if the input number is 0)
            if FPinput = 0 then F1(25) <= '0'; else F1(25) <= '1'; end if;
            done <= '0'; ovf <= '0'; unf <= '0'; State <= 1;
          end if;
        when 1 =>  -- load E2 and F2
          E2 <= FPinput(30 downto 23); S2 <= FPinput(31);
          F2(24 downto 0) <= FPinput(22 downto 0) & "00";
          if FPinput = 0 then F2(25) <= '0'; else F2(25) <= '1'; end if;
          State <= 2;
        when 2 =>  -- unnormalize fraction with smallest exponent
         if F1 = 0 or F2 = 0 then State <= 3;
         else
           if E1 = E2 then State <= 3;
```

```
              elsif E1 < E2 then
                F1 <= '0' & F1(25 downto 1); E1 <= E1 + 1;
              else
                F2 <= '0' & F2(25 downto 1); E2 <= E2 + 1;
              end if;
            end if;
          when 3 => -- add fractions and check for fraction overflow
            S1 <= Addout(27);
            if FV = '0' then F1 <= Fsum(25 downto 0);
            else F1 <= Fsum(26 downto 1); E1 <= E1 + 1; end if;
            State <= 4;
          when 4 => -- check for sum of fractions = 0
            if F1 = 0 then E1 <= "00000000"; State <= 6;
            else State <= 5; end if;
          when 5 => -- normalize
            if E1 = 0 then unf <= '1'; State <= 6;
            elsif FU = '0' then State <= 6;
            else F1 <= F1(24 downto 0) & '0'; E1 <= E1 - 1;
            end if;
          when 6 => -- check for exponent overflow
            if E1 = 255 then ovf <= '1'; end if;
            done <= '1'; State <= 0;
        end case;
      end if;
    end process;
  end FPADDER;
```

The fractions are added using 2's complement arithmetic, which is performed by concurrent statements. The input numbers are first converted to 2's complement representation. Two sign bits (00) are prefixed to $F_1$, and the 2's complement is formed if $S_1$ is 1 (negative). Two sign bits are used so that the sign is not lost if the fraction addition overflows into the first sign bit. $F_2$ is processed in a similar way. The resulting numbers, *F1comp* and *F2comp*, are added and the sum is assigned to *Addout*. The adder output is read in state 3. *Fsum* represents the magnitude of the fraction, so Addout must be complemented if it is negative. Normally the two sign bits of Fsum are "00", so they are discarded and the result is stored back in $F_1$, which serves as a floating-point accumulator. The sign bit is extracted from the MSB of *Addout*. Fraction overflow and underflow are indicated by *FV* and *FU*, respectively. Fraction overflow can be detected by exclusive-OR of the highest two bits of *Addout*. This is done as a concurrent statement. In case of fraction overflow, the sign bits of *Fsum* are "01", so *FV* = '1', Fsum is right shifted before it is stored in $F_1$, and $E_1$ is incremented. If the result of addition F1 = 0, E1 is set to 0 in state 4, and the floating-point addition is complete. If $F_1$ is unnormalized, it is normalized in state 5 by shifting $F_1$ left and decrementing $E_1$. Exponent overflow and underflow are represented by ovf and unf, respectively. Since the normal range of biased exponents is 1 to 254, an underflow occurs if $E_1$ is decremented to 0, and unf is set to '1' before exiting state 5. In state 6, if $E_1$ = 255, this indicates an exponent overflow, and ovf is set to '1'. The done signal is turned on before exiting state 6. $S_1$, $E_1$, and $F_1$ are merged by a concurrent statement to give the final sum, *FPsum*, in IEEE format.

The floating-point adder was tested for the following cases.

| Addend | | Augend | | Expected Result | |
|---|---|---|---|---|---|
| Number (Binary) | IEEE Single Precision | Number (Binary) | IEEE Single Precision | Number (Binary) | IEEE Single Precision |
| 0 | x00000000 | 0 | x00000000 | 0 | x00000000 |
| $1 \times 2^0$ | x3F800000 | $1 \times 2^0$ | x3F800000 | $1 \times 2^1$ | x40000000 |
| $-1 \times 2^0$ | xBF800000 | $-1 \times 2^0$ | xBF800000 | $-1 \times 2^1$ | xC0000000 |
| $1 \times 2^0$ | x3F800000 | $-1 \times 2^0$ | xBF800000 | 0 | x00000000 |
| $1.111 \ldots \times 2^{127}$ | x7F7FFFFF | $1 \times 2^0$ | x3F800000 | $1.111 \ldots \times 2^{127}$ | x7F7FFFFF |
| $-1.111 \ldots \times 2^{127}$ | xFF7FFFFF | $-1 \times 2^0$ | xBF800000 | $-1.111 \ldots \times 2^{127}$ | xFF7FFFFF |
| $1.111 \ldots \times 2^{127}$ | x7F7FFFFF | $1.111 \ldots \times 2^{127}$ | x7F7FFFFF | overflow | |
| $-1.111 \ldots \times 2^{127}$ | xFF7FFFFF | $-1.111 \ldots \times 2^{127}$ | xFF7FFFFF | overflow | |
| $1.11 \times 2^8$ | x43E00000 | $-1.11 \times 2^6$ | xC2E00000 | $1.0101 \times 2^8$ | x43A80000 |
| $-1.11 \times 2^8$ | xC3E00000 | $1.11 \times 2^6$ | x42E00000 | $-1.0101 \times 2^8$ | xC3A80000 |
| $1.111 \ldots \times 2^{127}$ | x7F7FFFFF | $0.0 \ldots 01 \times 2^{127}$ | x73800000 | overflow | |
| $-1.111 \ldots \times 2^{127}$ | xFF7FFFFF | $-0.0 \ldots 01 \times 2^{127}$ | xF3800000 | overflow | |
| $1.1 \ldots 10 \times 2^{127}$ | x7F7FFFFE | $0.0 \ldots 01 \times 2^{127}$ | x73800000 | $1.111 \ldots \times 2^{127}$ | x7F7FFFFF |
| $-1.1 \ldots 10 \times 2^{127}$ | xFF7FFFFE | $-0.0 \ldots 01 \times 2^{127}$ | xF3800000 | $-1.111 \ldots \times 2^{127}$ | xFF7FFFFF |
| $1.1 \times 2^{-126}$ | X00C00000 | $-1.0 \times 2^{-126}$ | x80800000 | underflow | |

● ● ● ● ● ● ● ● ● ● ● ●

# 7.4 **Other Floating-Point Operations**

### 7.4.1 **Subtraction**

Floating-point subtraction is the same as floating-point addition, except that we must subtract the fractions instead of adding them. The rest of the steps remain the same.

### 7.4.2 **Division**

The quotient of two floating-point numbers is

$$(F_1 \times 2^{E_1}) \div (F_2 \times 2^{E_2}) = (F_1/F_2) \times 2^{(E_1 - E_2)} = F \times 2^E$$

Thus, the basic rule for floating-point division is to divide the fractions and subtract the exponents. In addition to considering the same special cases as for multiplication, we must test for divide by 0 before dividing. If $F_1$ and $F_2$ are normalized, then the largest positive quotient ($F$) will be

$$0.1111 \ldots / 0.1000 \ldots = 01.111 \ldots$$

which is less than $10_2$, so the fraction overflow is easily corrected. For example,

$$(0.110101 \times 2^2) \div (0.101 \times 2^{-3}) = 01.010 \times 2^5 = 0.101 \times 2^6$$

Alternatively, if $F_1 \geq F_2$, we can shift $F_1$ right before dividing and avoid fraction overflow in the first place. In the IEEE format, when divide by 0 is involved, the result can be set to NaN (Not a Number).

In this chapter, we presented different representations of floating-point numbers. IEEE floating-point single precision and double precision formats were discussed. A floating-point format with 2's complement numbers was also presented. Then we presented a floating-point multiplier. We also presented a procedure to perform addition of floating-point numbers. In the process of designing the multiplier, we used the following steps:

**1.** Develop an algorithm for floating-point multiplication, taking all of the special cases into account.
**2.** Draw a block diagram of the system and define the necessary control signals.
**3.** Construct an SM chart (or state graph) for the control state machine using a separate linked state machine for controlling the fraction multiplier.
**4.** Write behavioral VHDL code.
**5.** Test the VHDL code to verify that the high-level design of the multiplier is correct.
**6.** Use the CAD software to synthesize the multiplier. Then implement the multiplier in the desired target technology (e.g., ASIC, FPGA, etc.).

● ● ● ● ● ● ● ● ● ● ● ●
# Problems

**7.1** **(a)** What is the biggest number that can be represented in the 8-bit 2's complement floating-point format with 4 bits for exponent and 4 for fraction?
**(b)** What is the smallest number that can be represented in the 8-bit 2's complement format with 4 bits for exponent and 4 for fraction?
**(c)** What is the biggest normalized number that can be represented in the IEEE single precision floating-point format?
**(d)** What is the smallest normalized number that can be represented in the IEEE single precision floating-point format?
**(e)** What is the biggest normalized number that can be represented in the IEEE double precision floating-point format?
**(f)** What is the smallest normalized number that can be represented in the IEEE double precision floating-point format?

**7.2** Convert the following decimal numbers in the IEEE single precision format.

**(i)** 25.25, **(ii)** 2000.25, **(iii)** 1, **(iv)** 0, **(v)** 1000, **(vi)** 8000, **(vii)** $10^6$, **(viii)** $-5.4$, **(ix)** $1.0 \times 2^{-140}$, **(x)** $1.5 \times 10^9$

**7.3** Convert the following decimal numbers to IEEE double precision format.

**(i)** 25.25, **(ii)** 2000.25, **(iii)** 1, **(iv)** 0, **(v)** 1000, **(vi)** 8000, **(vii)** $10^6$, **(viii)** $-5.4$, **(ix)** $1.0 \times 2^{-140}$, **(x)** $1.5 \times 10^9$

**7.4** What do the following hex representations mean if they are in IEEE single precision format?

**(i)** ABABABAB, **(ii)** 45454545, **(iii)** FFFFFFFF, **(iv)** 00000000, **(v)** 11111111, **(vi)** 01010101

7.5 What do the following hex representations mean if they are in IEEE double precision format?

**(i)** ABABABAB 00000000, **(ii)** 45454545 00000001, **(iii)** FFFFFFFF 10001000, **(iv)** 00000000 00000000, **(v)** 11111111 10001000, **(vi)** 01010101 01010101

7.6 **(a)** Represent $-35.25$ in IEEE single precision floating-point format.
   **(b)** What does the hex number ABCD0000 represent if it is in IEEE single precision floating-point format?

7.7 **(a)** Represent 25.625 in IEEE single precision floating-point format.
   **(b)** Represent $-15.6$ in IEEE single precision floating-point format.

7.8 This problem concerns the design of a digital system that converts an 8-bit signed integer (negative numbers are represented in 2's complement) to a floating-point number. Use a floating-point format similar to the ones used in Section 7.1.1 except the fraction should be 8 bits and the exponent 4 bits. The fraction should be properly normalized.

   **(a)** Draw a block diagram of the system and develop an algorithm for doing the conversion. Assume that the integer is already loaded into an 8-bit register, and when the conversion is complete the fraction should be in the same register. Illustrate your algorithm by converting –27 to floating point.
   **(b)** Draw a state diagram for the controller. Assume that the start signal is present for only one clock time. (Two states are sufficient.)
   **(c)** Write a VHDL description of the system.

7.9 **(a)** Multiply the following two floating-point numbers to give a properly normalized result. Assume 4-bit 2's complement format.

$$F_1 = 1.011, E_1 = 0101, F_2 = 1.001, E_2 = 0011$$

   **(b)** Repeat (a) for

$$F_1 = 1.011, E_1 = 1011, F_2 = 0.110, E_2 = 1101$$

7.10 A floating-point number system uses a 4-bit fraction and a 4-bit exponent with negative numbers expressed in 2's complement. Design an *efficient* system that will multiply the number by $-4$ (minus four). Take all special cases into account, and give a properly normalized result. Assume that the initial fraction is properly normalized or zero. *Note*: This system multiplies *only* by $-4$.

   **(a)** Give examples of the normal and special cases that can occur (for multiplication by $-4$).
   **(b)** Draw a block diagram of the system.
   **(c)** Draw an SM chart for the control unit. Define all signals used.

7.11 Redesign the floating-point multiplier in Figure 7-7 using a common 5-bit full adder connected to a bus instead of two separate adders for the exponents and fractions.

(a) Redraw the block diagram and be sure to include the connections to the bus and include all control signals.
(b) Draw a new SM chart for the new control.
(c) Write the VHDL description for the multiplier or specify what changes need to be made to an existing description.

7.12 This problem concerns the design of a circuit to find the square of a floating-point number, $F \times 2^E$. $F$ is a normalized 5-bit fraction, and $E$ is a 5-bit integer; negative numbers are represented in 2's complement. The result should be properly normalized. Take advantage of the fact that $(-F)^2 = F^2$.

(a) Draw a block diagram of the circuit. (Use only one adder and one complementer.)
(b) State your procedure, taking all special cases into account. Illustrate your procedure for

$$F = 1.0110 \qquad E = 00100$$

(c) Draw an SM chart for the main controller. You may assume that multiplication is carried out using a separate control circuit, which outputs $Mdone = 1$ when multiplication is complete.
(d) Write a VHDL description of the system.

7.13 Write a behavioral VHDL code for a floating-point multiplier using the IEEE single precision floating-point format. Use an overloaded multiplication operator instead of using an add-shift multiplier. Ignore special cases like infinity, denormalized, and not-a-number formats. Truncate the final result instead of rounding.

7.14 Write a test bench for the floating-point adder of Figure 7-14.

7.15 Add the following floating-point numbers (show each step). Assume that each fraction is 5 bits (including sign) and each exponent is 5 bits (including sign) with negative numbers in 2's complement.

$$F_1 = 0.1011 \qquad E_1 = 11111$$
$$F_2 = 1.0100 \qquad E_2 = 11101$$

7.16 Two floating-point numbers are added to form a floating-point sum:

$$(F_1 \times 2^{E_1}) + (F_2 \times 2^{E_2}) = F \times 2^E$$

Assume that $F_1$ and $F_2$ are normalized, and the result should be normalized.

(a) List the steps required to carry out floating-point addition, including all special cases.
(b) Illustrate these steps for $F_1 = 1.0101, E_1 = 1001, F_2 = 0.1010, E_2 = 1000$. Note that the fractions are 5 bits, including sign, and the exponents are 4 bits, including sign.
(c) Write a VHDL description of the system.

7.17 For the floating-point adder of Figure 7-14, modify the VHDL code so that

    **(a)** It handles IEEE standard single precision denormalized numbers both as input and output.

    **(b)** In state 2, it speeds up the processing when the exponents differ by more than 23.

    **(c)** It rounds up instead of truncating the resulting fraction.

7.18 **(a)** Add the floating-point numbers $0.111 \times 2^5 + 0.101 \times 2^3$ and normalize the result.

    **(b)** Draw an SM chart for a floating-point adder that adds $(F_1 \times 2^{E_1})$ and $(F_2 \times 2^{E_2})$. Assume that the fractions are initially normalized (or zero) and the final result should be normalized (or zero). A zero fraction should have an exponent of $-8$. Set an exponent overflow flag $(EV)$ if the final answer has an exponent overflow. Each number to be added consists of a 4-bit fraction and a 4-bit exponent, with negative numbers represented in 2's complement. Assume that all registers $(F_1, E_1, F_2,$ and $E_2)$ can be loaded in one clock time when a start signal $(St)$ is received. If $E_1 > E_2$, the control signal $GT = 1$, and if $E_1 < E_2$, the control signal $LT = 1$. Define all other control signals used. Include the special case where $|E_1 - E_2| > 3$.

7.19 **(a)** Draw a block diagram for a floating-point subtracter. Assume that the inputs to the subtracter are properly normalized, and the answer should be properly normalized. The fractions are 8 bits including sign, and the exponents are 5 bits including sign. Negative numbers are represented in 2's complement.

    **(b)** Draw an SM chart for the control circuit for the floating-point subtracter. Define the control signals used, and give an equation for each control signal used as an input to the control circuit.

    **(c)** Write the VHDL description of the floating-point subtracter.

7.20 **(a)** State the steps necessary to carry out floating-point subtraction, including special cases. Assume that the numbers are initially in normalized form, and the final result should be in normalized form.

    **(b)** Subtract the following (fractions are in 2's complement):

$$(1.0111 \times 2^{-3}) - (1.0101 \times 2^{-5})$$

    **(c)** Write a VHDL description of the system. Fractions are 5 bits including sign, and exponents are 4 bits including sign.

7.21 This problem concerns the design of a divider for floating point numbers:

$$(F_1 \times 2^{E_1}) / (F_2 \times 2^{E_2}) = F \times 2^E$$

Assume that $F_1$ and $F_2$ are properly normalized fractions (or 0), with negative fractions expressed in 2's complement. The exponents are integers with negative numbers expressed in 2's complement. The result should be properly normalized if it is not zero. Fractions are 8 bits including sign, and exponents are 5 bits including sign.

**(a)** Draw a flow chart for the floating-point divider. Assume that a divider is available that will divide two binary fractions to give a fraction as a result. Do not show the individual steps in the division of the fractions on your flowchart, just say "divide." The divider requires that $|F_2| > |F_1|$ before division is carried out.

**(b)** Illustrate your procedure by computing

$$0.111 \times 2^3 / 1.011 \times 2^{-2}$$

When you divide $F_1$ by $F_2$, you don't need to show the individual steps, just the result of the division.

**(c)** Write a VHDL description for the system.

7.22 Assume that $A$, $B$, and $C$ are floating-point numbers expressed in IEEE single precision floating-point format and that floating-point addition is performed.

$$\text{If } A = 2^{40}, B = -2^{40}, C = 1, \text{ then}$$

What is $A + (B + C)$? (i.e., $B + C$ done first and then $A$ added to it)
What is $(A + B) + C$? (i.e., $A + B$ done first and then $C$ added to it)

7.23 Assume that $A$, $B$, and $C$ are floating-point numbers expressed in IEEE double precision floating-point format and that floating-point addition is performed.

$$\text{If } A = 2^{40}, B = -2^{40}, C = 1, \text{ then}$$

What is $A + (B + C)$? (i.e., $B + C$ done first and then $A$ added to it)
What is $(A + B) + C$? (i.e., $A + B$ done first and then $C$ added to it)

7.24 Assume that $A$, $B$, and $C$ are floating-point numbers expressed in IEEE single precision floating-point format and that floating-point addition is performed.

$$\text{If } A = 2^{65}, B = -2^{65}, C = 1, \text{ then}$$

What is $A + (B + C)$? (i.e., $B + C$ done first and then $A$ added to it)
What is $(A + B) + C$? (i.e., $A + B$ done first and then $C$ added to it)

7.25 Assume that $A$, $B$, and $C$ are floating-point numbers expressed in IEEE double precision floating-point format and that floating-point addition is performed.

$$\text{If } A = 2^{65}, B = -2^{65}, C = 1, \text{ then}$$

What is $A + (B + C)$? (i.e., $B + C$ done first and then $A$ added to it)
What is $(A + B) + C$? (i.e., $A + B$ done first and then $C$ added to it)

# CHAPTER 8

# Additional Topics in VHDL

Up to this point, we have described the basic features of VHDL and how they can be used in the digital system design process. In this chapter, we describe additional features of VHDL that illustrate its power and flexibility. VHDL functions and procedures are presented. Several additional features, such as attributes, function overloading, and generic and generate statements, are also presented. The IEEE multivalued logic system and principles of signal resolution are described. A simple memory model is presented to illustrate the use of tristate signals.

• • • • • • • • • • • •

## 8.1 VHDL Functions

A key feature of VLSI circuits is the repeated use of similar structures. VHDL provides functions and procedures to easily express repeated invocation of the same functionality or the repeated use of structures. We describe functions in this section. Functions can return only a single value through a return statement. Procedures are more general and complex than functions. They can return any number of values using output parameters. Procedures are described in the next section.

A function executes a sequential algorithm and returns a single value to the calling program. When the following function is called, it returns a bit-vector equal to the input bit-vector (*reg*) rotated one position to the right:

```
function rotate_right (reg: bit_vector)
  return bit_vector is
begin
  return reg ror 1;
end rotate_right;
```

A function call can be used anywhere that an expression can be used. For example, if $A =$ "10010101", the statement

```
B <= rotate_right(A);
```

would set $B$ equal to "11001010", and leave $A$ unchanged.

The general form of a function declaration is

```
function function-name (formal-parameter-list)
  return return-type is
[declarations]
begin
  sequential statements    -- must include return return-value;
end function-name;
```

The general form of a function call is

```
function_name(actual-parameter-list)
```

The number and type of parameters on the `actual-parameter-list` must match the `formal-parameter-list` in the function declaration. The parameters are treated as input values and cannot be changed during the execution of the function.

**Example**

Write a VHDL function for generating an even parity bit for a 4-bit number. The input is a 4-bit number and the output is a code word that contains the data and the parity bit. Figure 8-1 shows the solution.

FIGURE 8-1: **Parity Generation Using a Function**

```
-- Function example code without a loop
-- This function takes a 4-bit vector
-- It returns a 5-bit code with even parity

function parity (A: bit_vector(3 downto 0))
  return bit_vector is

variable parity: bit;
variable B: bit_vector(4 downto 0);
begin
  parity := a(0) xor a(1) xor a(2) xor a(3);
  B := A & parity;
  return B;
end parity;
```

If parity circuits are used in several parts in a system, we could call the function each time it is desired.

Figure 8-2 illustrates a function using a **for** loop. In Figure 8-2, the loop index ($i$) will be initialized to 0 when the **for** loop is entered, and the sequential statements will be executed. Execution will be repeated for $i = 1$, $i = 2$, and $i = 3$; then the loop will terminate.

If $A$, $B$, and $C$ are integers, the statement C <= A + B will set $C$ equal to the sum of $A$ and $B$. However, if $A$, $B$, and $C$ are bit-vectors, this statement will not work, since the "+" operation is not defined for bit-vectors. However, we can write a function to perform bit-vector addition. The function given in Figure 8-2 adds two 4-bit

FIGURE 8-2: **Add Function**

```
-- This function adds two 4-bit vectors and a carry.
-- Illustrates function creation and use of loop
-- It returns a 5-bit sum

function add4 (A, B: bit_vector(3 downto 0); carry: bit)
  return bit_vector is

variable cout: bit;
variable cin: bit := carry;
variable sum: bit_vector(4 downto 0) := "00000";
begin
loop1: for i in 0 to 3 loop
  cout := (A(i) and B(i)) or (A(i) and cin) or (B(i) and cin);
  sum(i) := A(i) xor B(i) xor cin;
  cin := cout;
end loop loop1;
sum(4) := cout;
return sum;
end add4;
```

vectors plus a carry and returns a 5-bit vector as the sum. The function name is *add4*; the formal parameters are *A, B*, and *carry*; and the return type is a bit-vector. Variables *cout* and *cin* are defined to hold intermediate values during the calculation. The variable *sum* is used to store the value to be returned. When the function is called, *cin* will be initialized to the value of the carry. The **for** loop adds the bits of *A* and *B* serially in the same manner as a serial adder. The first time through the loop, *cout* and *sum*(0) are computed using $A(0)$, $B(0)$, and *cin*. Then the *cin* value is updated to the new *cout* value, and execution of the loop is repeated. During the second time through the loop, *cout* and *sum*(1) are computed using $A(1)$, $B(1)$, and the new *cin*. After four times through the loop, all values of *sum(i)* have been computed and *sum* is returned. The total simulation time required to execute the *add4* function is zero. Not even delta time is required, since all the computations are done using variables, and variables are updated instantaneously.

The function call is of the form

```
    add4(A, B, carry)
```

*A* and *B* may be replaced with any expressions that evaluate to bit-vectors with dimensions 3 **downto** 0, and *carry* may be replaced with any expression that evaluates to a bit. For example, the statement

```
    Z <= add4(X, not Y, '1');
```

calls the function *add4*. Parameters *A, B*, and *carry* are set equal to the values of *X*, **not** *Y*, and '1', respectively. *X* and *Y* must be bit-vectors dimensioned 3 **downto** 0. The function computes

$$Sum = A + B + carry = X + \textbf{not } Y + \text{'1'}$$

and returns this value. Since *Sum* is a variable, computation of *Sum* requires zero time. After delta time, $Z$ is set equal to the returned value of *Sum*. Since **not** $Y$ + '1' equals the 2's complement of $Y$, the computation is equivalent to subtracting by adding the 2's complement. If we ignore the carry stored in $Z(4)$, the result is $Z(3$ **downto** $0) = X - Y$.

Functions can be used to return an array. As an example, we will write a function that inputs an array of numbers and returns an array which contains the square of the input numbers. Figure 8-3 illustrates the function as well as the function call. The number of input numbers is provided as a parameter to the function. In the illustrated call to the function, the numbers are 4 bits wide.

**FIGURE 8-3: A Function to Compute Squares of an Array of Unsigned Numbers and Its Call**

```
library IEEE;
use IEEE.numeric_bit.all;

entity test_squares is
  port(CLK: in bit);
end test_squares;

architecture test of test_squares is
type FourBitNumbers is array (0 to 4) of unsigned (3 downto 0);
type squareNumbers is array (0 to 4) of unsigned (7 downto 0);
constant FN: FourBitNumbers := ("0001", "1000", "0011", "0010", "0101");
signal answer: squareNumbers;
signal length: integer := 4;

function squares (Number_arr: FourBitNumbers; length: positive)
  return squareNumbers is

variable SN: squareNumbers;
begin
   loop1: for i in 0 to length loop
         SN(i) := Number_arr(i) * Number_arr(i);
   end loop loop1;
return SN;
end squares;

begin
  process(CLK)
  begin
    if CLK = '1' and CLK'EVENT then
     answer <= squares(FN, length);
    end if;
  end process;
end test;
```

Functions are frequently used to do type conversions. We already came across type conversion functions in the IEEE numeric_bit library: `to_integer(A)` and `to_unsigned(B, N)`. The first one converts an unsigned-vector to an integer, and the second one converts an integer to an unsigned-vector with the specified number of bits.

● ● ● ● ● ● ● ● ● ● ● ●
## 8.2  VHDL Procedures

Procedures facilitate decomposition of VHDL code into modules. Unlike functions, which return only a single value through a return statement, procedures can return any number of values using output parameters. The form of a procedure declaration is

```
procedure procedure_name (formal-parameter-list) is
[declarations]
begin
  sequential statements
end procedure_name;
```

The `formal-parameter-list` specifies the inputs and outputs to the procedure and their types. A procedure call is a sequential or concurrent statement of the form

```
procedure_name(actual-parameter-list);
```

As an example we will write a procedure *Addvec*, which will add two *N*-bit vectors and a carry, and return an *N*-bit sum and a carry. We will use a procedure call of the form

```
Addvec(A, B, Cin, Sum, Cout, N);
```

where *A, B*, and *Sum* are *N*-bit vectors, *Cin* and *Cout* are bits, and *N* is an integer.

Figure 8-4 gives the procedure definition. *Add1, Add2*, and *Cin* are input parameters, and *Sum* and *Cout* are output parameters. *N* is a positive integer that specifies the number of bits in the bit-vectors. The addition algorithm is essentially the same as the one used in the *add4* function. *C* must be a variable, since the new value of *C* is needed each time through the loop; however, *Sum* can be a signal since *Sum* is not used within the loop. After *N* times through the loop, all the values of the signal *Sum* have been computed, but *Sum* is not updated until a delta time after exiting from the loop.

Within the procedure declaration, the class, mode, and type of each parameter must be specified in the `formal-parameter-list`. The class of each parameter can be **signal, variable**, or **constant**. If the class of an input parameter is omitted, **constant** is used as the default. If the class is a **signal**, then the actual parameter in the procedure call must be a **signal** of the same type. Similarly, for a formal parameter of class **variable**, the actual parameter must be a **variable** of the same type. However, for a **constant** formal parameter, the actual parameter can be any expression that evaluates to a constant of the proper type. This constant value is used inside the procedure and cannot be changed; thus, a **constant** formal parameter is always of mode **in**. Signals and variables can be of mode **in, out**, or **inout**. Parameters of mode **out** and **inout** can be changed in the procedure, so they are used to return values to the caller.

FIGURE 8-4: **Procedure for Adding Bit-Vectors**

```
-- This procedure adds two n-bit bit_vectors and a carry and
-- returns an n-bit sum and a carry. Add1 and Add2 are assumed
-- to be of the same length and dimensioned n-1 downto 0.

procedure Addvec (Add1, Add2: in bit_vector; Cin: in bit;
                  signal Sum: out bit_vector; signal Cout: out bit;
                  n: in positive) is

variable C: bit;
begin
  C := Cin;
  for i in 0 to n-1 loop
    Sum(i) <= Add1(i) xor Add2(i) xor C;
    C := (Add1(i) and Add2(i)) or (Add1(i) and C) or (Add2(i) and C);
  end loop;
  Cout <= C;
end Addvec;
```

In procedure *Addvec*, parameters *Add1, Add2*, and *Cin* are, by default, of class constant. Therefore, in the procedure call, *Add1, Add2*, and *Cin* can be replaced with any expressions that evaluate to constants of the proper type and dimension. Since *Sum* and *Cout* change within the procedure and are used to return values, they have been declared as class **signal**. Thus, in the procedure call, *Sum* and *Cout* can be replaced only with signals of the proper type and dimension.

The `formal-parameter-list` in a **function** declaration is similar to that of a **procedure**, except parameters of class **variable** are not allowed. Furthermore, all parameters must be of mode **in**, which is the default mode. Parameters of mode **out** or **inout** are not allowed, since a function returns only a single value, and this value cannot be returned through a parameter. Table 8-1 summarizes the modes and classes that may be used for procedure and function parameters. A procedure can have output parameters of mode **out** or **inout**. They can be signals or variables. They obviously cannot be constants because constants cannot be modified.

TABLE 8-1:
**Parameters for
Subprogram Calls**

| Mode | Class | Actual Parameter Procedure Call | Actual Parameter Function Call |
|---|---|---|---|
| In[1] | Constant[2] | Expression | Expression |
| | Signal | Signal | Signal |
| | Variable | Variable | n/a |
| Out/inout | Signal | Signal | n/a |
| | Variable[3] | Variable | n/a |

[1] Default mode for functions

[2] Default for in mode

[3] Default for out/inout mode

*NOTE:* n/a = "not applicable"

# 8.3 Attributes

An important feature of the VHDL language is attributes. Attributes can be associated with signals. They can also be associated with arrays.

### 8.3.1 Signal Attributes

You have already used a signal attribute, the `'EVENT` attribute, for creating edge-triggered clocks. As you know, `CLOCK'EVENT` (read as "CLOCK tick EVENT") returns a value of TRUE if a change in signal *CLOCK* has just occurred. VHDL has two types of attributes: (1) attributes that return a value and (2) attributes that return a signal.

Table 8-2 gives several examples of attributes that return a value. In this table, *S* represents a signal name, and *S* is separated from an attribute name by a tick mark (single quote). In VHDL, an event on a signal means a change in the signal. Thus, `S'ACTIVE` (read as "S tick ACTIVE") returns a value of TRUE if a transaction in *S* has just occurred. A transaction occurs on a signal every time it is evaluated, regardless of whether the signal changes or not. Consider the concurrent VHDL statement A <= B **and** C. If *B* = 0, then a transaction occurs on *A* every time *C* changes, since *A* is recomputed every time *C* changes. If *B* = 1, then an event and a transaction occur on *A* every time *C* changes. `S'ACTIVE` returns TRUE if *S* has just been re-evaluated, even if *S* does not change. In contrast, `S'EVENT` returns TRUE only if a change has occurred in *S*. If *S* changes at time *T*, then `S'EVENT` is true at time *T* but false at time $T + \Delta$.

| | Attribute | Returns |
|---|---|---|
| **TABLE 8-2:** | S'ACTIVE | True if a transaction occurred during the current delta, else false |
| **Signal Attributes** | S'EVENT | True if an event occurred during the current delta, else false |
| **That Return a** | S'LAST_EVENT | Time elapsed since the previous event on S |
| **Value** | S'LAST_VALUE | Value of S before the previous event on S |
| | S'LAST_ACTIVE | Time elapsed since previous transaction on S |

Table 8-3 gives signal attributes that create a signal. The brackets around (time) indicate that (time) is optional. If (time) is omitted, then one delta is used. The attribute `S'DELAYED(time)` creates a signal identical to *S*, except it is shifted by the amount of time specified. The example in Figure 8-5 illustrates use of the attributes listed in Table 8-3. The signal *C_delayed5* is the same as *C* shifted right by 5 ns. The signal *A_trans* toggles every time *B* or *C* changes, since *A* has a transaction whenever *B* or *C* changes. The initial computation of A <= B **and** C produces a transaction on *A* at time = $\Delta$, so *A_trans* changes to '1' at that time. The signal `A'STABLE(time)` is true if *A* has not changed during the preceding interval of length (time). Thus, *A_stable5* is false for 5 ns after *A* changes, and it is true otherwise. The signal `A'QUIET(time)` is true if *A* has had no transactions during the preceding interval of length (time). Thus, *A_quiet5* is false for 5 ns after *A* has had a

| TABLE 8-3: | Attribute | Creates |
|---|---|---|
| **Signal Attributes**<br>**That Create**<br>**a Signal** | S'DELAYED [(time)]* | Signal same as S delayed by specified time |
| | S'STABLE [(time)]* | Boolean signal that is true if S had no events for the specified time |
| | S'QUIET [(time)]* | Boolean signal that is true if S had no transactions for the specified time |
| | S'TRANSACTION | Signal of type bit that changes for every transaction on S |

*Delta is used if no time is specified.

FIGURE 8-5: **Examples of Signal Attributes**

```
entity attr_ex is
  port(B, C: in bit);
end attr_ex;

architecture test of attr_ex is
signal A, C_delayed5, A_trans: bit;
signal A_stable5, A_quiet5: boolean;
begin
  A <= B and C;
  C_delayed5 <= C'delayed(5 ns);
  A_trans <= A'transaction;
  A_stable5 <= A'stable(5 ns);
  A_quiet5 <= A'quiet(5 ns);
end test;
```

(a) VHDL code for attribute test

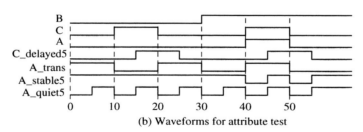

(b) Waveforms for attribute test

transaction. S'EVENT and **not** S'STABLE both return true if an event has occurred during the current delta; however, they cannot always be used interchangeably, since the former just returns a value and the latter returns a signal.

### 8.3.2 Array Attributes

Table 8-4 gives array attributes. In this table, *A* can either be an array name or an array type. In the examples, *ROM1* is a two-dimensional array for which the first index range is 0 **to** 15, and the second index range is 7 **downto** 0. ROM1'LEFT(2) is 7, since the left bound of the second index range is 7. Although *ROM1* is declared as

TABLE 8-4:
Array Attributes

```
type ROM is array (0 to 15, 7 downto 0) of bit;
signal ROM1 : ROM;
```

| Attribute | Returns | Examples |
|---|---|---|
| A'LEFT(N) | left bound of Nth index range | ROM1'LEFT(1) = 0<br>ROM1'LEFT(2) = 7 |
| A'RIGHT(N) | right bound of Nth index range | ROM1'RIGHT(1) = 15<br>ROM1'RIGHT(2) = 0 |
| A'HIGH(N) | largest bound of Nth index range | ROM1'HIGH(1) = 15<br>ROM1'HIGH(2) = 7 |
| A'LOW(N) | smallest bound of Nth index range | ROM1'LOW(1) = 0<br>ROM1'LOW(2) = 0 |
| A'RANGE(N) | Nth index range | ROM1'RANGE(1) = 0 to 15<br>ROM1'RANGE(2) = 7 downto 0 |
| A'REVERSE_RANGE(N) | Nth index range reversed | ROM1'REVERSE_RANGE(1) = 15 downto 0<br>ROM1'REVERSE_RANGE(2) = 0 to 7 |
| A'LENGTH(N) | size of Nth index range | ROM1'LENGTH(1) = 16<br>ROM1'LENGTH(2) = 8 |

a signal, the array attributes also work with array constants and array variables. In the examples, the results are the same if *ROM1* is replaced with its type, ROM. For a vector (a one-dimensional array), $N$ is 1 and can be omitted. If $A$ is a bit-vector dimensioned 2 **to** 9, then A'LEFT is 2 and A'LENGTH is 8.

### 8.3.3 Use of Attributes

Attributes are often used together with assert statements (see Section 2.19) for error checking. The assert statement checks to see if a certain condition is true and, if not, causes an error message to be displayed. We present two examples: one illustrating use of signal attributes and another one illustrating array attributes.

**Use of Signal Attributes**

Consider the process in Figure 8-6, which checks to see if the setup and hold times are satisfied for a D flip-flop. We will use attributes 'EVENT and 'STABLE. 'STABLE is an attribute that returns a Boolean signal if the signal has no events for a specified time (i.e., a TRUE signal returned by this indicates that the signal was stable for a specified time). For example, the signal A'STABLE(time) is true if $A$ has not changed during the preceding interval of length (time). Thus, A'stable(5) is false for 5 ns after $A$ changes, and it is true otherwise.

In the check process, after the active edge of the clock occurs, the $D$ input is checked to see if it has been stable for the specified *setup_time*. If not, a setup-time violation is reported as an error. Then, after waiting for the *hold_time*, D is checked to see if it has been stable during the hold-time period. If not, a hold-time violation is reported as an error.

FIGURE 8-6: **Process for Checking Setup and Hold Times**

```
check: process
begin
  wait until (Clk'event and CLK = '1');
  assert (D'stable(setup_time))
    report ("Setup time violation")
    severity error;
  wait for hold_time;
  assert (D'stable(hold_time))
    report ("Hold time violation")
    severity error;
end process check;
```

## Use of Array Attributes in Vector Addition

As an example of using the assert statement together with array attributes, consider the procedure illustrated in Figure 8-7 for adding bit-vectors. This procedure adds two vectors of arbitrary size. The vectors should, however, be of the same length. It is not required to pass the length of the arrays in the procedure call. Since vector lengths are not passed as a parameter to the procedure, the procedure uses array attributes and checks whether the lengths are equal. Figure 8-7 shows the code for the procedure *Addvec2*. The inputs to the procedure include the two input vectors and the carry in bit. The procedure creates a temporary variable, *C*, for the internal

FIGURE 8-7: **Procedure for Adding Bit-Vectors**

```
-- This procedure adds two bit_vectors and a carry and returns a sum
-- and a carry. Both bit_vectors should be of the same length.

procedure Addvec2 (Add1, Add2: in bit_vector; Cin: in bit;
                   signal Sum: out bit_vector;
                   signal Cout: out bit) is

variable C: bit := Cin;
alias n1: bit_vector(Add1'length-1 downto 0) is Add1;
alias n2: bit_vector(Add2'length-1 downto 0) is Add2;
alias S: bit_vector(Sum'length-1 downto 0) is Sum;
begin
  assert ((n1'length = n2'length) and (n1'length = S'length))
    report "Vector lengths must be equal!"
    severity error;
  for i in S'reverse_range loop   -- reverse range makes you start from LSB
    S(i) <= n1(i) xor n2(i) xor C;
    C := (n1(i) and n2(i)) or (n1(i) and C) or (n2(i) and C);
  end loop;
  Cout <= C;
end Addvec2;
```

carry and initializes it to the input carry, *Cin*. Then it creates aliases *n1, n2,* and *S*, which have the same length as *Add1, Add2,* and *Sum*, respectively. These aliases are dimensioned from their length minus 1 **downto** 0. Even though the ranges of *Add1, Add2,* and *Sum* might be **downto** or **to** and might not include 0, the ranges for the aliases are defined in a uniform manner to facilitate further computation. If the input vectors and *Sum* are not the same length, an error message is reported. The sum and carry are computed bit-by-bit in a loop. Since this loop must start with *i* = 0, the range of *i* is the reverse of the range for *S*. Finally, the carry output, *Cout*, is set equal to the corresponding temporary variable, *C*.

● ● ● ● ● ● ● ● ● ● ● ●
## 8.4 Creating Overloaded Operators

Let us understand how overloaded operators are created. Operator overloading means that we will extend the definition of the operator to other data types in addition to the default data types that have already been defined. The operator will implicitly call an appropriate function, which eliminates the need for an explicit function or procedure call. When the compiler encounters a function declaration in which the function name is an operator enclosed in double quotes, the compiler treats this function as an operator overloading function.

The VHDL arithmetic operators, + and −, are defined to operate on integers, but not on bit-vectors. We have been using the IEEE numeric_bit library in order to access the overloaded arithmetic operators for bit-vectors using the unsigned type. Let us create a "+" function for bit-vectors.

The package shown in Figure 8-8 illustrates the creation of a "+" function for bit-vectors. It adds two bit-vectors and returns a bit-vector. This function uses aliases so that it is independent of the ranges of the bit-vectors, but it assumes that the lengths of the vectors are the same. It uses a **for** loop to do the bit-by-bit addition. Without this overloaded function, the "+" function was not available for bit-vectors. The IEEE numeric_bit only provides it for the unsigned type.

FIGURE 8-8: **VHDL Package with Overloaded Operators for Bit-Vectors**

```
-- This package provides an overloaded function for the plus operator

package bit_overload is
  function "+" (Add1, Add2: bit_vector)
    return bit_vector;
end bit_overload;

package body bit_overload is
  -- This function returns a bit_vector sum of two bit_vector operands
  -- The add is performed bit by bit with an internal carry
  function "+" (Add1, Add2: bit_vector)
    return bit_vector is
```

```
    variable sum: bit_vector(Add1'length-1 downto 0);
    variable c: bit := '0';                    -- no carry in
    alias n1: bit_vector(Add1'length-1 downto 0) is Add1;
    alias n2: bit_vector(Add2'length-1 downto 0) is Add2;
    begin
      for i in sum'reverse_range loop
        sum(i) := n1(i) xor n2(i) xor c;
        c := (n1(i) and n2(i)) or (n1(i) and c) or (n2(i) and c);
      end loop;
      return (sum);
    end "+";
end bit_overload;
```

Overloading can also be applied to procedures and functions. Several procedures can have the same name, and the type of the actual parameters in the procedure call determines which version of the procedure is called. An examination of the IEEE numeric_bit library illustrates that several overloaded operators and functions are defined.

● ● ● ● ● ● ● ● ● ● ● ●
# 8.5 Multivalued Logic and Signal Resolution

In previous chapters, we have used 2-valued bit logic in our VHDL code. In order to represent tristate buffers and buses, it is necessary to be able to represent a third value, 'Z', which represents the high-impedance state. It is also at times necessary to have a fourth value, 'X', to represent an unknown state. This unknown state may occur if the initial value of a signal is unknown or if a signal is simultaneously driven to two conflicting values, such as '0' and '1'. If the input to a gate is 'Z', the gate output may assume an unknown value, 'X'.

We need multivalued logic in order to meet these requirements. The IEEE numeric_std and the IEEE standard logic use a 9-valued logic. Different CAD tool developers have defined 7-valued, 9-valued, and 11-valued logic conventions.

In this chapter, we will present two examples of multivalued logic, (1) a 4-valued logic system and (2) the **IEEE-1164 standard** 9-valued logic system. The 4-valued logic system is described in Section 8.5.1 and the 9-valued logic is explained in Section 8.6.

### 8.5.1 A 4-Valued Logic System

Signals in a 4-valued logic can assume the four values: 'X', '0', '1', and 'Z', where each of the symbols represent the following:

'X'   Unknown
'0'   0
'1'   1
'Z'   High impedance

The high-impedance state is used for modeling tristate buffers and buses. This unknown state can be used if the initial value of a signal is unknown or if a signal is simultaneously driven to two conflicting values, such as '0' and '1'.

Let us model tristate buffers using the 4-valued logic. Figure 8-9 shows two tristate buffers with their outputs tied together, and Figure 8-10 shows the corresponding VHDL representation. A new data type X01Z, which can assume the four values 'X', '0', '1', and 'Z' is assumed. The tristate buffers have an active-high output enable, so that when $b$ = '1' and $d$ = '0', $f = a$; when $b$ = '0' and $d$ = '1', $f = c$; and when $b = d$ = '0', the $f$ output assumes the high-Z state. If $b = d$ = '1', an output conflict can occur. Two VHDL architecture descriptions are shown. The first one uses two concurrent statements, and the second one uses two processes. In either case, $f$ is driven from two different sources, and VHDL uses a *resolution function* to determine the actual output. For example, if $a = c = d$ = '1' and $b$ = '0', $f$ is driven to 'Z' by one concurrent statement or process, and $f$ is driven to '1' by the other concurrent statement or process. The resolution function is automatically called to determine that the proper value of $f$ is '1'. The resolution function will supply a value of 'X' (unknown) if $f$ is driven to both '0' and '1' at the same time.

FIGURE 8-9:
**Tristate Buffers with Active-High Output Enable**

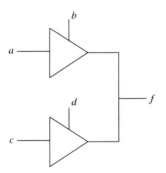

FIGURE 8-10: **VHDL Code for Tristate Buffers**

```
use WORK.fourpack.all; -- fourpack is a resolved package for 4-variable logic
                       -- more details on resolution in next subsection

entity t_buff_exmpl is
  port(a, b, c, d: in X01Z;  -- signals are four-valued
       f: out X01Z);
end t_buff_exmpl;

architecture t_buff_conc of t_buff_exmpl is
begin
  f <= a when b = '1' else 'Z';
  f <= c when d = '1' else 'Z';
end t_buff_conc;
```

```
architecture t_buff_bhv of t_buff_exmpl is
begin
  buff1: process(a, b)
  begin
    if (b = '1') then
      f <= a;
    else
      f <= 'Z';  -- "drive" the output high Z when not enabled
    end if;
  end process buff1;

  buff2: process(c, d)
  begin
    if (d = '1') then
      f <= c;
    else
      f <= 'Z';  -- "drive" the output high Z when not enabled
    end if;
  end process buff2;
end t_buff_bhv;
```

The code in Figure 8-10 utilizes a 4-valued logic package and corresponding signal resolution functions. Let us understand how to create signal resolution functions. A package, as described in the following subsection, is necessary to make the code in Figure 8-10 work.

### 8.5.2 Signal Resolution Functions

VHDL signals may either be resolved or unresolved. Signal resolution is necessary when different wires in a system are driving a common signal path. Signal resolution means arriving at a resulting value when two or more different signals are connected to the same point. VHDL with multivalued logic can be used to create resolutions when signals are connected.

Resolved signals have an associated resolution function, and unresolved signals do not. We have previously used signals of type bit, which are unresolved. With unresolved signals, if we drive a bit signal $B$ to two different values in two concurrent statements (or in two processes), the compiler will flag an error because there is no way to determine the proper value of $B$.

Consider the following three concurrent statements, where $R$ is a resolved signal of type X01Z:

```
R <= transport '0' after 2 ns, 'Z' after 6 ns;
R <= transport '1' after 4 ns;
R <= transport '1' after 8 ns, '0' after 10 ns;
```

Assuming that $R$ is initialized to 'Z', three drivers would be created for $R$, as shown in Figure 8-11. Each time one of the unresolved signals $s(0)$, $s(1)$, or $s(2)$ changes, the resolution function is automatically called to determine the value of the resolved signal, $R$.

FIGURE 8-11:
**Resolution of
Signal Drivers**

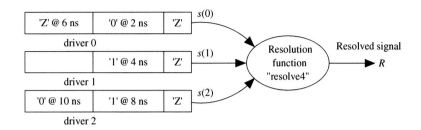

Since the X01Z logic has a symbol for high impedance, we can create resolution functions to model the wires when multiple signals are connected. Figure 8-12 shows how the resolution function for X01Z logic is defined in a package called *fourpack*. First, an unresolved logic type u_X01Z is defined, along with the corresponding

FIGURE 8-12:  **Resolution Function for X01Z Logic**

```
package fourpack is
  type u_x01z is ('X', '0', '1', 'Z');         -- u_x01z is unresolved
  type u_x01z_vector is array (natural range <>) of u_x01z;
  function resolve4 (s: u_x01z_vector) return u_x01z;
  subtype x01z is resolve4 u_x01z;
  -- x01z is a resolved subtype which uses the resolution function resolve4
  type x01z_vector is array (natural range <>) of x01z;
end fourpack;

package body fourpack is
  type x01z_table is array (u_x01z, u_x01z) of u_x01z;
  constant resolution_table: x01z_table := (
    ('X','X','X','X'),
    ('X','0','X','0'),
    ('X','X','1','1'),
    ('X','0','1','Z'));

  function resolve4 (s:u_x01z_vector)
    return u_x01z is

  variable result: u_x01z := 'Z';
  begin
    if (s'length = 1) then
      return s(s'low);
    else
      for i in s'range loop
        result := resolution_table(result, s(i));
      end loop;
    end if;
    return result;
  end resolve4;
end fourpack;
```

unconstrained array type, u_X01Z_vector. Then a resolution function, named *resolve4*, is declared. Resolved X01Z logic is defined as a subtype of u_X01Z. The subtype declaration contains the function name *resolve4*. This implies that whenever a signal of type X01Z is computed, function *resolve4* is called to compute the correct value.

The resolution function, which is based on the operation of a tristate bus, is specified by the following table:

|     | 'X' | '0' | '1' | 'Z' |
|-----|-----|-----|-----|-----|
| 'X' | 'X' | 'X' | 'X' | 'X' |
| '0' | 'X' | '0' | 'X' | '0' |
| '1' | 'X' | 'X' | '1' | '1' |
| 'Z' | 'X' | '0' | '1' | 'Z' |

This table gives the resolved value of a signal for each pair of input values: 'Z' resolved with any value returns that value, 'X' resolved with any value returns 'X', and '0' resolved with '1' returns 'X'. The function *resolve4* has an argument, $s$, which represents a vector of one or more signal values to be resolved. If the vector is of length 1, then the first (and only) element of the vector is returned. Otherwise, the return value (the resolved signal) is computed iteratively by starting with *result* = 'Z' and recomputing *result* by a table look-up using each element of the $s$ vector in turn. In the example of Figure 8-11, the $s$ vector has three elements, and *resolve4* would be called at 0, 2, 4, 6, 8, and 10 ns to compute $R$. The following table shows the result:

| Time | s(0) | s(1) | s(2) | R |
|------|------|------|------|-----|
| 0  | 'Z' | 'Z' | 'Z' | 'Z' |
| 2  | '0' | 'Z' | 'Z' | '0' |
| 4  | '0' | '1' | 'Z' | 'X' |
| 6  | 'Z' | '1' | 'Z' | '1' |
| 8  | 'Z' | '1' | '1' | '1' |
| 10 | 'Z' | '1' | '0' | 'X' |

In order to write VHDL code using X01Z logic, we need to define the required operations for this type of logic. For example, AND and OR may be defined using the following tables:

| AND | 'X' | '0' | '1' | 'Z' |
|-----|-----|-----|-----|-----|
| 'X' | 'X' | '0' | 'X' | 'X' |
| '0' | '0' | '0' | '0' | '0' |
| '1' | 'X' | '0' | '1' | 'X' |
| 'Z' | 'X' | '0' | 'X' | 'X' |

| OR  | 'X' | '0' | '1' | 'Z' |
|-----|-----|-----|-----|-----|
| 'X' | 'X' | 'X' | '1' | 'X' |
| '0' | 'X' | '0' | '1' | 'X' |
| '1' | '1' | '1' | '1' | '1' |
| 'Z' | 'X' | 'X' | '1' | 'X' |

The table on the left corresponds to the way an AND gate with 4-valued inputs would work. If one of the AND gate inputs is '0', the output is always '0'. If both inputs are '1', the output is '1'. In all other cases, the output is unknown ('X'), since a high-Z gate input may act like either a '0' or '1'. For an OR gate, if one of the inputs is '1', the output is always '1'. If both inputs are '0', the output is '0'. In all other cases, the output is 'X'. AND and OR functions based on these tables can be included in the package *fourpack* to overload the AND and OR operators.

While this section illustrated how resolved signals can be created, fortunately you do not have to create such signals. Standard libraries with resolved data types are available. The IEEE 1164 standard and IEEE_numeric_std are examples of such multivalued logic libraries.

## 8.6 The IEEE 9-Valued Logic System

The **IEEE 1164** standard specifies a 9-valued logic system with signal resolution. The 9 logic values defined in this standard are

| | |
|---|---|
| 'U' | Uninitialized |
| 'X' | Forcing unknown |
| '0' | Forcing 0 |
| '1' | Forcing 1 |
| 'Z' | High impedance |
| 'W' | Weak unknown |
| 'L' | Weak 0 |
| 'H' | Weak 1 |
| '–' | Don't care |

The unknown, '0', and '1' values come in two strengths—forcing and weak. A forcing '1' means that the signal is as perfect as the power supply voltage. A 'weak 1', represented by 'H', means that the signal is logically high, but there is a voltage drop (e.g., output of a pull-up resistor). A forcing '0' represents a perfect ground, whereas a 'weak 0' represents a signal which is logically '0', but not exactly the ground voltage (e.g., the output of a pull-down resistor). The 9-valued system has the representation 'U' for denoting uninitialized signals. Don't care states can be represented by '–'.

If a forcing signal and a weak signal are tied together, the forcing signal dominates. For example, if '0' and 'H' are tied together, the result is '0'. The 9-valued logic is useful in modeling the internal operation of certain types of ICs. In this text, we will normally use only a subset of the IEEE values—'X', '0', '1', and 'Z'.

The IEEE-1164 standard defines the AND, OR, NOT, XOR, and other functions for 9-valued logic. The package IEEE.std_logic_1164 defines a **std_logic type** that uses the 9-valued logic. It also specifies a number of subtypes of the 9-valued logic, such as the X01Z subtype, which we have already been using. Analogous to bit-vectors, when vectors are created with the std_logic type, they are called **std_logic vectors**. When bit-vectors are used, typically they are initialized to '0', whereas when the std_logic type is used, the uninitilized value 'U' is the default value.

Table 8-5 shows the resolution function table for the IEEE 9-valued logic. The row index values have been listed as comments to the right of the table. The resolution function table for X01Z logic is a subset of this table, as indicated by the black rectangle.

TABLE 8-5:
**Resolution
Function Table for
IEEE 9-Valued Logic**

```
CONSTANT resolution_table : stdlogic_table := (
--  -----------------------------------------------------------------
-- |  U    X    0    1    Z    W    L    H    -                |    |
--  -----------------------------------------------------------------
  ( 'U', 'U', 'U', 'U', 'U', 'U', 'U', 'U', 'U' ),  --  |  U  |
  ( 'U', 'X', 'X', 'X', 'X', 'X', 'X', 'X', 'X' ),  --  |  X  |
  ( 'U', 'X', '0', 'X', '0', '0', '0', '0', 'X' ),  --  |  0  |
  ( 'U', 'X', 'X', '1', '1', '1', '1', '1', 'X' ),  --  |  1  |
  ( 'U', 'X', '0', '1', 'Z', 'W', 'L', 'H', 'X' ),  --  |  Z  |
  ( 'U', 'X', '0', '1', 'W', 'W', 'W', 'W', 'X' ),  --  |  W  |
  ( 'U', 'X', '0', '1', 'L', 'W', 'L', 'W', 'X' ),  --  |  L  |
  ( 'U', 'X', '0', '1', 'H', 'W', 'W', 'H', 'X' ),  --  |  H  |
  ( 'U', 'X', 'X', 'X', 'X', 'X', 'X', 'X', 'X' )   --  |  -  |
);
```

Table 8-6 shows the AND function table for the IEEE 9-valued logic. The row index values have been listed as comments to the right of the table. The AND function table for X01Z logic is a subset of this table, as indicated by the black rectangle. The IEEE-1164 standard first defines **std_ulogic** (unresolved standard logic); then it defines the std_logic type as a subtype of std_ulogic with the associated resolution function.

TABLE 8-6:
**AND Table for IEEE
9-Valued Logic**

```
CONSTANT and_table : stdlogic_table := (
--  -----------------------------------------------------------------
-- |  U    X    0    1    Z    W    L    H    -                |    |
--  -----------------------------------------------------------------
  ( 'U', 'U', '0', 'U', 'U', 'U', '0', 'U', 'U' ),  --  |  U  |
  ( 'U', 'X', '0', 'X', 'X', 'X', '0', 'X', 'X' ),  --  |  X  |
  ( '0', '0', '0', '0', '0', '0', '0', '0', '0' ),  --  |  0  |
  ( 'U', 'X', '0', '1', 'X', 'X', '0', '1', 'X' ),  --  |  1  |
  ( 'U', 'X', '0', 'X', 'X', 'X', '0', 'X', 'X' ),  --  |  Z  |
  ( 'U', 'X', '0', 'X', 'X', 'X', '0', 'X', 'X' ),  --  |  W  |
  ( '0', '0', '0', '0', '0', '0', '0', '0', '0' ),  --  |  L  |
  ( 'U', 'X', '0', '1', 'X', 'X', '0', '1', 'X' ),  --  |  H  |
  ( 'U', 'X', '0', 'X', 'X', 'X', '0', 'X', 'X' )   --  |  -  |
);
```

The **and** functions given in Figure 8-13 use Table 8-6. These functions provide for operator overloading. This means that if we write an expression that uses the **and** operator, the compiler will automatically call the appropriate **and** function to evaluate the **and** operation depending on the type of the operands. If **and** is used with bit variables, the ordinary **and** function is used, but if **and** is used with std_logic variables, the std_logic **and** function is called. Operator overloading also automatically applies the appropriate **and** function to vectors. When **and** is used with bit-vectors, the ordinary bit-by-bit **and** is performed, but when **and** is applied to std_logic vectors, the std_logic **and** is applied on a bit-by-bit basis. The first **and** function in Figure 8-13 computes the **and** of the left (l) and right (r) operands by doing a table look-up. Although the **and** function is first defined for **std_ulogic**, it also works for std_logic since std_logic is a subtype of std_ulogic. The second **and** function works with std_logic vectors. Aliases are used to make sure the index range is the same direction for both operands. If the vectors are not the same length, the **assert** false always causes the message to be displayed. Otherwise, each bit in the result vector is computed by table look-up.

FIGURE 8-13: **AND Function for std_logic_vectors**

```
function "and" (l: std_ulogic; r: std_ulogic) return UX01 is
begin
  return (and_table(l, r));
end "and"; -- end of function for unresolved standard logic

function "and" (l, r: std_logic_vector) return std_logic_vector is
  alias lv: std_logic_vector (1 to l'LENGTH) is l; --alias makes index range
  alias rv: std_logic_vector (1 to r'LENGTH) is r; -- in same direction
  variable result: std_logic_vector ( 1 to l'LENGTH );
begin
  if (l'LENGTH /= r'LENGTH) then
    assert FALSE
      report "arguments of overloaded 'and' operator are not of the same length"
      severity FAILURE;
  else
    for i in result'RANGE loop
      result(i) := and_table(lv(i), rv(i));
    end loop;
  end if;
  return result;
end "and";
```

If multivalued logic is desired, we can use the IEEE standard numeric_std package instead of the numeric_bit package that we have been using so far. The IEEE.numeric_std package is similar to the IEEE.numeric_bit package, but it defines unsigned and signed types as vectors of std_logic type instead of as vectors of bits. It also defines the same set of overloaded operators and functions on unsigned and signed numbers as the numeric_bit package.

A VHDL program that used vectors with the unsigned type can be ported to use vectors with 9-valued logic by simply replacing the statement

```
use IEEE.numeric_bit.all;
```

with the statements

```
use IEEE.std_logic_1164.all; -- The IEEE.numeric_std package
                             -- uses the 1164 standard.
use IEEE.numeric_std.all;
```

The IEEE.numeric_std package uses the std_logic type from the 1164 standard. Hence, both the statements need to be included. With these statements, the unsigned type is considered to use 9-valued logic. No other changes in the program are required. If the original program used the type bit, they should be converted to the std_logic type.

Other popular VHDL package used for simulation and synthesis with multivalued logic are the std_logic_arith package and the std_logic_unsigned package, developed by Synopsis. These packages can be invoked by the following statements:

```
use IEEE.std_logic_unsigned.all;
use IEEE.std_logic_arith.all;
```

In examples from now on, we will use the IEEE numeric_std package because it is an IEEE standard and it is similar in functionality to the numeric_bit package that we have been using so far. We have chosen not to use the std_logic_arith and std_logic_unsigned packages because they are not IEEE standards and they have less functionality than the IEEE numeric_std package.

● ● ● ● ● ● ● ● ● ● ● ●
## 8.7 SRAM Model Using IEEE 1164

In this section, we develop a VHDL model to represent the operation of a static RAM (SRAM). *RAM* stands for random-access memory, which means that any word in the memory can be accessed in the same amount of time as any other word. Strictly speaking, ROM memories are also random access, but historically, the term *RAM* is normally applied only to read-write memories. This model also illustrates the usefulness of the multivalued logic system. Multivalued logic is used to model tristate conditions on the memory data lines.

Figure 8-14 shows the block diagram of a static RAM with $n$ address lines, $m$ data lines, and three control lines. This memory can store $2^n$ words, each $m$ bits wide. The data lines are bidirectional in order to reduce the required number of pins and the package size of the memory chip. When reading from the RAM, the data lines are outputs; when writing to the RAM, the data lines serve as inputs. The three control lines function as follows:

$\overline{CS}$    When asserted low, chip select selects the memory chip so that memory read and write operations are possible.

$\overline{OE}$    When asserted low, output enable enables the memory output onto an external bus.

$\overline{WE}$    When asserted low, write enable allows data to be written to the RAM.

FIGURE 8-14: **Block Diagram of Static RAM**

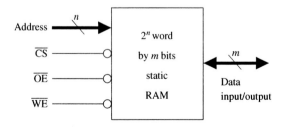

We say that a signal is asserted when it is in its active state. An active-low signal is asserted when it is low, and an active-high signal is asserted when it is high.

The truth table for the RAM (Table 8-7) describes its basic operation. High-Z in the I/O column means that the output buffers have high-Z outputs, and the data inputs are not used. In the read mode, the address lines are decoded to select $m$ of the memory cells, and the data comes out on the I/O lines after the memory access time has elapsed. In the write mode, input data is routed to the latch inputs in the selected memory cells when $\overline{WE}$ is low, but writing to the latches in the memory cells is not completed until either $\overline{WE}$ goes high or the chip is deselected. The truth table does not take memory timing into account.

| | $\overline{CS}$ | $\overline{OE}$ | $\overline{WE}$ | Mode | I/O pins |
|---|---|---|---|---|---|
| **TABLE 8-7:** | H | X | X | not selected | high-Z |
| **Truth Table for** | L | H | H | output disabled | high-Z |
| **Static RAM** | L | L | H | read | data out |
| | L | X | L | write | data in |

We now write a simple VHDL model for the memory that does not take timing considerations into account. In Figure 8-15, the RAM memory array is represented by an array of unsigned standard logic vectors (*RAM1*). This memory has 256 words, each of which are 8 bits. Since *Address* is typed as an unsigned bit-vector, it must be converted to an integer in order to index the memory array. The RAM process sets the I/O lines to high-Z if the chip is not selected. If *We_b* = '1', the RAM is in the read mode, and *IO* is the data read from the memory array. If *We_b* = '0', the memory is in the write mode, and the data on the I/O lines is stored in *RAM1* on the rising edge of *We_b*. If *Address* and *We_b* change simultaneously, the old value of *Address* should be used. *Address'delayed* is used as the array index to delay *Address* by one delta to make sure that the old address is used. *Address'delayed* uses one of the signal attributes described earlier in this chapter (Table 8-3). This is a RAM with asynchronous read and synchronous write.

**FIGURE 8-15: Simple Memory Model**

```
-- Simple memory model
library IEEE;
use IEEE.std_logic_1164.all;
use IEEE.numeric_std.all;

entity RAM6116 is
  port(Cs_b, We_b, Oe_b: in std_logic;
       Address: in unsigned(7 downto 0);
       IO: inout unsigned(7 downto 0));
end RAM6116;

architecture simple_ram of RAM6116 is
type RAMtype is array(0 to 255) of unsigned(7 downto 0);
```

```
signal RAM1: RAMtype := (others => (others =>'0'));
                        -- Initialize all bits to '0'
begin
  IO <= "ZZZZZZZZ" when Cs_b = '1' or We_b = '0' or Oe_b = '1'
    else RAM1(to_integer(Address));    -- read from RAM
  process(We_b, Cs_b)
  begin
    if Cs_b = '0' and rising_edge(We_b) then   -- rising-edge of We_b
      RAM1(to_integer(Address'delayed)) <= IO; -- write
    end if;
  end process;
end simple_ram;
```

● ● ● ● ● ● ● ● ● ● ● ●

## 8.8 Model for SRAM Read/Write System

To illustrate further the use of multivalued logic, we present an example with a **bidirectional tristate bus**. We will design a memory read-write system that reads the content of 32 memory locations from a RAM, increments each data value, and stores it back into the RAM. A block diagram of the system is shown in Figure 8-16. In order to hold the word that we read from memory, we use a **data register**. In order to hold the memory address that we are accessing, we use a memory address register (**MAR**). The system reads a word from the RAM, loads it into the data register, increments the data register, stores the result back in the RAM, and then increments the memory address register. This process continues until the memory address equals 32.

The data bus is used as a **bidirectional bus**. During the read operation, the memory output appears on the bus, and the data register output to the data bus will be in a tristate condition. During the write operation, the data register output is on the data bus and the memory will use it as input data.

**FIGURE 8-16: Block Diagram of RAM Read-Write System**

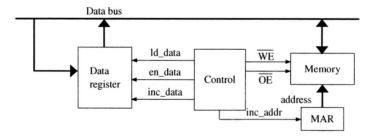

Control signals required to operate the system are

| | |
|---|---|
| *ld_data* | load data register from Data Bus |
| *en_data* | enable data register output onto Data Bus |
| *inc_data* | increment Data Register |
| *inc_addr* | increment MAR |
| $\overline{WE}$ | Write Enable for SRAM |
| $\overline{OE}$ | Output Enable for SRAM |

Figure 8-17 shows the SM chart for the system. The SM chart uses four states. In the first state, the SRAM drives the memory data onto the bus and the memory data is loaded into the Data Register. The control signal $\overline{OE}$ and *ld_data* are true in this state. The Data Register is incremented in $S_1$. The *en_data* control signal is true in state $S_2$, and hence the Data Register drives the bus. Write enable $\overline{WE}$ is an active-low signal, which is asserted low only in $S_2$, so that $\overline{WE}$ is high in the other states. The contents of the data register thus get written to the RAM at the transition from $S_2$ to $S_3$. The memory address is incremented. The process continues until the address is 32. State $S_3$ checks this and produces a done signal when the address reaches 32.

FIGURE 8-17: **SM Chart for RAM System**

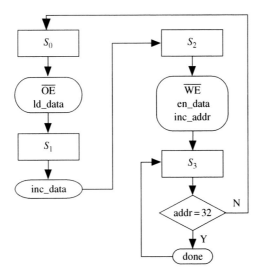

Figure 8-18 shows the VHDL code for the RAM system. The first process represents the SM chart, and the second process is used to update the registers on the rising edge of the clock. A short delay is added when the address is incremented to make sure the write to memory is completed before the address changes. A concurrent statement is used to simulate the tristate buffer, which enables the data register output onto the I/O lines.

FIGURE 8-18: **VHDL Code for RAM System**

```
-- SRAM Read-Write System model
library IEEE;
use IEEE.std_logic_1164.all;
use IEEE.numeric_std.all;

entity RAM6116_system is
end RAM6116_system;

architecture RAMtest of RAM6116_system is
component RAM6116 is
  port(Cs_b, We_b, Oe_b: in std_logic;
```

```
            Address: in unsigned(7 downto 0);
            IO: inout unsigned(7 downto 0));
end component RAM6116;

signal state, next_state: integer range 0 to 3;
signal inc_addr, inc_data, ld_data, en_data, Cs_b, clk, Oe_b, done:
       std_logic := '0';
signal We_b: std_logic := '1';                    -- initialize to read mode
signal Data: unsigned(7 downto 0);                -- data register
signal Address: unsigned(7 downto 0) := "00000000";  -- address register
signal IO: unsigned(7 downto 0);                  -- I/O bus
begin
  RAM1: RAM6116 port map (Cs_b, We_b, Oe_b, Address, IO);
  control: process(state, Address)
  begin
    --initialize all control signals (RAM always selected)
    ld_data <= '0'; inc_data <= '0'; inc_addr <= '0'; en_data <= '0';
    done <= '0'; We_b <= '1'; Cs_b <= '0'; Oe_b <= '1';

    --start SM chart here
    case state is
      when 0 => Oe_b <= '0'; ld_data <= '1'; next_state <= 1;
      when 1 => inc_data <= '1'; next_state <= 2;
      when 2 => We_b <= '0'; en_data <= '1'; inc_addr <= '1'; next_state <= 3;
      when 3 =>
        if (Address = "00100000") then done <= '1'; next_state <= 3;
        else next_state <= 0;
        end if;
    end case;
  end process control;

  --The following process is executed on the rising edge of a clock.
  register_update: process(clk)    -- process to update data register
  begin
    if rising_edge(clk) then
      state <= next_state;
      if (inc_data = '1') then data <= data + 1; end if;
                        -- increment data in data register
      if (ld_data = '1') then data <= IO; end if;
                        -- load data register from bus
      if (inc_addr = '1') then Address <= Address + 1 after 1 ns; end if;
                        -- delay added to allow completion of memory write
    end if;
  end process register_update;

  -- Concurrent statements
  clk <= not clk after 100 ns;
  IO <= data when en_data = '1'
    else "ZZZZZZZZ";
end RAMtest;
```

This system can be modified to include all memory locations for testing the correctness of the entire SRAM. Memory systems are often tested by writing checkerboard patterns (alternate 0's and 1's) in all locations. For instance, we can write 01010101 (55 hexadecimal) into all odd addresses and 10101010 (hexadecimal AA) into all even addresses. Then the odd and even locations can be swapped. Developing VHDL code for such a system is left as an exercise problem.

● ● ● ● ● ● ● ● ● ● ● ●

# 8.9   Generics

*Generics* are commonly used to specify parameters for a component in such a way that the parameter values may be specified when the component is instantiated. For example, the rise and fall times for a gate could be specified as generics, and different numeric values for these generics could be assigned for each instance of the gate. The example of Figure 8-19 describes a two-input NAND gate whose rise and fall delay times depend on the number of loads on the gate. In the entity declaration, *Trise, Tfall*, and *load* are generics that specify the no-load rise time, the no-load fall time, and the number of loads. In the architecture, an internal *nand_value* is computed whenever *a* or *b* changes. If *nand_value* has just changed to a '1', a rising output has occurred, and the gate delay time is computed as

$$Trise + 3\,ns * load$$

where 3 ns is the added delay for each load. Otherwise, a falling output has just occurred and the gate delay is computed as

$$Tfall + 2\,ns * load$$

where 2 ns is the added delay for each load.

FIGURE 8-19: **Rise/Fall Time Modeling Using Generic Statement**

```
entity NAND2 is
  generic(Trise, Tfall: time; load: natural);
  port(a, b: in bit;
       c: out bit);
end NAND2;

architecture behavior of NAND2 is
signal nand_value: bit;
begin
  nand_value <= a nand b;
  c <= nand_value after (Trise + 3 ns * load) when nand_value = '1'
    else nand_value after (Tfall + 2 ns * load);
end behavior;

entity NAND2_test is
  port(in1, in2, in3, in4: in bit;
```

```
       out1, out2: out bit);
end NAND2_test;

architecture behavior of NAND2_test is
component NAND2
  generic(Trise: time := 3 ns; Tfall: time := 2 ns; load: natural := 1);
  port(a, b: in bit; c: out bit);
end component;
begin
  U1: NAND2 generic map (2 ns, 1 ns, 2) port map (in1, in2, out1);
  U2: NAND2 port map (in3, in4, out2);
end behavior;
```

The entity *NAND2_test* tests the NAND2 component. The component declaration in the architecture specifies default values for *Trise, Tfall*, and *load*. When *U1* is instantiated, the generic map specifies different values for *Trise, Tfall*, and *load*. When *U2* is instantiated, no generic map is included, so the default values are used.

● ● ● ● ● ● ● ● ● ● ● ●

## 8.10   Named Association

Up to this point, we have used *positional association* in the port maps and generic maps that are part of an instantiation statement. For example, assume that the entity declaration for a full adder is

```
entity FullAdder is
  port(X, Y, Cin: in bit; Cout, Sum: out bit);
end FullAdder;
```

The statement

```
FA0: FullAdder port map (A(0), B(0), '0', open, S(0));
```

creates a full adder and connects $A(0)$ to the $X$ input of the adder, $B(0)$ to the $Y$ input, '0' to the *Cin* input, leaves the *Cout* output unconnected, and connects $S(0)$ to the *Sum* output of the adder. The first signal in the port map is associated with the first signal in the entity declaration, the second signal with the second signal, and so on. In order to indicate no connection, the keyword **open** is used.

As an alternative, we can use *named association*, in which each signal in the port map is explicitly associated with a signal in the port of the component entity declaration. For example, the statement

```
FA0: FullAdder port map (Sum=>S(0), X=>A(0), Y=>B(0), Cin=>'0');
```

makes the same connections as the previous instantiation statement (i.e., *Sum* connects to $S(0)$, $X$ connects to $A(0)$, etc). When named association is used, the order in which the connections are listed is not important, and any port signals not listed are left unconnected. Use of named association makes code easier to read, and it offers more flexibility in the order in which signals are listed.

When named association is used with a generic map, any unassociated generic parameter assumes its default value. For example, if we replace the statement in Figure 8-19 labeled U1 with

```
U1:NAND2 generic map (load => 3,Trise => 4ns) port map
(in1,in2,out1);
```

*Tfall* would assume its default value of 2 ns.

● ● ● ● ● ● ● ● ● ● ● ●

## 8.11 Generate Statements

In Chapter 2, we instantiated four full-adder components and interconnected them to form a 4-bit adder. Specifying the port maps for each instance of the full adder would become very tedious if the adder had 8 or more bits. When an iterative array of identical components is required, the **generate** statement provides an easy way of instantiating these components. The example of Figure 8-20 shows how a generate statement can be used to instantiate four 1-bit full adders to create a 4-bit adder. A 5-bit vector is used to represent the carries, with *Cin* the same as $C(0)$ and *Cout* the same as $C(4)$. The **for** loop generates four copies of the full adder, each with the appropriate **port map** to specify the interconnections between the adders.

Another example where the generate statement would have been very useful is the array multiplier. The VHDL code for the array multiplier (Chapter 4) used repeated use of **port map** statements in order to instantiate each component. They could have been replaced with generate statements.

FIGURE 8-20: Adder4 Using Generate Statement

```
entity Adder4 is
port(A, B: in bit_vector(3 downto 0); Ci: in bit;    -- Inputs
     S: out bit_vector(3 downto 0); Co: out bit);    -- Outputs
end Adder4;

architecture Structure of Adder4 is
component FullAdder
  port(X, Y, Cin: in bit;      -- Inputs
       Cout, Sum: out bit);    -- Outputs
end component;

signal C: bit_vector(4 downto 0);
begin
  C(0) <= Ci;
  -- generate four copies of the FullAdder
  FullAdd4: for i in 0 to 3 generate
  begin
    FAx: FullAdder port map (A(i), B(i), C(i), C(i+1), S(i));
  end generate FullAdd4;
  Co <= C(4);
end Structure;
```

In the preceding example, we used a generate statement of the form

```
generate_label: for identifier in range generate
[begin]
  concurrent statement(s)
end generate [generate_label];
```

At compile time, a set of concurrent statement(s) is generated for each value of the identifier in the given range. In Figure 8-20, one concurrent statement—a component instantiation statement—is used. A generate statement itself is defined to be a concurrent statement, so nested generate statements are allowed.

### 8.11.1 Conditional Generate

A generate statement with an if clause may be used to conditionally generate a set of concurrent statement(s). This type of generate statement has the form

```
generate_label: if condition generate
[begin]
  concurrent statement(s)
end generate [generate_label];
```

In this case, the concurrent statements(s) are generated at compile time only if the condition is true.

Figure 8-21 illustrates the use of conditional compilation using a generate statement with an **if** clause. An *N*-bit left-shift register is created if *Lshift* is true using the statement

```
genLS: if Lshift generate
  shifter <= Q(N-1 downto 1) & Shiftin;
end generate;
```

If *Lshift* is false, a right-shift register is generated using another conditional generate statement. The example also shows how generics and generate statements can be used together. It illustrates the use of generic parameters to write a VHDL model with parameters so that the size and function can be changed when it is instantiated.

FIGURE 8-21: **Shift Register Using Conditional Compilation**

```
entity shift_reg is
  generic(N: positive := 4; Lshift: Boolean := true);-- generic parameters used
  port(D: in bit_vector(N downto 1);
       Qout: out bit_vector(N downto 1);
       CLK, Ld, Sh, Shiftin: in bit);
end shift_reg;

architecture SRN of shift_reg is
signal Q, shifter: bit_vector(N downto 1);
begin
  Qout <= Q;
  genLS: if Lshift generate      -- conditional generate of left shift register
```

```
    shifter <= Q(N-1 downto 1) & Shiftin;
  end generate;
  genRS: if not Lshift generate -- conditional generate of right shift register
    shifter <= Shiftin & Q(N downto 2);
  end generate;
  process(CLK)
  begin
    if CLK'event and CLK = '1' then
      if LD = '1' then Q <= D;
      elsif Sh = '1' then Q <= shifter;
      end if;
    end if;
  end process;
end SRN;
```

* * * * * * * * * * * *

## 8.12    **Files and TEXTIO**

The ability to input files and text is very valuable while testing large VHDL designs. This section introduces file input and output in VHDL. Files are frequently used with test benches to provide a source of test data and to provide storage for test results. VHDL provides a standard *TEXTIO* package that can be used to read or write lines of text from or to a file.

Before a file is used, it must be declared using a declaration of the form

```
file file-name: file-type [open mode] is "file-pathname";
```

For example,

```
file test_data: text open read_mode is "c:\test1\test.dat";
```

declares a file named *test_data* of type text that is opened in the read mode. The physical location of the file is in the test1 directory on the c: drive.

A file can be opened in *read_mode, write_mode*, or *append_mode*. In read_mode, successive elements in the file can be read using the read procedure. When a file is opened in write_mode, a new empty file is created by the host computer's file system, and successive data elements can be written to the file using the write procedure. To write to an existing file, the file should be opened in the append_mode.

A file can contain only one type of object, such as integers, bit-vectors, or text strings, as specified by the file type. For example, the declaration

```
type bv_file is file of bit_vector;
```

defines *bv_file* to be a file type that can contain only bit-vectors. Each file type has an associated implicit endfile function. A call of the form

```
endfile(file_name)
```

returns TRUE if the file pointer is at the end of the file.

The standard *TEXTIO* package that comes with VHDL contains declarations and procedures for working with files composed of lines of text. The package specification for *TEXTIO* (see Appendix C) defines a file type named text:

```
type text is file of string;
```

The *TEXTIO* package contains procedures for reading lines of text from a file of type text and for writing lines of text to a file.

Procedure `readline` reads a line of text and places it in a buffer with an associated pointer. The pointer to the buffer must be of type line, which is declared in the *TEXTIO* package as

```
type line is access string;
```

When a variable of type line is declared, it creates a pointer to a string. The code

```
variable buff: line;
. . .
readline(test_data, buff);
```

reads a line of text from *test_data* and places it in a buffer that is pointed to by *buff*. After reading a line into the buffer, we must call a version of the read procedure one or more times to extract data from the line buffer. The *TEXTIO* package provides overloaded read procedures to read data of types bit, bit-vector, boolean, character, integer, real, string, and time from the buffer. For example, if *bv4* is a bit_vector of length four, the call

```
read(buff, bv4);
```

extracts a 4-bit vector from the buffer, sets *bv4* equal to this vector, and adjusts the pointer *buff* to point to the next character in the buffer. Another call to read then extracts the next data object from the line buffer.

A call to read may be of one of two forms:

```
read(pointer, value);
read(pointer, value, good);
```

where *pointer* is of type line and *value* is the variable into which we want to read the data. In the second form, *good* is a boolean that returns TRUE if the read is successful and FALSE if it is not. The size and type of *value* determines which of the read procedures in the *TEXTIO* package is called. For example, if *value* is a string of length 5, then a call to read reads the next five characters from the line buffer. If *value* is an integer, a call to read skips over any spaces and then reads decimal digits until a space or other nonnumeric character is encountered. The resulting string is then converted to an integer. Characters, strings, and bit-vectors within files of type text are not delimited by quotes.

To write lines of text to a file, we must call a version of the write procedure one or more times to write data to a line buffer and then call writeline to write the line of data to a file. The *TEXTIO* package provides overloaded write procedures to write data of types bit, bit-vector, boolean, character, integer, real, string, and time to the buffer. For example, the code

```
variable buffw: line;
variable int1: integer;
variable bv8: bit_vector(7 downto 0);
. . .
write(buffw, int1, right, 6);
write(buffw, bv8, right, 10);
writeline(buffw, output_file);
```

converts *int1* to a text string, writes this string to the line buffer pointed to by *buffw*, and adjusts the pointer. The text will be right justified in a field six characters wide. The second call to write puts the bit_vector *bv8* in a line buffer, and adjusts the pointer. The 8-bit vector will be right justified in a field 10 characters wide. Then writeline writes the buffer to the *output_file*. Each call to write has four parameters: (1) a buffer pointer of type line; (2) a value of any acceptable type; (3) justification (left or right), which specifies the location of the text within the output field; and (4) field width, an integer that specifies the number of characters in the field.

As an example, we write a procedure to read data from a file and store the data in a memory array. This procedure will later be used to load instruction codes into a memory module for a computer system. The computer system can then be tested by simulating the execution of the instructions stored in memory. The data in the file will be of the following format:

```
address N comments
byte1 byte2 byte3 . . . byteN comments
```

The address consists of four hexadecimal digits, and *N* is an integer that indicates the number of bytes of code that will be on the next line. Each byte of code consists of two hexadecimal digits. Each byte is separated by one space, and the last byte must be followed by a space. Anything following the last space will not be read and will be treated as a comment. The first byte should be stored in the memory array at the given address, the second byte at the next address, and so forth. For example, consider the following file:

```
12AC 7 (7 hex bytes follow)
AE 03 B6 91 C7 00 0C
005B 2 (2 hex bytes follow)
01 FC<space>
```

When the *fill_memory* procedure is called using this file as an input, AE is stored in 12AC, 03 in 12AD, B6 in 12AE, 91 in 12AF, and so on.

Figure 8-22 gives VHDL code that calls the procedure *fill_memory* to read data from a file and store it in an array named *mem*. Since *TEXTIO* does not include a read procedure for hex numbers, the procedure *fill_memory* reads each hex value as a string of characters and then converts the string to an integer. Conversion of a single hex digit to an integer value is accomplished by table look-up. The constant named *lookup* is an array of integers indexed by characters in the range '0' to 'F'. This range includes the 23 ASCII characters: '0', '1', '2', . . . , '9', ':', ';', '<', '=', '>',

'?', '@', 'A', 'B', 'C', 'D', 'E', 'F'. The corresponding array values are $0, 1, 2, \ldots, 9, -1,$ $-1, -1, -1, -1, -1, -1, 10, 11, 12, 13, 14, 15$. The $-1$ could be replaced with any integer value, since the seven special characters in the index range should never occur in practice. Thus, *lookup*('2') is the integer value 2, *lookup*('C') is 12, and so forth.

FIGURE 8-22: **VHDL Code to Fill a Memory Array from a File**

```
library IEEE;
use IEEE.numeric_bit.all;                 -- to use TO_UNSIGNED(int, size)
use std.textio.all;

entity testfill is
end testfill;

architecture fillmem of testfill is
type RAMtype is array (0 to 8191) of unsigned(7 downto 0);
signal mem: RAMtype := (others => (others => '0'));

procedure fill_memory(signal mem: inout RAMType) is
type HexTable is array (character range <>) of integer;
-- valid hex chars: 0, 1, . . . A, B, C, D, E, F (upper-case only)
constant lookup: HexTable('0' to 'F'): =
  (0, 1, 2, 3, 4, 5, 6, 7, 8, 9, -1, -1, -1,
   -1, -1, -1, -1, 10, 11, 12, 13, 14, 15);
file infile: text open read_mode is "mem1.txt"; -- open file for reading
-- file infile: text is in "mem1.txt";          -- VHDL '87 version
variable buff: line;
variable addr_s: string(4 downto 1);
variable data_s: string(3 downto 1);            -- data_s(1) has a space
variable addr1, byte_cnt: integer;
variable data: integer range 255 downto 0;
begin
  while (not endfile(infile)) loop
    readline(infile, buff);
    read(buff, addr_s);                       -- read addr hexnum
    read(buff, byte_cnt);                     -- read number of bytes to read
    addr1 := lookup(addr_s(4)) * 4096 + lookup(addr_s(3)) * 256
             + lookup(addr_s(2)) * 16 + lookup(addr_s(1));
    readline(infile, buff);
    for i in 1 to byte_cnt loop
      read(buff, data_s);                     -- read 2 digit hex data and a space
      data := lookup(data_s(3)) * 16 + lookup(data_s(2));
      mem(addr1) <= TO_UNSIGNED(data, 8);
      addr1:= addr1 + 1;
    end loop;
  end loop;
end fill_memory;
```

```
begin
  testbench: process
  begin
    fill_memory(mem);
    -- insert code which uses memory data
  end process;
end fillmem;
```

Procedure *fill_memory* calls *readline* to read a line of text that contains a hex address and an integer. The first call to read reads the address string from the line buffer, and the second call to read reads an integer, which is the byte count for the next line. The integer *addr1* is computed using the look-up table for each character in the address string. The next line of text is read into the buffer, and a loop is used to read each byte. Since *data_s* is three characters long, each call to read reads two hex characters and a space. The hex characters are converted to an integer and then to an unsigned vector, which is stored in the memory array. The address is incremented before reading and storing the next byte. The procedure exits when the end of file is reached.

This chapter has introduced several important features of VHDL. Functions and procedures were introduced first. Attributes were presented next. Attributes associated with signals allow checking of setup and hold times and other timing specifications. Attributes associated with arrays allow us to write procedures that do not depend on the manner in which the arrays are indexed. Operator overloading can be used to extend the definition of VHDL operators so that they can be used with different types of operands. The IEEE Standard 1164 defines a system of 9-valued logic that is widely used with VHDL. Multivalued logic and the associated resolution functions allow us to model tristate buses and other systems where a signal is driven from more than one source. Generics enable us to specify parameter values for a component when the component is instantiated. Generate statements provide an efficient way to describe systems that have an iterative structure. The *TEXTIO* package provides a convenient way of doing file input and output.

● ● ● ● ● ● ● ● ● ● ● ●
# Problems

8.1 Write a VHDL function that converts a 5-bit bit_vector to an integer. Note that the integer value of the binary number $a_4a_3a_2a_1a_0$ can be computed as

$$((((0 + a_4)*2 + a_3)*2 + a_2)*2 + a_1)*2 + a_0$$

How much simulated time will it take for your function to execute?

8.2 Write a VHDL function that will create the 2's complement of an *n*-bit vector. Use a call of the form `comp2(bit_vec, N)` where *N* is the length of the vector. State any assumptions you make about the range of bit_vec. Do the complement on a bit-by-bit basis using a loop.

8.3 Write a VHDL function which will return the largest integer in an array of $N$ integers. The function call should be of the form LARGEST(ARR, N).

8.4 $A$ and $B$ are bit-vectors that represent unsigned binary numbers. Write a VHDL function that returns TRUE if $A > B$. The function call should be of the form GT(A, B, N), where $N$ is the length of the bit vectors. Do not call any functions or procedures from within your code. *Hint*: Start comparing the most significant bits of $A$ and $B$ first and proceed from left to right. As soon as you find a pair of unequal bits you can determine whether or not $A > B$. For example, if $A = 1011010$ and $B = 1010110$, you can determine that $A > B$ when you make the fourth comparison.

8.5 What are the major differences between VHDL functions and VHDL procedures?

8.6 Write a VHDL procedure that counts the number of ones in an input bit-vector that is $N$ bits long ($N \leq 31$). The output should be an unsigned vector that is 5 bits long.

8.7 $X$ and $Y$ are bit-vectors of length $N$ that represent signed binary numbers, with negative numbers represented in 2's complement. Write a VHDL procedure that will compute $D = X - Y$. This procedure should also return the borrow from the last bit position ($B$) and an overflow flag ($V$). Do not call any other functions or procedures in your code. The procedure call should be of the form SUBVEC(X, Y, D, B, V, N);.

8.8 Write a VHDL module that implements a 4-digit BCD adder with accumulator (see block diagram below). If $LD = 1$, then the contents of $BCDacc$ are replaced with $BCDacc + BCDin$. Each four-digit BCD signal should be represented by an array of the following type:

```
type BCD4 is array (3 downto 0) of unsigned (3 downto 0);
```

Write a procedure that adds two BCD digits and a carry and returns a BCD digit and a carry. Call this procedure concurrently four times in your code.

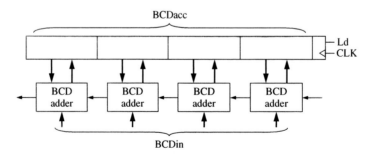

8.9 For the following VHDL code, list the values of $B$ and $C$ at each time a change occurs. Include all deltas, and stop your listing when time > 8 ns. Assume that $B$ is changed to "0110" at time 5 ns. Indicate the times at which procedure $P1$ is called.

```
entity Q1 is
  port(B, C: inout bit_vector(3 downto 0));
end Q1;

architecture Q1 of Q1 is
  procedure P1(signal A: inout bit_vector) is
  begin
    for i in 1 to 3 loop
      A(i) <= A(i-1);
    end loop;
    A(0) <= A(3);
  end P1;
begin
  process
  begin
    wait until B'event;
    P1(B);
    wait for 1 ns;
    P1(B);
  end process;
  C <= B;
end Q1;
```

8.10 The following VHDL code is part of a process. Assume that $A = B = $ '0' before the code is executed. Give the values of the variables $X1$, $X2$, $X3$, and $X4$ immediately after the code is executed.

```
wait until clock'event and clock = '1';
A <= not B;
A <= transport B after 5 ns;
wait for 5 ns;
X1 := A'event;
X2 := A'delayed'event;
X3 := A'last_event;
X4 := A'delayed'last_event;
```

8.11 Write a VHDL function that will take two integer vectors, $A$ and $B$, and find the dot product $C = \Sigma \, a_i * b_i$. The function call should be of the form DOT(A,B), where $A$ and $B$ are integer vector signals. Use attributes inside the function to determine the length and ranges of the vectors. Make no assumptions about the high and low values of the ranges. For example,

$A(3 \text{ downto } 1) = (1, 2, 3), B(3 \text{ downto } 1) = (4, 5, 6), C = 3 * 6 + 2 * 5 + 1 * 4 = 32$

Output a warning if the ranges are not the same.

8.12 Write a VHDL procedure that will add two $n \times m$ matrices of integers, $C <= A + B$. The procedure call should be of the form addM(A, B, C). The procedure should report an error if the number of rows in $A$ and $B$ are not the same or if the number of columns in $A$ and $B$ are not the same. Make no assumptions about the high and low values or direction of the ranges for either dimension.

8.13 Write a VHDL procedure that will add two bit-vectors that represent signed binary numbers. Negative numbers are represented in 2's complement. If the vectors are of different lengths, the shorter one should be sign-extended during the addition. Make no assumptions about the range for either vector. The procedure call should be of the form Add2(A, B, Sum, V), where $V = 1$ if the addition produces a 2's complement overflow.

8.14 A VHDL entity has inputs $A$ and $B$, and outputs $C$ and $D$. $A$ and $B$ are initially high. Whenever $A$ goes low, $C$ will go high 5 ns later, and if $A$ changes again, $C$ will change 5 ns later. $D$ will change if $B$ does not change for 3 ns after $A$ changes.

   **(a)** Write the VHDL architecture with a process that determines the outputs $C$ and $D$.
   **(b)** Write another process to check that $B$ is stable 2 ns before and 1 ns after $A$ goes high. The process should also report an error if $B$ goes low for a time interval less than 10 ns.

8.15 Write an overloading function for the "<" operator for bit-vectors. Return a boolean TRUE if $A$ is less than $B$, otherwise return FALSE. Report an error if the bit-vectors are of different lengths.

8.16 Write an overloading function for the unary "−" operator for bit-vectors. If $A$ is a bit-vector −A should return the 2's complement of $A$.

8.17 Consider the following three concurrent statements, where $R$ is a resolved signal of type X01Z:

```
R <= transport '0' after 2 ns, 'Z' after 8 ns;
R <= transport '1' after 10 ns;
R <= transport '1' after 4 ns, '0' after 6 ns;
```

Draw the multiple drivers that will be created and the resolved output signal $R$ from time 0 until time 12 ns.

8.18 Write a VHDL description of an address decoder/address match detector. One input to the address decoder is an 8-bit address, which can have any range with a length of 8 bits; for example, bit_vector addr(8 to 15). The second input is check: x01z_vector(5 downto 0). The address decoder will output $Sel = $ '1' if the upper 6 bits of the 8-bit address match the 6-bit check vector. For example, if $addr = $ "10001010" and $check = $ "1000XX", then $Sel = $ '1'. Only the six leftmost bits of $addr$ will be compared; the remaining bits are ignored. An 'X' in the check vector is treated as a don't care.

8.19 Write a VHDL model for one flip-flop in a 74HC374 (octal D-type flip-flop with three-state outputs). Use the IEEE-standard nine-valued logic package. Assume that all logic values are 'x', '0', '1' or 'z'. Check setup, hold, and pulse width specs using assert statements. Unless the output is 'z', the output should be 'x' if $CLK$ or $OC$ is 'x', or if an 'x' has been stored in the flip-flop.

8.20 Write a VHDL function to compare two IEEE std_logic_vectors to see if they are equal. Report an error if any bit in either vector is not '0', '1', or '−' (don't care), or if the lengths of the vectors are not the same. The function call should pass only the vectors. The function should return TRUE if the vectors are equal, else FALSE. When comparing the vectors, consider that '0' = '−', and '1' = '−'. Make no assumptions about the index range of the two vectors (for example, one could be 1 **to** 7 and the other 8 **downto** 0).

8.21 Consider the following concurrent statements, where *A, B*, and *C* are of type std_logic:

```
A <= transport '1' after 5 ns, '0' after 10 ns, 'Z' after 15 ns;
B <= transport '0' after 4 ns, 'Z' after 10 ns;
C <= A after 6 ns;
C <= transport A after 5 ns;
C <= reject 3 ns B after 4 ns;
```

**(a)** Draw drivers (see Figure 2-27) for signals *A* and *B*.
**(b)** Draw the three drivers $s(0)$, $s(1)$, and $s(2)$ for *C* (similar to Figure 8-11).
**(c)** List the value for *C* each time it is resolved by the drivers, and draw a timing chart for *C*.

8.22 Subtype X01LH of std_logic has values of 'X', '0', '1', 'L', and 'H'. Complete the following table for a resolution function of this subtype.

|   | 'X' | '0' | '1' | 'L' | 'H' |
|---|-----|-----|-----|-----|-----|
| 'X' |  |  |  |  |  |
| '0' |  |  |  |  |  |
| '1' |  |  |  |  |  |
| 'L' |  |  |  |  |  |
| 'H' |  |  |  |  |  |

8.23 Write an overloading function for "not", where the input and returned value are standard logic vectors. The "not" function should basically simulate a group of inverters. The output bits should be one of the following: 'U', '0', '1', or 'X'. An uninitialized input should give an uninitialized output.

8.24 In the following code, all signals are 1-bit std_logic. Draw a logic diagram that corresponds to the code. Assume that a D flip-flop with CE is available.

```
F <= A when EA = '1' else B when EB = '1' else 'Z';
process(CLK)
begin
  if CLK'event and CLK = '1' then
    if Ld = '1' then A <= B; end if;
    if Cm = '1' then A <= not A; end if;
  end if;
end process;
```

8.25 Design a memory-test system to test the first 256 bytes of a static RAM memory. The system consists of simple controller, an 8-bit counter, a comparator, and a memory as shown below. The counter is connected to both the address and data (IO) bus so that 0 will be written to address 0, 1 to address 1, 2 to address 2, . . . , and 255 to address 255. Then the data will be read back from address 0, address 1, . . . , address 255 and compared with the address. If the data does not match, the controller goes to the fail state as soon as a mismatch is detected; otherwise, it goes to a pass state after all 256 locations have been matched. Assume that $OE\_b = 0$ and $CS\_b = 0$.

(a) Draw an SM chart or a state graph for the controller (five states). Assume that the clock period is long enough so that one word can be read every clock period.

(b) Write VHDL code for the memory-test system.

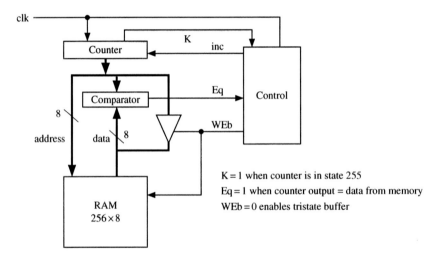

8.26 Design a memory-test system similar to that of Problem 8.25, except write a checkerboard pattern into memory (01010101 into address 0, 10101010 into address 1, etc.). Draw the block diagram and SM chart.

8.27 Design a memory tester that verifies the correct operation of a 6116 static RAM (Figure 8-15). The tester should store a checkerboard pattern (alternating 0's and 1's in the even addresses, and alternating 1's and 0's in the odd addresses) in all memory locations and then read it back. The tester should then repeat the test using the reverse pattern.

(a) Draw a block diagram of the memory tester. Show and explain all control signals.

(b) Draw an SM chart or state graph for the control unit. Use a simple RAM model and disregard timing.

(c) Write VHDL code for the tester and use a test bench to verify its operation.

8.28 A clocked T flip-flop has propagation delays from the rising edge of *CLK* to the changes in $Q$ and $Q'$ as follows: If $Q$ (or $Q'$) changes to 1, $t_{plh}$ = 8 ns, and if $Q$ (or $Q'$) changes to 0, $t_{phl}$ = 10 ns. The minimum clock pulse width is $t_{ck}$ = 15 ns, the setup time for the $T$ input is $t_{su}$ = 4 ns, and the hold time is $t_h$ = 2 ns. Write a VHDL model for the flip-flop that includes the propagation delay and that reports if any timing specification is violated. Write the model using generic parameters with default values.

8.29 **(a)** Write a model for a D flip-flop with a direct clear input. Use the following generic timing parameters: $t_{plh}, t_{phl}, t_{su}, t_h$, and $t_{cmin}$. The minimum allowable clock period is $t_{cmin}$. Report appropriate errors if timing violations occur.
**(b)** Write a test bench to test your model. Include tests for every error condition.

8.30 Write a VHDL model for an *N*-bit comparator using an iterative circuit. In the entity, use the generic parameter *N* to define the length of the input bit-vectors *A* and *B*. The comparator outputs should be *EQ* = '1' if *A* = *B*, and *GT* = '1' if *A* > *B*. Use a for loop to do the comparison on a bit-by-bit basis, starting with the high-order bits. Even though the comparison is done on a bit-by-bit basis, the final values of *EQ* and *GT* apply to *A* and *B* as a whole.

8.31 Four RAM memories are connected to CPU busses as shown below. Assume that the following RAM component is available:

```
component SRAM
  port(cs_b, we_b, oe_b: in bit;
       address: in bit_vector(14 downto 0);
       data: inout std_logic_vector(7 downto 0));
end component;
```

Write a VHDL code segment which will connect the four RAMs to the busses. Use a generate statement and named association.

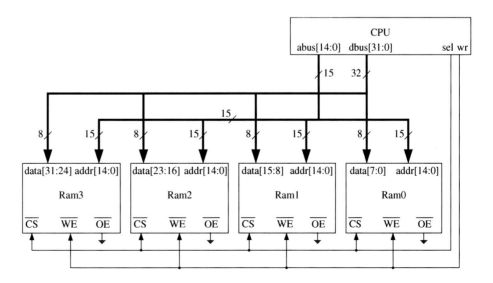

8.32 Write structural VHDL code for a module that is an $N$-bit serial-in, serial-out right-shift register. Inputs to the shift register are bit signals: $SI$ (serial input), $Sh$ (shift enable), and $CLK$. Your module should include a generic in the entity declaration, and a generate statement in the architecture. Assume that a component for a D flip-flop with clock enable ($CE$) is available.

8.33 Write structural VHDL code for a module that has two inputs: an $N$-bit vector $A$, and a control signal $B$ (1 bit). The module has an $N$-bit output vector, $C$. When $B = 1$, $C \leq A$. When $B = 0$, $C$ is all 0's. Use a generic to specify the value of $N$ (default = 4). To implement the logic, use a generate statement that instantiates $N$ 2-input AND gates.

8.34 Create a $4 \times 4$ array multiplier using generate statements. Use full adder, half adder, and AND gate components as in Chapter 4.

8.35 $B$ is an integer array with range 0 to 4. Write a VHDL code segment which will read a line of text from a file named "FILE2" and then read five integers into array $B$. Assume that TEXTIO libraries are available.

8.36 Write a procedure that has an integer signal and a file name as parameters. Each line of the file contains a delay value and an integer. The procedure reads a line from the file, waits for the delay time, assigns the integer value to the signal, and then reads the next line. The procedure should return when end-of-file is reached.

8.37 Write a procedure that logs the history of values of a bit-vector signal to a text file. Each time the signal changes, write the current time and signal value to the file. VHDL has a built in function called NOW that returns the current simulation time when it is called.

# Design of a RISC Microprocessor

**CHAPTER 9**

A microprocessor is an example of a complex digital system. In this chapter, we will describe a microprocessor from MIPS Technologies, the MIPS R2000, and implement a subset of the MIPS processor's **instruction set architecture (ISA)**. The term *instruction set architecture* denotes the instructions that are visible to the assembly language programmer, the number of registers, the addressing modes, and the operations (opcodes) available in the particular processor. An introduction to the RISC philosophy is presented first followed by a description of the MIPS ISA. The arithmetic, memory access, and control transfer instructions of the MIPS are presented. A design to implement a subset of the ISA is presented. A synthesizeable VHDL model for the MIPS subset is then presented. Use of a test bench for testing the design is illustrated.

● ● ● ● ● ● ● ● ● ● ● ● ●

## 9.1 The RISC Philosophy

Many early microprocessors, such as the Intel 8086 and Motorola 68000, incorporated a variety of powerful instructions and addressing modes. A natural consequence of this was the complexity of the design, especially the control unit complexity. These microprocessors included a microprogrammed control unit because it was difficult to design and debug a hardwired control unit for such complex digital systems. (See Chapter 5 for a discussion of tradeoffs between microprogramming and hardwiring.)

The value of simplicity became clearer in the late 1970s and early 1980s. The result was the advent of **RISC** or the *Reduced Instruction Set Computing* philosophy. RISC processors are a type of microprocessor that use a small and simple set of instructions rather than a variety of complex instructions and versatile addressing modes. The first RISC projects came from IBM, Stanford University, and The University of California–Berkeley in the late 1970s and early 1980s. The IBM 801, Stanford MIPS, and Berkeley RISC 1 and 2 were all designed with a similar philosophy, which has become known as RISC. In contrast, earlier processors such as the Intel 8086 and the Motorola 68000/68020 started to be called **CISC** (*Complex Instruction Set Computing*) processors, after the advent of the RISC philosophy. The first generation of RISC processors included MIPS R2000 from MIPS, SPARC from Sun Microsystems, and RS/6000 from IBM. The IBM RS/6000 has evolved into the POWERPC and POWER architecture.

**MIPS**

MIPS Technologies is a computer manufacturer that has designed and sold several RISC microprocessors starting with the MIPS R2000 processor in the 1980s. The term *MIPS* was commonly known to computer designers as a performance metric, the Millions of Instructions Per Second metric. The MIPS in the name of the MIPS Corporation, however, does not stand for that. It stands for **M**icroprocessor without hardware **I**nterlock **P**ipeline **S**tages. In a pipelined processor, there must exist a mechanism to enforce dependencies between instructions. So, if one instruction needs the result of the previous one, the second instruction should not proceed. Enforcing of this type of dependency is usually done by hardware interlocks. The first MIPS processor, however, did not have hardware interlocks. It reflected the early RISC idealism that anything that can be done in software should be done in software. Pipeline interlocks were implemented by software by inserting the appropriate number of *nop* (no operation) instructions.

Certain design features have been characteristic of most RISC processors:

- *Uniform instruction length*: All instructions have the same length (e.g., 32 bits). This is in sharp contrast to previous microprocessors, which contained instructions as small as a byte and as large as 16 bytes.

- *Few instruction formats*: The RISC ISAs emphasized having as few instruction formats as possible and encoding the different fields in the instruction as uniformly as possible. This greatly simplifies instruction decoding.

- *Few addressing modes*: Most RISC processors support only one or two memory addressing modes. Addressing modes offer different ways an instruction can indicate the memory address to be accessed. Examples are direct addressing, immediate addressing, base plus offset addressing, based indexed addressing, and indirect addressing. Many RISC processors support only one addressing mode. Typically, this addressing mode specifies addresses with a register and an offset.

- *Large number of registers*: The RISC design philosophy generally incorporates a larger number of registers to prevent the loss of performance by frequently accessing memory. RISC processors are also often called *register-register architectures*. All arithmetic operations operate on register operands. CISC architectures typically contained 8 or 12 registers, whereas most RISC architectures contained 32 registers.

- *Load/store architecture*: RISC architectures are also called load/store architectures. The key idea is the absence of arithmetic instructions that directly operate on memory operands (i.e., arithmetic instructions that take one or more operands from memory). The only instructions that are allowed to access memory are load and store instructions. The load instructions bring the data to registers and arithmetic operations operate on the data in the registers. These architectures are also called register-register architectures because input and output operands for computation operations are in registers. A load/store architecture inherently means that it is also a register-register architecture.

- *No implied operands or side-effects*: Most earlier ISAs contained implied operands, such as accumulators, or implied results (*side-effects*), such as flags (condition codes), to indicate conditions such as carry, overflow, and negative. Implied operands and side-effects can cause difficulties/challenges in pipelined and parallel implementations. A principle behind RISC architectures is to have minimal implied operands/operations and side-effects.

The RISC philosophy has been to adhere to the above features and embrace simplicity of design. The terms *RISC* and *CISC* are used very often as antonyms, but perhaps it is not clear how *reduced* is the opposite of *complex*. It is not even clear that RISC processors have a smaller instruction set than prior CISC processors. Some RISC ISAs have 100+ instructions, whereas some CISC processors have only 80 instructions. However, these 80 CISC instructions could assume several addressing modes. A CISC processor, the Motorola 68020, supported up to 20 different addressing modes. Considering all the different forms an instruction could take, most RISC ISAs do contain fewer instructions than CISC ISAs. The key point in the RISC philosophy has been the emphasis on simplicity: having only simple basic operations, simplifying instruction formats, reducing the number of addressing modes, and eliminating complex operations. This computing paradigm could have been called Simple Instruction Set Computing (SISC); however, SISC sounds like CISC.

CISC architectures are not without advantages. Instruction encoding is denser in CISC than RISC. The fixed instruction width in RISC leads to using more bits than necessary for some instructions. In CISC ISAs, every instruction is just as wide as it needs to be. Hence, code size is smaller in the CISC case. If instruction memory size has to be kept small, as in embedded environments, CISC ISAs have an advantage.

Most modern microprocessors have RISC ISAs. Some examples are the MIPS R14000, Sun UltraSPARC, IBM PowerPC, and HP PA-RISC. The Pentium 4 or the x86 processors in general are examples of modern processors with a CISC ISA. (The term x86 is used to refer to the different processors that have used the ISA that originated with Intel 8086. This list includes Intel 8086, 80286, 80386, 80486, Pentium and AMD K5, K6, Opteron, etc.)

Whether RISC or CISC is better was a topic of intense debate in the 1980s and 1990s. It has now become understood that decoding and processing is easy with a RISC ISA; however, it also has been shown that hardware can translate complex CISC-style instructions into RISC-style instructions and process them. Pentium 4 and other high-end x86 processors of today have a CISC ISA; however, they use hardware to convert each CISC instruction to one or more RISC-type instructions or microoperations (called *uops or R-ops*) that can be pipelined easily. In spite of all arguments that have taken place, there is no disagreement about the ease of implementation of RISC ISAs.

The MIPS instruction set architecture is one of the earliest RISC ISAs and is one of the simplest ones. It only has one memory addressing mode. In contrast, another early RISC architecture, the SPARC, has two memory addressing modes. The MIPS ISA is described in detail in a book by Gerry Kane, *MIPS RISC Architecture* [26].

It is also described in the book *Computer Organization and Design: The Hardware Software Interface*, by Patterson and Hennessey [37]. We provide a very concise description of the MIPS ISA here.

---

**The Single-Instruction Computer**

It has also been shown in the past that a microprocessor can be designed with a single instruction. This single instruction should be able to access memory operands, do arithmetic operations, and do control transfers. A subtract instruction that operates on memory operands, writes results to memory, and branches to an address if the result of the subtraction is negative can be used to write any program. Will such a single-instruction microprocessor qualify to be called a RISC? Probably not. Although it is a single-instruction computer, it is not a register-register architecture, and it is not an ISA that supports simple operations. We would classify it under a CISC category since every instruction is a complex branch and memory access instruction. More discussion of such a computer and illustration of a program written using the single instruction can be found in [37].

---

## 9.2 The MIPS ISA

The MIPS ISA contains a set of simple arithmetic, logical, memory access, branch, and jump instructions. The architecture emphasizes simplicity and excludes instructions that could take longer than the most common instructions.

There are 32 general-purpose registers in the MIPS architecture. Each register is 32 bits wide. The MIPS registers are referred to as $0, $1, $2, ..., and $31 with a $ sign as in [37]. The MIPS instructions follow a **three-address format** for ALU instructions, meaning they specify two source addresses and one destination address. For example, an add instruction that adds registers $3 and $4 and writes the result to $5 is written as

```
add $5, $3, $4
```

We will describe each group of instructions.

### 9.2.1 Arithmetic Instructions

The MIPS ISA contains instructions for performing addition, subtraction, multiplication, and division of integers. The various arithmetic instructions are summarized in Table 9-1. Addition and subtraction of signed or unsigned quantities can be accomplished using the *add, addu, sub*, and *subu* instructions. Signed arithmetic instructions detect overflows, whereas unsigned arithmetic instructions do not detect overflows. For example, the instruction

```
sub $5, $3, $4
```

will subtract the value in register $4 from the value in register $3 and write the result to register $5. It is a signed instruction and overflow will be detected.

When an overflow is detected, it is handled as an exception. The address of the instruction that caused the exception is saved and control is transferred to the operating system, which handles the exception.

Addition of the contents of a register with an immediate value specified in the instruction can be done using the *addi* and *addiu* instructions. The instruction

```
addi $5, $3, 400
```

will add the value in register $3 to the immediate constant 400 and write the result to register $5. The immediate constant is sign-extended before the addition. The action of the *addiu* instruction is similar, except that the *addiu* instruction never causes an overflow exception.

Multiplication of two 32-bit quantities results in a 64-bit result that cannot be contained in one MIPS register. Hence, two special registers called HI and LO are used by the MIPS processors to hold the products. Use of implied HI and LO registers is certainly a deviation from the RISC philosophy. Table 9-1 illustrates the multiply and divide instructions in the MIPS ISA and the use of the HI and LO registers. The use of these special registers also necessitates special instructions to transfer

| TABLE 9-1: Arithmetic Instructions in the MIPS ISA | Instruction | Assembly Code | | Operation | Comments |
|---|---|---|---|---|---|
| | add | add | $s1, $s2, $s3 | $s1 = $s2 + $s3 | Overflow detected |
| | subtract | sub | $s1, $s2, $s3 | $s1 = $s2 − $s3 | Overflow detected |
| | add immediate | addi | $s1, $s2, k | $s1 = $s2 + k | k, a 16-bit constant, is sign-extended and added; 2's complement overflow detected |
| | add unsigned | addu | $s1, $s2, $s3 | $s1 = $s2 + $s3 | Overflow not detected |
| | subtract unsigned | subu | $s1, $s2, $s3 | $s1 = $s2 − $s3 | Overflow not detected |
| | add immediate unsigned | addiu | $s1, $s2, k | $s1 = $s2 + k | Same as addi except no overflow |
| | move from co-processor register | mfc0 | $s1, $epc | $s1 = $epc | epc is exception program counter |
| | multiply | mult | $s2, $s3 | Hi, Lo = $s2 × $s3 | 64-bit signed product in Hi, Lo |
| | multiply unsigned | multu | $s2, $s3 | Hi, Lo = $s2 × $s3 | 64-bit unsigned product in Hi, Lo |
| | divide | div | $s2, $s3 | Lo = $s2 / $s3 Hi = $s2 mod $s3 | Lo = quotient, Hi = remainder |
| | divide unsigned | divu | $s2, $s3 | Lo = $s2 / $s3 Hi = $s2 mod $s3 | Unsigned quotient and remainder |
| | move from Hi | mfhi | $s1 | $s1 = Hi | Copy Hi to $s1 |
| | move from Lo | mflo | $s1 | $s1 = Lo | Copy Lo to $s1 |

data from these registers to the required destination registers. The **mfhi** and **mflo** instructions accomplish this task.

### 9.2.2 Logical Instructions

The logical instructions in the MIPS ISA are presented in Table 9-2. The MIPS ISA contains logical instructions for performing bit-wise AND and OR of register contents. The *and* and *or* instructions perform these operations for register operands. The *andi* and *ori* instructions can be used when one operand is in a register and the other operand is an immediate constant. The *sll* and *srl* instructions are provided to perform logical left and right shifts of register contents (with zero fill). The number of shifts is encoded as an immediate value in the instruction.

TABLE 9-2: **Logical Instructions in the MIPS ISA**

| Instruction | Assembly Code | Operation | Comments |
|---|---|---|---|
| and | and $s1, $s2, $s3 | $s1 = $s2 AND $s3 | logical AND |
| or | or $s1, $s2, $s3 | $s1 = $s2 OR $s3 | logical OR |
| and immediate | andi $s1, $s2, k | $s1 = $s2 AND k | k is a 16-bit constant; k is 0-extended first |
| or immediate | ori $s1, $s2, k | $s1 = $s2 OR k | k is a 16-bit constant; k is 0-extended first |
| shift left logical | sll $s1, $s2, k | $s1 = $s2 << k | Shift left by 5-bit constant k |
| shift right logical | srl $s1, $s2, k | $s1 = $s2 >> k | Shift right by 5-bit constant k |

### 9.2.3 Memory Access Instructions

The only instructions in the MIPS ISA to access the memory are load and store instructions. A load instruction transfers data from memory to the specified register. A store instruction transfers data from a register to the specified memory address.

The RISC researchers investigated the number of addressing modes that are needed to efficiently code high-level language programs such as those in C. They concluded that one addressing mode with a base register and an offset was sufficient. The only addressing mode that is supported for memory instructions in the MIPS processor is this addressing mode with one base register and a signed offset. The memory address is computed as the sum of the register contents and the offset specified in the instruction.

Consider the MIPS load instruction

```
lw $5, 100($4)
```

This instruction computes the memory address as the sum of the value in register $4 and the offset 100. So if register $4 contains 4000, the effective address is 4100. The content of memory location 4100 is moved to register $5 in the processor. In the case of sw $6, 100($8), the content of register $6 is written to the memory location pointed to by the sum of the contents of register $8 and 100.

A group of 32 bits is called a word in the MIPS world. MIPS has instructions to load and store words, halfwords (16 bits), or bytes (8 bits). These instructions are summarized in Table 9-3.

**TABLE 9-3:**
**Memory Access Instructions in the MIPS ISA**

| Instruction | Assembly Code | Operation | Comments |
|---|---|---|---|
| load word | lw $s1, k($s2) | $s1 = Memory[$s2 + k] | Read 32 bits from memory; memory address = register content + *k*; *k* is 16-bit offset |
| store word | sw $s1, k($s2) | Memory[$s2 + k] = $s1 | Write 32 bits to memory; memory address = register content + *k*; *k* is 16-bit offset; |
| load halfword | lh $s1, k($s2) | $s1 = Memory[$s2 + k] | Read 16 bits from memory; sign-extend and load into register |
| store halfword | sh $s1, k($s2) | Memory[$s2 + k] = $s1 | Write 16 bits to memory |
| load byte | lb $s1, k($s2) | $s1 = Memory[$s2 + k] | Read byte from memory; sign-extend and load to register |
| store byte | sb $s1, k($s2) | Memory[$s2 + k] = $s1 | Write byte to memory |
| load byte unsigned | lbu $s1, k($s2) | $s1 = Memory[$s2 + k] | Read byte from memory; byte is 0-extended |
| load upper immediate | lui $s1, k | $s1 = k * $2^{16}$ | Loads constant *k* to upper 16 bits of register |

### 9.2.4 Control Transfer Instructions

Typically program execution proceeds in a sequential fashion, but loops, procedures, functions, and subroutines change the program control flow. A microprocessor needs branch and jump instructions in order to accomplish transfer of control whenever nonsequential control flow is required. The MIPS ISA includes two conditional branch instructions, *branch on equal (beq)* and *branch on not equal (bne)*, as illustrated in Table 9-4.

The MIPS instruction

```
beq $5, $4, 25
```

will compare the contents of $5 and $4 and branch to PC + 4 + 100 if $4 and $5 are equal. The constant offset provided in the branch instruction is specified in

TABLE 9-4:
Conditional Control
Related
Instructions in the
MIPS ISA

| Instruction | Assembly Code | Operation | Comments |
|---|---|---|---|
| branch on equal | beq  $s1, $s2, k | If ($s1 == $s2) go to PC + 4 + k * 4 | Branch if registers are equal; PC-relative branch; Target = PC + 4 + Offset * 4; k is sign-extended |
| branch on not equal | bne  $s1, $s2, k | If ($s1 / = $s2) go to PC + 4 + k * 4 | Branch if registers are not equal; PC-relative branch; Target = PC + 4 + Offset * 4; k is sign-extended |
| set on less than | slt   $s1, $s2, $s3 | If ($s2 < $s3) $s1 = 1; else $s1 = 0; | Compare and set (2's complement) |
| set on less than immediate | slti  $s1, $s2, k | If ($s2 < k) $s1 = 1; else $s1 = 0; | Compare and set; k is 16-bit constant; sign-extended and compared |
| set on less than unsigned | sltu $s1, $s2, $s3 | If ($s2 < $s3) $s1 = 1; else $s1 = 0; | Compare and set; natural numbers |
| set on less than immediate unsigned | sltiu $s1, $s2, k | If ($s2 < k) $s1 = 1;  else $s1 = 0; | Compare and set; natural numbers; k, the16-bit constant, is sign-extended; no overflow |

terms of the number of instructions from the current PC (program counter). MIPS uses byte addressing, and hence the offset in words is multiplied by 4 to get the offset in bytes. The program counter is assumed to point to the next instruction at PC + 4 already; hence the target address is computed as PC + 4 + 4 * offset. The offset is 16 bits long, however one bit is used for sign. Branching is thus possible to only +/−32K.

Having only two conditional branch instructions is in contrast to CISC processors that provide branch on less than, branch on greater than, branch on higher than, branch on lower than, branch on carry, branch on overflow, branch on negative, and several such conditional branch instructions. The MIPS philosophy was that only two conditional branch instructions are necessary and that checking of other conditions can be accomplished using combinations of instructions. In order to facilitate checking of less than and greater than, MIPS ISA provides the set on less than (*slt*) instructions. These are explicit compare instructions that will set an explicit destination register to 1 or 0 depending on the results of the compare. The *slt* instruction is used along with a *bne* or *beq* instruction to create the effect of branch on less than, branch on greater than, and so on. These instructions are used for implementing *loop* and *if-then-else* statements from high-level languages.

The MIPS ISA also includes three unconditional jump instructions as illustrated in Table 9-5. These instructions are used for implementing function and procedure calls and returns.

TABLE 9-5:
Unconditional
Control Transfer
Instructions in the
MIPS ISA

| Instruction | Assembly Code | Operation | Comments |
|---|---|---|---|
| jump | j  addr | Go to addr * 4; i.e., PC = addr * 4 | Target address = Imm offset * 4; addr is 26 bits |
| jump register | jr  $reg | Go to $reg; i.e., PC = $reg | $reg contains 32-bit target address |
| jump and link | jal  addr | return address = PC + 4; go to addr * 4 | For procedure call, return address saved in the link register $31 |

The *jump* instruction transfers control to the address specified in the instruction. Since the MIPS instruction is 32 bits wide, the number of bits available for encoding the address will be (32 − number of opcode bits). In the MIPS, the opcode consumes 6 bits; therefore, only 26 bits are available for the address in the jump instruction. In order to increase the range of addresses to which control can be transferred, MIPS designers consider the specified address as a word address (instead of a byte address) and multiply the specified address by 4 to get the resulting byte address.

The *jump register (jr)* instruction is an *indirect jump*. In contrast, the jump instruction described in the previous paragraph is called a *direct* jump because the jump address is directly specified in the instruction itself. In the case of the *jump register* instruction, the content of the register is used as the address to which program should transfer control to. This type of branch instruction is very useful for implementing case statements from high-level languages.

The *jump and link (jal)* instruction is specifically designed for procedure calls. It computes the target address from the offset specified in the instruction, but in addition to transferring control to that address, it also saves the return address in link register *$31*. The return address means the address control should return to after the subroutine or procedure call is completed. The return address is equal to the current PC + 4, since every instruction is four bytes wide and PC + 4 is the address of the instruction following the current instruction (the *jal* instruction).

We have described the major classes of instructions in the MIPS ISA. In order to become familiar with the instructions, let us practice some assembly language programming.

**Example**

Write a MIPS assembly language program for the following program that adds two arrays x(i) and y(i), each of which has 100 elements.

$$\text{for } i = 1, 100, i++ \qquad \text{; repeat 100 times}$$
$$y(i) = x(i) + y(i) \qquad \text{; add ith element of the arrays}$$

Assume that the *x* and *y* arrays start at locations 4000 and 8000 (decimal).

**Answer**

|  |  |  |  |
|---|---|---|---|
|  | andi | $3, $3, 0 | ; initialize loop counter $3 to 0 |
|  | andi | $2, $2, 0 | ; clear register for loop bound |
|  | addi | $2, $2, 400 | ; loop bound |
| $label: | lw | $15, 4000($3) | ; load x(i) to R15 |
|  | lw | $14, 8000($3) | ; load y(i) to R14 |
|  | add | $24, $15, $14 | ; x(i) + y(i) |
|  | sw | $24, 8000($3) | ; save new y(i) |
|  | addi | $3, $3, 4 | ; update address register, address= address + 4 |
|  | bne | $3, $2, $label | ; check if loop counter=loop bound |

Several microprocessors with the MIPS ISA have been designed since the MIPS R2000 was designed in the 1980s. In those days, the main processor could not integrate the floating-point unit. Hence, the floating-point units were implemented as a math coprocessor, the MIPS R2010. Nowadays, the floating-point unit is integrated with the main CPU. The MIPS R2000 was followed by MIPS R3000, R4000, R8000, R10000, R12000, and the R14000. They all have the MIPS ISA but different implementations with different levels of pipelining and different techniques to obtain high performance.

● ● ● ● ● ● ● ● ● ● ● ●

# 9.3 MIPS Instruction Encoding

Adhering to the RISC philosophy, all instructions in the MIPS processor have the same width, 32 bits. In a move toward simplicity, there are only three different instruction formats for the MIPS instructions. The three formats are called **R-format**, **I-format**, and **J-format**, as illustrated in Table 9-6.

TABLE 9-6: **Instruction Formats in the MIPS ISA**

| Format | Fields | | | | | | Used by |
|---|---|---|---|---|---|---|---|
|  | 6 bits 31–26 | 5 bits 25–21 | 5 bits 20–16 | 5 bits 15–11 | 5 bits 10–6 | 6 bits 5–0 |  |
| **R-format** | opcode | rs | rt | rd | shamt | F_code (funct) | ALU instructions except immediate, Jump Register (JR) |
| **I-format** | opcode | rs | rt | offset/immediate | | | Load, store, Immediate ALU, beq, bne |
| **J-format** | opcode | target address | | | | | Jump (J), Jump and Link (JAL) |

The **R-format** is primarily for ALU instructions which require three operands. These ALU instructions have two source operands (input registers) and one destination address (result register) to be specified. The jump register instruction (jr) also uses this format. The instruction consists of six fields, the first of which is the 6-bit **opcode** field. The opcode field is followed by the three register fields *rs, rt*, and *rd*, each of which takes 5 bits. The first two are the source register fields, and the third one is the destination register field. The next field is called shift amount (*shamt*) field, which is used to specify the amount of shifting to be done in shift instructions. Any number between 0 and 31 can be specified as the shift amount. This field is used only in shift instructions. The last field is an additional opcode field, called the function field *funct* or *F_code*. The first opcode field can encode only $2^6$ or 64 instructions. The MIPS processor does have more than 64 instructions considering the different variations of loads (byte load, halfword load, word load, floating-point loads, etc.). Hence, more than 6 bits are required to fully specify an instruction. The MIPS designers chose a scheme in which the first 6 bits are 0 for the R-format instructions, and then an additional field (the last 6 bits of the instruction) is used to further identify the instruction.

The **I-format** is for arithmetic instructions, load/store instructions, and branch instructions that need an immediate constant to be specified in the instruction. These instructions need only two registers to be specified in addition to the immediate constant. The opcode field takes 6 bits, and the two register fields take 5 bits each. The remaining 16 bits are used as an immediate constant to specify an operand for instructions such as *addi*, or to specify the offset in a load/store instruction, or to specify the branch offset in a conditional branch instruction.

The **J-format** is for jump instructions. The first 6 bits of the instruction word are used for the opcode, and the remaining 26 bits are used to specify the jump offset. Since the jump offset is specified as a word address rather than byte address, the offset is first multiplied by 4 and then concatenated to the highest 4 bits of the PC to get the 32-bit target address. MIPS uses byte addressing for accessing instructions and data.

Table 9-7 illustrates the instruction encoding for the MIPS instructions we have discussed. The opcode, source, and destination are assigned the same field in the instruction format as much as possible. The first 6 bits (bits 31–26) are for the opcode in all the three different formats. The source and destination register fields are in similar positions (bits 25–21, bits 20–16, and bits 15–11) as much as possible. This greatly simplifies decoding.

The encoding is very regular; however, compromises had to be made to accommodate various instructions into the same width. For instance, the destination register appears in different fields in three-register and two-register formats. Similarly, in a load instruction, the second register field is a destination register; whereas in a store instruction, it is the source of the data to be stored. In spite of these irregularities, we can say that the encoding is largely regular.

To increase the familiarity with the MIPS instruction encoding, let us practice some machine coding.

TABLE 9-7: **Instruction Encoding for the MIPS Instructions**

| | | Fields | | | | | | | |
|---|---|---|---|---|---|---|---|---|---|
| Name | Format | Bits 31–26 | Bits 25–21 | Bits 20–16 | Bits 15–11 | Bits 10–6 | Bits 5–0 | Instruction (operation dest, src1, src2) | |
| add | R | 0 | 2 | 3 | 1 | 0 | 32 | add | $1, $2, $3 |
| sub | R | 0 | 2 | 3 | 1 | 0 | 34 | sub | $1, $2, $3 |
| addi | I | 8 | 2 | 1 | | 100 | | addi | $1, $2, 100 |
| addu | R | 0 | 2 | 3 | 1 | 0 | 33 | addu | $1, $2, $3 |
| subu | R | 0 | 2 | 3 | 1 | 0 | 35 | subu | $1, $2, $3 |
| addiu | I | 9 | 2 | 1 | | 100 | | addiu | $1, $2, 100 |
| mfc0 | R | 16 | 0 | 1 | 14 | 0 | 0 | mfc0 | $1, $epc |
| mult | R | 0 | 2 | 3 | 0 | 0 | 24 | mult | $2, $3 |
| multu | R | 0 | 2 | 3 | 0 | 0 | 25 | multu | $2, $3 |
| div | R | 0 | 2 | 3 | 0 | 0 | 26 | div | $2, $3 |
| divu | R | 0 | 2 | 3 | 0 | 0 | 27 | divu | $2, $3 |
| mfhi | R | 0 | 0 | 0 | 1 | 0 | 16 | mfhi | $1 |
| mflo | R | 0 | 0 | 0 | 1 | 0 | 18 | mflo | $1 |
| and | R | 0 | 2 | 3 | 1 | 0 | 36 | and | $1, $2, $3 |
| or | R | 0 | 2 | 3 | 1 | 0 | 37 | or | $1, $2, $3 |
| andi | I | 12 | 2 | 1 | | 100 | | andi | $1, $2, 100 |
| ori | I | 13 | 2 | 1 | | 100 | | ori | $1, $2, 100 |
| sll | R | 0 | 0 | 2 | 1 | 10 | 0 | sll | $1, $2, 10 |
| srl | R | 0 | 0 | 2 | 1 | 10 | 2 | srl | $1, $2, 10 |
| lw | I | 35 | 2 | 1 | | 100 | | lw | $1, 100($2) |
| sw | I | 43 | 2 | 1 | | 100 | | sw | $1, 100($2) |
| lui | I | 15 | 0 | 1 | | 100 | | lui | $1, 100 |
| beq | I | 4 | 1 | 2 | | 25 | | beq | $1, $2, 25 |
| bne | I | 5 | 1 | 2 | | 25 | | bne | $1, $2, 25 |
| slt | R | 0 | 2 | 3 | 1 | 0 | 42 | slt | $1, $2, $3 |
| slti | I | 10 | 2 | 1 | | 100 | | slti | $1, $2, 100 |
| sltu | R | 0 | 2 | 3 | 1 | 0 | 43 | sltu | $1, $2, $3 |
| sltiu | I | 11 | 2 | 1 | | 100 | | sltiu | $1, $2, 100 |
| j | J | 2 | | 2500 | | | | j | 2500 |
| jr | R | 0 | 31 | 0 | 0 | 0 | 8 | jr | $31 |
| jal | J | 3 | | 2500 | | | | jal | 2500 |

**Example**

Create the machine code equivalent of the following assembly language program.

```
         andi   $3, $3, 0        ; initialize loop counter $3 to 0
         andi   $2, $2, 0        ; clear register for loop bound
         addi   $2, $2, 4000     ; loop bound register
$label:  lw     $15, 4000($3)    ; load x(i) to R15
         lw     $14, 8000($3)    ; load y(i) to R14
         add    $24, $15, $14    ; x(i) + y(i)
```

```
sw      $24, 8000($3)      ; save new y(i)
addi    $3, $3, 4          ; update address register, address=
                             address + 4
bne     $3, $2, $label     ; Check if loop counter=
                             loop bound
```

**Answer**

The first instruction

```
andi    $3, $3, 0
```

can be translated as follows. Table 9-7 shows that the opcode for **andi** is 12. Hence, the first 6 bits for the first instruction will be 001100, as indicated in row 1 (after the header row) of Table 9-8. The source register field is next. It should be 00011 because the source register is $3. The destination register field is next. It should be 00011 because the destination register is $3. The immediate constant is 0 and leads to sixteen 0's in bits 0 to 15. This explains the contents of row 1. In hex representation, it becomes 3063 0000.

We will also explain the encoding of the last instruction, bne $3, $2, label. The opcode is 5 (i.e., 000101). The next field corresponds to register $3, so it is 00011. The next field is 00010 to indicate the register $2. The byte offset should be −24, but the instruction is supposed to contain the word offset which is −24 divided by 4 (i.e., −6). In 2's complement representation, it is 1010. Sign extending to fill the 16 bits, we get 1111111111111010, which will occupy bits 0 to 15.

Machine code corresponding to all the instructions is shown in Table 9-8.

TABLE 9-8: MIPS Machine Code for Example. Binary as Well as Hex Representations Shown

| Instruction | Bits 31–26 | Bits 25–21 | Bits 20–16 | Bits 15–11 | Bits 10–6 | Bits 5–0 | Equivalent Hex |
|---|---|---|---|---|---|---|---|
| andi $3, $3, 0 | 001100 | 00011 | 00011 | 00000 | 00000 | 000000 | 3063 0000 |
| andi $2, $2, 0 | 001100 | 00010 | 00010 | 00000 | 00000 | 000000 | 3042 0000 |
| addi $2, $2, 4000 | 001000 | 00010 | 00010 | 00001 | 11110 | 100000 | 2042 0FA0 |
| lw   $15, 4000($3) | 100011 | 00011 | 01111 | 00001 | 11110 | 100000 | 8C6F 0FA0 |
| lw   $14, 8000($3) | 100011 | 00011 | 01110 | 00011 | 11101 | 000000 | 8C6E 1F40 |
| add  $24, $15, $14 | 000000 | 01111 | 01110 | 11000 | 00000 | 100000 | 01EE C020 |
| sw   $24, 8000($3) | 101011 | 00011 | 11000 | 00011 | 11101 | 000000 | AC78 1F40 |
| addi $3, $3, 4 | 001000 | 00011 | 00011 | 00000 | 00000 | 000100 | 2063 0004 |
| bne  $3, $2, −6 | 000101 | 00011 | 00010 | 11111 | 11111 | 111010 | 1462 FFFA |

● ● ● ● ● ● ● ● ● ● ● ●

# 9.4 Implementation of a MIPS Subset

In this section, we describe a simple implementation of a subset of the MIPS ISA. This subset, illustrated in Table 9-9, includes most of the important instructions, including ALU, memory access, and branch instructions. What we present in this

| TABLE 9-9: Subset of MIPS Instructions Implemented in This Chapter | | |
|---|---|---|
| | Arithmetic | add<br>subtract<br>add immediate |
| | Logical | and<br>or<br>and immediate<br>or immediate<br>shift left logical<br>shift right logical |
| | Data Transfer | load word<br>store word |
| | Conditional branch | branch on equal<br>branch on not equal<br>set on less than |
| | Unconditional branch | jump<br>jump register |

section is a naïve implementation of this instruction set. Modern microprocessors implement features such as multiple instruction issue, out-of-order execution, branch prediction, and pipelining. For the sake of simplicity, what is presented here is a simple in-order, nonpipelined implementation. Some of the exercise problems describe other implementations that will provide better performance.

### 9.4.1 Design of the Data Path

In order to design a microprocessor, first we examine the sequence of operations during execution of instructions, and then we describe the nature of the hardware required to accomplish the instruction execution. In general, any microprocessor works in the following manner:

1. The processor **fetches** an instruction.
2. It **decodes** the instruction that was fetched. *Decoding* means identifying what the instruction is.
3. It reads the operands and **executes** the instruction. For a RISC ISA, for arithmetic instructions, the operands are in registers. The registers that contain the input operands are called source registers. For memory access instructions, addresses are computed using registers, and memory is accessed. After execution, the processor **writes** the result of the instruction execution into the destination. The destination is a register for all instructions other than the store instruction, which has to write the result into the memory.

Hence, the design must contain a **unit to fetch** the instructions, a **unit to decode** the instructions, an arithmetic and logic unit (**ALU**) to execute the instructions, a **register file** to hold the operands, and the **memory** that stores instructions and data. These components are described in the following subsections.

### Instruction Fetch Unit

In general, a microprocessor has a special register called the program counter (PC), which points to the next instruction in the instruction memory. The PC sends this address to the instruction memory (or the instruction caches), which sends the instruction back. The processor increments the PC to point to the next instruction to be fetched. A block diagram for this unit is shown in Figure 9-1.

**FIGURE 9-1: Block Diagram for Instruction Fetch**

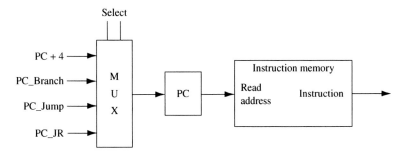

The next PC is one of the following, depending on the current instruction:

a. **PC + 4**: For instructions other than branch and jump instructions, the next instruction is at address PC + 4, since four bytes are needed for the current instruction.

b. **PC_Branch**: In the branch (bne and beq) instructions, the next PC is obtained by adding the offset in the instruction to the current PC. In the MIPS ISA, the branch offset is provided as a signed-word offset (number of words to jump forward or backward). First the word offset is sign-extended, converted to a byte offset by multiplying by 4, and then it is added to the current PC. Thus, the next PC for branch instructions is

$$PC\_Branch = PC + 4 + Offset * 4$$

c. **PC_Jump**: In the jump (J) instruction, the new target is provided in the instruction. In the MIPS ISA, the opcode takes 6 out of the 32 bits. Hence, the biggest jump address that can be encoded is only 26 bits. In order to compute the 32-bit jump address, first, the 26-bit word address in the instruction is shifted twice to the left, resulting in a 28-bit address, which is a byte address. Then it is concatenated with the four highest bits of the PC, yielding a 32-bit address. Thus, the next PC for jump instructions is

$$PC\_Jump = PC31..28 \parallel Address * 4$$

where $\parallel$ stands for concatenation.

d. **PC_JR**: In the jump register (JR) instruction, the jump target is obtained from the register specified in the instruction. Thus, the next PC for a JR instruction is

$$PC\_JR = [REG]$$

where [REG] indicates contents of the register.

The appropriate target addresses are computed and fed to the PC. A multiplexer is used to select among the branch target, jump target, jump register target, or PC + 4, depending on the instruction.

There are several choices as to when the target addresses are computed. The default target, PC + 4, can be computed at instruction fetch itself, since it needs no other information other than the PC itself. In conditional branch instructions, the branch target (PC_Branch) computation can be done as soon as the instruction is read; however, whether the branch is taken or not will not be known until the registers are read and compared. In the case of the jump instruction, the target (PC_Jump) can be computed as soon as the instruction is fetched, since the information for the target is available in the instruction itself. In a jump register (jr) instruction, the branch target (PC_JR) can be computed after the register is read.

## Instruction Decode Unit

Decoding is fairly simple due to the simplicity of the RISC ISA. We can observe from Table 9-7 that the instruction formats in the MIPS ISA are very regular and uniform. The first 6 bits of the instruction specify the opcode in most cases. But, as described in Section 9.3, for the R-format ALU instructions, the first 6 bits are 0, and the last 6 bits of the instruction, called **F_code**, need to be used to further identify the instruction.

The opcode is used to identify the instruction and the instruction format used by the instruction. The uniformity of the instruction format allows many of the instruction fields to be directly used for register addressing and control signal generation. The instruction opcode bits are fed to a control unit that generates the various control signals.

## Instruction Execution Unit

Once the instruction is identified at the decode stage, the next task is to read the operands and perform the operation. In RISC instruction sets, the operands are in registers. The MIPS architecture contains 32 registers, and these registers are collectively referred to as the **register file**. The register file should have at least two read ports to support reading two operands at the same time, and it should have one write port.

The operation of the register file is as follows. The registers that hold the input operands are called **source registers**, and the register that should receive the result is called the **destination register**. The source register addresses are applied to the register file. The register file will produce the data from the corresponding registers on the output data lines. This data is fed to the arithmetic and logic unit (**ALU**), which executes the instruction. The ALU contains functional units such as adders and shifters. It may also include more complex units such as multipliers, although our restricted design here does not include multiplication.

In most instructions, the result from the ALU should be written into the destination register. To accomplish this, the ALU result is applied to the input data lines of the register file. The destination register name and the *register write (RegW) command* is applied to the register file. That causes the input data to get written into the destination register.

Figure 9-2 shows a block diagram of the datapath that is required to execute the ALU and memory instructions. The data path includes an ALU, which will perform the following operations: add, sub, and, and or. In the case of R-format instructions, both operands for the ALU are read from the register file. In the case of the I-format instructions, the immediate constant in the instruction is sign-extended to create the second operand. Since one of the ALU operands comes from either the register file or the sign extender, a multiplexer is required to select the appropriate operand.

The ALU is also required for nonarithmetic instructions. For memory access instructions, we have to first calculate the address to be accessed. The ALU can be used for calculating the address. For address calculation for load and store instructions, the first operand is obtained from the register specified in the instruction and the second operand is obtained by sign-extending the immediate offset specified in the instruction.

The ALU is required for conditional branch instructions also. As you know, MIPS has only two branch instructions, branch on equal (beq) and branch not equal (bne). The comparison for determining whether the registers are equal can be done by the ALU. Both operands for this comparison can be obtained from the register file. The data path also has to include a **data memory unit** because load and store instructions have to access the data memory unit. Modern microprocessors contain on-chip data caches. We will not be designing a data cache memory; however, we will assume the presence of on-chip data memory that can be accessed by the instructions in one cycle after the data address is provided to the memory.

## Overall Data Path

The overall data path is shown in Figure 9-3. It integrates the fetch and execute hardware from Figures 9-1 and 9-2 and adds other required elements for correct operation. In addition, control signals are shown.

**FIGURE 9-2: Required Data Path for Computation and Memory Instructions**

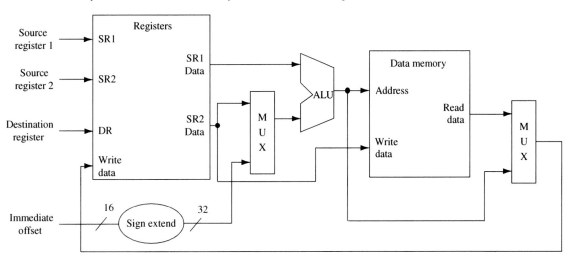

Figure 9-3 also shows use of several multiplexers and how the different bits of the instruction are connected to the register file. As Table 9-6 illustrates, bits 21 to 25 of the instruction contains one of the source register addresses in all ALU instructions. Hence, these bits can be directly connected to the first source register address of the register file. Any instruction with a second register source contains the register address in bits 16 to 20. Hence, these bits can also be directly connected to the source register address of the register file. However, the destination register address appears in different fields in different instructions. In R-format instructions, the destination register address appears in bits 11 to 15. In I-format instructions, however, the destination address is in bits 16 to 20. Hence, a multiplexer is required to choose the appropriate destination register address. Another multiplexer chooses between the immediate operand or register operand for the ALU. A third multiplexer is used to select whether ALU output or memory data will be written to the destination register.

Figure 9-3 also illustrates the details of the computation of the target addresses in the various kinds of instructions. Default next address of PC + 4 is calculated

FIGURE 9-3: **Overall Data Path**

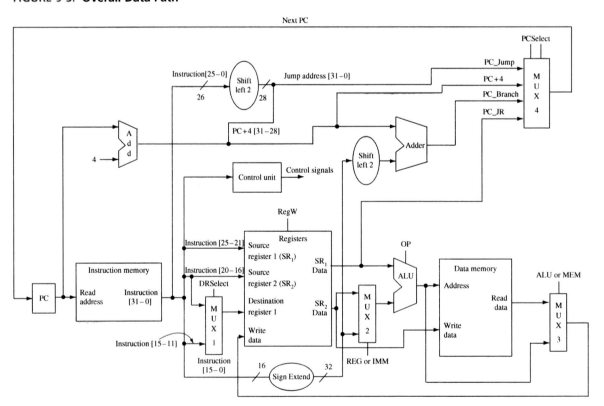

with an adder. Addition of the branch offset to the PC is also done using a separate adder. A multiplexer is used to select the appropriate PC.

In summary:

- MUX 1 selects a destination register address from an appropriate register field depending on the instruction format. For R-format instructions, bits 20–16 yield the destination address, and for I-format instructions, bits 15–11 of instruction provide the destination address.
- MUX 2 selects whether the second operand for ALU comes from a register or an immediate constant. For R-format ALU instructions and conditional branch instructions, the register is chosen. For I-format ALU instructions, the immediate constant provides the operand.
- MUX 3 selects between the memory or the ALU output for data to go into the destination register. For load instructions, the memory data is chosen.
- MUX 4 selects between the four possible next PC values depending on the type of instruction.

### 9.4.2 Instruction Execution Flow

Figure 9-4 illustrates the flow of execution for a possible implementation.

The first step is fetch for all instructions. The address in the program counter (PC) is sent to the instruction memory unit. All instructions also need to update the PC to point to the next instruction. While PC should be updated differently for branch or jump instructions, the vast majority of instructions are in sequence, and hence PC can be updated to point to the next instruction in sequence. Branch and jump instructions can later modify the PC appropriately.

The second step is decode. Depending on the opcode that is encountered, different actions follow. For R-type instructions, and for some I-type instructions (e.g., *bne* and *beq*), both ALU operands are read from registers. For other I-type instructions, one operand is read from the register file and the immediate constant in the instruction is sign-extended as the other operand. Reading of a register source satisfies requirements for a jump register (*jr*) instruction, which is an R-type instruction. The ALU operation required for each instruction is identified during the decode step. For instance, the *bne* and *beq* instructions need a subtract operation. The load and store instructions require an add operation. If the jump opcode is encountered, a jump target is calculated. Since the jump instruction does not need any further action, flow of control can go to step 1.

Step 3 is the actual execution of the instructions. Depending on the instruction, different ALU operations are performed during this step. The different actions are shown in boxes labeled 3a, 3b, and so on for the different types of instructions. Each instruction goes through only one of these operations, depending on what type of instruction it is. All instructions other than the jump instruction must come to this step. The jump register (*jr*) instruction does not need any arithmetic operation, but the content of the register fetched during step 2 must be loaded into the PC. For load and store instructions, the ALU performs an addition to calculate the memory address.

FIGURE 9-4: **Flow Chart for Instruction Processing**

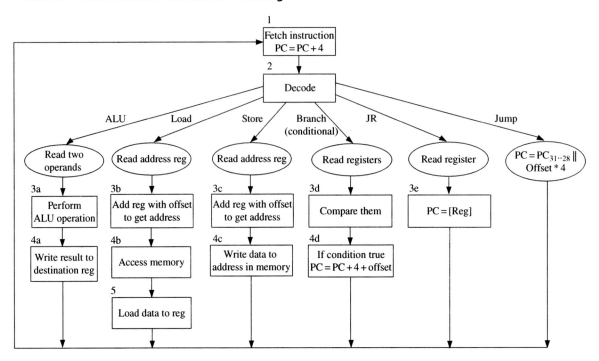

Step 4 varies widely between the instructions. Arithmetic and logic instructions (of R-type and I-type) can write their computation result to the destination register. Branch instructions must examine their condition and decide to take the branch or not. If the branch is to be taken, the branch target address is calculated. For load instructions, a memory read operation is initiated. For memory store instructions, the data from the second source register is steered to the memory, and a memory write operation is initiated. This is the final step for all instructions other than load instructions.

Step 5 is required only for load instructions. The data output from memory is written into the destination register.

We can implement this instruction flow in a variety of ways. In the most naïve implementation, we can have a very slow clock and the processor performs all operations required for each instruction in one clock cycle. The disadvantage with this scheme is that all instructions will be as slow as the slowest instruction because the clock cycle has to be long enough for the slowest instruction. Another option is to do an implementation where each instruction takes multiple cycles, but just enough cycles to finish all operations for each class of instruction. For instance, Figure 9-4 can be considered as an SM chart with each box taking one cycle. In this case, a jump instruction can finish in two cycles, while an ALU instruction needs four cycles and a load instruction takes five cycles. In the next section, we present the VHDL model of such an implementation.

• • • • • • • • • • •
# 9.5 VHDL Model

The VHDL model for the processor is organized as in Figure 9-5. The instruction memory, data memory, and register file are created as components with their architecture and entity descriptions. The main code, the MIPS entity embeds the control sequencing the instructions through the various stages of its operation. For simplicity we combined the instruction and data memory units to be a single memory and illustrate the use of the address and data buses. Later, when we use a test bench, we allow the test bench to directly write into the instruction memory in order to deposit instructions to be tested.

**FIGURE 9-5:**
**Organization of the VHDL Model for the Processor**

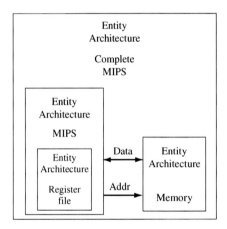

Let us model the register and memory components first.

### 9.5.1 VHDL Model for the Register File

Figure 9-6 shows the VHDL model for the register file. The REG entity is used to represent the 32 MIPS registers. Each register is 32 bits long. The destination register address is *DR*, and the source register addresses are *SR1* and *SR2*. Since there are 32 registers, *DR, SR1*, and *SR2* are 5 bits each. The outputs *ReadReg1* and *ReadReg2* are the contents of the registers specified by *SR1* and *SR2*. *ReadReg1* is fed straight to the ALU. *ReadReg2* can be used as a second ALU input, or as the input to data memory in the case of store instructions. The control signal *RegW* is used to control the write operation to the register file. If *RegW* is true, the data on lines *Reg_In* is written into the register pointed to by *DR*.

If this code is synthesized for a Xilinx Spartan FPGA, the reads have to be performed asynchronously as in the provided code in order to get the register file mapped into distributed RAM. As you know from Chapter 6, the Xilinx Spartan/Virtex FPGAs contain dedicated block RAM. It is desirable to perform reads synchronously; however, then the register file gets synthesized into BlockRAM with current Xilinx synthesis tools. We used the asynchronous reads to allow generation of distributed RAM for the register file.

FIGURE 9-6: **VHDL Code for Register File**

```
library IEEE;
use IEEE.std_logic_1164.all;
use IEEE.numeric_std.all;

entity REG is
port(CLK: in std_logic;
     RegW: in std_logic;
     DR, SR1, SR2: in unsigned(4 downto 0);
     Reg_In: in unsigned(31 downto 0);
     ReadReg1, ReadReg2: out unsigned(31 downto 0));
end REG;

architecture Behavioral of REG is
  type RAM is array (0 to 31) of unsigned(31 downto 0);
  signal Regs: RAM := (others => (others => '1')); -- set all reg bits to '1'
begin
  process(clk)
  begin
    if CLK = '1' and CLK'event then
      if RegW = '1' then
        Regs(to_integer(DR)) <= Reg_In;
      end if;
    end if;
  end process;
  ReadReg1 <= Regs(to_integer(SR1)); -- asynchronous read
  ReadReg2 <= Regs(to_integer(SR2)); -- asynchronous read
end Behavioral;
```

### 9.5.2 VHDL Model for Memory

Figure 9-7 illustrates the VHDL code for the memory unit. The VHDL model is similar to the SRAM model that we did in Chapter 8. This SRAM model has tristated input-output lines and allows easy testing with a test bench, where the test bench can write instructions into the memory and the processor can read instruction and read/write data. The test bench and the processor can drive the data bus of the memory. Although Figure 9-3 illustrated separate instruction and data memories, for convenience and for illustrating the use of address and data buses, we have used a unified memory module which stores both instructions and data. The memory consists of 128 locations, each 32 bits wide. We assume that the instructions are the first 64 words in the array, and the other 64 words are allocated for data memory. The signal *Address* specifies the location in memory to be read from or stored to. The address bus is actually 32 bits wide, but we only use the seven lower bits since we implement only a small memory.

The address bus will be driven by the processor appropriately for instruction and data access. The address input may come from the program counter for reading the instruction, or from the ALU that computes the address to access the data portion of the memory. The chip select (*CS*) and write enable (*WE*) signals

FIGURE 9-7: **VHDL Code for the Unified Instruction/Data Memory**

```vhdl
library IEEE;
use IEEE.std_logic_1164.all;
use IEEE.numeric_std.all;

entity Memory is
  port(CS, WE, Clk: in std_logic;
       ADDR: in unsigned(31 downto 0);
       Mem_Bus: inout unsigned(31 downto 0));
end Memory;

architecture Internal of Memory is
  type RAMtype is array (0 to 127) of unsigned(31 downto 0);
  signal RAM1: RAMtype := (others => (others => '0'));
  signal output: unsigned(31 downto 0);
begin
  Mem_Bus <= (others => 'Z') when CS = '0' or WE = '1'
    else output;
  process(Clk)
  begin
    if Clk = '0' and Clk'event then
      if CS = '1' and WE = '1' then
        RAM1(to_integer(ADDR(6 downto 0))) <= Mem_Bus;
      end if;
    output <= RAM1(to_integer(ADDR(6 downto 0)));
    end if;
  end process;
end Internal;
```

allow the processor to control the reads and writes. When *CS* and *WE* are true, the data on *Mem_Bus* gets written to the memory location pointed to by address *ADDR*.

For simplicity, **the address is shown as a word address in the VHDL code for the memory**. Hence, branch and jump offsets are used as such in Figure 9-8 without multiplying by 4. In the actual MIPS processor, the memory is byte-addressable. Therefore, each instruction memory access should obtain the data found in the specified location concatenated with the next three memory locations. For example, if address = 0, the instruction register must be loaded with the contents of MEM[0], MEM[1], MEM[2], and MEM[3]. The instructions are stored depending on the endianness of the machine (See sidebar). Many modern microprocessors support both big-endian and little-endian approaches.

### 9.5.3 VHDL Code for the Processor CPU

In this section, we present the VHDL code for the central processing unit (CPU) of the microprocessor. The register module that was created in the earlier section is

---

**Little-Endian and Big-Endian**

When we store 16-bit or 32-bit data into byte-addressable memory, there are two possible ways to store the data: little-endian and big-endian. In a little-endian system, the least significant byte in the sequence is stored first. In a big-endian system, the most significant byte in the sequence is stored at the lowest storage address (i.e., first). Let us consider how a MIPS instruction will be stored into byte-addressable memory in the two systems. The MIPS instruction **andi $3, $3, 0** will be encoded as **30630000 (hex)**. When this instruction is stored at address 2000, depending on whether big-endian or little-endian system is used, the memory will look as follows:

| Address | Big-Endian Representation of 30630000hex | Little-Endian Representation of 30630000hex |
|---|---|---|
| 2000 | 30 | 00 |
| 2001 | 63 | 00 |
| 2002 | 00 | 63 |
| 2003 | 00 | 30 |

---

used here. Figure 9-8 shows a VHDL model for the MIPS instructions in Table 9-9. The VHDL model generally follows the flow in Figure 9-4, implementing the fetch, decode, and execute phases of an instruction. In order to increase the readability of the code, several aliases are defined. The most significant 6 bits of the instruction are denoted by the alias *Opcode*. The lowest 6 bits of the instruction are denoted with the alias *F_Code*. The shift amount in shift instructions is denoted using *NumShift*. The two register source fields are aliased to *SR1* and *SR2*. The following statements accomplish this aliasing:

```
alias opcode: unsigned(5 downto 0) is Instr(31 downto 26);
alias SR1: unsigned(4 downto 0) is Instr(25 downto 21);
alias SR2: unsigned(4 downto 0) is Instr(20 downto 16);
alias F_Code: unsigned(5 downto 0) is Instr(5 downto 0);
alias NumShift: unsigned(4 downto 0) is Instr(10 downto 6);
```

For readability of the code, we have also used constant declarations to associate the various opcodes with the corresponding codes from Table 9-7. For instance, the load instruction *lw* has 35 as its opcode, and the store instruction *sw* has 43 as its opcode. Several statements, such as the following, are used in order to denote the various opcodes:

```
constant lw  : unsigned(5 downto 0) := "100011";   -- 35
constant sw  : unsigned(5 downto 0) := "101011";   -- 43
```

Sign extension of the immediate quantity is accomplished by the following statement:

```
Imm_Ext <= x"FFFF"&Instr(15 downto 0) when Instr(15) = '1' else
           x"0000"&Instr(15 downto 0);
```

Following are the signals used in the VHDL model:

MIPS Processor Model Signals:

| | |
|---|---|
| *CLK (input)* | Clock. |
| *Rst (input)* | Synchronous reset. |
| *CS (output)* | Memory chip select. When CS is active and WE is inactive, the memory module outputs the memory contents at the address specified by Addr to mem_bus. |
| *WE (output)* | Memory write enable. When WE and CS are active, the memory module stores the contents of mem_bus to the location specified by Addr during the falling edge of the clock. |
| *Addr (ouput)* | Memory address. During state 0 (fetch instruction from memory), Addr is connected to the PC. Otherwise, it is connected to the ALU result (32 bits). |
| *Mem_Bus (in/out)* | Tristate memory bus; carries data to and from the memory module. The MIPS module outputs to the bus during memory writes. The memory module outputs to the bus during memory reads. When not in use, the bus is at 'hi-Z' (32 bits). |
| *Op* | ALU operation select; determines the specific operation (e.g., add, and, or) to be performed by ALU. Determined during decode. |
| *Format* | Indicates whether the current instruction is of R, I, or J format. |
| *Instr* | The current instruction (32 bits). |
| *Imm_Ext* | Sign-extended immediate constant from the instruction (32 bits). |
| *PC* | Current program counter (32 bits). |
| *NPC* | Next program counter (32 bits). |
| *ReadReg1* | Contents of the first source register (SR1) (32 bits). |
| *ReadReg2* | Contents of the second source register (SR2) (32 bits). |
| *Reg_In* | Data input to registers. When executing a load instruction, Reg_In is connected to the memory bus. Otherwise, it is connected to the ALU result (32 bits). |
| *ALU_InA* | First operand for the ALU (32 bits). |
| *ALU_InB* | Second operand for the ALU. ALU_InB is connected to Imm_Ext during immediate mode instructions. Otherwise, it is connected to ReadReg2 (32 bits). |
| *ALU_Result* | Output of ALU (32 bits). |
| *ALUorMEM* | Select signal for the Reg_In multiplexer; indicates if the register input should come from the memory, or the ALU. |
| *REGorIMM* | Select signal for the ALU_InB multiplexer; determines if the second ALU operand is a register output or sign extended immediate constant. |
| *RegW* | Indicates if the destination register should be written to. Some instructions do not write any results to a register (e.g., branch, store). |
| *FetchDorI* | Select signal for the Address multiplexer; determines if Addr is the location of an instruction to be fetched, or the location of data to be read or written. |
| *Writing* | Control signal for the MIPS processor output to the memory bus. Except during memory writes, the output is 'hi-Z' so the bus can be used by other modules. Note Writing cannot be replaced with WE, because WE is of mode *out*. *Writing* is used in mode *in* too. |
| *DR* | Address of destination register (5 bits). |
| *State* | Current state. |
| *nState* | Next state. |

Two processes are used in the code. Since we have used separate clock cycles for the fetch operation, decode operation, execute operation, and so on, it is necessary to save signals created during each stage for later use. The statements such as

```
OpSave <= Op;
REGorIMM_Save <= REGorIMM;
ALUorMEM_Save <= ALUorMEM;
ALU_Result_Save <= ALU_Result;
```

are used in the clocked process (the second process) for saving (explicit latching) of the relevant signals.

FIGURE 9-8: **VHDL Code for the MIPS Subset Implementation**

```vhdl
library IEEE;
use IEEE.std_logic_1164.all;
use IEEE.numeric_std.all;

entity MIPS is
  port(CLK, RST: in std_logic;
       CS, WE: out std_logic;
       ADDR: out unsigned (31 downto 0);
       Mem_Bus: inout unsigned(31 downto 0));
end MIPS;

architecture structure of MIPS is
  component REG is
    port(CLK: in std_logic;
         RegW: in std_logic;
         DR, SR1, SR2: in unsigned(4 downto 0);
         Reg_In: in unsigned(31 downto 0);
         ReadReg1, ReadReg2: out unsigned(31 downto 0));
  end component;
  type Operation is (and1, or1, add, sub, slt, shr, shl, jr);
  signal Op, OpSave: Operation := and1;
  type Instr_Format is (R, I, J);  -- (Arithmetic, Addr_Imm, Jump)
  signal Format: Instr_Format := R;
  signal Instr, Imm_Ext: unsigned (31 downto 0);
  signal PC, nPC, ReadReg1, ReadReg2, Reg_In: unsigned(31 downto 0);
  signal ALU_InA, ALU_InB, ALU_Result: unsigned(31 downto 0);
  signal ALU_Result_Save: unsigned(31 downto 0);
  signal ALUorMEM, RegW, FetchDorI, Writing, REGorIMM: std_logic := '0';
  signal REGorIMM_Save, ALUorMEM_Save: std_logic := '0';
  signal DR: unsigned(4 downto 0);
  signal State, nState: integer range 0 to 4 := 0;
  constant addi: unsigned(5 downto 0) := "001000"; -- 8
  constant andi: unsigned(5 downto 0) := "001100"; -- 12
  constant ori:  unsigned(5 downto 0) := "001101"; -- 13
  constant lw:   unsigned(5 downto 0) := "100011"; -- 35
```

```
     constant sw:    unsigned(5 downto 0) := "101011"; -- 43
     constant beq:   unsigned(5 downto 0) := "000100"; -- 4
     constant bne:   unsigned(5 downto 0) := "000101"; -- 5
     constant jump: unsigned(5 downto 0) := "000010"; -- 2
     alias opcode: unsigned(5 downto 0) is Instr(31 downto 26);
     alias SR1: unsigned(4 downto 0) is Instr(25 downto 21);
     alias SR2: unsigned(4 downto 0) is Instr(20 downto 16);
     alias F_Code: unsigned(5 downto 0) is Instr(5 downto 0);
     alias NumShift: unsigned(4 downto 0) is Instr(10 downto 6);
     alias ImmField: unsigned (15 downto 0) is Instr(15 downto 0);
begin
  A1: Reg port map (CLK, RegW, DR, SR1, SR2, Reg_In, ReadReg1, ReadReg2);
  Imm_Ext <= x"FFFF" & Instr(15 downto 0) when Instr(15) = '1'
     else x"0000" & Instr(15 downto 0);  -- Sign extend immediate field
  DR <= Instr(15 downto 11) when Format = R
     else Instr(20 downto 16);           -- Destination Register MUX (MUX1)
  ALU_InA <= ReadReg1;
  ALU_InB <= Imm_Ext when REGorIMM_Save = '1' else ReadReg2; -- ALU MUX (MUX2)
  Reg_in <= Mem_Bus when ALUorMEM_Save = '1' else ALU_Result_Save; -- Data MUX
  Format <= R when Opcode = 0 else J when Opcode = 2 else I;
  Mem_Bus <= ReadReg2 when Writing = '1' else
     "ZZZZZZZZZZZZZZZZZZZZZZZZZZZZZZZZ"; -- drive memory bus only during writes
  ADDR <= PC when FetchDorI = '1' else ALU_Result_Save; --ADDR Mux

  process(State, PC, Instr, Format, F_Code, opcode, Op, ALU_InA, ALU_InB, Imm_Ext)
  begin
    FetchDorI <= '0'; CS <= '0'; WE <= '0'; RegW <= '0'; Writing <= '0';
    ALU_Result <= "00000000000000000000000000000000";
    nPC <= PC; Op <= jr; REGorIMM <= '0'; ALUorMEM <= '0';
    case state is
      when 0 =>  --fetch instruction
        nPC <= PC + 1; CS <= '1'; nState <= 1; -- increment by 1 since word address
        FetchDorI <= '1';
      when 1 =>
        nState <= 2; REGorIMM <= '0'; ALUorMEM <= '0';
        if Format = J then
          nPC <= "000000" & Instr(25 downto 0); nState <= 0; --jump, and finish
          -- offset not multiplied by 4 since mem is word address
        elsif Format = R then  -- register instructions
          if    F_code = "100000" then Op <= add;  -- add
          elsif F_code = "100010" then Op <= sub;  -- subtract
          elsif F_code = "100100" then Op <= and1; -- and
          elsif F_code = "100101" then Op <= or1;  -- or
          elsif F_code = "101010" then Op <= slt;  -- set on less than
          elsif F_code = "000010" then Op <= shr;  -- shift right
          elsif F_code = "000000" then Op <= shl;  -- shift left
          elsif F_code = "001000" then Op <= jr;   -- jump register
          end if;
        elsif Format = I then -- immediate instructions
          REGorIMM <= '1';
```

```
                    if Opcode = lw or Opcode = sw or Opcode = addi then Op <= add;
                    elsif Opcode = beq or Opcode = bne then Op <= sub; REGorIMM <= '0';
                    elsif Opcode = andi then Op <= and1;
                    elsif Opcode = ori then Op <= or1;
                    end if;
                    if Opcode = lw then ALUorMEM <= '1'; end if;
                 end if;
              when 2 =>
                 nState <= 3;
                 if   OpSave = and1 then ALU_Result <= ALU_InA and ALU_InB;
                 elsif OpSave = or1 then ALU_Result <= ALU_InA or ALU_InB;
                 elsif OpSave = add then ALU_Result <= ALU_InA + ALU_InB;
                 elsif OpSave = sub then ALU_Result <= ALU_InA - ALU_InB;
                 elsif OpSave = shr then ALU_Result <= ALU_InB srl to_integer(numshift);
                 elsif OpSave = shl then ALU_Result <= ALU_InB sll to_integer(numshift);
                 elsif OpSave = slt then -- set on less than
                    if ALU_InA < ALU_InB then ALU_Result <= X"00000001";
                    else ALU_Result <= X"00000000";
                    end if;
                 end if;
                 if ((ALU_InA = ALU_InB) and Opcode = beq) or
                    ((ALU_InA /= ALU_InB) and Opcode = bne) then
                    nPC <= PC + Imm_Ext; nState <= 0;
                 elsif opcode = bne or opcode = beq then nState <= 0;
                 elsif OpSave = jr then nPC <= ALU_InA; nState <= 0;
                 end if;
              when 3 =>
                 nState <= 0;
                 if Format = R or Opcode = addi or Opcode = andi or Opcode = ori then
                    RegW <= '1';
                 elsif Opcode = sw then CS <= '1'; WE <= '1'; Writing <= '1';
                 elsif Opcode = lw then CS <= '1'; nState <= 4;
                 end if;
              when 4 =>
                 nState <= 0; CS <= '1';
                 if Opcode = lw then RegW <= '1'; end if;
        end case;
end process;

process(CLK)
begin
   if CLK = '1' and CLK'event then
      if rst = '1' then
         State <= 0;
         PC <= x"00000000";
      else
         State <= nState;
         PC <= nPC;
      end if;
      if State = 0 then Instr <= Mem_Bus; end if;
```

```
        if State = 1 then
            OpSave <= Op;
            REGorIMM_Save <= REGorIMM;
            ALUorMEM_Save <= ALUorMEM;
        end if;
        if State = 2 then ALU_Result_Save <= ALU_Result; end if;
    end if;
  end process;
end structure;
```

The multiplexer at the input of the program counter is not explicitly coded. The various data transfers are coded behaviorally in the various states. A good synthesizer will be able to generate the multiplexer to accomplish the various data transfers. Similarly, the multiplexer to select the destination register address is also not explicitly coded. If the synthesis tool generates inefficient hardware for this multiplexed data transfer, we can code the multiplexer into the data path and generate control signals for the select signals.

### 9.5.4 Complete MIPS

The processor module and the memory are integrated to yield the complete MIPS model (Figure 9-9). Component descriptions are created for the processor and the memory units. These components are integrated by using port-map statements. The high-level entity is called Complete_MIPS. We have also brought out the address and data buses as outputs from the high-level entity. If no outputs are shown in an entity, when the code is synthesized, it results in empty blocks. Depending on the synthesis tool, unused signals (and corresponding nets) may be deleted from the synthesized circuit.

FIGURE 9-9: **VHDL Code Integrating the Processor and Memory Modules**

```
library IEEE;
use IEEE.std_logic_1164.all;
use IEEE.numeric_std.all;

entity Complete_MIPS is
  port(CLK, RST: in std_logic;
       A_Out, D_Out: out unsigned(31 downto 0));
end Complete_MIPS;

architecture model of Complete_MIPS is
  component MIPS is
    port(CLK, RST: in std_logic;
         CS, WE: out std_logic;
         ADDR: out unsigned(31 downto 0);
         Mem_Bus: inout unsigned(31 downto 0));
  end component;
```

```
  component Memory is
    port(CS, WE, Clk: in std_logic;
         ADDR: in unsigned(31 downto 0);
         Mem_Bus: inout unsigned(31 downto 0));
  end component;
  signal CS, WE: std_logic;
  signal ADDR, Mem_Bus: unsigned(31 downto 0);
begin
  CPU: MIPS port map (CLK, RST, CS, WE, ADDR, Mem_Bus);
  MEM: Memory port map (CS, WE, CLK, ADDR, Mem_Bus);
  A_Out <= ADDR;
  D_Out <= Mem_Bus;
end model;
```

We synthesized the model shown in Figure 9-9. The Xilinx ISE tools targeted for a Spartan 3 FPGA yield 1108 four-input LUTs, 660 slices, 111 flip-flops, and 1 block RAM. The register file takes 194 four-input LUTs. Since one LUT can give 16 bits of storage, thirty-two 32-bit registers would need the storage from 64 LUTs. Since the register file has two read ports, it would need 128 LUTs. Additional LUTs are required for the address decoder and the control signals. In order to implement the design on a prototyping board, interface to the input and display modules should be added.

### 9.5.5 Testing the Processor Model

The overall MIPS VHDL model is tested using a test bench illustrated in Figure 9-10. The test bench must verify the proper operation of each implemented instruction. The test bench consists of a MIPS program with test instructions and VHDL code to load the program into memory and verify the program's output. We use a constant array of instructions that we want to write into the memory and a constant array of expected outputs to which we will compare the processor execution result.

However, note that now the memory is connected to the processor and test bench, and that means both our test bench and the processor will try to control the two signals at the same time. One way to resolve this is to put muxes at the input ports of the memory. There are a few muxes for that purpose: Address_Mux (for choosing the address), CS_Mux for choosing the CS signal, and WE_Mux (for choosing the WE signal). The select signal for the muxes is *init*. When the signal is '1', the three muxes select the address and CS and WE signals from the test bench. Otherwise, these signals from the processor module are chosen. We also assert the reset of our CPU throughout the initialization process to make sure the CPU does not run until the test bench finishes writing the instructions into the memory. When *init* is '0', the CPU and memory are connected for normal operation.

As the MIPS program executes, each test instruction stores its result in a different register. After all of the test instructions have been executed, the program performs a series of store instructions. Each of these instructions places the contents of a different register onto the bus as it executes. So if there are 10 instructions that we want to verify, we also have 10 store word instructions. During each store, the value on the bus is compared to the expected result for that register with an assert statement. In the MIPS processor, register $0 is always 0. We did not implement that in the register file. Hence we clear register $0 using an instruction. The first instruction in the test sequence does that. In normal MIPS processor code, you will not find instructions with register $0 as the destination. Essentially, writes to register $0 are ignored in MIPS.

FIGURE 9-10: **Test Bench for the Processor Model**

```
library IEEE;
use IEEE.std_logic_1164.all;
use IEEE.numeric_std.all;

entity MIPS_Testbench is
end MIPS_Testbench;

architecture test of MIPS_Testbench is
  component MIPS
    port(CLK, RST: in std_logic;
         CS, WE: out std_logic;
         ADDR: out unsigned (31 downto 0);
         Mem_Bus: inout unsigned(31 downto 0));
  end component;
  component Memory
    port(CS, WE, CLK: in std_logic;
         ADDR: in unsigned(31 downto 0);
         Mem_Bus: inout unsigned(31 downto 0));
  end component;

  constant N: integer : = 8;
  constant W: integer : = 26;
  type Iarr is array(1 to W) of unsigned(31 downto 0);
  constant Instr_List: Iarr : = (
    x"30000000", -- andi $0, $0, 0   => $0 = 0
    x"20010006", -- addi $1, $0, 6   => $1 = 6
    x"34020012", -- ori $2, $0, 18   => $2 = 18
    x"00221820", -- add $3, $1, $2   => $3 = $1 + $2 = 24
    x"00412022", -- sub $4, $2, $1   => $4 = $2 - $1 = 12
    x"00222824", -- and $5, $1, $2   => $5 = $1 and $2 = 2
    x"00223025", -- or $6, $1, $2    => $6 = $1 or $2 = 22
    x"0022382A", -- slt $7, $1, $2   => $7 = 1 because $1<$2
    x"00024100", -- sll $8, $2, 4    => $8 = 18 * 16 = 288
    x"00014842", -- srl $9, $1, 1    => $9 = 6/2 = 3
    x"10220001", -- beq $1, $2, 1    => should not branch
    x"8C0A0004", -- lw $10, 4($0)    => $10 = 5th instr = x"00412022" = 4268066
    x"14620001", -- bne $1, $2, 1    => must branch to PC+1+1
    x"30210000", -- andi $1, $1, 0   => $1 = 0 (skipped if bne worked correctly)
    x"08000010", -- j 16             => PC = 16
```

```
    x"30420000", -- andi $2, $2, 0   => $2 = 0 (skipped if j 16 worked correctly)
    x"00400008", -- jr $2            => PC = $2 = 18 = PC+1+1. $3 wrong if fails
    x"30630000", -- andi $3, $3, 0   => $3 = 0 (skipped if jr $2 worked correctly)
    x"AC030040", -- sw $3, 64($0)    => Mem(64) = $3
    x"AC040041", -- sw $4, 65($0)    => Mem(65) = $4
    x"AC050042", -- sw $5, 66($0)    => Mem(66) = $5
    x"AC060043", -- sw $6, 67($0)    => Mem(67) = $6
    x"AC070044", -- sw $7, 68($0)    => Mem(68) = $7
    x"AC080045", -- sw $8, 69($0)    => Mem(69) = $8
    x"AC090046", -- sw $9, 70($0)    => Mem(70) = $9
    x"AC0A0047"  -- sw $10, 71($0)   => Mem(71) = $10
);
    -- The last instructions perform a series of sw operations that store
    -- registers 3-10 to memory. During the memory write stage, the testbench
    -- will compare the value of these registers (by looking at the bus value)
    -- with the expected output. No explicit check/assertion for branch
    -- instructions, however if a branch does not execute as expected, an error
    -- will be detected because the assertion for the instruction after the
    -- branch instruction will be incorrect.
  type output_arr is array(1 to N) of integer;
  constant expected: output_arr: = (24, 12, 2, 22, 1, 288, 3, 4268066);
  signal CS, WE, CLK: std_logic : = '0';
  signal Mem_Bus, Address, AddressTB, Address_Mux: unsigned(31 downto 0);
  signal RST, init, WE_Mux, CS_Mux, WE_TB, CS_TB: std_logic;
begin
  CPU: MIPS port map (CLK, RST, CS, WE, Address, Mem_Bus);
  MEM: Memory port map (CS_Mux, WE_Mux, CLK, Address_Mux, Mem_Bus);

  CLK <= not CLK after 10 ns;
  Address_Mux <= AddressTB when init = '1' else Address;
  WE_Mux <= WE_TB when init = '1' else WE;
  CS_Mux <= CS_TB when init = '1' else CS;

  process
  begin
    rst <= '1';
    wait until CLK = '1' and CLK'event;

    --Initialize the instructions from the testbench
    init <= '1';
    CS_TB <= '1'; WE_TB <= '1';
    for i in 1 to W loop
      wait until CLK = '1' and CLK'event;
      AddressTB <= to_unsigned(i-1,32);
      Mem_Bus <= Instr_List(i);
    end loop;
    wait until CLK = '1' and CLK'event;
    Mem_Bus <= "ZZZZZZZZZZZZZZZZZZZZZZZZZZZZZZZZ";
    CS_TB <= '0'; WE_TB <= '0';
    init <= '0';
```

```
    wait until CLK = '1' and CLK'event;
    rst <= '0';

    for i in 1 to N loop
      wait until WE = '1' and WE'event;  -- When a store word is executed
      wait until CLK = '0' and CLK'event;
      assert(to_integer(Mem_Bus) = expected(i))
        report "Output mismatch:" severity error;
    end loop;

    report "Testing Finished:";
  end process;
end test;
```

The following command file was used to test the VHDL model. All the signals that we are interested in are not available in the topmost entity, which here is the test bench. In such cases, the full path describing the signal (specifically pointing to the component in which the signal is appearing) must be provided for correct simulation. The configure list -delta collapse command removes outputs at intermediate deltas.

```
add list -hex sim:/mips_testbench/cpu/instr
add list -unsigned sim:/mips_testbench/cpu/npc
add list -unsigned sim:/mips_testbench/cpu/pc
add list -unsigned sim:/mips_testbench/cpu/state
add list -unsigned sim:/mips_testbench/cpu/alu_ina
add list -unsigned sim:/mips_testbench/cpu/alu_inb
add list -signed sim:/mips_testbench/cpu/alu_result
add list -signed sim:/mips_testbench/cpu/addr
configure list -delta collapse
run 2330
```

The simulation results are illustrated as follows:

| MIPS Instruction | ns | Instr | PC | State | ALU_InA | ALU_InB | ALU_Result | Addr |
|---|---|---|---|---|---|---|---|---|
| andi $0, $0, 0 | 570 | 30000000 | 0 | 0 | – | – | 0 | 0 |
| | 908 | 30000000 | 1 | 1 | – | – | 0 | X |
| | 610 | 30000000 | 1 | 2 | – | 0 | 0 | X |
| | 630 | 30000000 | 1 | 3 | – | 0 | 0 | 0 |
| addi $1, $0, 6 | 650 | 30000000 | 1 | 0 | 0 | 0 | 0 | 1 |
| | 670 | 20010006 | 2 | 1 | 0 | 6 | 0 | 0 |
| | 690 | 20010006 | 2 | 2 | 0 | 6 | 6 | 0 |
| | 710 | 20010006 | 2 | 3 | 0 | 6 | 0 | 6 |
| ori $2, $0, 18 | 730 | 20010006 | 2 | 0 | 0 | 6 | 0 | 2 |
| | 750 | 34020012 | 3 | 1 | 0 | 18 | 0 | 6 |
| | 770 | 34020012 | 3 | 2 | 0 | 18 | 18 | 6 |
| | 790 | 34020012 | 3 | 3 | 0 | 18 | 0 | 18 |

| | | | | | | | | |
|---|---|---|---|---|---|---|---|---|
| add $3, $1, $2 | 810 | 34020012 | 3 | 0 | 0 | 18 | 0 | 3 |
| | 830 | 00221820 | 4 | 1 | 6 | 6176 | 0 | 18 |
| | 850 | 00221820 | 4 | 2 | 6 | 18 | 24 | 18 |
| | 870 | 00221820 | 4 | 3 | 6 | 18 | 0 | 24 |
| sub $4, $2, $1 | 890 | 00221820 | 4 | 0 | 6 | 18 | 0 | 4 |
| | 910 | 00412022 | 5 | 1 | 18 | 6 | 0 | 24 |
| | 930 | 00412022 | 5 | 2 | 18 | 6 | 12 | 24 |
| | 950 | 00412022 | 5 | 3 | 18 | 6 | 0 | 12 |
| and $5, $1, $2 | 970 | 00412022 | 5 | 0 | 18 | 6 | 0 | 5 |
| | 990 | 00222824 | 6 | 1 | 6 | 18 | 0 | 12 |
| | 1010 | 00222824 | 6 | 2 | 6 | 18 | 2 | 12 |
| | 1030 | 00222824 | 6 | 3 | 6 | 18 | 0 | 2 |
| or $6, $1, $2 | 1050 | 00222824 | 6 | 0 | 6 | 18 | 0 | 6 |
| | 1070 | 00223025 | 7 | 1 | 6 | 18 | 0 | 2 |
| | 1090 | 00223025 | 7 | 2 | 6 | 18 | 22 | 2 |
| | 1110 | 00223025 | 7 | 3 | 6 | 18 | 0 | 22 |
| slt $7, $1, $2 | 1130 | 00223025 | 7 | 0 | 6 | 18 | 0 | 7 |
| | 1150 | 0022382A | 8 | 1 | 6 | 18 | 0 | 22 |
| | 1170 | 0022382A | 8 | 2 | 6 | 18 | 1 | 22 |
| | 1190 | 0022382A | 8 | 3 | 6 | 18 | 0 | 1 |
| sll $8, $2, 4 | 1210 | 0022382A | 8 | 0 | 6 | 18 | 0 | 8 |
| | 1230 | 00024100 | 9 | 1 | 0 | 18 | 0 | 1 |
| | 1250 | 00024100 | 9 | 2 | 0 | 18 | 288 | 1 |
| | 1270 | 00024100 | 9 | 3 | 0 | 18 | 0 | 288 |
| srl $9, $1, 1 | 1290 | 00024100 | 9 | 0 | 0 | 18 | 0 | 9 |
| | 1310 | 00014842 | 10 | 1 | 0 | 6 | 0 | 288 |
| | 1330 | 00014842 | 10 | 2 | 0 | 6 | 3 | 288 |
| | 1350 | 00014842 | 10 | 3 | 0 | 6 | 0 | 3 |
| beq $1, $2, 1 | 1370 | 00014842 | 10 | 0 | 0 | 6 | 0 | 10 |
| | 1390 | 10220001 | 11 | 1 | 6 | 18 | 0 | 3 |
| | 1410 | 10220001 | 11 | 2 | 6 | 18 | −12 | 3 |
| lw $10, 4($0) | 1430 | 10220001 | 11 | 0 | 6 | 18 | 0 | 11 |
| | 1450 | 8C0A0004 | 12 | 1 | 0 | − | 0 | −12 |
| | 1470 | 8C0A0004 | 12 | 2 | 0 | 4 | 4 | −12 |
| | 1490 | 8C0A0004 | 12 | 3 | 0 | 4 | 0 | 4 |
| | 1510 | 8C0A0004 | 12 | 4 | 0 | 4 | 0 | 4 |
| bne $1, $2, 1 | 1530 | 8C0A0004 | 12 | 0 | 0 | 4 | 0 | 12 |
| | 1550 | 14620001 | 13 | 1 | 24 | 1 | 0 | 4 |
| | 1570 | 14620001 | 13 | 2 | 24 | 18 | 6 | 4 |
| j 16 | 1590 | 14620001 | 14 | 0 | 24 | 18 | 0 | 14 |
| | 1610 | 08000010 | 15 | 1 | 0 | 0 | 0 | 6 |
| jr $2 | 1630 | 08000010 | 16 | 0 | 0 | 0 | 0 | 16 |
| | 1650 | 00400008 | 17 | 1 | 18 | 0 | 0 | 6 |
| | 1670 | 00400008 | 17 | 2 | 18 | 0 | 0 | 6 |
| sw $3, 64($0) | 1690 | 00400008 | 18 | 0 | 18 | 0 | 0 | 18 |
| | 1710 | AC030040 | 19 | 1 | 0 | 24 | 0 | 0 |
| | 1730 | AC030040 | 19 | 2 | 0 | 64 | 64 | 0 |
| | 1750 | AC030040 | 19 | 3 | 0 | 64 | 0 | 64 |

The initial cycles that are used to load the instructions into the memory module are not shown. The presented data corresponds to the cycles once the instruction fetch by the processor begins. Only the first store instruction is shown here. But all store instructions are tested in the test bench. More comprehensive tests can be devised by reading the data from the stored locations in the memory.

In this chapter, we have presented a popular RISC instruction set, the MIPS. We presented a design for a subset of the MIPS instruction set starting from the instruction set specification. We presented a synthesizable VHDL model. We illustrated the use of a test bench to test the processor model.

● ● ● ● ● ● ● ● ● ● ● ● ●
# Problems

9.1 What does the term *ISA* mean? Do the Pentium 4 and Pentium 3 have the same ISA?

9.2 Microprocessor $X$ has 30 instructions in its instruction set and microprocessor $Y$ has 45 instructions in its instruction set. You are told that $Y$ is a RISC processor. Can you conclusively say that $X$ is a RISC processor? Why or why not?

9.3 List four important characteristics that make a processor RISC type.

9.4 What is the difference between the MIPS `addi` instruction and `addiu` instruction?

9.5 What is the machine language encoding for the following MIPS instructions? Give the answers in hexadecimal (hex). All offsets are in decimal.

```
(i)   add   $6, $7, $8
(ii)  lw    $5, 4($6)
(iii) addiu $3, $2, -2000
(iv)  sll   $3, $7, 12
(v)   beq   $6, $5, -16
(vi)  j     4000
```

9.6 What is the machine language encoding for the following MIPS instructions? Give the answers in hexadecimal (hex). All offsets are in decimal.

```
(i)   addi  $5, $4, 4000
(ii)  sw    $5, 20($3)
(iii) addu  $4, $5, $3
(iv)  bne   $2, $3, 32
(v)   jr    $5
(vi)  jal   8000
```

9.7 What MIPS instruction do the following hexadecimal (hex) numbers correspond to? If it is not any instruction in Table 9-7, denote as an illegal opcode.

    **(i)** 33333300
    **(ii)** 8D8D8D8D
    **(iii)** 1777FF00
    **(iv)** BDBD00BD
    **(v)** 01010101

9.8 What MIPS instruction do the following hexadecimal (hex) numbers correspond to? If it is not any instruction in Table 9-7, denote as an illegal opcode.

    **(i)** 20202020
    **(ii)** 00E70018
    **(iii)** 13D300C8
    **(iv)** 0192282A
    **(v)** 0F6812A4

9.9 Write a MIPS assembly language program for the following pseudo code segment. Assume the x and y arrays start at locations 4000 and 8000 (decimal).

```
for(i = 0; i < 100; i++)
    x(i) = x(i) * y(i)
```

9.10 Write a MIPS assembly language program for the following pseudo code segment. Assume the x and y arrays start at locations 4000 and 8000 (decimal).

```
for(i = 1; i < 100; i++)
    x(i) = x(i) + x(i-1)
```

9.11 Write a MIPS assembly language program for the following pseudo code segment. Assume the x and y arrays start at locations 4000 and 8000 (decimal), and *a* is at location 12000 (decimal).

```
for(i = 0; i < 100; i++)
    y(i) = a * x(i) + y(i)
```

9.12 Figure 9-8 presents a model for a subset of MIPS instructions. Synthesize the model using current Xilinx software with a state of the art Xilinx FPGA as the target. How many logic blocks, flip-flops, and memory blocks are used? (*Note*: Substitute a different FPGA company and its software to create variations of this question that suit your environment.)

9.13 **(a)** Figure 9-8 presents a model for a subset of MIPS instructions. Enhance the model by adding modules to interface the model to input switches and LEDs/displays on an FPGA prototyping board. Your interface must be able to halt operation of the MIPS processor and display the lower 8 bits of $1 on eight LEDs. Your interface must also divide the prototyping board's internal clock to provide the model with a slow clock (e.g., 100-Hz clock). You may display additional information using other LEDs or display devices, depending on the capabilities of your prototyping board. Synthesize the model and implement it on a prototyping board.

**(b)** For this question, use the model in part (a). Write a MIPS assembly language program to create a rotating light (implemented using eight LEDs on the prototyping board). The light rotates from one LED to the next at a one second interval.

**(c)** For this question, you'll use the model in part (a). Write a MIPS assembly language program to create a traffic light controller. Implement your traffic light with the following pattern:

| Street A | | | Street B | | | |
|---|---|---|---|---|---|---|
| Red | Yellow | Green | Red | Yellow | Green | |
| 0 | 0 | 1 | 1 | 0 | 0 | (5 seconds) |
| 0 | 1 | 0 | 1 | 0 | 0 | (2 seconds) |
| 1 | 0 | 0 | 1 | 0 | 0 | (1 second) |
| 1 | 0 | 0 | 0 | 0 | 1 | (5 seconds) |
| 1 | 0 | 0 | 0 | 1 | 0 | (2 seconds) |
| 1 | 0 | 0 | 1 | 0 | 0 | (1 second), then repeats |

9.14 Many microprocessors perform input-output operations by memory mapping. Assume that memory location F0002F2F is a parallel port for the processor. Write a MIPS program to generate a square wave with approximate frequency 8MHz on LSB of the parallel port. Assume that you have a MIPS processor prototype based on Figure 9-8, running with a 100-MHz clock.

9.15 **(a)** Add overflow detection to the add and addi instructions in the MIPS subset VHDL code (Figure 9-8).
**(b)** Write a test bench to test your code from part (a).

9.16 **(a)** Add overflow detection to all overflow-capable instructions in the MIPS subset that is implemented in Figure 9-8.
**(b)** Write a test bench to test your code from part (a).

9.17 **(a)** Add the MIPS instruction JAL (jump and link) to the MIPS subset VHDL code (Figure 9-8). JAL is used for procedure calls. JAL jumpaddr puts the return address (PC + 1) in register file $31 and then goes to jumpaddr for the next instruction. (*Note:* The original MIPS used (PC + 4) and jumpaddr*4; however, the implementation in Chapter 9 uses word addressing instead of byte addressing so the "4" is replaced with "1".) The JAL instruction uses the J format; therefore, the first 6 bits are the opcode (3) and the remaining 26 bits are jumpaddr. Make as few changes to the VHDL code as you need.
**(b)** Create a test bench to test this instruction.

9.18 **(a)** Add an instruction that multiplies two 16-bit numbers stored in the lower half of two general-purpose registers and deposits the product into another 32-bit register to the processor model in Figure 9-8. (*Note:* Such an instruction does not exist in MIPS.)
**(b)** Create a test bench to test this instruction.

9.19 (This problem can be used as a term project. More information on pipelining can be obtained from Reference 37.) Modern microprocessors employ pipelining to improve instruction throughput. Consider a five-stage pipeline consisting of fetch, decode and read registers, execute, memory access, and register write-back stages. During the first stage, an instruction is fetched from the instruction memory. During the second stage, the fetched instruction is decoded. The operand registers are also read during this stage. During the third stage, the arithmetic or logic operation is performed on the register data read during the second stage. During the fourth stage, in load/store instructions, data memory is read/written into memory. Arithmetic instructions do not perform any operation during this stage. During the fifth stage, arithmetic instructions write the results to the destination register.

**(a)** Design a pipelined implementation of the MIPS design in Figure 9-8. Draw a block diagram indicating the general structure of the pipeline. Write VHDL code, synthesize it for an FPGA target, and implement it on an FPGA proto-typing board. Assume that each stage takes one clock cycle. While implementing on the prototyping board, use an 8-Hz clock.

Assume that instruction memory access and data memory access take only one cycle. Instruction and data memories need to be separated (or must have two ports) in order to allow simultaneous access from the first stage and fourth stage.

An instruction can read the operands in second stage from the register file, as long as there are no dependencies with an incomplete instruction (ahead of it in the pipeline). If such a dependency exists, the current instruction in decode stage must wait until the register data is ready. Each instruction should test for dependencies with previous instructions. This can be done by comparing source registers of the current instruction with destination registers of the incomplete instructions ahead of the current instruction.

The register file is written into during stage 5 and read from during stage 2. A reasonable assumption to make is that the write is performed during the first half of the cycle and the read is performed during the second half of the cycle. Assume that data written into the destination register during the first half of a cycle can be read by another instruction during the second half of the same cycle.

**(b)** How many cycles does it take to execute $N$ instructions with no dependencies?

**(c)** How many cycles does it take to execute the following instruction sequence through this pipeline?

```
add  $5,$4,$3
add  $6,$5,$4
add  $7,$6,$5
add  $8,$7,$6
```

9.20 (This problem can be used as a term project. More information on pipelining and data forwarding can be obtained from Reference 37.) In Problem 9.19, it is assumed that data should be written into the register file during the write-back stage of an instruction before a subsequent instruction can read it. This introduces two idle cycles if instruction $i + 1$ is dependent on instruction $i$. A technique that many processors use to solve this problem is called data forwarding. If an instruction needs the result from an instruction ahead of it, the result is forwarded to the current instruction. This can be done by having multiplexers at the input of the ALU which take the operand either from the register file, the forwarding path from the output of the ALU, or the output of the memory access stage (fourth stage). The dependencies between instructions are clearly identified and then the multiplexers are appropriately controlled to forward the correct data.

**(a)** Design a pipelined implementation of the MIPS design in Figure 9-8 with data forwarding. Draw a block diagram indicating the forwarding hardware. Write VHDL code, synthesize it for an FPGA target, and implement it on an FPGA prototyping board. While implementing on the prototyping board, use an 8-Hz clock.

**(b)** Compare the number of cycles taken by the code in Problems 9.10 and 9.11 for this design, the design in Problem 9.19, and the design in Figure 9-8.

# Hardware Testing and Design for Testability

This chapter introduces digital system testing and design methods that make the systems easier to test. We have already discussed the use of testing during the design process. We have written VHDL test benches to verify that the overall design and algorithms used are correct. We have used simulation at the logic level to verify that a design is logically correct and that it meets specifications. After the logic level design of an IC is completed, additional testing can be done by simulating it at the circuit level to verify that the design has been correctly implemented and that the timing is correct.

When a digital system is manufactured, further testing is required to verify that it functions correctly. When multiple copies of an IC are manufactured, each copy must be tested to verify that it is free from manufacturing defects. This testing process can become very expensive and time consuming. With today's complex ICs, the cost of testing is a major component of the manufacturing cost. Therefore, it is very important to develop efficient methods of testing digital systems and to design the systems so that they are easy to test. **Design for testability (DFT)** is thus an important issue in modern IC design.

In this chapter, we first discuss methods of testing combinational logic for the basic types of faults that can occur. Then we describe methods for determining test sequences for sequential logic. **Automatic test pattern generators (ATPGs)** are employed in order to generate test sequences required for testing circuits and systems. One of the problems encountered is that normally we have access only to the inputs and outputs of the circuit being tested and not to the internal state. To remedy this problem, internal test points may be brought out to additional pins on the IC. To reduce the number of test pins required, we introduce the concept of **scan design**, in which the state of the system can be stored in a shift register and shifted out serially. Finally, we discuss the concept of **built-in self-test (BIST)**. By adding more components to the IC, we can generate test sequences and verify the response to these sequences internally without the need for expensive external testing.

● ● ● ● ● ● ● ● ● ● ● ●
## 10.1  Testing Combinational Logic

Two common types of faults are short circuits and open circuits. If the input to a gate is shorted to ground, the input acts as if it is stuck at a logic 0. If the input to a gate is shorted to a positive power supply voltage, the gate input acts as if it is stuck at a logic

1. If the input to a gate is an open circuit, the input may act as if it is stuck at 0 or stuck at 1, depending on the type of logic being used. Thus, it is common practice to model faults in logic circuits as stuck-at-1 (s-a-1) or stuck-at-0 (s-a-0) faults. To test a gate input for s-a-0, the gate input must be 1 so a change to 0 can be detected. Similarly, to test a gate input for s-a-1, the normal gate input must be 0 so a change to 1 can be detected.

We can test an AND gate for s-a-0 faults by applying 1's to all inputs, as shown in Figure 10-1(a). The normal gate output is then 1, but if any input is s-a-0, the output becomes 0. The notation $1 \rightarrow 0$ on the gate input $a$ means that the normal value of $a$ is 1, but the value has changed to 0 because of the s-a-0 fault. The notation $1 \rightarrow 0$ at the gate output indicates that this change has propagated to the gate output. We can test an AND gate input for s-a-1 by applying 0 to the input being tested and 1's to the other inputs, as shown in Figure 10-1(b). The normal gate output then is 0, but if the input being tested is s-a-1, the output becomes 1. To test OR gate inputs for s-a-1, we apply 0's to all inputs, and if any input is s-a-1, the output will change to 1 (Figure 10-1(c)). To test an OR gate input for s-a-0, we apply a 1 to the input under test and 0's to the other inputs. If the input under test is s-a-0, the output will change to 0 (Figure 10-1(d)). In the process of testing the inputs to a gate for s-a-0 and s-a-1, we also can detect s-a-0 and s-a-1 faults at the gate output.

**FIGURE 10-1:**
**Testing AND and OR Gates for Stuck-At Faults**

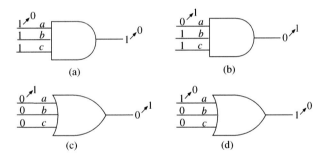

The two-level AND-OR circuit of Figure 10-2 has nine inputs and one output. We assume that the OR gate inputs ($p$, $q$, and $r$) are not accessible, so the gates cannot be tested individually. One approach to testing the circuit would be to apply all $2^9 = 512$ different input combinations and observe the output. A more efficient approach is based on testing for all s-a-0 and s-a-1 faults, as shown in Table 10-1. To test the $abc$ AND gate inputs for s-a-0, we must apply 1's to $a$, $b$, and $c$, as shown in Figure 10-2(a). Then, if any gate input is s-a-0, the gate output ($p$) will become 0. In order to transmit the change to the OR gate output, the other OR gate inputs must be 0. To achieve this, we can set $d = 0$ and $g = 0$ ($e$, $f$, $h$, and $i$ are then don't cares). This test vector will detect $p0$ ($p$ stuck-at-0) as well as $a0$, $b0$, and $c0$. In a similar manner, we can test for $d0$, $e0$, $f0$, and $q0$ by setting $d = e = f = 1$ and $a = g = 0$. A third test with $g = h = i = 1$ and $a = d = 0$ will test the remaining s-a-0 faults. To test $a$ for s-a-1 ($a1$), we must set $a = 0$ and $b = c = 1$, as shown in Figure 10-2(b). Then, if $a$ is s-a-1, $p$ will become 1. In order to transmit this change to the output, we must have $q = r = 0$, as before. However, if we set $d = g = 0$ and $e = f = h = i = 1$, we can test for $d1$ and $g1$ at the same time as $a1$. This same test vector also tests for $p1, q1$, and $r1$. As shown in the table, we can test for $b1, e1$, and $h1$ with a single test vector and test similarly for $c1, f1$, and $i1$. Thus, we can test all s-a-0 and s-a-1 faults with only six tests, whereas the brute-force approach would require 512 tests. When we apply the six tests, we can determine whether or not a fault is present, but we cannot determine the exact location of the fault. In the preceding analysis, we have assumed that only one fault occurs at a time. In many cases the presence of multiple faults will also be detected.

**FIGURE 10-2:**
**Testing an AND-OR Circuit**

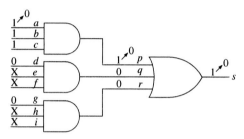

(a) stuck-at-0 test

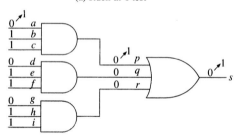

(b) stuck-at-1 test

**TABLE 10-1: Test Vectors for Figure 10-2**

| $a$ | $b$ | $c$ | $d$ | $e$ | $f$ | $g$ | $h$ | $i$ | Faults Tested |
|---|---|---|---|---|---|---|---|---|---|
| 1 | 1 | 1 | 0 | X | X | 0 | X | X | $a0, b0, c0, p0$ |
| 0 | X | X | 1 | 1 | 1 | 0 | X | X | $d0, e0, f0, q0$ |
| 0 | X | X | 0 | X | X | 1 | 1 | 1 | $g0, h0, i0, r0$ |
| 0 | 1 | 1 | 0 | 1 | 1 | 0 | 1 | 1 | $a1, d1, g1, p1, q1, r1$ |
| 1 | 0 | 1 | 1 | 0 | 1 | 1 | 0 | 1 | $b1, e1, h1, p1, q1, r1$ |
| 1 | 1 | 0 | 1 | 1 | 0 | 1 | 1 | 0 | $c1, f1, i1, p1, q1, r1$ |

Testing multilevel circuits is considerably more complex than testing two-level circuits. In order to test for an internal fault in a circuit, we must choose a set of inputs that will excite that fault and then propagate the effect of that fault to the circuit output. In Figure 10-3, $a, b, c, d$, and $e$ are circuit inputs. If we want

**FIGURE 10-3: Fault Detection Using Path Sensitization**

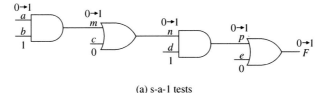

(a) s-a-1 tests

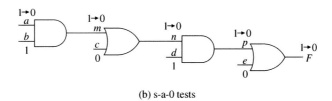

(b) s-a-0 tests

to test for gate input $n$ s-a-1, $n$ must be 0. This can be achieved if we make $c = 0$, $a = 0$, and $b = 1$, as shown. In order to propagate the fault $n$ s-a-1 to the output $F$, we must make $d = 1$ and $e = 0$. With this set of inputs, if $a$, $m$, $n$, or $p$ is s-a-1, the output $F$ will have the incorrect value and the fault can be detected. Furthermore, if we change $a$ to 1 and gate input $a$, $m$, $n$, or $p$ is s-a-0, the output $F$ will change from 1 to 0. We say that the path through $a$, $m$, $n$, and $p$ has been sensitized, since any fault along that path can be detected. The method of **path sensitization** allows us to test for a number of different stuck-at faults using one set of circuit inputs.

Next, we try to determine a minimum set of test vectors to test the circuit of Figure 10-4 for all single stuck-at-1 and stuck-at-0 faults. We assume that we can apply inputs to $A$, $B$, $C$, and $D$ and observe the output $F$ and that the internal gate inputs and outputs cannot be accessed. The general procedure to determine the test vectors is the following:

1. Select an untested fault.
2. Determine the required $ABCD$ inputs.
3. Determine the additional faults that are tested.
4. Repeat this procedure until tests are found for all of the faults.

Let us start by testing input $p$ for s-a-1. In order to do this, we must choose inputs $A$, $B$, $C$, and $D$ such that $p = 0$, and if $p$ is s-a-1, we must propagate this fault to the output $F$ so it can be observed. In order to propagate the fault, we must make $c = 0$ and $w = 1$. We can make $w = 1$ by making $t = 1$ or $u = 1$. To make $u = 1$, we must have both $D$ and $r = 1$. Fortunately, our choice of $C = 0$ makes $r = 1$. To make $p = 0$, we choose $A = 0$. By choosing $B = 1$, we can sensitize the path $A$-$a$-$p$-$v$-$f$-$F$ so that the set of inputs $ABCD = 0101$ will test for faults $a1, p1, v1$, and $f1$. This set of inputs also tests for $c$ s-a-1. We assume that $c$ s-a-1 is a fault internal to the gate, so it is still possible to have $q = 0$ and $r = 1$ if $c$ s-a-1 occurs.

FIGURE 10-4:
**Example Circuit
for Stuck-At Fault
Testing (*p* stuck
at 1)**

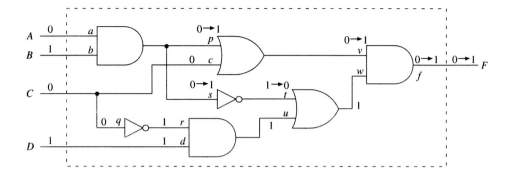

To test for s-a-0 inputs along the path *A-a-p-v-f-F*, we can use the inputs *ABCD* =
1101. In addition to testing for faults *a*0, *p*0, *v*0, and *f*0, this input vector also tests the
following faults: *b*0, *w*0, *u*0, *r*0, *q*1, and *d*0. To determine tests for the remaining
stuck-at faults, we select an untested fault, determine the required *ABCD* inputs,
and then determine the additional faults that are tested. Then we can repeat this
procedure until tests are found for all of the faults. Table 10-2 lists a set five test vectors that will test for all single stuck-at faults in Figure 10-4.

TABLE 10-2: **Tests
for Stuck-At Faults
in Figure 10-4**

| Test Vectors | | | | Normal Gate Inputs | | | | | | | | | | | | Faults Tested |
|---|---|---|---|---|---|---|---|---|---|---|---|---|---|---|---|---|
| A | B | C | D | a | b | p | c | q | r | d | s | t | u | v | w | F | |
| 0 | 1 | 0 | 1 | 0 | 1 | 0 | 0 | 0 | 1 | 1 | 0 | 1 | 1 | 0 | 1 | 0 | *a*1 *p*1 *c*1 *v*1 *f*1 |
| 1 | 1 | 0 | 1 | 1 | 1 | 1 | 0 | 0 | 1 | 1 | 1 | 0 | 1 | 1 | 1 | 1 | *a*0 *b*0 *p*0 *q*1 *r*0 *d*0 *u*0 *v*0 *w*0 *f*0 |
| 1 | 0 | 1 | 1 | 1 | 0 | 0 | 1 | 1 | 0 | 1 | 0 | 1 | 0 | 1 | 1 | 1 | *b*1 *c*0 *s*1 *t*0 *v*0 *w*0 *f*0 |
| 1 | 1 | 0 | 0 | 1 | 1 | 1 | 0 | 0 | 1 | 0 | 1 | 0 | 0 | 1 | 0 | 0 | *a*0 *b*0 *d*1 *s*0 *t*1 *u*1 *w*1 *f*1 |
| 1 | 1 | 1 | 1 | 1 | 1 | 1 | 1 | 0 | 1 | 1 | 0 | 0 | 1 | 0 | 0 | 0 | *a*0 *b*0 *q*0 *r*1 *s*0 *t*1 *u*1 *w*1 *f*1 |

In addition to stuck-at faults, other types of faults, such as bridging faults, may
occur. A bridging fault occurs when two unconnected signal lines are shorted
together. For a large combinational circuit, finding a minimum set of test vectors
that will test for all possible faults is very difficult and time consuming. For circuits
that contain redundant gates, testing for some of the faults may be impossible.
Even if a comprehensive set of test vectors can be found, applying all of the vectors may take too much time and cost too much. For these reasons, it is common
practice to use a relatively small set of test vectors that will test most of the faults.
In general, determining such a set of vectors is a difficult and computationally
intensive problem. Many algorithms and corresponding computer programs have
been developed to generate such sets of test vectors. Computer programs have
also been developed to simulate faulty circuits. Such programs allow the user to
determine what percentage of possible faults are tested by a given set of input vectors. The percentage of possible faults that can be tested by a set of input vectors
is called the **coverage** of the test vectors.

● ● ● ● ● ● ● ● ● ● ●
## 10.2  Testing Sequential Logic

Testing sequential logic is generally much more difficult than testing combinational logic, because we must use sequences of inputs for testing. If we can observe only the input and output sequences and not the state of the flip-flops in a sequential circuit, a very large number of test sequences may be required. Basically, the problem is to determine if the circuit under test is equivalent to a correctly functioning circuit. We will assume that the sequential circuit being tested has a reset input so we can reset it to a known initial state. If we attempted to test the circuit using the brute-force approach, we would reset the circuit to the initial state, apply a test sequence, and observe the output sequence. If the output sequence was correct, then we would repeat the test for another sequence. This process has to be repeated for all possible input sequences. A large number of tests are required to test exhaustively all states and all state transitions in the machine. Since the brute-force approach is totally impractical, the question arises: Can we derive a relatively small set of test sequences that will adequately test the circuit?

One way to derive test sequences for a sequential circuit is to convert it to an iterative circuit. The iterative circuit means that the combinational part of the sequential circuit is repeated several times to indicate the condition of the combinational part of the circuit at each time. Since the iterative circuit is a combinational circuit, we could derive test vectors for the iterative circuit using one of the standard methods for combinational circuits.

As an example, Figure 10-5 shows a standard Mealy sequential circuit and the corresponding iterative circuit. In these figures, $X$, $Z$, and $Q$ can either be single variables or vectors. The iterative circuit has $k + 1$ identical copies of the combinational network used in the sequential circuit, where $k + 1$ is the length of the sequence used to test the sequential circuit. For the sequential circuit, $X(t)$ represents a sequence of

FIGURE 10-5:
**Sequential and
Iterative Circuits**

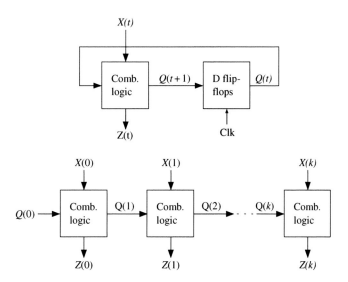

inputs in time. In the iterative circuit, $X(0)$ $X(1)$ ... $X(k)$ represents the same sequence in space. Each cell of the iterative circuit computes $Z(t)$ and $Q(t + 1)$ in terms of $Q(t)$ and $X(t)$. The leftmost cell computes the values for $t = 0$, the next cell for $t = 1$, and so on. After the test vectors have been derived for the iterative circuit, these vectors become the input sequences used to test the original sequential circuit. The number of cells in the iterative circuit depends on the length of the sequences required to test the sequential circuit.

Derivation of a small set of test sequences that will adequately test a sequential circuit is generally difficult to do. Consider the state graph shown in Figure 10-6 and the corresponding state table (Table 10-3). We assume that we can reset the circuit to state $S_0$. It is necessary that the test sequence cause the circuit to go through all possible state transitions, but this is not an adequate test. For example, the input sequence

$$X = 0\ 1\ 0\ 1\ 1\ 0\ 0\ 1\ 1$$

traverses all the arcs connecting the states and produces the output sequence

$$Z = 0\ 0\ 1\ 0\ 1\ 1\ 1\ 1\ 0$$

If we replace the arc from $S_3$ to $S_0$ with a self-loop, as shown by the dashed line, the output sequence will be the same, but the new sequential machine is not equivalent to the old one.

FIGURE 10-6: **State Graph for Test Example**

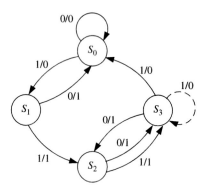

TABLE 10-3: **State Table for Figure 10-6**

| Q1Q2 | State | Next State X = 0 | 1 | Output X = 0 | 1 |
|------|-------|------------------|-----|--------------|-----|
| 00 | $S_0$ | $S_0$ | $S_1$ | 0 | 0 |
| 10 | $S_1$ | $S_0$ | $S_2$ | 1 | 1 |
| 01 | $S_2$ | $S_3$ | $S_3$ | 1 | 1 |
| 11 | $S_3$ | $S_2$ | $S_0$ | 1 | 0 |

A state graph in which every state can be reached from every other state is referred to as **strongly connected**. A general test strategy for a sequential circuit with a strongly connected state graph and no equivalent states is first to find an input sequence that will distinguish each state from the other states. Such an input sequence is referred to as a **distinguishing sequence**. Two states of a state machine $M$ are distinguishable if and only if there exists at least one finite input sequence, which, when applied to $M$, causes different output sequences. If the output sequence is identical for every possible input sequence, then obviously the states are equivalent. It has been proved that if two states of machine $M$ are distinguishable, they can be distinguished by a sequence of length $n - 1$ or less, where $n$ is the number of states in $M$ [28]. Given a distinguishing sequence, each entry in the state table can be verified.

For the example of Figure 10-6, one distinguishing sequence is 11. This distinguishing sequence can be obtained as follows. Divide the states $S_0, S_1, S_2,$ and $S_3$ into two groups, where the states in each group are equivalent if the test sequence is only one-bit long. For instance, Table 10-3 shows that by applying a one bit test sequence, we can distinguish between groups $\{S_0, S_3\}$ and $\{S_1, S_2\}$. If the input is 1, output is 0 for $\{S_0, S_3\}$ and 1 for $\{S_1, S_2\}$. States inside each partition are equivalent if the test sequence is only a 1. Now, from Table 10-3, we can see that if we applied a test input of 1 again, states in group $\{S_0, S_3\}$ can be distinguished. The states in group $\{S_1, S_2\}$ can also be distinguished by the test input 1. Hence, the sequence 11 is sufficient to distinguish among the four states. In the worst case, a sequence of three bits would have been sufficient since there are only four states in the machine. If we start in $S_0$, the input sequence 11 gives the output sequence 01; for $S_1$ the output is 11; for $S_2$, 10; and for $S_3$, 00. Thus, we can distinguish the four states by using the input sequence 11. We can then verify every entry in the state table using the following sequences, where $R$ means reset to state $S_0$:

| Input | Output | Transition Verified |
|---|---|---|
| R 0 1 1 | 0 0 1 | ($S_0$ to $S_0$) |
| R 1 1 1 | 0 1 1 | ($S_0$ to $S_1$) |
| R 1 0 1 1 | 0 1 0 1 | ($S_1$ to $S_0$) |
| R 1 1 1 1 | 0 1 1 0 | ($S_1$ to $S_2$) |
| R 1 1 0 1 1 | 0 1 1 0 0 | ($S_2$ to $S_3$) |
| R 1 1 1 1 1 | 0 1 1 0 0 | ($S_2$ to $S_3$) |
| R 1 1 0 0 1 1 | 0 1 1 1 1 0 | ($S_3$ to $S_2$) |
| R 1 1 0 1 1 1 | 0 1 1 0 0 1 | ($S_3$ to $S_0$) |

Another approach to deriving test sequences is based on testing for stuck-at faults. Figure 10-7 shows the realization of Figure 10-6 using the following state assignment: $S_0$, 00; $S_1$, 10; $S_2$, 01; $S_3$, 11. If we want to test for $a$ s-a-1, we must first excite the fault by going to state $S_1$, in which $Q1Q2 = 10$ and then setting $X = 0$. In normal operation, the next state will be $S_0$. However, if $a$ is s-a-1, then next state is $Q1Q2 = 01$, which is $S_2$. This test sequence can then be constructed as follows:

- To go to $S_1$: reset followed by $X = 1$.
- To test $a$ s-a-1: $X = 0$.
- To distinguish the state that is reached: $X = 11$.

The final sequence is R1011. The normal output is 0101, and the faulty output is 0110.

FIGURE 10-7:
Realization of
Figure 10-6

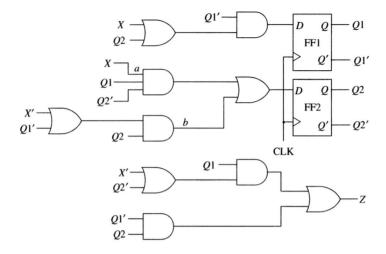

We have shown some simple examples that illustrate some of the methods used to derive test sequences for sequential circuits. As the number of inputs and states in the circuit increases, the number and length of the required test sequence increases rapidly, and the derivation of these test sequences becomes much more difficult. This, in turn, means that the time and expense required to test the circuits increases rapidly with the number of inputs and states.

• • • • • • • • • • • •

## 10.3 Scan Testing

The problem of testing a sequential circuit is greatly simplified if we can observe the state of all the flip-flops instead of just observing the circuit outputs. For each state of the flip-flops and for each input combination, we need to verify that the circuit outputs are correct and that the circuit goes to the correct next state. One approach would be to connect the output of each flip-flop within the IC being tested to one of the IC pins. Since the number of pins on the IC is very limited, this approach is not very practical. So the question arises: How can we observe the state of all the flip-flops without using up a large number of pins on the IC? If the flip-flops were arranged to form a shift register, then we could shift out the state of the flip-flops bit by bit using a single serial output pin on the IC. This leads to the concept of **scan path testing**.

Figure 10-8 shows a method of scan path testing based on two-port flip-flops. In the usual way, the sequential circuit is separated into a combinational logic part and a state register composed of flip-flops. Each of the flip-flops has two $D$ inputs and two clock inputs. When $C1$ is pulsed, the $D1$ input is stored in the flip-flop. When $C2$ is pulsed, $D2$ is stored in the flip-flop. The $Q$ output of each flip-flop is connected to the $D2$ input of the next flip-flop to form a shift register. The next state $(Q_1^+ Q_2^+ \ldots Q_k^+)$ generated by the combinational logic is loaded into the flip-flops when $C1$ is pulsed, and the new state $(Q_1 Q_2 \ldots Q_k)$ feeds back into the combinational logic. When the circuit is not being tested, the system clock ($SCK = C1$) is used. A set of inputs ($X_1 X_2 \ldots X_n$) is applied, the outputs ($Z_1 Z_2 \ldots Z_m$) are generated, $SCK$ is pulsed, and the circuit goes to the next state.

**FIGURE 10-8: Scan Path Test Circuit Using Two-Port Flip-Flops**

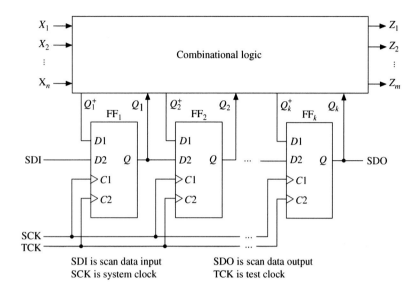

SDI is scan data input
SCK is system clock

SDO is scan data output
TCK is test clock

When the circuit is being tested, the flip-flops are set to a specified state by shifting the state code into the register using the scan data input ($SDI$) and the test clock ($TCK$). A test input vector ($X_1 X_2 \ldots X_n$) is applied, the outputs ($Z_1 Z_2 \ldots Z_m$) are verified, and $SCK$ is pulsed to take the circuit to the next state. The next state is then verified by pulsing $TCK$ to shift the state code out of the scan data register via the scan data output ($SDO$). This method reduces the problem of testing a sequential circuit to that of testing a combinational circuit. Any of the standard methods can be used to generate a set of test vectors for the combinational logic. Each test vector contains ($n + k$) bits, since there are $n$ $X$ inputs and $k$ state inputs to the combinational logic. The $X$ part of the test vector is applied directly, and the $Q$ part is shifted in via the $SDI$. In summary, the test procedure is as follows:

1. Scan in the test vector $Q_i$ values via $SDI$ using the test clock $TCK$.
2. Apply the corresponding test values to the $X_i$ inputs.
3. After sufficient time for the signals to propagate through the combinational circuit, verify the output $Z_i$ values.
4. Apply one clock pulse to the system clock $SCK$ to store the new values of $Q_i^+$ into the corresponding flip-flops.
5. Scan out and verify the $Q_i$ values by pulsing the test clock $TCK$.
6. Repeat steps 1 through 5 for each test vector.

Steps 5 and 1 can overlap, since it is possible to scan in one test vector while scanning out the previous test result.

We will apply this method to test a sequential circuit with two inputs, three flip-flops, and two outputs. The circuit is configured as in Figure 10-8 with inputs $X_1X_2$, flip-flops $Q_1Q_2Q_3$, and outputs $Z_1Z_2$. One row of the state transition table is as follows:

| $Q_1Q_2Q_3$ | $X_1X_2 =$ | $Q_1^+ Q_2^+ Q_3^+$ 00 | 01 | 11 | 10 | $Z_1Z_2$ 00 | 01 | 11 | 10 |
|---|---|---|---|---|---|---|---|---|---|
| 101 | | 010 | 110 | 011 | 111 | 10 | 11 | 00 | 01 |

Figure 10-9 shows the timing diagram for testing this row of the transition table. First, 101 is shifted in using $TCK$, least significant bit ($Q_3$) first. The input $X_1X_2 = 00$ is applied, and $Z_1Z_2 = 10$ is then read. $SCK$ is pulsed and the circuit goes to state 010. As 010 is shifted out using $TCK$, 101 is shifted in for the next test. This process continues until the test is completed.

**FIGURE 10-9: Timing Chart for Scan Test**

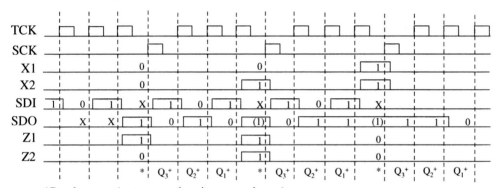

*Read output (output at other times not shown)

In general, a digital system implemented by an IC consists of flip-flop registers separated by blocks of combinational logic, as shown in Figure 10-10(a). In order to apply scan test to the IC, we need to replace the flip-flops with two-port flip-flops

**FIGURE 10-10: System with Flip-Flop Registers and Combinational Logic Blocks**

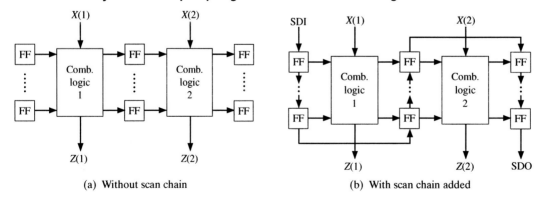

(a) Without scan chain          (b) With scan chain added

(or other types of scannable flip-flops) and link all the flip-flops into a scan chain, as shown in Figure 10-10(b). Then we can scan test data into all the registers, apply the test clock, and scan out the results.

When multiple ICs are mounted on a PC board, it is possible to chain together the scan registers in each IC so that the entire board can be tested using a single serial access port (Figure 10-11).

**FIGURE 10-11: Scan Test Configuration with Multiple ICs**

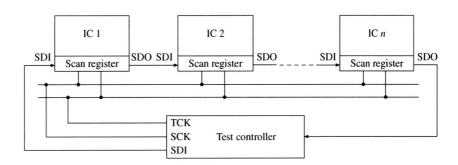

• • • • • • • • • • • •
## 10.4    Boundary Scan

As ICs have become more complex, with more and more pins, printed circuit boards have become denser, with multiple layers and very fine traces. Testing these PC boards after they have been loaded with complex ICs has become very difficult. Testing a board by means of its edge connector does not provide adequate testing and may require very long test sequences. When PC boards were less dense with wider traces, testing was often done using a **bed-of-nails test fixture**. This method used sharp probes to contact the traces on the board so test data could be applied to and read from various ICs on the board. Bed-of-nails testing is not practical for high-density PC boards with fine traces and complex ICs.

Boundary scan test methodology was introduced to facilitate the testing of complex PC boards. It is an integrated method for testing circuit boards with many ICs. A standard for boundary scan testing was developed by the Joint Test Action Group (**JTAG**), and this standard has been adopted as **ANSI/IEEE Standard 1149.1**, "Standard Test Access Port and Boundary-Scan Architecture." Many IC manufacturers make ICs that conform to this standard. Such ICs can be linked together on a PC board so that they can be tested using only a few pins on the PC board edge connector.

Figure 10-12 shows an IC with added boundary scan logic according to the IEEE standard. One cell of the boundary scan register (BSR) is placed between each input or output pin and the internal core logic. Four or five pins of the IC are devoted to the **test-access port**, or **TAP**. The TAP controller and additional test logic are also

**FIGURE 10-12: IC with Boundary Scan Register and Test-Access Port**

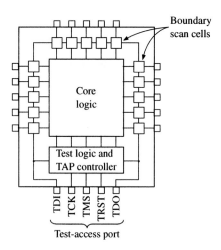

added to the core logic on the IC. The functions of the TAP pins (according to the standard) are as follows:

*TDI*    Test data input (this data is shifted serially into the BSR)
*TCK*    Test clock
*TMS*    Test mode select
*TDO*    Test data output (serial output from the BSR)
*TRST*   Test reset (resets the TAP controller and test logic; optional pin)

A PC board with several boundary scan ICs is shown in Figure 10-13. The boundary scan registers in the ICs are linked serially in a single chain with input *TDI* and output *TDO. TCK, TMS*, and *TRST* (if used) are connected in parallel to all of the ICs. Using these signals, test instructions and test data can be clocked into every IC on the board.

**FIGURE 10-13: PC Board with Boundary Scan ICs**

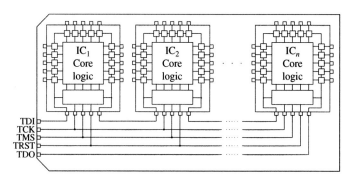

Figure 10-12 illustrated the boundary scan cells on the periphery of each IC that conforms to the boundary scan standard. The structure of a typical boundary scan cell is shown in Figure 10-14. A boundary scan cell has two inputs, TDI serial input and the parallel input pin. Similarly, it has two outputs, the serial out and the parallel data out. When in the normal mode, data from the parallel input pin is routed to the internal

FIGURE 10-14:
**Typical Boundary
Scan Cell**

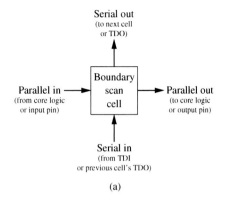

(a)

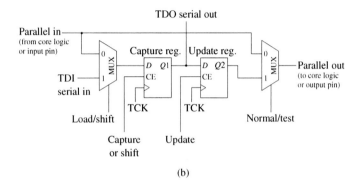

(b)

core logic in the IC, or data from the core logic is routed to the output pin. When in the shift mode, serial data from the previous cell is clocked into flip-flop $Q1$ at the same time as the data stored in $Q1$ is clocked into the next boundary scan cell. After $Q2$ is updated, test data can be supplied to the internal logic or to the output pin.

Figure 10-15 shows the basic boundary scan architecture that is implemented on each boundary scan IC. The boundary scan register is divided into two parts. BSR1 represents the shift register, which consists of the $Q1$ flip-flops in the boundary scan cells. BSR2 represents the $Q2$ flip-flops, which can be parallel-loaded from BSR1 when an update signal is received. The serial input data ($TDI$) can be shifted into the boundary scan register (BSR1), through a bypass register, or into the instruction register. The TAP controller on each IC contains a state machine (Figure 10-16). The input to the state machine is $TMS$, and the sequence of 0's and 1's applied to $TMS$ determines whether the $TDI$ data is shifted into the instruction register or through the boundary scan cells. The TAP controller and the instruction register control the operation of the boundary scan cells.

The TAP controller state machine has 16 states. States 9 through 15 are used for loading and updating the instruction register, and states 2 through 8 are used for loading and updating the data register (BSR1). The TRST signal, if used, resets the state to Test-Logic-Reset. The state graph has the interesting property that, regardless of the initial state, a sequence of five 1's on the TMS input will always reset the machine to state 0.

FIGURE 10-15:
**Basic Boundary
Scan Architecture**

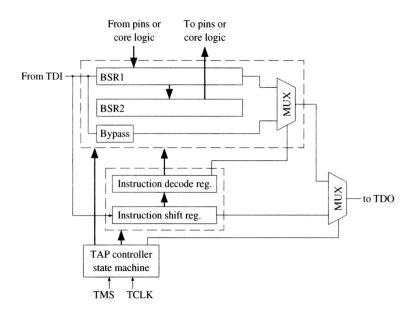

FIGURE 10-16:
**State Machine for
TAP Controller**

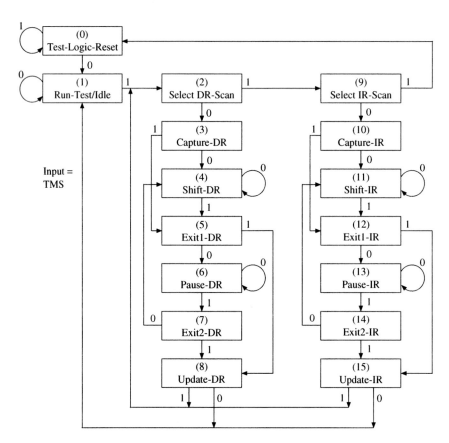

The following instructions are defined in the IEEE standard:

- BYPASS: This instruction allows the *TDI* serial data to go through a 1-bit bypass register on the IC instead of through the boundary scan register. In this way, one or more ICs on the PC board may be bypassed while other ICs are being tested.

- SAMPLE/PRELOAD: This instruction is used to scan the boundary scan register without interfering with the normal operation of the core logic. Data is transferred to or from the core logic from or to the IC pins without interference. Samples of this data can be taken and scanned out through the boundary scan register. Test data can be shifted into the BSR.

- EXTEST: This instruction allows board-level interconnect testing, and it also allows testing of clusters of components that do not incorporate the boundary scan test features. Test data is shifted into the BSR and then it goes to the output pins. Data from the input pins is captured by the BSR.

- INTEST (optional): This instruction allows testing of the core logic by shifting test data into the boundary scan register. Data shifted into the BSR takes the place of data from the input pins, and output data from the core logic is loaded into the BSR.

- RUNBIST (optional): This instruction causes special built-in self-test (BIST) logic within the IC to execute. (Section 10.5 explains how BIST logic can be used to generate test sequences and check the test results.)

Several other optional and user-defined instructions may also be included.

The data paths between the IC pins, the boundary scan registers, and the core logic depend on the instruction being executed as well as the state of the TAP controller. Figures 10-17, 10-18, and 10-19 highlight the data paths for the Sample/Preload, Extest, and Intest instructions. In each case, the boundary scan registers BSR1 and BSR2 have been split into two sections—one associated with the input pins and one associated with the output pins. Test data can be shifted into BSR1 from TDI and shifted out to TDO.

For the Sample/Preload instruction (Figure 10-17) the core logic operates in the normal mode with inputs from the input pins of the IC and outputs going to the output pins. When the controller is in the CaptureDR state, BSR1 is parallel-loaded from the input pins and from the outputs of the core logic. In the UpdateDR state, BSR2 is loaded from BSR1.

**FIGURE 10-17:**
**Signal Paths for Sample/Preload Instruction (highlighted)**

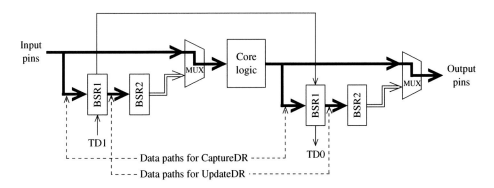

For the Extest instruction (Figure 10-18) the core logic is not used. In the UpdateDR state, BSR1 is loaded into BSR2 and the data is routed to the output pins of the IC. In the CaptureDR state, data from the input pins is loaded into BSR1.

**FIGURE 10-18:**
**Signal Paths for**
**Extest Instruction**
**(highlighted)**

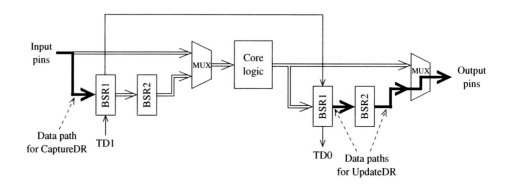

For the Intest instruction (Figure 10-19) the IC pins are not used. In the UpdateDR state, test data that has previously been shifted into BSR1 is loaded into BSR2 and routed to the core logic inputs. In the CaptureDR state, data from the core logic is loaded into BSR1.

**FIGURE 10-19:**
**Signal Paths for**
**Intest Instruction**
**(highlighted)**

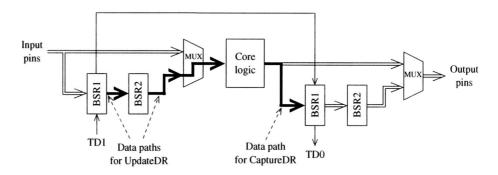

The following simplified example illustrates how the connections between two ICs can be tested using the SAMPLE/PRELOAD and EXTEST instructions. The test is intended to check for shorts and opens in the PC board traces. Both ICs have two input pins and two output pins, as shown in Figure 10-20. Test data is shifted into the BSRs via *TDI*. Then data from the input pins is parallel-loaded into the BSRs and shifted out via *TDO*. We assume that the instruction register on each IC is three bits long with EXTEST coded as 000 and SAMPLE/PRELOAD as 001. The core logic in IC1 is an inverter connected as a clock oscillator and two flip-flops. The core logic in IC2 is an inverter and XOR gate. The two ICs are interconnected to form a 2-bit counter.

FIGURE 10-20:
**Interconnection
Testing Using
Boundary Scan**

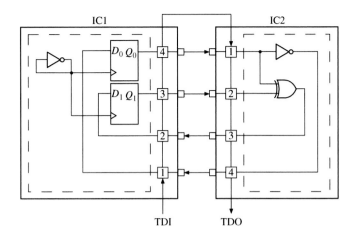

The steps required to test the connections between the ICs are as follows:

1. Reset the TAP state machine to the Test-Logic-Reset state by inputting a sequence of five 1's on *TMS*.

2. Scan in the SAMPLE/PRELOAD instruction to both ICs using the sequences for *TMS* and *TDI* given here. The state numbers refer to Figure 10-16.

```
State:  0  1  2  9  10  11  11  11  11  11  11  12  15  2
TMS:    0  1  1  0  0   0   0   0   0   0   0   1   1   1
TDI:    -  -  -  -  -   1   0   0   1   0   0   -   -
```

The *TMS* sequence 01100 takes the TAP controller to the Shift-IR state. In this state, copies of the SAMPLE/PRELOAD instruction (code 001) are shifted into the instruction registers on both ICs. In the Update-IR state, the instructions are loaded into the instruction decode registers. Then the TAP controller goes back to the Select DR-scan state.

3. Preload the first set of test data into the ICs using the following sequences for *TMS* and *TDI*:

```
State:  2  3  4  4  4  4  4  4  4  4  5  8  2
TMS:    0  0  0  0  0  0  0  0  0  1  1  1
TDI:    -  -  0  1  0  0  0  1  0  0  -  -
```

Data is shifted into BSR1 in the Shift-DR state, and it is transferred to BSR2 in the Update-DR state. The result is as follows:

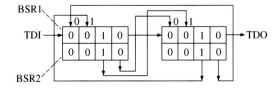

4. Scan in the EXTEST instruction to both ICs using the following sequences:

| State: | 2 | 9 | 10 | 11 | 11 | 11 | 11 | 11 | 11 | 12 | 15 | 2 |
|--------|---|---|----|----|----|----|----|----|----|----|----|---|
| TMS: | 1 | 0 | 0 | 0 | 0 | 0 | 0 | 0 | 1 | 1 | 1 | |
| TDI: | – | – | – | 0 | 0 | 0 | 0 | 0 | 0 | – | – | |

The EXTEST instruction (000) is scanned into the instruction register in state Shift-IR and loaded into the instruction decode register in state Update-IR. At this point, the preloaded test data goes to the output pins, and it is transmitted to the adjacent IC input pins via the printed circuit board traces.

5. Capture the test results from the IC inputs. Scan this data out to *TDO* and scan the second set of test data in using the following sequences:

| State: | 2 | 3 | 4 | 4 | 4 | 4 | 4 | 4 | 4 | 4 | 5 | 8 | 2 |
|--------|---|---|---|---|---|---|---|---|---|---|---|---|---|
| TMS: | 0 | 0 | 0 | 0 | 0 | 0 | 0 | 0 | 0 | 1 | 1 | 1 | |
| TDI: | – | – | 1 | 0 | 0 | 0 | 1 | 0 | 0 | 0 | – | – | |
| TDO: | – | – | x | x | 1 | 0 | x | x | 1 | 0 | – | – | |

The data from the input pins is loaded into BSR1 in state Capture-DR. At this time, if no faults have been detected, the BSRs should be configured as shown below, where the X's indicate captured data that is not relevant to the test.

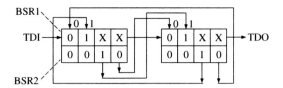

The test results are then shifted out of BSR1 in state Shift-DR as the new test data is shifted in. The new data is loaded into BSR2 in the Update-IR state.

6. Capture the test results from the IC inputs. Scan this data out to *TDO* and scan all 0's in using the following sequences:

| State: | 2 | 3 | 4 | 4 | 4 | 4 | 4 | 4 | 4 | 4 | 5 | 8 | 2 | 9 | 0 |
|--------|---|---|---|---|---|---|---|---|---|---|---|---|---|---|---|
| TMS: | 0 | 0 | 0 | 0 | 0 | 0 | 0 | 0 | 0 | 1 | 1 | 1 | 1 | 1 | |
| TDI: | – | – | 0 | 0 | 0 | 0 | 0 | 0 | 0 | 0 | – | – | – | – | |
| TDO: | – | – | x | x | 0 | 1 | x | x | 0 | 1 | – | – | – | – | |

The data from the input pins is loaded into BSR1 in state Capture-DR. Then it is shifted out in state Shift-DR as all 0's are shifted in. The 0's are loaded into BSR2 in the Update-DR state. The controller then returns to the Test-Logic-Reset state, and normal operation of the ICs can then occur. The interconnection test passes if the observed *TDO* sequences match the ones given above.

VHDL code for the basic boundary scan architecture of Figure 10-15 is given in Figure 10-21. Only the three mandatory instructions (EXTEST, SAMPLE/PRELOAD, and BYPASS) are implemented using a 3-bit instruction register. These instructions are coded as 000, 001, and 111, respectively. The number of cells in the *BSR* is a generic parameter. A second generic parameter, *CellType*, is a bit_vector that specifies whether each cell is an input cell or output cell. The case statement implements the TAP controller state machine. The instruction code is scanned in and loaded into *IDR* in states Capture-IR, Shift-IR, and Update-IR. The instructions are executed in states Capture-DR, Shift-DR, and Update-DR. The actions taken in these states depend on the instruction being executed. The register updates and state changes all occur on the rising edge of *TCK*. The VHDL code implements most of the functions required by the IEEE boundary scan standard, but it does not fully comply with the standard.

FIGURE 10-21: **VHDL Code for Basic Boundary Scan Architecture**

```
-- VHDL for Boundary Scan Architecture of Figure 10-15

entity BS_arch is
  generic(NCELLS: natural range 2 to 120 := 2);
          -- number of boundary scan cells
  port(TCK, TMS, TDI: in bit;
       TDO: out bit;
       BSRin: in bit_vector(1 to NCELLS);
       BSRout: inout bit_vector(1 to NCELLS);
       CellType: bit_vector(1 to NCELLS));
       -- '0' for input cell, '1' for output cell
end BS_arch;

architecture behavior of BS_arch is
  signal IR, IDR: bit_vector(1 to 3);        -- instruction registers
  signal BSR1, BSR2: bit_vector(1 to NCELLS); -- boundary scan cells
  signal BYPASS: bit;                        -- bypass bit
  type TAPstate is (TestLogicReset, RunTest_Idle,
    SelectDRScan, CaptureDR, ShiftDR, Exit1DR, PauseDR, Exit2DR, UpdateDR,
    SelectIRScan, CaptureIR, ShiftIR, Exit1IR, PauseIR, Exit2IR, UpdateIR);
  signal St: TAPstate;                       -- TAP Controller State
begin
  process (TCK)
  begin
    if TCK'event and TCK='1' then
      -- TAP Controller State Machine
      case St is
        when TestLogicReset =>
          if TMS='0' then St <= RunTest_Idle; else St<=TestLogicReset; end if;
        when RunTest_Idle =>
          if TMS='0' then St <= RunTest_Idle; else St <= SelectDRScan; end if;
        when SelectDRScan =>
          if TMS='0' then St <= CaptureDR; else St <= SelectIRScan; end if;
        when CaptureDR =>
          if IDR = "111" then BYPASS <= '0';
          elsif IDR = "000" then  -- EXTEST (input cells capture pin data)
```

```
              BSR1 <= (not CellType and BSRin) or (CellType and BSR1);
           elsif IDR = "001" then - SAMPLE/PRELOAD
              BSR1 <= BSRin;
           end if;  -- all cells capture cell input data
           if TMS='0' then St <= ShiftDR; else St <= Exit1DR; end if;
         when ShiftDR =>
           if IDR = "111" then BYPASS <= TDI; -- shift data through bypass reg.
           else BSR1 <= TDI & BSR1(1 to NCELLS-1); end if;
              -- shift data into BSR
           if TMS='0' then St <= ShiftDR; else St <= Exit1DR; end if;
         when Exit1DR =>
           if TMS='0' then St <= PauseDR; else St <= UpdateDR; end if;
         when PauseDR =>
           if TMS='0' then St <= PauseDR; else St <= Exit2DR; end if;
         when Exit2DR =>
           if TMS='0' then St <= ShiftDR; else St <= UpdateDR; end if;
         when UpdateDR =>
           if IDR = "000" then -- EXTEST (update output reg. for output cells)
              BSR2 <= (CellType and BSR1) or (not CellType and BSR2);
           elsif IDR = "001" then    -- SAMPLE/PRELOAD
              BSR2 <= BSR1;           -- update output reg. in all cells
           end if;
           if TMS='0' then St <= RunTest_Idle; else St <= SelectDRScan; end if;
         when SelectIRScan =>
           if TMS='0' then St <= CaptureIR; else St <= TestLogicReset; end if;
         when CaptureIR =>
           IR <= "001"; -- load 2 LSBs of IR with 01 as required by the standard
           if TMS='0' then St <= ShiftIR; else St <= Exit1IR; end if;
         when ShiftIR =>
           IR <= TDI & IR(1 to 2);  -- shift in instruction code
           if TMS='0' then St <= ShiftIR; else St <= Exit1IR; end if;
         when Exit1IR =>
           if TMS='0' then St <= PauseIR; else St <= UpdateIR; end if;
         when PauseIR =>
           if TMS='0' then St <= PauseIR; else St <= Exit2IR; end if;
         when Exit2IR =>
           if TMS='0' then St <= ShiftIR; else St <= UpdateIR; end if;
         when UpdateIR =>
           IDR <= IR;   -- update instruction decode register
           if TMS='0' then St <= RunTest_Idle; else St <= SelectDRScan; end if;
      end case;
    end if;
  end process;

  TDO <= BYPASS when St = ShiftDR and IDR = "111"  -- BYPASS
    else BSR1(NCELLS) when St = ShiftDR   -- EXTEST or SAMPLE/PRELOAD
    else IR(3) when St = ShiftIR;

  BSRout <= BSRin when (St = TestLogicReset or not (IDR = "000"))
    else BSR2;                          -- define cell outputs
end behavior;
```

VHDL code that implements the interconnection test example of Figure 10-20 is given in Figure 10-22. The *TMS* and *TDI* test patterns are the concatenation of the test patterns used in steps 2 through 6. A copy of the basic boundary scan architecture is instantiated for IC1 and for IC2. The external connections and internal logic for each IC are then specified. The internal clock frequency was arbitrarily chosen to be different than the test clock frequency. The test process runs the internal logic, then runs the scan test, and then runs the internal logic again. The test results verify that the IC logic runs correctly and that the scan test produces the expected results.

FIGURE 10-22: **VHDL Code for Interconnection Test Example**

```
-- Boundary Scan Tester

entity system is
end system;

architecture IC_test of system is
  component BS_arch is
    generic(NCELLS:natural range 2 to 120 := 4);
    port(TCK, TMS, TDI: in bit;
         TDO: out bit;
         BSRin: in bit_vector(1 to NCELLS);
         BSRout: inout bit_vector(1 to NCELLS);
         CellType: in bit_vector(1 to NCELLS));
         -- '0' for input cell, '1' for output cell
  end component;

  signal TCK, TMS, TDI, TDO, TDO1: bit;
  signal Q0, Q1, CLK1: bit;
  signal BSR1in, BSR1out, BSR2in, BSR2out: bit_vector(1 to 4);
  signal count: integer := 0;

  constant TMSpattern: bit_vector(0 to 62) :=
    "0110000000111000000000111100000001110000000000111000000000111111";
  constant TDIpattern: bit_vector(0 to 62) :=
    "0000010010000000100010000000000000000100010000000000000000000000";
begin
  BS1: BS_arch port map(TCK, TMS, TDI, TDO1, BSR1in, BSR1out, "0011");
  BS2: BS_arch port map(TCK, TMS, TDO1, TDO, BSR2in, BSR2out, "0011");
  -- each BSR has two input cells and two output cells
  BSR1in(1) <= BSR2out(4);              -- IC1 external connections
  SR1in(2) <= BSR2out(3);
  BSR1in(3) <= Q1;                      -- IC1 internal logic
  BSR1in(4) <= Q0;
  CLK1 <= not CLK1 after 7 ns;          -- internal clock
```

```
  process(CLK1)
  begin
    if CLK1 = '1' then                    -- D flip-flops
      Q0 <= BSR1out(1);
      Q1 <= BSR1out(2);
    end if;
  end process;

  BSR2in(1) <= BSR1out(4);                -- IC2 external connections
  BSR2in(2) <= BSR1out(3);
  BSR2in(3) <= BSR2out(1) xor BSR2out(2); -- IC2 internal logic
  BSR2in(4) <= not BSR2out(1);

  TCK <= not TCK after 5 ns;                         -- test clock

  process
  begin
    TMS <= '1';
    wait for 70 ns;                       -- run internal logic
    wait until TCK = '1';
    for i in TMSpattern'range loop        -- run scan test
      TMS <= TMSpattern(i);
      TDI <= TDIpattern(i);
      wait for 0 ns;
      count <= i + 1;                        -- count triggers listing output
      wait until TCK = '1';
    end loop;
    wait for 70 ns;                       -- run internal logic
    wait;                                 -- stop
  end process;
end IC_test;
```

● ● ● ● ● ● ● ● ● ● ● ●

## 10.5  Built-In Self-Test

As digital systems become more and more complex, they become much harder and more expensive to test. One solution to this problem is to add logic to the IC so that it can test itself. This is referred to as built-in self-test, or BIST. Figure 10-23 illustrates the general method for using BIST. An on-chip test generator applies test patterns to the circuit under test. The resulting output is observed by the response monitor, which produces an error signal if an incorrect output pattern is detected.

BIST is often used for testing memory. The regular structure of a memory chip makes it easy to generate test patterns. Figure 10-24 shows a block diagram of a self-test circuit for a RAM. The BIST controller enables the write-data generator and address counter so that data is written to each location in the RAM. Then the address counter and read-data generator are enabled, and the data read from each

FIGURE 10-23:
**Generic BIST
Scheme**

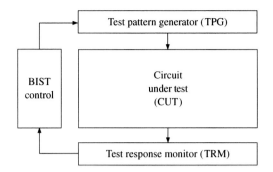

RAM location is compared with the output of the read-data generator to verify that it is correct. Memory is often tested by writing **checkerboard patterns** (alternating 0's and 1's) in all memory locations and reading them back. For instance, we could first write alternating 0's and 1's in all even addresses and alternating 1's and 0's in all odd addresses. After reading these back, the odd and even address patterns can be swapped to complete the test. In another test, the **March test**, each cell is read and then the complemented value is written. This process is continued until the entire memory array has been traversed. Then the process is repeated in the reverse order of addresses.

FIGURE 10-24:
**Self-Test Circuit
for RAM**

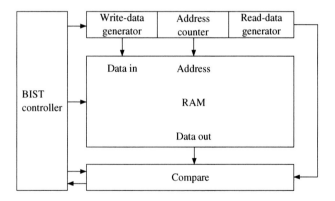

The test circuit can be simplified by using a signature register. The signature register compresses the output data into a short string of bits called a **signature**, and this signature is compared with the signature for a correctly functioning component. A **multiple-input signature register (MISR)** combines and compresses several output streams into a single signature. Figure 10-25 shows a simplified version of the RAM self-test circuit. The read-data generator and comparator have been eliminated and replaced with a MISR. One type of MISR simply forms a check sum by adding up all the data bytes stored in the RAM. When testing a ROM, Figure 10-25 can be simplified further, since no write-data generator is needed.

FIGURE 10-25:
**Self-Test Circuit
for RAM with
Signature Register**

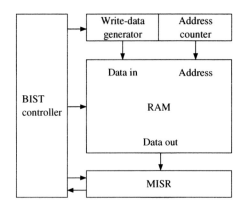

**Linear feedback shift registers (LFSRs)** are often used to generate test patterns and to compress test outputs into signatures. An LFSR is a shift register whose serial input bit is a linear function of some bits of the current shift register content. The bit positions that affect the serial input are called **taps**. The general form of a LFSR is a shift register with two or more flip-flop outputs XOR'ed together and fed back into the first flip-flop. The name **linear** comes from the fact that exclusive OR is equivalent to modulo-2 addition, and addition is a linear operation. Figure 10-26 shows an example of a LFSR. The outputs from the first and fourth flip-flops are XOR'ed together and fed back into the $D$ input of the first flip-flop; the taps are positions 1 and 4.

FIGURE 10-26:
**Four-Bit Linear
Feedback Shift
Register (LFSR)**

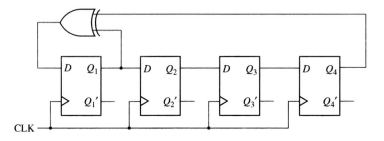

By proper choice of the outputs that are fed back through the exclusive OR gate, it is possible to generate $2^n - 1$ different bit patterns using an $n$-bit shift register. All possible patterns can be generated except for all 0's. The patterns generated by the LFSR of Figure 10-26 are

$$1000, 1100, 1110, 1111, 0111, 1011, 0101, 1010, 1101, 0110, 0011,$$
$$1001, 0100, 0010, 0001, 1000, \ldots$$

These patterns have no obvious order, and they have certain randomness properties. Such an LFSR is often referred to as a **pseudo-random pattern generator**, or **PRPG**. PRPGs are obviously very useful for BIST, since they can generate a large number of test patterns with a small amount of logic circuitry. Table 10-4 gives a feedback combination that will generate all $2^n - 1$ bit patterns for some LFSRs with lengths in the range $n = 4$ to 32.

| $n$ | Feedback |
|---|---|
| 4, 6, 7 | $Q_1 \oplus Q_n$ |
| 5 | $Q_2 \oplus Q_5$ |
| 8 | $Q_2 \oplus Q_3 \oplus Q_4 \oplus Q_8$ |
| 12 | $Q_1 \oplus Q_4 \oplus Q_6 \oplus Q_{12}$ |
| 14, 16 | $Q_3 \oplus Q_4 \oplus Q_5 \oplus Q_n$ |
| 24 | $Q_1 \oplus Q_2 \oplus Q_7 \oplus Q_{24}$ |
| 32 | $Q_1 \oplus Q_2 \oplus Q_{22} \oplus Q_{32}$ |

TABLE 10-4:
**Feedback for Maximum-Length LFSR Sequence**

If the all-0s test pattern is required, an $n$-bit LFSR can be modified by adding an AND gate with $n - 1$ inputs, as shown in Figure 10-27 for $n = 4$. When in state 0001, the next state is 0000; when in state 0000, the next state is 1000; otherwise, the sequence is the same as for Figure 10-26.

FIGURE 10-27:
**Modified LFSR with 0000 State**

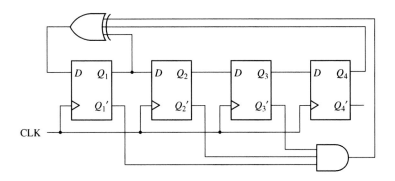

An MISR can be constructed by modifying a LFSR by adding XOR gates, as shown in Figure 10-28. The test data $(Z_1 Z_2 Z_3 Z_4)$ is XOR'ed into the register with each clock, and the final result represents a signature that can be compared with the signature for a known correctly functioning component. This type of signature analysis will catch many, but not all, possible errors. An $n$-bit signature register maps all possible input streams into one of the $2^n$ possible signatures. One of these is the correct signature, and the others indicate that errors have occurred. The probability that an incorrect input sequence will map to the correct signature is of the order of $1/2^n$.

FIGURE 10-28:
**Multiple-Input Signature Register (MISR)**

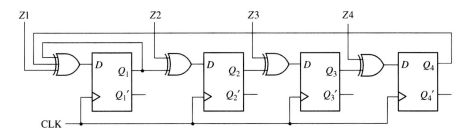

For the MISR of Figure 10-28, assume that the correct input sequence is 1010, 0001, 1110, 1111, 0100, 1011, 1001, 1000, 0101, 0110, 0011, 1101, 0111, 0010, 1100. This sequence maps to the signature 0010, assuming the initial contents of the MISR to be 0000. Any input sequence that differs in one bit will map to a different signature. For example, if 0001 in the sequence is changed to 1001, the resulting sequence maps to 0000. Most sequences with two errors will be detected, but if we change 0001 to 1001 and 0010 to 0110 in the original sequence, the result maps to 0010, which is the correct signature, so the errors would not be detected.

Several types of architectures have been proposed for BIST. Two popular examples are the STUMPS architecture and the BILBO architecture.

**STUMPS** stands for **S**elf-**T**esting **U**sing an **M**ISR and **P**arallel **S**RSG. SRSG, in turn, stands for Shift Register Sequence Generator. STUMPS is a BIST architecture that uses scan chains. An overview of the STUMPS architecture is shown in Figure 10-29. A pseudo-random pattern generator feeds test stimulus to the scan chains, and after a capture cycle, the test response analyzer receives the test responses. The test procedure in STUMPS is the following:

1. Scan in patterns from the test pattern generator (LFSR) into all scan chains.
2. Switch to normal function mode and clock once with system clock.
3. Shift out scan chain into test response analyzer (MISR) where test signature is generated.

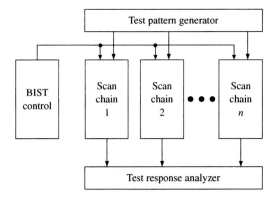

FIGURE 10-29:
**The STUMPS Architecture**

If the scan chain contains 100 scan cells, steps 1 and 3 will take 100 clocks. All scan chains should first be filled by the pseudo-random generator; hence, long scan chains necessitate long testing times. Since one test is done per scan, the STUMPS architecture is called a **test-per-scan** scheme. In order to reduce the testing time, a large number of parallel scan chains can be used, which reduces the time for filling the scan chains with the test since all scan chains can be loaded in parallel.

The STUMPS architecture was originally developed for self-testing of multi-chip modules [7]. The scan chain on each logic chip (module) is loaded in parallel from the pseudo-random pattern source. The number of clock cycles required is equal to the number of flip-flops in the longest scan chain. If there are $m$ scan cells in the longest scan chain, it will take $2m + 1$ cycles to perform one test ($m$ cycles for scan-in, one for capture, and $m$ cycles for scan out). The shorter scan chains will overflow into the MISR, but that will not affect the final correct signature.

In order to reduce test-times, steps 1 and 3 can be overlapped. When the scan chain is unloaded into the MISR after one test, simultaneously the next pseudo-random pattern set from the SRSG can be loaded into the scan chain (i.e., when test response from test $I$ is being shifted out, test pattern for test $I + 1$ can be shifted in). Assuming overlap between scan-out of a test and scan-in of the following test, each test vector will take $m + 1$ cycles, and it will take $n(m + 1) + m$ cycles to apply $n$ test vectors, including the $m$ cycles taken for the last scan-out.

As opposed to the test-per-scan scheme just discussed, a **test-per-clock** scheme can be used for faster testing. One such scheme is called the **BILBO (Built-In Logic Block Observer)** technique. In BILBO schemes, the scan register is modified so that parts of the scan register can serve as a state register, pattern generator, signature register, or shift register. When used as a shift register, the test data can be scanned in and out in the usual way. During testing, part of the scan register can be used as a pattern generator (PRPG) and part as a signature register (MISR) to test one of the combinational blocks. The roles can then be changed to test another combinational block. When the testing is finished, the scan register is placed in the state register mode for normal operation. After the BILBO registers are initialized, since there is no loading of test patterns as in the case of scan chains, a test can be applied in each clock cycle. Hence, this is categorized as a test-per-clock BIST scheme. BILBO involves shorter test lengths, but more test hardware.

Figure 10-30 shows the placement of BILBO registers for testing a circuit with two combinational blocks. Combinational circuit 1 is tested when the first BILBO is used as a PRPG and the second as an MISR. The roles of the registers are reversed to test combinational circuit 2. In the normal operating mode, both BILBOs serve as registers for the associated combinational logic. To scan data in and out, both BILBOs operate in the shift register mode.

**FIGURE 10-30: BIST Using BILBO Registers**

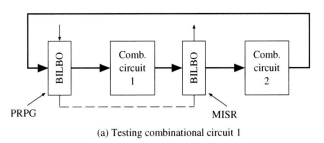

(a) Testing combinational circuit 1

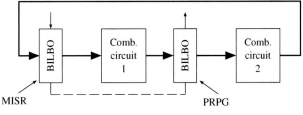

(b) Testing combinational circuit 2

Figure 10-31 shows the structure of one version of a 4-bit BILBO register. The control inputs $B_1$ and $B_2$ determine the operating mode. *Si* and *So* are the serial input and output for the shift register mode. The $Z$'s are inputs from the combinational logic. The equations for this BILBO register are

$$D_1 = Z_1 B_1 \oplus (Si\ B_2' + FB\ B_2)\ (B_1' + B_2)$$
$$D_i = Z_i B_1 \oplus Q_{i-1}\ (B_1' + B_2)\ (i > 1)$$

When $B_1 = B_2 = 0$, these equations reduce to

$$D_1 = Si \text{ and } D_i = Q_{i-1}\ (i > 1)$$

which corresponds to the shift register mode. When $B_1 = 0$ and $B_2 = 1$, the equations reduce to

$$D_1 = FB, \qquad\qquad D_i = Q_{i-1}$$

which corresponds to the PRPG mode, and the BILBO register is equivalent to Figure 10-26. When $B_1 = 1$ and $B_2 = 0$, the equations reduce to

$$D_1 = Z_1, \qquad\qquad D_i = Z_i$$

which corresponds to the normal operating mode. When $B_1 = B_2 = 1$, the equations reduce to

$$D_1 = Z_1 \oplus FB, \qquad D_i = Z_i \oplus Q_{i-1}$$

**FIGURE 10-31:**
**Four-Bit BILBO**
**Register**

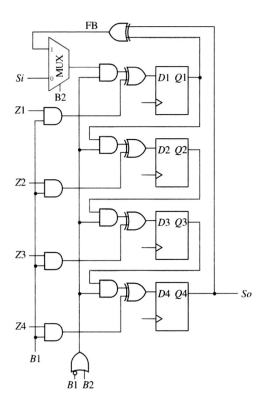

which corresponds to the MISR mode, and the BILBO register is equivalent to Figure 10-28. In summary, the BILBO operating modes are as follows:

| B1B2 | Operating Mode |
|------|----------------|
| 00 | Shift register |
| 01 | PRPG |
| 10 | Normal |
| 11 | MISR |

Figure 10-32 shows the VHDL description of an $n$-bit BILBO register. NBITS, which equals the number of bits, is a generic parameter in the range 4 through 8. The

FIGURE 10-32: **VHDL Code for BILBO Register of Figure 10-31**

```vhdl
entity BILBO is                    -- BILBO Register
  generic (NBITS: natural range 4 to 8 := 4);
  port (Clk, CE, B1, B2, Si: in bit;
        So: out bit;
        Z: in bit_vector(1 to NBITS);
        Q: inout bit_vector(1 to NBITS));
end BILBO;

architecture behavior of BILBO is
  signal FB: bit;
begin
  Gen8: if NBITS = 8 generate
    FB <= Q(2) xor Q(3) xor Q(NBITS); end generate;
  Gen5: if NBITS = 5 generate
    FB <= Q(2) xor Q(NBITS); end generate;
  GenX: if not(NBITS = 5 or NBITS = 8) generate
    FB <= Q(1) xor Q(NBITS); end generate;
  process(Clk)
    variable mode: bit_vector(1 downto 0);
  begin
    if (Clk = '1' and CE = '1') then
      mode := B1 & B2;
      case mode is
        when "00" =>      -- Shift register mode
          Q <= Si & Q(1 to NBITS-1);
        when "01" =>      -- Pseudo Random Pattern Generator mode
          Q <= FB & Q(1 to NBITS-1);
        when "10" =>      -- Normal Operating mode
          Q <= Z;
        when "11" =>      -- Multiple Input Signature Register mode
          Q <= Z(1 to NBITS) xor (FB & Q(1 to NBITS-1));
      end case;
    end if;
  end process;
  So <= Q(NBITS);
end behavior;
```

register is functionally equivalent to Figure 10-31, except that we have added a clock enable (*CE*). The feedback (*FB*) for the LFSR depends on the number of bits.

The system shown in Figure 10-33 illustrates the use of BILBO registers. In this system, registers *A* and *B* can be loaded from the Dbus using the *LDA* and *LDB* signals. Then the registers are added and the sum and carry are stored in register *C*. When B1 & B2 = 10, the registers are in the normal mode (*Test* = 0), and loading of the registers is controlled by *LDA, LDB*, and *LDC*. To test the adder, we first set B1 & B2 = 00 to place the registers in the shift register mode and scan in initial values for *A, B*, and *C*. Then we set B1 & B2 = 01, which places registers *A* and *B* in PRPG mode and register *C* in MISR mode. After 15 clocks, the test is complete. Then we can set B1 & B2 = 00 and scan out the signature.

**FIGURE 10-33:**
**System with BILBO Registers and Tester**

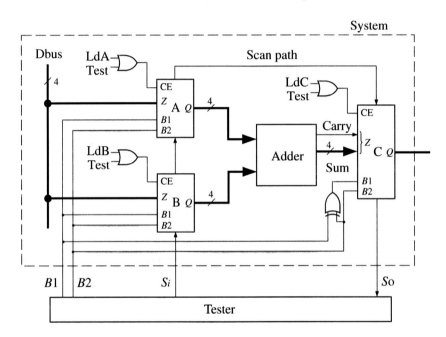

The VHDL code for the system is given in Figure 10-34, and a test bench is given in Figure 10-35. The system uses three BILBO registers and the 4-bit adder of Figure 8-20. The test bench scans in a test vector to initialize the BILBO registers; then it runs the test with registers *A* and *B* used as PRPGs and register *C* as a MISR. The resulting signature is shifted out and compared with the correct signature.

FIGURE 10-34: **VHDL Code for System with BILBO Registers and Tester**

```
entity BILBO_System is
  port(Clk, LdA, LdB, LdC, B1, B2, Si: in bit;
       So: out bit;
       DBus: in bit_vector(3 downto 0);
       Output: inout bit_vector(4 downto 0));
end BILBO_System;
```

```
architecture BSys1 of BILBO_System is
  component Adder4 is
    port(A, B: in bit_vector(3 downto 0); Ci: in bit;
         S: out bit_vector(3 downto 0); Co: out bit);
  end component;
  component BILBO is
    generic(NBITS: natural range 4 to 8 := 4);
    port(Clk, CE, B1, B2, Si : in bit;
         So: out bit;
         Z: in bit_vector(1 to NBITS);
         Q: inout bit_vector(1 to NBITS));
  end component;

  signal Aout, Bout: bit_vector(3 downto 0);
  signal Cin: bit_vector(4 downto 0);
  alias Carry: bit is Cin(4);
  alias Sum: bit_vector(3 downto 0) is Cin(3 downto 0);
  signal ACE, BCE, CCE, CB1, Test, S1, S2: bit;
begin
  Test <= not B1 or B2;
  ACE <= Test or LdA;
  BCE <= Test or LdB;
  CCE <= Test or LdC;
  CB1 <= B1 xor B2;
  RegA: BILBO generic map (4) port map(Clk, ACE, B1, B2, S1, S2, DBus, Aout);
  RegB: BILBO generic map (4) port map(Clk, BCE, B1, B2, Si, S1, DBus, Bout);
  RegC: BILBO generic map (5) port map(Clk, CCE, CB1, B2, S2, So, Cin, Output);
  Adder: Adder4 port map(Aout, Bout, '0', Sum, Carry);
end BSys1;
```

FIGURE 10-35: **Test Bench for BILBO System**

```
-- System with BILBO test bench

entity BILBO_test is
end BILBO_test;

architecture Btest of BILBO_test is
  component BILBO_System is
    port(Clk, LdA, LdB, LdC, B1, B2, Si: in bit;
         So: out bit;
         DBus: in bit_vector(3 downto 0);
         Output: inout bit_vector(4 downto 0));
  end component;
  signal Clk: bit := '0';
  signal LdA, LdB, LdC, B1, B2, Si, So: bit := '0';
```

```vhdl
  signal DBus: bit_vector(3 downto 0);
  signal Output: bit_vector(4 downto 0);
  signal Sig: bit_vector(4 downto 0);

  constant test_vector: bit_vector(12 downto 0) := "1000110000000";
  constant test_result: bit_vector(4 downto 0) := "01011";
begin
  clk <= not clk after 25 ns;
  Sys: BILBO_System port map(Clk,Lda,LdB,LdC,B1,B2,Si,So,DBus,Output);
  process
  begin
    B1 <= '0'; B2 <= '0';          -- Shift in test vector
    for i in test_vector'right to test_vector'left loop
      Si <= test_vector(i);
      wait until clk = '1';
    end loop;

    B1 <= '0'; B2 <= '1';          -- Use PRPG and MISR
    for i in 1 to 15 loop
      wait until clk = '1';
    end loop;

    B1 <= '0'; B2 <= '0';          -- Shift signature out
    for i in 0 to 5 loop
      Sig <= So & Sig(4 downto 1);
      wait until clk = '1';
    end loop;

    if (Sig = test_result) then    -- Compare signature
      report "System passed test.";
    else
      report "System did not pass test!";
    end if;

    wait;
  end process;
end Btest;
```

In this chapter, we introduced the subject of testing hardware, including combinational circuits, sequential circuits, complex ICs, and PC boards. Use of scan techniques for testing and built-in self-test has become a necessity as digital systems have become more complex. It is very important that design for testability be considered early in the design process so that the final hardware can be tested efficiently and economically.

● ● ● ● ● ● ● ● ● ● ●

# Problems

10.1 **(a)** Determine the necessary inputs to the following circuit to test for *u* stuck-at-0.
   **(b)** For this set of inputs, determine which other stuck-at faults can be tested.
   **(c)** Repeat (a) and (b) for *r* stuck-at-1.

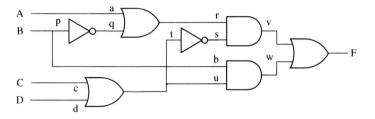

10.2 For the following circuit,

   **(a)** Determine the values of *A*, *B*, *C*, and *D* necessary to test for *e* s-a-1. Specify the other faults tested by this input vector.
   **(b)** Repeat (a) for *g* s-a-0.

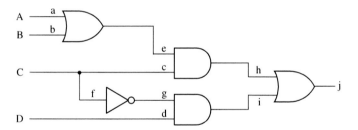

10.3 Find a minimum set of tests that will test all single stuck-at-0 and stuck-at-1 faults in the following circuit. For each test, specify which faults are tested for s-a-0 and for s-a-1.

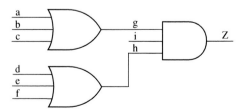

10.4 Give a minimum set of test vectors that will test for all stuck-at faults in the following circuit. List the faults tested by each test vector.

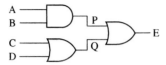

10.5 For the following circuit, specify a minimum set of test vectors for $a, b, c, d,$ and $e$ that will test for all stuck-at faults. Specify the faults tested by each vector.

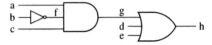

10.6 For the following circuit, find a minimum number of test vectors that will test all s-a-0 and s-a-1 faults at the AND and OR gate inputs. For each test vector, specify the values of $A, B, C$ and $D$, and the stuck-at faults that are tested.

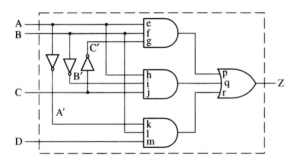

10.7 Find a test sequence to test for $b$ s-a-0 in the sequential circuit of Figure 10-7.

10.8 A sequential circuit has the following state graph:

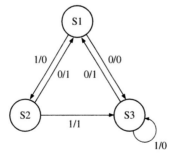

The three states can be distinguished using the input sequence 11 and observing the output. The circuit has a reset input, $R$, that resets the circuit to state $S_1$. Give a set of test sequences that will test every state transition and give the transition tested by each sequence. (When you test a state transition, you must verify that the output and the next state are correct by observing the output sequence.)

10.9 State graphs for two sequential machines are given below. The first graph represents a correctly functioning machine, and the second represents the same machine with a malfunction. Assuming that the two machines can be reset to their starting states ($S_0$ and $T_0$), determine the shortest input sequence that will distinguish the two machines.

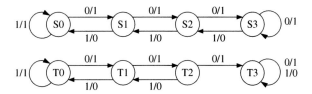

10.10 When testing a sequential circuit, what are the major advantages of using scan-path testing compared to applying input sequences and observing output sequences?

10.11 A scan path test circuit of the type shown in Figure 10-8 has three flip-flops, two inputs, and two outputs. One row of the state table of the sequential circuit to be tested is as follows:

| $Q_1Q_2Q_3$ | $X_1X_2 =$ | $Q_1^+Q_2^+Q_3^+$ | | | | $Z_1Z_2$ | | | |
|---|---|---|---|---|---|---|---|---|---|
| | | 00 | 01 | 11 | 10 | 00 | 01 | 11 | 10 |
| 011 | | 010 | 110 | 011 | 111 | 10 | 11 | 00 | 01 |

For this row of the table, complete a timing chart similar to Figure 10-9 to show how the circuit can be tested to verify the next states and outputs for inputs 00, 01, and 10. Show the expected $Z_1$ and $Z_2$ outputs only at the time when they should be read.

10.12 **(a)** Redraw the code converter circuit of Figure 1-26 in the form of Figure 10-8 using dual-port flip-flops.
   **(b)** Determine a test sequence that will verify the first two rows of the transition table of Figure 1-24(b). Draw a timing diagram similar to Figure 10-9 for your test sequence.

10.13 **(a)** Write VHDL code for a dual-port flip-flop.
   **(b)** Write VHDL code for your solution to Problem 10.12(a).
   **(c)** Write a test bench that applies the test sequence from Problem 10.12(b), and compare the resulting waveforms with your solution to Problem 10.12(b).

10.14 Instead of using dual-port flip-flops of the type shown in Figure 10-8, scan testing can be accomplished using standard D flip-flops with a mux on each D input to select $D_1$ or $D_2$. Redraw the circuit of Figure 1-22 to establish a scan chain using D flip-flops and muxes. A test signal ($T$) should control the muxes.

10.15 Referring to Figure 10-16, determine the sequence of *TMS* and *TDI* inputs required to load the instruction register with 011 and the boundary scan register BSR2 with 1101. Start in state 0 and end in state 1. Give the sequence of states along with the *TMS* and *TDI* inputs.

10.16 The INTEST instruction (code 010) allows testing of the core logic by shifting test data into the boundary scan register (BSR1) and then updating BSR2 with this test data. For input cells this data takes the place of data from the input pins. Output data from the core logic is captured in BSR1 and then shifted out. For this problem, assume that the BSR has three cells.

(a) Referring to Figure 10-16, give the sequence for *TMS* and *TDI* that will load the instruction register with 010 and BSR2 with 011. Also give the state sequence, starting in state 0.

(b) In the code of Figure 10-21, what changes or additions must be made in the last BSRout assignment statement, in the CaptureDR state, and in the UpdateDR state to implement the INTEST instruction?

10.17 Based on the VHDL code of Figure 10-21, design a two-cell boundary scan register. The first cell should be an input cell, and the second cell an output cell. Do not design the TAP controller; just assume that the necessary control signals like *shift-DR*, *capture-DR*, and *update-DR* are available. Do not design the instruction register or instruction decoding logic; just assume that the following signals are available: *EXT* (EXTEST instruction is being executed), *SPR* (Sample/Preload instruction is being executed), and *BYP* (Bypass instruction is being executed). Use two flip-flops for BSR1, two flip-flops for BSR2, and one BYPASS flip-flop. In addition to the control signals mentioned above, the inputs are *Pin*1 (from a pin), *Core*2 (from the core logic), *TDI*, and *TCK*; outputs are *Core*1 (to core logic), *Pin*2 (to a pin), and *TDO*. Use *TCK* as the clock input for all of the flip-flops. Draw a block diagram showing the flip-flops, muxes, and so on. Then give the logic equations or connections for each flip-flop D input, each CE (clock enable), and each MUX control input.

10.18 Simulate the boundary scan tester of Figure 10-22 and verify that the results are as expected. Change the code to represent the case where the lower input to IC1 is shorted to ground, simulate again, and interpret the results.

10.19 Write VHDL code for the boundary scan cell of Figure 10-14(b). Rewrite the VHDL code of Figure 10-21 to use this boundary scan cell as a component in place of some of the behavioral code for the BSR. Use a generate statement to instantiate NCELLS copies of this component. Test your new code using the boundary scan tester example of Figure 10-22.

10.20 **(a)** Draw a circuit diagram for an LFSR with $n = 5$ that generates a maximum length sequence.
   **(b)** Add logic so that 00000 is included in the state sequence.
   **(c)** Determine the actual state sequence.

10.21 **(a)** Draw a circuit diagram for an LFSR with $n = 6$ that generates a maximum length sequence.
   **(b)** Add logic so that 000000 is included in the sequence.
   **(c)** Determine the 10 elements of the sequence starting in 101010.

10.22 **(a)** Write VHDL for an 8-bit MISR that is similar to Figure 10-28.
   **(b)** Design a self-test circuit, similar to Figure 10-25, for a 6116 static RAM (see Figure 8-15). The write-data generator should store data in the following sequence: 00000000, 10000000, 11000000, ..., 11111111, 01111111, 00111111, ..., 00000000.
   **(c)** Write VHDL code to test your design. Simulate the system for at least one example with no errors, one error, two errors, and three errors.

10.23 In the system of Figure 10-33, $A$, $B$, and $C$ are BILBO registers. The $B_1$ and $B_2$ inputs to each of the registers determine its BILBO operating mode as follows:

$B_1B_2 = 00$, shift register; $B_1B_2 = 01$, PRPG (pattern generator);
$B_1B_2 = 10$, normal system mode; $B_1B_2 = 11$, MISR (signature register).

The shifting into $A$, $B$, and $C$ is always LSB first. When in the test mode, the Dbus is not used. Specify the sequence of the Tester outputs ($B_1$, $B_2$, and $S_i$) needed to perform the following operations:

**(1)** Load $A$ with 1011 and $B$ with 1110, clear $C$.
**(2)** Test the system by using $A$ and $B$ as pattern generators and $C$ as a signature register for four clock times.
**(3)** Shift the $C$ register output into the tester.
**(4)** Return to the normal system mode.

$B_1 \ B_2 \ S_i = 0 \ 0 \ 0, \dots$

10.24 Given the BILBO register shown below, specify $B_1$ and $B_0$ for each of the following modes:

   normal mode
   shift register mode
   PRPG (LSFR) mode
   MISR mode

When in the PRPG mode, what sequence of states would be generated for $Q_1$, $Q_2$, and $Q_3$, assuming that the initial state is 001?

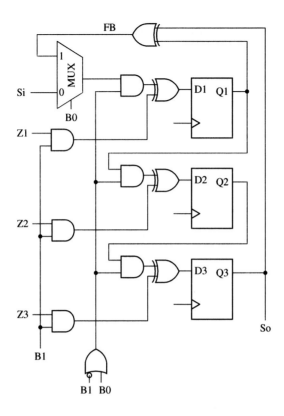

# Additional Design Examples

In this chapter, we present additional examples that show how VHDL, together with synthesis tools, can be used to simulate and design complex digital systems. We first design a wristwatch with alarm and stopwatch functions. Simulation models for memory chips with specific timing specifications are presented next. Finally, a receiver-transmitter for a serial data port is presented.

• • • • • • • • • • • •

## 11.1 Design of a Wristwatch

In this section, we will design a multifunction wristwatch that has time-keeping, alarm, and stopwatch functions. The wristwatch has three buttons (B1, B2, and B3) that are used to change the mode, set the time, set the alarm, start and stop the stopwatch, and so on. Pushing button B1 changes the mode from **Time** to **Alarm** to **Stopwatch** and back to **Time**. The functions of buttons B2 and B3 vary depending on the mode and are explained in the following paragraphs.

### 11.1.1 Specifications

**Operation in time mode:** Display indicates the time and whether it is A.M. or P.M. using the format hh:mm:ss (A or P). When in time mode, the alarm can be shut off manually by pressing B3. Pushing B2 changes the state to Set Hours or Set Minutes and back to Time mode. When in the Set Hours or Set Minutes state, each press of B3 advances the hours or minutes by 1.

**Operation in alarm mode:** Display indicates the alarm time and whether it is A.M. or P.M. using the format hh:mm (A or P). Pushing B2 changes the state to Set Alarm Hours or Set Alarm Minutes and then back to Alarm. When in the Set Alarm Hours or Set Alarm Minutes state, each press of B3 advances the alarm hours or minutes by 1. When in the Alarm state, pressing B3 sets or resets the alarm. Once the alarm starts ringing, it will ring for 50 seconds and then shut itself off. It can also be shut off manually by pressing B3 in time mode.

**Operation in the stopwatch mode:** Display indicates stopwatch time in the format mm:ss.cc (where cc is hundredths of a second). Pressing B2 starts the time counter, pressing B2 again stops it, and then pressing B2 restarts it, and so on. Pressing B3 resets the time. Once the stopwatch is started, it will keep running even when the wristwatch is in time or alarm mode.

### 11.1.2 Design Implementation

Figure 11-1 shows a block diagram for the design. The **input module** divides the system clock down to a 100-Hz clock, CLK. It debounces the input buttons (PB1, PB2, and PB3) and synchronizes them with CLK. Each time PB1, PB2, or PB3 is pressed, the corresponding signal, B1, B2, or B3, will be 1 for exactly one clock time. The single pulser circuitry that we designed in Section 4.7 can be used to build this module.

**FIGURE 11-1:**
**Block Diagram of**
**Wristwatch Design**

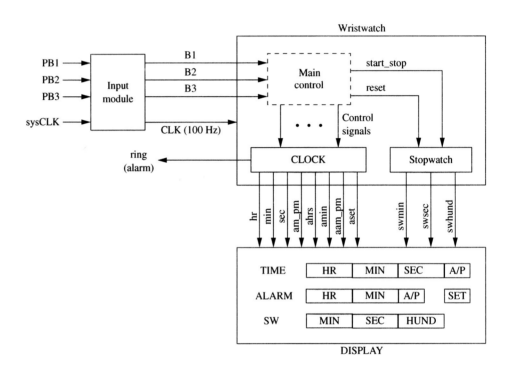

The wristwatch module contains the **main control** for the wristwatch; the **clock module**, which implements the timekeeping and alarm functions; and the **stopwatch module**, which implements the stopwatch functions. The 100-Hz clock (CLK) synchronizes operation of the control unit and time registers. Figure 11-2 shows the state graph for the controller. This state machine generates the following control signals in response to pressing the buttons:

| | |
|---|---|
| *inch* | increments hours in the set_hours state |
| *incm* | increments minutes in the set_minutes state |
| *alarm_off* | turns off the alarm when it is ringing |
| *incha* | increments hours for the alarm |
| *incma* | increments minutes for the alarm |
| *set_alarm* | toggles the alarm set on and off |
| *start_stop* | starts or stops the stopwatch counter |
| *reset* | resets the stopwatch counter |

FIGURE 11-2:
**State Graph of
Wristwatch
Module**

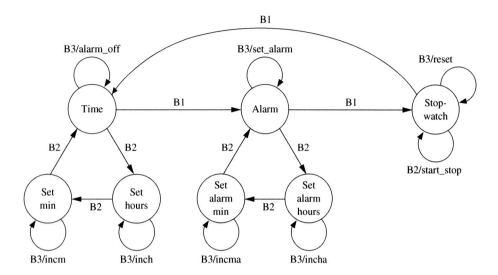

Figure 11-3 shows the VHDL code for the wristwatch module. This module instantiates the clock and stopwatch modules, and it implements the state machine. This state machine tests the B1, B2, and B3 button signals and generates the control signals. Following are some of the signal names used in the VHDL code:

| | |
|---|---|
| *am_pm* | A.M. or P.M. in time mode |
| *aam_pm* | A.M. or P.M. in alarm mode |
| *alarm_set* | indicates that alarm is set |
| *ring* | indicates that alarm setting matches time counters, if alarm is set |
| *hours* | hours in time mode |
| *ahours* | hours in alarm mode |
| *minutes* | minutes in the time mode |
| *aminutes* | minutes in the alarm mode |
| *seconds* | seconds in the time mode |
| *swhundredths* | hundredths of a second during stopwatch mode |
| *swseconds* | seconds in stopwatch mode |
| *swminutes* | minutes in stopwatch mode |

FIGURE 11-3: **VHDL Code for the Wristwatch Module**

```
library IEEE;
use IEEE.numeric_bit.all;

entity wristwatch is
  port(B1, B2, B3, clk: in bit;
       am_pm, aam_pm, ring, alarm_set: inout bit;
       hours, ahours, minutes, aminutes, seconds: inout unsigned(7 downto 0);
       swhundreths, swseconds, swminutes: out unsigned(7 downto 0));
end wristwatch;
```

```vhdl
architecture wristwatch1 of wristwatch is
  component clock is
    port(clk, inch, incm, incha, incma, set_alarm, alarm_off: in bit;
         hours, ahours, minutes, aminutes, seconds: inout unsigned(7 downto 0);
         am_pm, aam_pm, ring, alarm_set: inout bit);
  end component;
  component stopwatch is
    port(clk, reset, start_stop: in bit;
         swhundreths, swseconds, swminutes: out unsigned(7 downto 0));
  end component;
  type st_type is (time1, set_min, set_hours, alarm, set_alarm_hrs,
                   set_alarm_min, stop_watch);
  signal state, nextstate: st_type;
  signal inch, incm, alarm_off, set_alarm, incha, incma,
         start_stop, reset: bit;
begin
  clock1: clock port map(clk, inch, incm, incha, incma, set_alarm, alarm_off,
                         hours, ahours, minutes, aminutes, seconds, am_pm,
                         aam_pm, ring, alarm_set);
  stopwatch1: stopwatch port map(clk, reset, start_stop, swhundreths,
                                 swseconds, swminutes);
process(state, B1, B2, B3)
begin
  alarm_off <= '0'; inch <= '0'; incm <= '0'; set_alarm <= '0'; incha <= '0';
  incma <= '0'; start_stop <= '0'; reset <= '0';
  case state is
    when time1 =>
      if B1 = '1' then nextstate <= alarm;
      elsif B2 = '1' then nextstate <= set_hours;
      else nextstate <= time1;
      end if;
      if B3 = '1' then alarm_off <= '1';
      end if;
    when set_hours =>
      if B3 = '1' then inch <= '1'; nextstate <= set_hours;
      else nextstate <= set_hours;
      end if;
      if B2 = '1' then nextstate <= set_min;
      end if;
    when set_min =>
      if B3 = '1' then incm <= '1'; nextstate <= set_min;
      else nextstate <= set_min;
      end if;
      if B2 = '1' then nextstate <= time1;
      end if;
    when alarm =>
      if B1 = '1' then nextstate <= stop_watch;
      elsif B2 = '1' then nextstate <= set_alarm_hrs;
```

```
      else nextstate <= alarm;
      end if;
      if B3 = '1' then set_alarm <= '1'; nextstate <= alarm;
      end if;
    when set_alarm_hrs =>
      if B2 = '1' then nextstate <= set_alarm_min;
      else nextstate <= set_alarm_hrs;
      end if;
      if B3 = '1' then incha <= '1';
      end if;
    when set_alarm_min =>
      if B2 = '1' then nextstate <= alarm;
      else nextstate <= set_alarm_min;
      end if;
      if B3 = '1' then incma <= '1';
      end if;
    when stop_watch =>
      if B1 = '1' then nextstate <= time1;
      else nextstate <= stop_watch;
      end if;
      if B2 = '1' then start_stop <= '1';
      end if;
      if B3 = '1' then reset <= '1';
      end if;
    end case;
  end process;
  process(clk)
  begin
    if clk'event and clk = '1' then
      state <= nextstate;
    end if;
  end process;
end wristwatch1;
```

The clock module contains the counters that keep track of time (*hours*, *minutes*, and *seconds*) as well as the counters that are used to store the hour and minute settings for the alarm (*ahours* and *aminutes*). Each of these counters stores a two-digit BCD (binary-coded-decimal) number that is incremented at the appropriate time. The module also contains a counter that divides the 100-Hz clock by 100 and provides a signal to increment the seconds counter.

The VHDL code in Figure 11-4 instantiates three counters labeled *sec*1, *min*1, and *hrs*1. When the divide-by-100 counter is in state 99, it outputs a signal *c*99, which is used as an increment signal to the *sec*1 counter. *Sec*1 counts the seconds, and when the divide-by-100 counter rolls over, the seconds are incremented by 1 because *c*99 = 1. *Sec*1 is a divide-by-60 counter, and when it reaches 59, it outputs a signal *s*59. *Min*1 counts the minutes. It is incremented when *s*59 and *c*99 are both 1, and also when *incm* is 1 in the *set_minutes* state. A signal *incmin* is used to denote the condition when minutes has to be incremented, whether due to pressing of a button or due to a control signal while counting. When *min1* reaches 59, it outputs a signal *m59*. *Hrs1* counts the hours and also toggles the *am_pm* flip-flop when time changes from 11:59:59:99 to 12:00:00:00. *Hrs1* is incremented when $m59 = s59 = c99 = 1$, and also when *inch* is 1 in the *set_hours* state. A signal *inchr* denotes the condition when the hour has to be incremented, whether due to pressing of a button or due to a control signal while counting.

FIGURE 11-4: **VHDL Code of Clock Module**

```vhdl
library IEEE;
use IEEE.numeric_bit.all;

entity clock is
  port(clk, inch, incm, incha, incma, set_alarm, alarm_off: in bit;
      hours, ahours, minutes, aminutes, seconds: inout unsigned(7 downto 0);
      am_pm, aam_pm, ring, alarm_set: inout bit);
end clock;

architecture clock1 of clock is
  component CTR_59 is
    port(clk, inc, reset: in bit; dout: out unsigned(7 downto 0);
        t59: out bit);
  end component;
  component CTR_12 is
    port(clk, inc: in bit; dout: out unsigned(7 downto 0); am_pm: inout bit);
  end component;
  signal s59, m59, inchr, incmin, c99: bit;
  signal alarm_ring_time: integer range 0 to 50;
  signal div100: integer range 0 to 99;
  begin
    sec1: ctr_59 port map(clk, c99, '0', seconds, s59);
    min1: ctr_59 port map(clk, incmin, '0', minutes, m59);
    hrs1: ctr_12 port map(clk, inchr, hours, am_pm);
    incmin <= (s59 and c99) or incm;
    inchr <= (m59 and s59 and c99) or inch;
    alarm_min: ctr_59 port map(clk, incma, '0', aminutes, open);
    alarm_hr: ctr_12 port map(clk, incha, ahours, aam_pm);
    c99 <= '1' when div100 = 99 else '0';
    process(clk)
    begin
      if clk'event and clk = '1' then
        if c99 = '1' then div100 <= 0;    -- divide by 100 counter
        else div100 <= div100 + 1;
        end if;
        if set_alarm = '1' then
        alarm_set <= not alarm_set;
      end if;
```

```
        if ((minutes = aminutes) and (hours = ahours) and (am_pm = aam_pm)) and
            seconds = 0 and alarm_set = '1' then
          ring <= '1';
        end if;
        if ring = '1' and c99 = '1' then
          alarm_ring_time <= alarm_ring_time + 1;
        end if;
        if alarm_ring_time = 50 or alarm_off = '1' then
          ring <= '0'; alarm_ring_time <= 0;
        end if;
      end if;
  end process;
end clock1;
```

The clock VHDL code also implements the alarm functions. It instantiates counters for setting the alarm minutes and hours. The *alarm_set* flip-flop is toggled when *alarm_set* is 1. The *ring* flip-flop is set to 1 when the alarm setting matches the time counters and the alarm is set. *Alarm_ring_time* is a counter that counts seconds when the alarm is ringing. The ring flip-flop is cleared after 50 seconds or when the *alarm_off* signal is received.

The VHDL code in Figure 11-5 implements the stopwatch functions. It instantiates counters for hundredths of a second, seconds, and minutes. When a *start_stop* signal is received, the counting flip-flop is toggled. *Ctr2* is a divide-by-100 BCD counter that is incremented every clock when counting = 1. It generates a signal *swc99* when it is in state 99. VHDL code for the divide-by-100 counter is shown in Figure 11-6. *Sec2* is the seconds counter that is incremented when *swc99* = 1. *Sec2* generates a signal *s59* when it is in state 59. The minutes counter, *min2*, is incremented when both *s59* and *swc99* are 1.

**FIGURE 11-5: VHDL Code of the Stopwatch Module**

```
library IEEE;
use IEEE.numeric_bit.all;

entity stopwatch is
  port(clk, reset, start_stop: in bit;
       swhundreths, swseconds, swminutes: out unsigned(7 downto 0));
end stopwatch;

architecture stopwatch1 of stopwatch is
  component CTR_59 is
    port(clk, inc, reset: in bit; dout: out unsigned(7 downto 0); t59: out bit);
  end component;
  component CTR_99 is
    port(clk, inc, reset: in bit; dout: out unsigned(7 downto 0); t59: out bit);
  end component;
  signal swc99, s59, counting, swincmin: bit;
begin
  ctr2: ctr_99 port map(clk, counting, reset, swhundreths, swc99);
```

```
      --counts hundreths of seconds
  sec2: ctr_59 port map(clk, swc99, reset, swseconds, s59);
     --counts seconds
  min2: ctr_59 port map(clk, swincmin, reset, swminutes, open);
     --counts minutes
  swincmin <= s59 and swc99;
  process(clk)
  begin
     if clk'event and clk = '1' then
       if start_stop = '1' then
         counting <= not counting;
       end if;
     end if;
  end process;
end stopwatch1;
```

FIGURE 11-6: VHDL Code for Divide-by-100 Counter

```
library IEEE;
use IEEE.numeric_bit.all;
--divide by 100 BCD counter
entity CTR_99 is
  port(clk, inc, reset: in bit; dout: out unsigned(7 downto 0); t59: out bit);
end CTR_99;

architecture count99 of CTR_99 is
  signal dig1, dig0: unsigned(3 downto 0);
begin
  process(clk)
  begin
    if clk'event and clk = '1' then
      if reset = '1' then dig0 <= "0000"; dig1 <= "0000";
      else
        if inc = '1' then
          if dig0 = 9 then dig0 <= "0000";
            if dig1 = 9 then dig1 <= "0000";
            else dig1 <= dig1 + 1;
            end if;
          else dig0 <= dig0 + 1;
          end if;
        end if;
      end if;
    end if;
  end process;
  t59 <= '1' when (dig1 = 9 and dig0 = 9) else '0';
  dout <= dig1 & dig0;
end count99;
```

VHDL code for the divide-by-60 counter (Figure 11-7) is straightforward. The counter counts to 59 and then resets.

FIGURE 11-7: **VHDL Code of Divide-by-60 Counter**

```vhdl
library IEEE;
use IEEE.numeric_bit.all;
--this counter counts seconds or minutes 0 to 59
entity CTR_59 is
  port(clk, inc, reset: in bit; dout: out unsigned(7 downto 0); t59: out bit);
end CTR_59;

architecture count59 of CTR_59 is
  signal dig1, dig0: unsigned(3 downto 0);
begin
  process(clk)
  begin
    if clk'event and clk = '1' then
      if reset = '1' then dig0 <= "0000"; dig1 <= "0000";
      else
        if inc = '1' then
          if dig0 = 9 then dig0 <= "0000";
            if dig1 = 5 then dig1 <= "0000";
            else dig1 <= dig1 + 1;
            end if;
          else dig0 <= dig0 + 1;
          end if;
        end if;
      end if;
    end if;
  end process;
  t59 <= '1' when (dig1 = 5 and dig0 = 9) else '0';
  dout <= dig1 & dig0;
end count59;
```

The hours counter (Figure 11-8) counts to 12 and then changes to 1 the next time the increment signal is 1. It toggles the *am_pm* signal when the count changes from 11 to 12.

FIGURE 11-8: **VHDL Code for Hours Counter**

```vhdl
library IEEE;
use IEEE.numeric_bit.all;
--this counter counts hours 1 to 12 and toggles am_pm
entity CTR_12 is
  port(clk, inc: in bit; dout: out unsigned(7 downto 0); am_pm: inout bit);
end CTR_12;

architecture count12 of CTR_12 is
  signal dig0: unsigned(3 downto 0);
  signal dig1: bit;
```

```
begin
  process(clk)
  begin
    if clk'event and clk = '1' then
      if inc = '1' then
        if dig1 = '1' and dig0 = 2 then
          dig1 <= '0'; dig0 <= "0001";
        else
          if dig0 = 9 then dig0 <= "0000"; dig1 <= '1';
          else dig0 <= dig0 + 1;
          end if;
          if dig1 = '1' and dig0 = 1 then am_pm <= not am_pm;
          end if;
        end if;
      end if;
    end if;
  end process;
  dout <= "000" & dig1 & dig0;
end count12;
```

### 11.1.3 Testing the Wristwatch

Next we will write a test bench for the wristwatch module (Figure 11-9). The test bench must generate a series of button pushes as well as the 100-Hz clock, and it must display the time, alarm settings, and stopwatch counters. In effect, the test bench takes the place of the input and display modules in the overall design. To simplify writing the test bench code, we have written two procedures. Procedure `wait1(N1)` waits for *N1* clocks each time it is called. Procedure `push(button, N)` simulates pushing a button *N* times each time it is called. Thus `push(B2, 23)` simulates pushing *B2* 23 times. The push procedure simulates the output from the Input Module. Therefore, each button signal is on for exactly one clock time, and it is synchronized with CLK. The procedure waits 1.2 seconds after each button push. Testing should also be done with longer and shorter wait times between button pushes. Because we are using unsigned numbers and the numeric_bit package, all registers will be clear when the test bench is started. If we used numeric_std instead, we would have to reset all registers before running the simulation.

The test sequence we used is as follows:

**1.** Set the time to 11:58 P.M.
**2.** Set the alarm time to 12:00 A.M.
**3.** Set the alarm and change to time mode, and wait until the time rolls over at midnight.
**4.** Turn off the alarm 5 seconds later.
**5.** Change to stopwatch mode and start the stopwatch
**6.** Switch to time mode and wait for 10 seconds (stopwatch keeps running)
**7.** Switch back to stopwatch mode and wait until it reads 1 minute and 2 seconds
**8.** Stop the stopwatch, reset it, and return to time mode.

FIGURE 11-9: **Test Bench for Wristwatch**

```
library IEEE;
use IEEE.numeric_bit.all;

entity testww is    -- test bench for wristwatch
  port(hours, ahours, minutes, aminutes, seconds,
       swhundreths, swseconds, swminutes: inout unsigned(7 downto 0);
       am_pm, aam_pm, ring, alarm_set: inout bit);
end testww;

architecture testww1 of testww is
  component wristwatch is
    port(B1, B2, B3, clk: in bit;
         am_pm, aam_pm, ring, alarm_set: inout bit;
         hours, ahours, minutes, aminutes, seconds: inout unsigned(7 downto 0);
         swhundreths, swseconds, swminutes: out unsigned(7 downto 0));
  end component;
  signal B1, B2, B3, clk: bit;
begin
  wristwatch1: wristwatch port map(B1, B2, B3, clk, am_pm, aam_pm, ring,
                                   alarm_set, hours, ahours, minutes, aminutes,
                                   seconds, swhundreths, swseconds, swminutes);
  clk <= not clk after 5 ms;    -- generate 100 hz clock
  process
  procedure wait1    -- waits for N1 clocks
    (N1: in integer)
    variable count: integer;
  begin
    count := N1;
    while count /= 0 loop
      wait until clk'event and clk = '1';
      count := count - 1;
      wait until clk'event and clk = '0';
    end loop;
  end procedure wait1;
  procedure push    -- simulates pushing a button N times
    (signal button: out bit; N: in integer) is
  begin
    for i in 1 to N loop
      button <= '1';
      wait1(1);
      button <= '0';
      wait1(120);    -- wait 1200 ms between pushes
    end loop;
  end procedure push;
  begin
    wait1(10);    -- set time to 11:58 pm
    push(b2, 1); push(b3, 23); push(b2, 1); push(b3, 57); push(b2, 1);
```

```
      report "time should be 11:58 P.M.";
      push(b1, 1);    -- set alarm to 12:00 am
      push(b2, 1); push(b3, 24); push(b2, 2); push(b3, 1); push(b1, 2);
      report "alarm should be set to 12:00 A.M.";
      wait until hours = "00010010" and seconds = "00000101";
      push(b3, 1);    -- turn alarm off at 12 hours and 5 seconds
      push(b1, 2);    -- run stopwatch, go to time mode, go back to stopwatch
      push(b2, 1); wait1(120); push(b1, 1); wait1(1000); push(b1, 2);
      wait until swminutes = "00000001" and swseconds = "00000010";
        --stop stopwatch after 1 min. and 2 sec., then reset
      report "stopwatch should read 1 min. 2 sec.";
      push(b2, 1); push(b3, 1); push(b1, 1);
      wait;
   end process;
end testww1;
```

We used the following commands to run the simulation with the preceding test sequence.

```
vsim -t 1ms testww    -- set simulator resolution to 1 ms
add list -hex hours minutes seconds am_pm
               ahours aminutes aam_pm ring
add list b1 b2 b3 wristwatch1/state
add list -hex swminutes swseconds -notrigger swhundredths
run 300000 ms
```

The test results showed that the wristwatch module functions according to the specifications. When the wristwatch module is implemented using the Xillinx Spartan 3 FPGA, it requires 87 slices, 80 flip-flops, and 158 four-input LUTs. To complete the design, we still need to write code for the Input and Display modules.

# 11.2 Memory Timing Models

When we design a complex digital system with several components, many timing constraints must be satisfied in order for the system components to function together properly. For example, if we are interfacing memory components to a microprocessor bus, all bus interface timing specifications must be satisfied. In order to simulate such a system using VHDL, we must develop accurate timing models for each component. In this section we will develop a timing model for a small static RAM. This will illustrate the process of going from manufacturer's specifications to a VHDL model that takes timing parameters into account. These types of timing models are very useful when developing system on a chip (SoC) designs.

Figure 11-10 illustrates the block diagram of a 6116 static RAM, which can store 2048 eight-bit words of data. This memory has 16,384 cells, arranged in a $128 \times 128$ memory matrix. The RAM contains address decoders and a memory array. The address decoder is typically split into the column decoder and the row decoder. The 11 address lines, which are needed to address the $2^{11}$ bytes of data, are divided into two groups, one for the column decoder and the other for the row decoder. Lines A0 through A3 select eight columns in the matrix at a time, since there are eight data lines for each address. Lines A4 through A10 select one of the 128 rows in the matrix. The data outputs from the matrix go through tristate buffers before connecting to the data I/O lines. These buffers are disabled except when reading from the memory.

Figure 11-10 also illustrates the connections of the chip select ($\overline{\text{CS}}$), output enable ($\overline{\text{OE}}$), and write enable ($\overline{\text{WE}}$) signals. The functions of these signals were explained in Table 8-7. When $\overline{\text{CS}}$ is high (i.e., not asserted), a 0 input reaches the two AND gates; hence, the tristate control voltage is low, resulting in a high-$Z$ output. Similarly, when $\overline{\text{OE}}$ is high, even if chip select is asserted, the tristate control is inactive, resulting in high-$Z$ output. When chip select and $\overline{\text{WE}}$ are asserted, a write operation happens, and the data on the I/O lines gets written into the RAM. When chip select and $\overline{\text{OE}}$ are asserted and $\overline{\text{WE}}$ is not asserted, a read operation happens, and the RAM contents appear on the I/O lines.

**FIGURE 11-10:**
**Block Diagram of 6116 Static RAM**

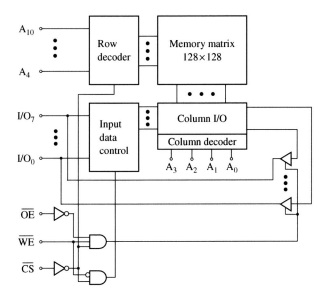

We presented some static RAM memory models in Section 8.7; however, the models do not take into account timing specifications. Memory timing diagrams and specifications must be considered when designing systems using the memory chips. In this section, we present simulation models of memory chips with particular timing specifications.

Let us consider a CMOS static RAM 6116, whose timing parameters are defined in Table 11-1 for both read and write cycles. Specifications are given for the 6116 SA-15 RAM, which has a 15-ns access time. A dash in the table indicates that either the specification was not relevant or that the manufacturer did not provide the specification.

**TABLE 11-1:**
**Timing Specifications for CMOS Static RAM 6116 SA-15**

| Parameter | Symbol | Timing Specification min(ns) | max(ns) |
|---|---|---|---|
| Read cycle time | $t_{RC}$ | 15 | — |
| Address access time | $t_{AA}$ | — | 15 |
| Chip select access time | $t_{ACS}$ | — | 15 |
| Chip selection to output in low-Z | $t_{CLZ}$ | 5 | — |
| Output enable to output valid | $t_{OE}$ | — | 10 |
| Output enable to output in low-Z | $t_{OLZ}$ | 0 | — |
| Chip deselection to output in high-Z | $t_{CHZ}$ | 2* | 10 |
| Output disable to output in high-Z | $t_{OHZ}$ | 2* | 8 |
| Output hold from address change | $t_{OH}$ | 5 | — |
| Write cycle time | $t_{WC}$ | 15 | — |
| Chip selection to end of write | $t_{CW}$ | 13 | — |
| Address valid to end of write | $t_{AW}$ | 14 | — |
| Address setup time | $t_{AS}$ | 0 | — |
| Write pulse width | $t_{WP}$ | 12 | — |
| Write recovery time | $t_{WR}$ | 0 | — |
| Write enable to output in high-Z | $t_{WHZ}$ | — | 7 |
| Data valid to end of write | $t_{DW}$ | 12 | — |
| Data hold from end of write | $t_{DH}$ | 0 | — |
| Output active from end of write | $t_{OW}$ | 0 | — |

*Estimated value, not specified by manufacturer.

Figure 11-11(a) shows the read cycle timing for the case where $\overline{CS}$ and $\overline{OE}$ are both low before the address changes. In this case, after the address changes, the old data remains at the memory output for a time $t_{OH}$; then there is a transition period during which the data may change (as indicated by the cross-hatching). The new data is stable at the memory output after the address access time, $t_{AA}$. The address must be stable for the read cycle time, $t_{RC}$.

Figure 11-11(b) shows the timing for the case where $\overline{OE}$ is low and the address is stable before $\overline{CS}$ goes low. When $\overline{CS}$ is high, *Dout* is in the high-Z state, as indicated by a line halfway between '0' and '1'. When $\overline{CS}$ goes low, *Dout* leaves the high-Z state after time $t_{CLZ}$, there is a transition period during which the data may change, and the new data is stable at time $t_{ACS}$ after $\overline{CS}$ changes. *Dout* returns to high-Z at time $t_{CHZ}$ after $\overline{CS}$ goes high.

**FIGURE 11-11:**
**Read Cycle Timing**

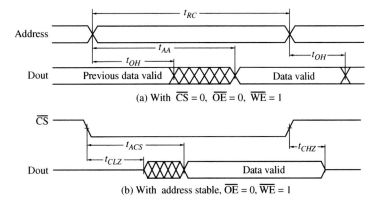

(a) With $\overline{CS} = 0$, $\overline{OE} = 0$, $\overline{WE} = 1$

(b) With address stable, $\overline{OE} = 0$, $\overline{WE} = 1$

Figure 11-12 shows the write cycle timing for the case where $\overline{OE}$ is low during the entire cycle and where writing to memory is controlled by $\overline{WE}$. In this case, it is assumed that $\overline{CS}$ goes low before or at the same time as $\overline{WE}$ goes low, and goes high before or at the same time as $\overline{CS}$ does. The cross-hatching on $\overline{CS}$ indicates the interval in which it can go from high to low (or from low to high). The address must be stable for the address setup time, $t_{AS}$, before $\overline{WE}$ goes low. After time $t_{WHZ}$, the data out from the tristate buffers go to the high-Z state and input data may be placed on the I/O lines. The data into the memory must be stable for the setup time $t_{DW}$ before $\overline{WE}$ goes high, and then it must be kept stable for the hold time $t_{DH}$. The address must be stable for $t_{WR}$ after $\overline{WE}$ goes high. When $\overline{WE}$ goes high, the memory switches back to the read mode. After $t_{OW}$ (min) and during region (a), *Dout* goes through a transition period and then becomes the same as the data just stored in the memory. Further change in *Dout* may occur if the address changes or if $\overline{CS}$ goes high. To avoid bus conflicts during region (a), *Din* should either be high-Z or the same as *Dout*.

**FIGURE 11-12:**
**$\overline{WE}$-Controlled**
**Write Cycle Timing**
**($\overline{OE} = 0$)**

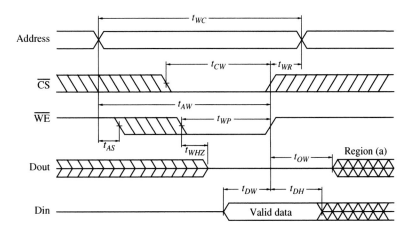

Figure 11-13 shows the write cycle timing for the case where $\overline{OE}$ is low during the entire cycle and where writing to memory is controlled by $\overline{CS}$. In this case, it is assumed that $\overline{WE}$ goes low before or at the same time as $\overline{CS}$ goes low, and $\overline{CS}$ goes high before or at the same time as $\overline{WE}$ does. The address must be stable for the address setup time, $t_{AS}$, before $\overline{CS}$ goes low. The data into the memory must be stable for the setup time $t_{DW}$ before $\overline{CS}$ goes high, and then it must be kept stable for the hold time $t_{DH}$. The address must be stable for $t_{WR}$ after $\overline{CS}$ goes high. Note that this write cycle is very similar to the $\overline{WE}$-controlled cycle. In both cases, writing to memory occurs when both $\overline{CS}$ and $\overline{WE}$ are low, and writing is completed when either one goes high.

**FIGURE 11-13:**
**$\overline{CS}$-Controlled**
**Write Cycle Timing**
**($\overline{OE}$ = 0)**

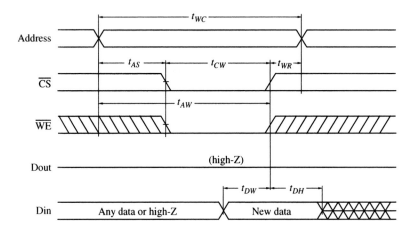

Next, we revise the RAM model presented in Figure 8-15 to include timing information based on the read and write cycles shown in Figures 11-11, 11-12, and 11-13. We assume that $\overline{OE}$ = '0'. The VHDL RAM timing model in Figure 11-14 uses a generic declaration to define default values for the important timing parameters. Transport delays are used throughout to avoid cancellation problems, which can occur with inertial delays. The RAM process waits for a change in $CS\_b$, $WE\_b$, or the address. If a rising edge of $WE\_b$ occurs when $CS\_b$ is '0', or a rising edge of $CS\_b$ occurs when $WE\_b$ is '0', this indicates the end of write, so the data is stored in the RAM, and then the data is read back out after $t_{OW}$. If a falling edge of $WE\_b$ occurs when $CS\_b$ = '0', the RAM switches to write mode and the data output goes to high-Z.

**FIGURE 11-14: Timing Simulation Model for 6116 Static CMOS RAM**

```
-- memory model with timing (OE_b=0)
library IEEE;
use IEEE.std_logic_1164.all;
use IEEE.numeric_std.all;

entity static_RAM is
  generic(constant tAA:  time := 15 ns;   -- 6116 static CMOS RAM
          constant tACS: time := 15 ns;
```

```vhdl
          constant tCLZ: time :=  5 ns;
          constant tCHZ: time :=  2 ns;
          constant tOH:  time :=  5 ns;
          constant tWC:  time := 15 ns;
          constant tAW:  time := 14 ns;
          constant tWP:  time := 12 ns;
          constant tWHZ: time :=  7 ns;
          constant tDW:  time := 12 ns;
          constant tDH:  time :=  0 ns;
          constant tOW:  time :=  0 ns);
  port(CS_b, WE_b, OE_b: in std_logic;
       Address: in unsigned(7 downto 0);
       IO: inout unsigned(7 downto 0) := (others => 'Z'));
end Static_RAM;

architecture SRAM of Static_RAM is
  type RAMtype is array(0 to 255) of unsigned(7 downto 0);
  signal RAM1: RAMtype := (others => (others => '0'));
begin
  RAM: process (CS_b, WE_b, Address)
  begin
    if CS_b='0' and WE_b='1' and Address'event then
    -- read when address changes
      IO <= transport "XXXXXXXX" after tOH,
            Ram1(to_integer(Address)) after tAA; end if;
    if falling_edge(CS_b)and WE_b='1' then
    -- read when CS_b goes low
      IO <= transport "XXXXXXXX" after tCLZ,
            Ram1(to_integer(Address)) after tACS; end if;
    if rising_edge(CS_b) then    -- deselect the chip
      IO <= transport "ZZZZZZZZ" after tCHZ;
      if We_b='0' then -- CS-controlled write
        Ram1(to_integer(Address'delayed)) <= IO; end if;
    end if;
    if falling_edge(WE_b) and CS_b='0' then    -- WE-controlled write
      IO <= transport "ZZZZZZZZ" after tWHZ; end if;
    if rising_edge(WE_b) and CS_b='0' then
      Ram1(to_integer(Address'delayed)) <= IO'delayed;
      IO <= transport IO'delayed after tOW;    -- read back after write
        -- IO'delayed is the value of IO just before the rising edge
    end if;
  end process RAM;

  check: process
  begin
    if NOW /= 0 ns then
      if address'event then
        assert (address'delayed'stable(tWC))    -- tRC = tWC assumed
          report "Address cycle time too short"
          severity WARNING;
```

```
      end if;
  -- The following code only checks for a WE_b controlled write:
    if rising_edge(WE_b) and CS_b'delayed = '0' then
      assert (address'delayed'stable(tAW))
        report "Address not valid long enough to end of write"
        severity WARNING;
      assert (WE_b'delayed'stable(tWP))
        report "Write pulse too short"
        severity WARNING;
      assert (IO'delayed'stable(tDW))
        report "IO setup time too short"
        severity WARNING;
      wait for tDH;
        assert (IO'last_event >= tDH)
        report "IO hold time too short"
        severity WARNING;
    end if;
  end if;
  wait on CS_b, WE_b, Address;
 end process check;
end SRAM;
```

If a rising edge of $CS\_b$ has occurred, the RAM is deselected, and the data output goes to high-Z after the specified delay. Otherwise, if a falling edge of $CS\_b$ has occurred and $WE\_b$ is '1', the RAM is in the read mode. The data bus can leave the high-Z state after time $t_{CLZ}$ (min), but it is not guaranteed to have valid data out until time $t_{ACS}$ (max). The region in between is a transitional region where the bus state is unknown, so we model this region by outputting 'X' on the I/O lines. If an address change has just occurred and the RAM is in the read mode (Figure 11-11(a)), the old data holds its value for time $t_{OH}$. Then the output is in an unknown transitional state until valid data has been read from the RAM after time $t_{AA}$.

The check process, which runs concurrently with the RAM process, tests to see if some of the memory timing specifications are satisfied. *NOW* is a predefined variable that equals the current time. (VHDL provides NOW in order to access the current simulation time. It is actually a predefined function. It returns different values when called at different times during the course of a simulation.) To avoid false error messages, checking is not done when $NOW = 0$ or when the chip is not selected. When the address changes, the process checks to see if the address has been stable for the write cycle time ($t_{WC}$) and outputs a warning message if it is not. Since an address event has just occurred when this test is made, *Address'stable*($t_{WC}$) would always return FALSE. Therefore, *Address'delayed* must be used instead of *Address* so that *Address* is delayed one delta and the stability test is made just before *Address* changes. Next, the timing specifications for write are checked. First, we verify that the address has been stable for $t_{AW}$. Then we check to see that $WE\_b$ has been low for $t_{WP}$. Finally, we check the setup and hold times for the data.

VHDL code for a partial test of the RAM timing model is shown in Figure 11-15. This code runs a write cycle followed by two read cycles. The RAM is deselected between cycles. Figure 11-16 shows the test results. We also tested the model for cases where simultaneous input changes occur and cases where timing specifications are violated, but these test results are not included here.

FIGURE 11-15: **VHDL Code for Testing the RAM Timing Model**

```vhdl
library IEEE;
use IEEE.std_logic_1164.all;
use IEEE.numeric_std.all;

entity RAM_timing_tester is
end RAM_timing_tester;

architecture test1 of RAM_timing_tester is
  component static_RAM is
    port(CS_b, WE_b, OE_b: in std_logic;
         Address: in unsigned(7 downto 0);
         IO: inout unsigned(7 downto 0));
  end component Static_RAM;
  signal Cs_b, We_b: std_logic := '1';    -- active low signals
  signal Data: unsigned(7 downto 0) := "ZZZZZZZZ";
  signal Address: unsigned(7 downto 0):= "00000000";
begin
  SRAM1: Static_RAM port map(Cs_b, We_b, '0', Address, Data);
  process
  begin
    wait for 20 ns;
    Address <= "00001000";              -- WE-controlled write
    Cs_b <= transport '0', '1' after 50 ns;
    We_b <= transport '0' after 8 ns, '1' after 40 ns;
    Data <= transport "11100011" after 25 ns, "ZZZZZZZZ" after 55 ns;

    wait for 60 ns;
    Address <= "00011000";              -- RAM deselected
    wait for 40 ns;
    Address <= "00001000";              -- Read cycles
    Cs_b <= '0';
    wait for 40 ns;
    Address <= "00010000";
    Cs_b <= '1' after 40 ns;
    wait for 40 ns;
    Address <= "00011000";              -- RAM deselected
    wait for 40 ns;
    report "DONE";
  end process;
end test1;
```

FIGURE 11-16:
**Test Results for
RAM Timing Model**

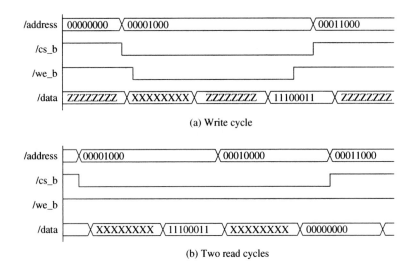

(a) Write cycle

(b) Two read cycles

● ● ● ● ● ● ● ● ● ● ● ● ●

# 11.3 A Universal Asynchronous Receiver Transmitter

Most computers and microcontrollers have one or more serial data ports used to communicate with serial input/output devices such as keyboards and serial printers. By using a modem (modulator-demodulator) connected to a serial port, serial data can be transmitted to and received from a remote location via telephone lines (see Figure 11-17). The serial communication interface, which receives and transmits serial data, is often called a UART (universal asynchronous receiver-transmitter). In Figure 11-17, $RxD$ is the received serial data signal and $TxD$ is the transmitted data signal.

FIGURE 11-17:
**Serial Data
Transmission**

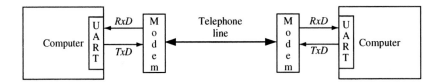

Figure 11-18 shows the standard format for serial data transmission. Since there is no clock line, the data ($D$) is transmitted asynchronously, one byte at a time. When no data is being transmitted, $D$ remains high. To mark the start of transmission, $D$ goes low for one bit time, which is referred to as the start bit. Then eight data bits are transmitted, least significant bit first. When text is being transmitted, ASCII code is usually used. In ASCII code, each alphanumeric character is represented by a 7-bit code. The eighth bit may be used as a parity check bit. In the example, the letter U, coded as 1010101, is transmitted followed by a 0 parity bit, so that the total number of 1's is even (even parity). After 8 bits are transmitted, $D$ must go high for at least one bit time, which is referred to as the stop bit. Then transmission of another character can start at any time.

The number of bits transmitted per second is frequently referred to as the *baud rate*.

FIGURE 11-18:
**Standard Serial
Data Format**

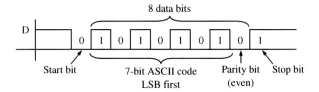

When transmitting, the UART takes 8 bits of parallel data and converts the data to a serial bit stream that consists of a start bit (logic '0'), 8 data bits (least significant bit first), and one or more stop bits (logic '1'). When receiving, the UART detects the start bit, receives the 8 data bits, and converts the data to parallel form when it detects the stop bit. Since no clock is transmitted, the UART must synchronize the incoming bit stream with the local clock.

We now design a simplified version of a UART similar to the one used within the microcontroller MC6805, MC6811, and other microcontrollers. Figure 11-19 shows the UART connected to the 8-bit data bus. The following six 8-bit registers are used:

| | |
|---|---|
| *RSR* | Receive shift register |
| *RDR* | Receive data register |
| *TDR* | Transmit data register |
| *TSR* | Transmit shift register |
| *SCCR* | Serial communications control register |
| *SCSR* | Serial communications status register |

The following discussion assumes that the UART is connected to a microcontroller data and address bus so that the CPU can read and write to the registers. *RDR*,

FIGURE 11-19: **UART Block Diagram**

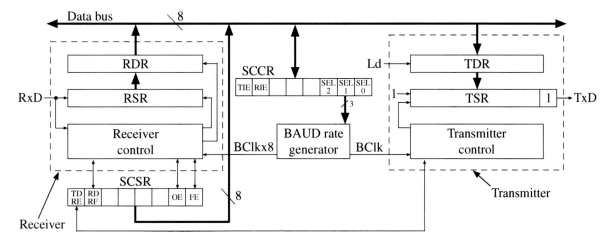

*TDR, SCCR,* and *SCSR* are memory-mapped; that is, each register is assigned an address in the microcontroller memory space. *RDR, SCSR,* and *SCCR* can drive the data bus through tristate buffers; *TDR* and *SCCR* can be loaded from the data bus.

Besides the registers, the three main components of the UART are the baud rate generator, the receiver controller, and the transmitter controller. The baud rate generator divides down the system clock to provide the bit clock (*BClk*) with a period equal to one bit time and also *BClkX8*, which has a frequency eight times the *BClk* frequency.

The *TDRE* (Transmit Data Register Empty) bit in the *SCSR* is set when *TDR* is empty. When the microcontroller is ready to transmit data, the following occurs:

1. The microcontroller waits until *TDRE* = '1' and then loads a byte of data into *TDR* and clears *TDRE*.
2. The UART transfers data from *TDR* to *TSR* and sets *TDRE*.
3. The UART outputs a start bit ('0') for one bit time and then shifts *TSR* right to transmit the eight data bits followed by a stop bit ('1').

Figure 11-20 shows the SM chart for the transmitter. The corresponding sequential machine (SM) is clocked by the microcontroller system clock (*CLK*). In the IDLE state, the SM waits until *TDR* has been loaded and *TDRE* is cleared. In the SYNCH state, the SM waits for the rising edge of the bit clock (*Bclk*↑) and then clears the low-order bit of *TSR* to transmit a '0' for one bit time. In the TDATA state, each time *Bclk*↑ is detected, *TSR* is shifted right to transmit the next data bit,

**FIGURE 11-20:**
**SM Chart for UART Transmitter**

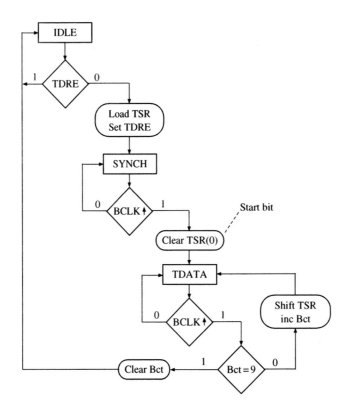

and the bit counter (*Bct*) is incremented. When *Bct* = 9, 8 data bits and a stop bit have transmitted. *Bct* is then cleared and the SM goes back to IDLE.

The VHDL code for the UART transmitter (Figure 11-21) is based on the SM chart of Figure 11-20. The `use IEEE.numeric_std.all` statement is not necessary if std_logic_vector type is used. We use the unsigned type here and in the other modules of the UART. The transmitter contains the *TDR* and *TSR* registers and the transmit control. It interfaces with *TDRE* and the data bus (DBUS). The first process represents the combinational network, which generates the nextstate and control signals. The second process updates the registers on the rising edge of the clock. The signal *Bclk_rising* is '1' for one system clock time following the rising edge of *Bclk*. To generate *Bclk_rising*, *Bclk* is stored in a flip-flop named *Bclk_Dlayed*. Then *Bclk_rising* is '1' if the current value of *Bclk* is '1' and the previous value (stored in *Bclk_Dlayed*) is '0'. Thus,

```
Bclk_rising <= Bclk and not Bclk_Dlayed;
```

FIGURE 11-21: **VHDL Code for UART Transmitter**

```
library IEEE;
use IEEE.std_logic_1164.all;
use IEEE.numeric_std.all; -- use this if unsigned type is used.

entity UART_Transmitter is
  port(Bclk, sysclk, rst_b, TDRE, loadTDR: in std_logic;
       DBUS: in unsigned(7 downto 0);
       setTDRE, TxD: out std_logic);
end UART_Transmitter;

architecture xmit of UART_Transmitter is
  type stateType is (IDLE, SYNCH, TDATA);
  signal state, nextstate: stateType;
  signal TSR: unsigned(8 downto 0);    -- Transmit Shift Register
  signal TDR: unsigned(7 downto 0);    -- Transmit Data Register
  signal Bct: integer range 0 to 9;    -- counts number of bits sent
  signal inc, clr, loadTSR, shftTSR, start: std_logic;
  signal Bclk_rising, Bclk_Dlayed: std_logic;
begin
  TxD <= TSR(0);
  setTDRE <= loadTSR;
  Bclk_rising <= Bclk and (not Bclk_Dlayed);
  -- indicates the rising edge of bit clock

  Xmit_Control: process(state, TDRE, Bct, Bclk_rising)
  begin
    inc <= '0'; clr <= '0'; loadTSR <= '0'; shftTSR <= '0'; start <= '0';
    -- reset control signals
    case state is
      when IDLE =>
        if (TDRE = '0') then
```

```
              loadTSR <= '1'; nextstate <= SYNCH;
          else nextstate <= IDLE;
          end if;
        when SYNCH =>   -- synchronize with the bit clock
          if (Bclk_rising = '1') then
            start <= '1'; nextstate <= TDATA;
          else nextstate <= SYNCH;
          end if;
        when TDATA =>
          if (Bclk_rising = '0') then nextstate <= TDATA;
          elsif (Bct /= 9) then
            shftTSR <= '1'; inc <= '1'; nextstate <= TDATA;
          else clr <= '1'; nextstate <= IDLE;
          end if;
    end case;
end process;

Xmit_update: process(sysclk, rst_b)
begin
  if (rst_b = '0') then
    TSR <= "111111111"; state <= IDLE; Bct <= 0; Bclk_Dlayed <= '0';
  elsif (sysclk'event and sysclk = '1') then
    state <= nextstate;
    if (clr = '1') then Bct <= 0;
    elsif (inc = '1') then
      Bct <= Bct + 1;
    end if;
    if (loadTDR = '1') then TDR <= DBUS;
    end if;
    if (loadTSR = '1') then TSR <= TDR & '1';
    end if;
    if (start = '1') then TSR(0) <= '0';
    end if;
    if (shftTSR = '1') then TSR <= '1' & TSR(8 downto 1);
    end if;
    -- shift out one bit
    Bclk_Dlayed <= Bclk;     -- Bclk delayed by 1 sysclk
  end if;
end process;
end xmit;
```

The operation of the UART receiver is as follows:

**1.** When the UART detects a start bit, it reads in the remaining bits serially and shifts them into the *RSR*.
**2.** When all the data bits and the stop bit have been received, the *RSR* is loaded into the *RDR*, and the Receive Data Register Full (*RDRF*) flag in the *SCSR* is set.
**3.** The microcontroller checks the *RDRF* flag, and if it is set, the *RDR* is read and the flag is cleared.

The bit stream coming in on *RxD* is not synchronized with the local bit clock (*Bclk*). If we attempted to read *RxD* at the rising edge of *Bclk*, we would have a problem if *RxD* changed near the clock edge. We could have setup and hold time problems. If the bit rate of the incoming signal differed from *Bclk* by a small amount, we could end up reading some bits at the wrong time. To avoid these problems, we will sample *RxD* eight times during each bit time. (Some systems sample 16 times per bit.) We will sample on the rising edge of *BclkX8*. The arrows in Figure 11-22 indicate the rising edge of *BclkX8*. Ideally, we should read the bit value at the middle of each bit time for maximum reliability. When *RxD* first goes to '0', we will wait for four *BclkX8* periods, and we should be near the middle of the start bit. Then we will wait eight more *BclkX8* periods, which should take us near the middle of the first data bit. We continue reading once every eight *BclkX8* clocks until we have read the stop bit.

**FIGURE 11-22:**
**Sampling RxD with BclkX8**

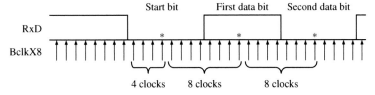

*Read data at these points

Figure 11-23 shows an SM chart for the UART receiver. Two counters are used. *Ct*1 counts the number of *BclkX8* clocks. *Ct*2 counts the number of bits received after the start bit. In the IDLE state, the SM waits for the start bit (*RxD* = '0') and then goes to the Start Detected state. The SM waits for the rising edge of *BclkX8* (*BclkX8↑*) and then samples *RxD* again. Since the start bit should be '0' for eight *BclkX8* clocks, we should read '0'. *Ct*1 is still 0, so *Ct*1 is incremented and the SM waits for *BclkX8↑*. If *RxD* = '1', this is an error condition and the SM clears *Ct*1 and resets to the IDLE state. Otherwise, the SM keeps looping. When *RxD* is '0' for the fourth time, *Ct*1 = 3, so *Ct*1 is cleared and the state goes to Receive Data. In this state, the SM increments *Ct*1 after every rising edge of *BclkX8*. After the eighth clock, *Ct*1 = 7 and *Ct*2 is checked. If it is not 8, the current value of *RxD* is shifted into *RSR*, *Ct*2 is incremented, and *Ct*1 is cleared. If *Ct*2 = 8, all 8 bits have been read and we should be in the middle of the stop bit. If *RDRF* = '1', the microcontroller has not yet read the previously received data byte, and an overrun error has occurred, in which case the *OE* flag in the status register is set and the new data is ignored. If *RxD* = '0', the stop bit has not been detected properly, and the framing error (*FE*) flag in the status register is set. If no errors have occurred, *RDR* is loaded from *RSR*. In all cases, *RDRF* is set to indicate that the receive operation is completed and the counters are cleared.

**FIGURE 11-23: SM Chart for UART Receiver**

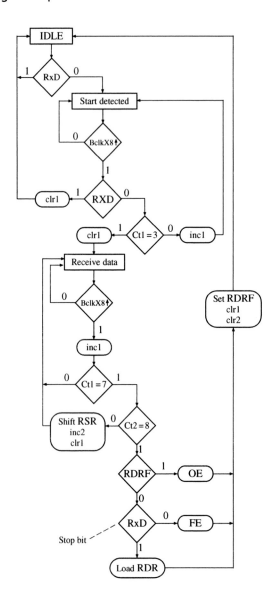

The VHDL code for the UART receiver (Figure 11-24) is based on the SM chart of Figure 11-23. The receiver contains the *RDR* and *RSR* registers and the receive control. The control interfaces with *SCSR*, and *RDR* can drive data onto the data bus. The first process represents the combinational network, which generates the nextstate and control signals. The second process updates the registers on the rising edge of the clock. The signal *BclkX8_rising* is '1' for one system clock time following the rising edge of *BclkX8*. *BclkX8_rising* is generated the same manner as *Bclk_rising*.

FIGURE 11-24: **VHDL Code for UART Receiver**

```
library IEEE;
use IEEE.std_logic_1164.all;
use IEEE.numeric_std.all; -- to use unsigned type

entity UART_Receiver is
port(RxD, BclkX8, sysclk, rst_b, RDRF: in std_logic;
     RDR: out unsigned(7 downto 0);
     setRDRF, setOE, setFE: out std_logic);
end UART_Receiver;

architecture rcvr of UART_Receiver is
  type stateType is (IDLE, START_DETECTED, RECV_DATA);
  signal state, nextstate: stateType;
  signal RSR: unsigned(7 downto 0);   -- receive shift register
  signal ct1 : integer range 0 to 7;  -- indicates when to read the RxD input
  signal ct2 : integer range 0 to 8;  -- counts number of bits read
  signal inc1, inc2, clr1, clr2, shftRSR, loadRDR: std_logic;
  signal BclkX8_Dlayed, BclkX8_rising: std_logic;
begin
  BclkX8_rising <= BclkX8 and (not BclkX8_Dlayed);
    -- indicates the rising edge of bitX8 clock
  Rcvr_Control: process(state, RxD, RDRF, ct1, ct2, BclkX8_rising)
  begin
    -- reset control signals
    inc1 <= '0'; inc2 <= '0'; clr1 <= '0'; clr2 <= '0';
    shftRSR <= '0'; loadRDR <= '0'; setRDRF <= '0'; setOE <= '0'; setFE <= '0';
    case state is
      when IDLE =>
        if (RxD = '0') then nextstate <= START_DETECTED;
        else nextstate <= IDLE;
        end if;
      when START_DETECTED =>
        if (BclkX8_rising = '0') then nextstate <= START_DETECTED;
        elsif (RxD = '1') then clr1 <= '1'; nextstate <= IDLE;
        elsif (ct1 = 3) then clr1 <= '1'; nextstate <= RECV_DATA;
        else inc1 <= '1'; nextstate <= START_DETECTED;
        end if;
      when RECV_DATA =>
        if (BclkX8_rising = '0') then nextstate <= RECV_DATA;
        else inc1 <= '1';
          if (ct1 /= 7) then nextstate <= RECV_DATA;
            -- wait for 8 clock cycles
          elsif (ct2 /= 8) then
            shftRSR <= '1'; inc2 <= '1'; clr1 <= '1'; -- read next data bit
            nextstate <= RECV_DATA;
          else
            nextstate <= IDLE;
            setRDRF <= '1'; clr1 <= '1'; clr2 <= '1';
```

```
                if (RDRF = '1') then setOE <= '1';       -- overrun error
                elsif (RxD = '0') then setFE <= '1';      -- framing error
                else loadRDR <= '1';                      -- load recv data register
                end if;
              end if;
            end if;
      end case;
  end process;

  Rcvr_update: process(sysclk, rst_b)
  begin
    if (rst_b = '0') then state <= IDLE; BclkX8_Dlayed <= '0';
      ct1 <= 0; ct2 <= 0;
    elsif (sysclk'event and sysclk = '1') then
      state <= nextstate;
      if (clr1 = '1') then ct1 <= 0; elsif (inc1 = '1') then
        ct1 <= ct1 + 1;
      end if;
      if (clr2 = '1') then ct2 <= 0; elsif (inc2 = '1') then
        ct2 <= ct2 + 1;
      end if;
      if (shftRSR = '1') then RSR <= RxD & RSR(7 downto 1);
      end if;
      -- update shift reg.
      if (loadRDR = '1') then RDR <= RSR;
      end if;
      BclkX8_Dlayed <= BclkX8;    -- BclkX8 delayed by 1 sysclk
    end if;
  end process;
end rcvr;
```

Figure 11-25 shows the result of synthesizing the UART receiver using the Xilinx Spartan 3 device FPGA series as a target. The resulting implementation requires 26 flip-flops, 21 slices, and 32 four-input LUTs.

Next we will design a programmable baud rate generator. Three bits in the *SCCR* are used to select any one of eight baud rates. We will assume that the system clock is 8 MHz and we want baud rates 300, 600, 1200, 2400, 4800, 9600, 19,200, and 38,400. The maximum *BclkX*8 frequency needed is $38,400 \times 8 = 307,200$. To get this frequency, we should divide 8 MHz by 26.04. Since we can divide only by an integer, we need to either accept a small error in the baud rate or adjust the system clock frequency downward to 7.9877 MHz to compensate.

Figure 11-26 shows a block diagram for the baud rate generator. The 8-MHz system clock is first divided by 13 using a counter. This counter output goes to an 8-bit binary counter. The outputs of the flip-flops in this counter correspond to divide by 2, divide by 4, . . . , and divide by 256. One of these outputs is selected by a multiplexer. The MUX select inputs come from the lower 3 bits of the *SCCR*. The MUX

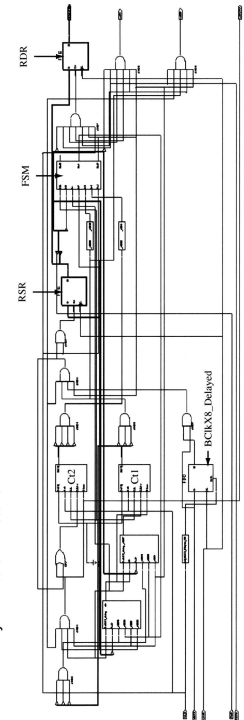

**FIGURE 11-25:  Synthesized UART Receiver**

output corresponds to *BclkX*8, which is further divided by 8 to give *Bclk*. Assuming an 8-MHz clock, the frequencies generated are given by the following table:

| Select Bits | BAUD Rate (*Bclk*) |
|---|---|
| 000 | 38,462 |
| 001 | 19,231 |
| 010 | 9615 |
| 011 | 4808 |
| 100 | 2404 |
| 101 | 1202 |
| 110 | 601 |
| 111 | 300.5 |

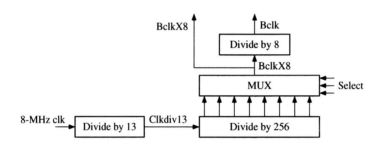

FIGURE 11-26:
Baud Rate
Generator

The VHDL code for the baud rate generator is given in Figure 11-27. The first process increments the divide-by-13 counter on the rising edge of the system clock. The second process increments the divide-by-256 counter on the rising edge of *Clkdiv*13. A concurrent statement generates the MUX output, *BclkX*8. The third process increments the divide-by-8 counter on the rising edge of *BclkX*8 to generate *Bclk*.

FIGURE 11-27: **VHDL Code for Baud Rate Generator**

```
library IEEE;
use IEEE.std_logic_1164.all;
use IEEE.numeric_std.all;    -- for overloaded + operator and conversion functions

entity clk_divider is
  port(Sysclk, rst_b: in std_logic;
       Sel: in unsigned(2 downto 0);
       BclkX8: buffer std_logic;
       Bclk: out std_logic);
end clk_divider;

architecture baudgen of clk_divider is
  signal ctr1: unsigned(3 downto 0) := "0000";      -- divide by 13 counter
  signal ctr2: unsigned(7 downto 0) := "00000000";  -- div by 256 ctr
  signal ctr3: unsigned(2 downto 0) := "000";       -- divide by 8 counter
  signal Clkdiv13: std_logic;
```

```
begin
  process(Sysclk)   -- first divide system clock by 13
  begin
    if (Sysclk'event and Sysclk = '1') then
      if (ctr1 = "1100") then ctr1 <= "0000";
      else ctr1 <= ctr1 + 1;
      end if;
    end if;
  end process;
  Clkdiv13 <= ctr1(3);    -- divide Sysclk by 13

  process(Clkdiv13)   -- ctr2 is an 8-bit counter
  begin
    if (Clkdiv13'event and Clkdiv13 = '1') then
      ctr2 <= ctr2 + 1;
    end if;
  end process;

  BclkX8 <= ctr2(to_integer(sel));   -- select baud rate
  process(BclkX8)
  begin
    if (BclkX8'event and BclkX8 = '1') then
      ctr3 <= ctr3 + 1;
    end if;
  end process;
  Bclk <= ctr3(2);    -- Bclk is BclkX8 divided by 8
end baudgen;
```

To complete the UART design, we need to interconnect the three components we have designed, connect them to the control and status registers, and add the interrupt generation logic and the bus interface. Figure 11-28 gives the VHDL code for the complete UART.

FIGURE 11-28: **VHDL Code for Complete UART**

```
library IEEE;
use IEEE.std_logic_1164.all;
use IEEE.numeric_std.all;

entity UART is
  port(SCI_sel, R_W, clk, rst_b, RxD: in std_logic;
       ADDR2: in unsigned(1 downto 0);
       DBUS: inout unsigned(7 downto 0);
       SCI_IRQ, TxD: out std_logic);
end UART;

architecture uart1 of UART is
  component UART_Receiver
    port(RxD, BclkX8, sysclk, rst_b, RDRF: in std_logic;
```

```vhdl
            RDR: out unsigned(7 downto 0);
            setRDRF, setOE, setFE: out std_logic);
    end component;
    component UART_Transmitter
      port(Bclk, sysclk, rst_b, TDRE, loadTDR: in std_logic;
            DBUS: in unsigned(7 downto 0);
            setTDRE, TxD: out std_logic);
    end component;
    component clk_divider
      port(Sysclk, rst_b: in std_logic;
            Sel: in unsigned(2 downto 0);
            BclkX8: buffer std_logic; Bclk: out std_logic);
    end component;
    signal RDR: unsigned(7 downto 0);    -- Receive Data Register
    signal SCSR: unsigned(7 downto 0);   -- Status Register
    signal SCCR: unsigned(7 downto 0);   -- Control Register
    signal TDRE, RDRF, OE, FE, TIE, RIE: std_logic;
    signal BaudSel: unsigned(2 downto 0);
    signal setTDRE, setRDRF, setOE, setFE, loadTDR, loadSCCR: std_logic;
    signal clrRDRF, Bclk, BclkX8, SCI_Read, SCI_Write: std_logic;
begin
    RCVR: UART_Receiver port map(RxD, BclkX8, clk, rst_b, RDRF, RDR,
                                setRDRF, setOE, setFE);
    XMIT: UART_Transmitter port map(Bclk, clk, rst_b, TDRE, loadTDR,
                                DBUS, setTDRE, TxD);
    CLKDIV: clk_divider port map(clk, rst_b, BaudSel, BclkX8, Bclk);
    -- This process updates the control and status registers
    process(clk, rst_b)
    begin
      if (rst_b='0') then
        TDRE <= '1'; RDRF <= '0'; OE <= '0'; FE <= '0';
        TIE <= '0'; RIE <= '0';
      elsif (rising_edge(clk)) then
        TDRE <= (setTDRE and not TDRE) or (not loadTDR and TDRE);
        RDRF <= (setRDRF and not RDRF) or (not clrRDRF and RDRF);
        OE <= (setOE and not OE) or (not clrRDRF and OE);
        FE <= (setFE and not FE) or (not clrRDRF and FE);
        if (loadSCCR = '1') then TIE <= DBUS(7); RIE <= DBUS(6);
          BaudSel <= DBUS(2 downto 0);
        end if;
      end if;
    end process;

    -- IRQ generation logic
    SCI_IRQ <= '1' when ((RIE='1' and (RDRF='1' or OE='1')) or
                        (TIE='1' and TDRE='1'))
                    else '0';
    -- Bus Interface
    SCSR <= TDRE & RDRF & "0000" & OE & FE;
    SCCR <= TIE & RIE & "000" & BaudSel;
```

```
    SCI_Read <= '1' when (SCI_sel = '1' and R_W = '0') else '0';
    SCI_Write <= '1' when (SCI_sel = '1' and R_W = '1') else '0';
    clrRDRF <= '1' when (SCI_Read = '1' and ADDR2 = "00") else '0';
    loadTDR <= '1' when (SCI_Write = '1' and ADDR2 = "00") else '0';
    loadSCCR <= '1' when (SCI_Write = '1' and ADDR2 = "10") else '0';
    DBUS <= "ZZZZZZZZ" when (SCI_Read = '0')    -- tristate bus when not reading
            else RDR when (ADDR2 = "00")
    -- write appropriate register to the bus
            else SCSR when (ADDR2 = "01")
            else SCCR;    -- dbus = sccr, if ADDR2 is "10" or "11"
end uart1;
```

*SCI_IRQ* is an interrupt signal that interrupts the CPU when the UART receiver or transmitter needs attention. When the *RIE* (receive interrupt enable) is set in *SCCR, SCI_IRQ* is generated whenever *RDRF* or *OE* is '1'. When *TIE* (transmit interrupt enable) is set in *SCCR, SCI_IRQ* is generated whenever *TDRE* is '1'.

The UART is interfaced to microcontroller address and data buses so that the CPU can read and write to the UART registers when the UART is selected by *SCIsel* = '1'. The last two bits of the address (*ADDR2*), together with the *R_W* signal, are used for register selection as follows:

| ADDR2 | R_W | Action |
|-------|-----|--------|
| 00 | 0 | DBUS ← RDR |
| 00 | 1 | TDR ← DBUS |
| 01 | 0 | DBUS ← SCSR |
| 01 | 1 | DBUS ← hi-Z |
| 1– | 0 | DBUS ← SCCR |
| 1– | 1 | SCCR ← DBUS |

When the UART is not selected for reading, the data bus is driven to high-Z.

The VHDL code in Figure 11-28 was synthesized using the Xilinx SPARTAN 3 series FPGA as a target. The resulting implementation required 62 slices, 109 four-input LUTs, and 74 flip-flops.

This chapter presented three examples for the use of VHDL in design and simulation of digital systems. Two design examples, a wristwatch and a UART, and a simulation example, a memory chip, were presented. In the design examples, we first developed a block diagram for the design and state machine charts representing the controller of the system. Then, we presented behavioral VHDL models for the various blocks in the system. Use of test benches is illustrated. The VHDL code was then synthesized for FPGAs. Designs were downloaded and operation verified.

We also presented a simulation model for a memory chip. This model included timing parameters for the memory chip and built-in checks to verify that setup and hold times and other timing specifications are met. Such models are helpful when third party cores/chips are utilized during system on a chip (SoC) design.

● ● ● ● ● ● ● ● ● ● ● ●

# Problems

**11.1** Assume that you are implementing the wristwatch design from Section 11.1 on an FPGA board. Design the input module for the wristwatch for the FPGA board that you have and write the VHDL code.

**11.2** Assume that you are implementing the wristwatch design from Section 11.1 on an FPGA board. Design the display module for the wristwatch and write the VHDL code. Use an FPGA board with an LCD display. Display the time, the alarm setting, or the stopwatch time depending on which mode the wristwatch is in.

**11.3** **(a)** Add a count-down timer mode to the wristwatch module of Figures 11-2 and 11-3. The timer should count seconds, minutes, and hours. When in the timer state, B2 should change states to allow setting the hours, minutes, and seconds with B3. When setting is complete and the wristwatch is back in the main timer state, B3 should start the count-down. If B3 is pressed again, it should stop the count-down; otherwise, the count-down stops when it reaches 00:00:00, in which case the timer beeps for one second.
**(b)** Modify the test bench and test your timer.

**11.4** The problem concerns the design of a simple calculator for adding unsigned binary numbers. Operation is similar to a simple hand-held calculator, except all inputs and outputs are in binary, and the only operation is +. The calculator displays 8 bits with a binary point. The calculator has only five keys: 0, 1, . , +, and reset. Reset clears all registers and resets the calculator to the starting state. After entering the first number, the + key terminates that entry and allows a second number to be entered. When + is pushed again, the sum is put in the accumulator, and another number can be entered. This continues until the calculator is reset. Note that there is no equals key. You may assume that only normal input sequences occur, that is, a number will always be entered each time before + is pressed. Before addition can be done, the binary points of the numbers to be added must be aligned by shifting. If addition produces an overflow, the overflow should be corrected if possible. If not, set $E = 1$ to indicate an error.

The keys are not encoded. The calculator has six input signals: *zero, one, dot, plus, reset*, and *V*. Assume that all input signals are debounced, and $V = 1$ for one clock time whenever a key is pressed. Outputs to the display are 8 bits from the *A* register, *RCTA* (the number of bits to the right of the binary point), and *E*.

**(a)** Draw a block diagram for the calculator showing required registers, counters, adders, and so on. Show the necessary control signals and tell what they mean. For example, *RSHA* means right shift *A*. Specify the size of each register.
**(b)** Draw an SM chart for the main calculator code. Include inputting the binary numbers, aligning the binary points, adding, and correcting for overflow if possible. Define all control signals used.

(c) Write VHDL code for the main calculator module.

(d) Write a test bench for your VHDL module.

11.5 This question refers to static RAM read and write cycles (refer to Figures 11-11 and 11-12). Answer this question in general, not for any specific set of numeric values.

(a) If $\overline{WE} = 1$, and the address changes at the same time $\overline{CS}$ goes to 0, what is the maximum time before valid data is available at the RAM output? (*Note*: The timing diagrams are not drawn to scale).

(b) What determines the maximum number of bytes per second that can be read from the RAM? State any assumptions which you make.

(c) For a $\overline{WE}$-controlled write cycle, what is the normal sequence of events which occur when writing to RAM?

(d) State clearly what timing conditions must be satisfied in order to correctly write data to the RAM. For example, $\overline{WE}$ must be 0 for at least $t_{wp}$.

11.6 Answer the following questions for the 6116 SA-15 static CMOS RAM. Refer to the timing specifications in Table 11-1.

(a) What is the maximum clock frequency that can be used?

(b) What is the minimum time after a change in address or $\overline{CS}$ at which valid data can be read?

(c) For a $\overline{WE}$-controlled write cycle, what is the earliest time new data can be driven after $\overline{WE}$ goes low?

(d) For a write cycle, what is the minimum time that valid data must be driven onto the data bus?

11.7 This problem concerns a simplified memory model for a 6116 CMOS RAM. Assume that both $\overline{CS}$ and $\overline{OE}$ are always low, so memory operation depends only on the address and $\overline{WE}$.

(a) Write a simple VHDL model for the memory that ignores all timing information. (Your model should not contain $\overline{CS}$ or $\overline{OE}$.)

(b) Add the following timing specs to your model: $t_{AA}$, $t_{OH}$, $t_{WHZ}$, and $t_{OW}$. For reads, *Dout* should go to "XXXXXXXX" (unknown) after $t_{OH}$ and then to valid data out after $t_{AA}$. For writes, *Dout* should go to high-Z after $t_{WHZ}$, and it should go to the value just stored after $t_{OW}$.

(c) Add another process that gives appropriate error messages if any of the following specs are not satisfied: $t_{WP}$, $t_{DW}$, and $t_{DH}$.

11.8 A VHDL model that describes the operation of the 6116 memory is given in Figure 11-14.

(a) Verify that the code will report a warning if the data setup time for writing to memory is not met, if the data hold time for writing to memory is not met, or if the minimum pulse width spec for *WE_b* is not met.

**(b)** Indicate the changes and additions to the original VHDL code that are necessary if *OE_b* ($\overline{\text{OE}}$) is taken into account. Note that for reads, if *OE_b* goes low after *CS_b* goes low, the $t_{OE}$ access time must be considered. Also note that when *OE_b* goes high, the data bus will go high-Z after time $t_{OHZ}$.

**11.9** What modifications must be made in the check process in the VHDL 6116 RAM timing model (Figure 11-14) in order to verify the address setup time ($t_{AS}$) and the write recovery time ($t_{WR}$) specifications?

**11.10** Consider the *CS*-controlled write cycle for a static CMOS RAM (Figure 11-13). What VHDL code is needed in the check process in the timing model (Figure 11-14) to verify the correct operation of a *CS*-controlled write? You must check timing specifications such as $t_{CW}$, $t_{DW}$, and $t_{DH}$.

**11.11** A ROM (read-only memory) has an 8-bit address input, an output enable (*OE*), and an 8-bit data output. When *OE* = 0, *data* = hi-Z; when *OE* = 1, *data* is read from the ROM. Timing diagrams are shown below. Write a VHDL model for the ROM that includes the timing specifications.

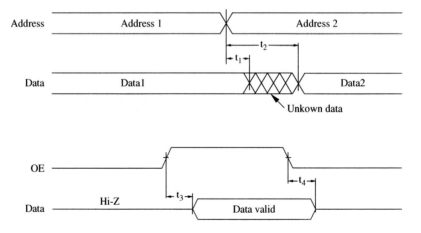

**11.12** A static RAM memory uses a $\overline{\text{WE}}$ controlled write cycle as shown in the figure. This memory has a negative data hold time with a magnitude $t_{hn}$. This means that as long as the setup time ($t_{dw}$) is satisfied, it is okay for the input data to change anytime during the interval $t_{hn}$ before the rising edge of $\overline{\text{WE}}$. Write a process that will report an error if the input data (*Dbus*) changes at any time during the time interval $t_{dw}$ to $t_{hn}$ before $\overline{\text{WE}}$ rises.

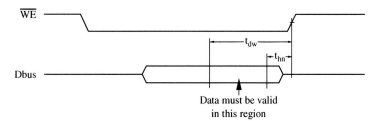

Data must be valid
in this region

11.13 Make necessary changes in the UART receiver VHDL code so that it uses a 16X bit clock instead of an 8X bit clock. Using a faster sampling clock can improve the noise immunity of the receiver.

11.14 **(a)** Write a VHDL test bench for the UART. Include cases to test overrun error, framing error, noise causing a false start, change of BAUD rate, and so on. Simulate the VHDL code.

**(b)** If suitable hardware is available, write a simpler test bench to allow a loop-back test with *TxD* externally connected to *RxD*. Synthesize the test bench along with the UART, download to the target device, and verify correct operation of the hardware.

11.15 Make necessary changes to the VHDL code to add a parity option to the UART described in Section 11.3. Add 2 bits $(P_1 P_0)$ to the SCCR that select the parity mode as follows:

$P_1 P_0 = 00$    8 data bits, no parity bit
$P_1 P_0 = 01$    7 data bits, 8th bit makes parity even
$P_1 P_0 = 10$    7 data bits, 8th bit makes parity odd
$P_1 P_0 = 11$    7 data bits, 8th bit is always '0'

The transmitter should generate the even, odd, or '0' parity bit as specified. The receiver should check the parity bit to verify that it is correct. If not, it should set a PE (parity error) flag in the SCSR.

11.16 The operation of a synchronous receiver is somewhat similar to the UART receiver discussed in Section 11.3, except both data (*RxD*) and a data clock (*Dclk*) are transmitted so there is no need to synchronize data with a local clock, and no start and stop bits are required. As shown below, when 8 bits of data are transmitted, the clock is actually active for nine clock times and then it becomes inactive. On the first eight clocks data is shifted into the receive shift register (RSR), and on the ninth clock, the data is transferred to the receive data register (RDR) and the RDRF flag is set.

**(a)** Draw a block diagram for the synchronous receiver, including a counter. (*Note*: A state machine is not necessary, but generation of control signals *Load* and *Shift* is required.)

**(b)** Write synthesizable VHDL code that corresponds to (a). Signals *Load* and *Shift* should appear explicitly in your code.

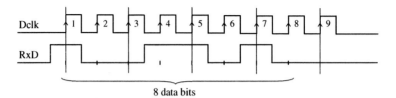

8 data bits

11.17 Write a test bench for the UART that performs a loop-back test. The test bench connects the *TxD* output of the UART to the *RxD* input so that any data loaded into TDR will automatically be transmitted from *TxD*, received into *RxD*, and loaded into RDR. The test bench should simulate the action of a CPU that writes "01010101" to TDR, reads the status register in a loop until RDRF = '1', and then reads from RDR.

# VHDL Language Summary

Disclaimer: This VHDL summary is not complete and contains some special cases. Only VHDL statements used in this text are listed. For a complete description of VHDL syntax, refer to References 6, 9, and 23.

Notes:

- VHDL is not case sensitive.
- Signal names and other identifiers may contain letters, numbers, and the underscore (_) character.
- An identifier must start with a letter.
- An identifier cannot end with an underscore.
- Every VHDL statement must be terminated with a semicolon.
- VHDL is a strongly typed language. In general, mixing of data types is not allowed.

**LEGEND**

| | |
|---|---|
| **bold** | reserved word |
| [ ] | optional items |
| { } | repeated zero or more times |
| \| | or |

## 1. Predefined Types

| | |
|---|---|
| bit | '0' or '1' |
| boolean | FALSE or TRUE |
| integer | an integer in the range $-(2^{31} - 1)$ to $+(2^{31} - 1)$ (some implementations support a wider range) |
| real | floating-point number in the range $-1.0E38$ to $+1.0E38$ |
| character | any legal VHDL character including upper- and lowercase letters, digits, and special characters (each printable character must be enclosed in single quotes; e.g., 'd', '7', '+') |
| time | an integer with units fs, ps, ns, us, ms, sec, min, or hr |
| natural | integers $\geq 0$ |
| positive | integers $> 0$ |

| | |
|---|---|
| bit_vector | array of bits |
| string | array of characters |
| delay_length | time $\geq 0$ |

## 2. Operators By Increasing Precedence

**1.** Binary logical operators:    **and or nand nor xor xnor**
**2.** Relational operators:      $= /= < <= > >=$
**3.** Shift operators:        **sll srl sla sra rol ror**
**4.** Adding operators:       $+ - \&$ (concatenation)
**5.** Unary sign operators:     $+ -$
**6.** Multiplying operators:     **\* / mod rem**
**7.** Miscellaneous operators:    **not abs \*\***

## 3. Predefined Attributes

*Signal attributes that return a value*:

| Attribute | Returns |
|---|---|
| S'ACTIVE | true if a transaction occurred during the current delta, else false |
| S'EVENT | true if an event occurred during the current delta, else false |
| S'LAST_EVENT | time elapsed since the previous event on S |
| S'LAST_VALUE | value of S before the previous event on S |
| S'LAST_ACTIVE | time elapsed since previous transaction on S |

*Signal attributes that create a signal*:

| Attribute | Creates |
|---|---|
| S'DELAYED [(time)]\* | signal same as S delayed by specified time |
| S'STABLE [(time)]\* | boolean signal that is true if S had no events for the specified time |
| S'QUIET [(time)]\* | boolean signal that is true if S had no transactions for the specified time |
| S'TRANSACTION | signal of type bit that changes for every transaction on S |

\*Delta is used if no time is specified.

*Array attributes*:
```
type ROM is array (0 to 15, 7 downto 0) of bit;
signal ROM1 : ROM;
```

| Attribute | Returns | Examples |
|---|---|---|
| A'LEFT(N) | left bound of Nth index range | ROM1'LEFT(1) = 0<br>ROM1'LEFT(2) = 7 |
| A'RIGHT(N) | right bound of Nth index range | ROM1'RIGHT(1) = 15<br>ROM1'RIGHT(2) = 0 |
| A'HIGH(N) | largest bound of Nth index range | ROM1'HIGH(1) = 15<br>ROM1'HIGH(2) = 7 |
| A'LOW(N) | smallest bound of Nth index range | ROM1'LOW(1) = 0<br>ROM1'LOW(2) = 0 |
| A'RANGE(N) | Nth index range | ROM1'RANGE(1) = 0 to 15<br>ROM1'RANGE(2) = 7 downto 0 |
| A'REVERSE_RANGE(N) | Nth index range<br><br>reversed | ROM1'REVERSE_RANGE(1) =<br>    15 downto 0<br>ROM1'REVERSE_RANGE(2) =<br>    0 to 7 |
| A'LENGTH(N) | size of Nth index range | ROM1'LENGTH(1) = 16<br>ROM1'LENGTH(2) = 8 |

## 4. Predefined Functions

| | |
|---|---|
| NOW | returns current simulation time |
| FILE_OPEN([status], FileID, string, mode) | open file |
| FILE_CLOSE(FileID) | close file |

## 5. Declarations

*entity declaration*:
```
entity entity-name is
        [generic (list-of-generics-and-their-types);]
        [port (interface-signal-declaration);]
        [declarations]
end [entity] [entity-name];
```

*interface-signal declaration*:
```
list-of-interface-signals: mode type [:= initial-value]
{; list-of-interface-signals: mode type [:= initial-value]}
```

*Note*: An interface signal can be of mode in, out, inout, or buffer.

*architecture declaration*:
```
architecture architecture-name of entity-name is
        [declarations]      -- variable declarations not allowed
begin
        architecture-body
end [architecture] [architecture-name];
```

*Note*: The architecture body may contain component-instantiation statements, processes, blocks, assignment statements, procedure calls, etc.

*integer type declaration*:
```
type type_name is range integer_range;
```

*enumeration type declaration*:
```
type type_name is (list-of-names-or-characters);
```

*subtype declaration*:
```
subtype subtype_name is type_name [index-or-range-constraint];
```

*variable declaration*:
```
variable list-of-variable-names: type_name [:= initial_value];
```

*signal declaration*:
```
signal list-of-signal-names: type_name [:= initial_value];
```

*constant declaration*:
```
constant constant_name: type_name := constant_value;
```

*alias declaration*:
```
alias identifier[:identifier-type] is item-name;
```
*Note*: Item-name can be a constant, signal, variable, file, function name, type name, etc.

*array type and object declaration*:
```
type array_type_name is array index_range of element_type;
signal|variable|constant array_name: array_type_name
[:= initial_values];
```

*procedure declaration*:
```
procedure procedure-name (parameter list) is
        [declarations]
begin
        sequential statements
end procedure-name;
```
*Note*: Parameters may be signals, variables, or constants.

*function declaration*:
```
function function-name (parameter-list) return return-type is
        [declarations]
begin
        sequential statements           -- must include return
                                        return-value;
end function-name;
```
*Note*: Parameters may be signals or constants.

*library declaration*:
```
library list-of-library-names;
```


*use statement*:
```
use library_name.package_name.item;            (.item may be .all)
```


*package declaration*:
```
package package-name is
        package declarations
end [package][package-name];
```


*package body*:
```
package body package-name is
        package body declarations
end [package body][package name];
```


*component declaration*:
```
component component-name
        [generic (list-of-generics-and-their-types);]
        port (list-of-interface-signals-and-their-types);
end component;
```


*file type declaration*:
```
type file_name is file of type_name;
```


*file declaration*:
```
file file_name: file_type [open mode] is "file_pathname";
```

*Note*: Mode may be read_mode, write_mode, or append_mode.


## 6. Concurrent Statements

*signal assignment statement*:
```
signal <= [transport | [reject pulse-width] inertial] expression
        [after delay_time];
```

*Note*: If signal assignment done as concurrent statement, signal value is recomputed every time a change occurs on the right-hand side. If [**after** delay_time] is omitted, signal is updated after delta time.


*conditional assignment statement*:
```
signal <= expression1 when condition1
        else expression2 when condition2
        . . .
        [else expression];
```

*selected signal assignment statement:*
```
    with expression select
            signal <= expression1 [after delay_time] when choice1,
                      expression2 [after delay_time] when choice2,
                      . . .
                      [expression [after delay_time] when others];
```

*assert statement:*
```
    assert boolean-expression
            [report string-expression]
            [severity severity-level];
```

*component instantiation:*
```
    label: component-name
            [generic map (generic-association-list);]
            port map (list-of-actual-signals);
```

*Note*: Use **open** if a component output has no connection

*generate statements:*
```
    generate_label: for identifier in range generate
    [begin]
            concurrent statement(s)
    end generate [generate_label];
```
```
    generate_label: if condition generate
    [begin]
            concurrent statement(s)
    end generate [generate_label];
```

*process statement (with sensitivity list):*
```
    [process-label:] process (sensitivity-list)
            [declarations]        -- signal declarations not allowed
    begin
            sequential statements
    end process [process-label];
```
*Note*: This form of process is executed initially and thereafter only when an item on the sensitivity list changes value. The sensitivity list is a list of signals. No wait statements are allowed.

*process statement (without sensitivity list):*
```
    [process-label:] process
            [declarations]        -- signal declarations not allowed
    begin
            sequential statements
    end process [process-label];
```
*Note*: This form of process must contain one or more wait statements. It starts execution immediately and continues until a wait statement is encountered.

*procedure call*:
```
procedure-name (actual-parameter-list);
```

*Note*: An expression may be used for an actual parameter of mode in; types of the actual parameters must match the types of the formal parameters; open cannot be used.

*function call*:
```
function-name (actual-parameter list)
```

*Note*: A function call is used within (or in place of) an expression. Function call is not a statement by itself, it is part of a statement.

## 7. Sequential Statements

*signal assignment statement*:
```
signal <= [transport | [reject pulse-width] inertial] expression
          [after delay_time];
```

*Note*: If [**after** delay_time] is omitted, signal is updated after delta time.

*variable assignment statement*:
```
variable := expression;
```

*Note*: This can be used only within a process, function, or procedure. The variable is always updated immediately.

*wait statements can be of the form*:
```
wait on sensitivity-list;
wait until boolean-expression;
wait for time-expression;
```

*if statement*:
```
if condition then
        sequential statements
{elsif condition then
        sequential statements}   -- 0 or more elsif clauses may
                                     be included
[else sequential statements]
end if;
```

*case statement*:
```
case expression is
        when choice1 => sequential statements
        when choice2 => sequential statements
        . . .
        [when others => sequential statements]
end case;
```

*for loop statement*:
```
[loop-label:] for identifier in range loop
        sequential statements
end loop [loop-label];
```

*Note*: You may use **exit** to exit the current loop.

*while loop statement*:
```
[loop-label:] while boolean-expression loop
        sequential statements
end loop [loop-label];
```

*exit statement*:
```
exit [loop-label] [when condition];
```

*assert statement*:
```
assert boolean-expression
        [report string-expression]
        [severity severity-level];
```

*report statement*:
```
report string-expression
        [severity severity-level];
```

*procedure call*:
```
procedure-name (actual-parameter-list);
```

*Note*: An expression may be used for an actual parameter of mode in; types of the actual parameters must match the types of the formal parameters; open cannot be used.

*function call*:
```
function-name (actual-parameter list)
```

*Note*: A function call is used within (or in place of) an expression. Function call is not a statement by itself, it is part of a statement.

# IEEE Standard Libraries

The two packages from the IEEE libraries that we have used in the book are NUMERIC_BIT and NUMERIC_STD. The headers of these packages read as follows:

**Standard VHDL Synthesis Package (1076.3, NUMERIC_BIT)**

```
-- Developers: IEEE DASC Synthesis Working Group, PAR 1076.3
-- Purpose: This package defines numeric types and arithmetic functions
--          :for use with synthesis tools. Two numeric types are defined:
--          :--> UNSIGNED: represents an UNSIGNED number in vector form
--          :--> SIGNED: represents a SIGNED number in vector form
--          :The base element type is type BIT.
--          :The leftmost bit is treated as the most significant bit.
--          :Signed vectors are represented in two's complement form.
--          :This package contains overloaded arithmetic operators on
--          :the SIGNED and UNSIGNED types. The package also contains
--          :useful type conversions functions, clock detection
--          :functions, and other utility functions.
```

**Standard VHDL Synthesis Package (1076.3, NUMERIC_STD)**

```
-- Developers: IEEE DASC Synthesis Working Group, PAR 1076.3
-- Purpose: This package defines numeric types and arithmetic functions
--          :for use with synthesis tools. Two numeric types are defined:
--          :--> UNSIGNED: represents UNSIGNED number in vector form
--          :--> SIGNED: represents a SIGNED number in vector form
--          :The base element type is type STD_LOGIC.
--          :The leftmost bit is treated as the most significant bit.
--          :Signed vectors are represented in two's complement form.
--          :This package contains overloaded arithmetic operators on
--          :the SIGNED and UNSIGNED types. The package also contains
--          :useful type conversions functions.
```

The entire package listings can be viewed at

http://www.eda.org/rassp/vhdl/models/standards/numeric_bit.vhd
http://www.eda.org/rassp/vhdl/models/standards/numeric_std.vhd

**Useful conversion functions in the numeric_bit package:**

TO_INTEGER(A): converts an unsigned (or signed) vector $A$ to an integer
TO_UNSIGNED(B,N): converts an integer to an unsigned vector of length $N$
TO_SIGNED(B,N): converts an integer to an signed vector of length $N$
UNSIGNED(A): causes the compiler to treat a bit_vector $A$ as an unsigned vector
SIGNED(A): causes the compiler to treat a bit_vector $A$ as a signed vector
BIT_VECTOR(B): causes the compiler to treat an unsigned (or signed) vector $B$ as a bit_vector

The same conversion functions are available in the numeric_std package, except replace bit_vector with std_logic_vector.

**Notes:**

**1.** The numeric_bit package provides an overloaded operator to add an integer to an unsigned, but not to add a bit to an unsigned type. Thus, if $A$ and $B$ are unsigned, A+B+1 is allowed, but a statement of the form

```
Sum <= A + B + carry;
```

is not allowed when carry is of type bit. The carry must be converted to unsigned before it can be added to the unsigned vector A+B. The notation **unsigned'(0=>carry)** will accomplish the necessary conversion. Use the statement

```
Sum <= A + B + unsigned'(0=>carry);
```

**2.** If we want more bits in the sum than there are in the numbers being added, we must extend the numbers by concatenating '0'. For example, if $X$ and $Y$ are 4 bits, and a 5-bit sum including the carry out is desired, extend $X$ to 5 bits by concatenating '0' and $X$. ($Y$ will automatically be extended to match.) Hence:

```
Sum5 <= '0' & X + Y;
```

accomplishes the addition of two 4-bit numbers and provides a 5-bit sum.

# Textio Package

```
package TEXTIO is
  -- Type definitions for text I/O:
  type LINE is access STRING; -- A LINE is a pointer to a STRING value.
  -- The predefined operators for this type are as follows:
  -- function "=" (anonymous, anonymous: LINE) return BOOLEAN;
  -- function "/=" (anonymous, anonymous: LINE) return BOOLEAN;
  type TEXT is file of STRING; -- A file of variable-length ASCII records.
  -- The predefined operators for this type are as follows:
  -- procedure FILE_OPEN (file F: TEXT; External_Name; in STRING;
  -- Open_Kind: in FILE_OPEN_KIND := READ_MODE);
  -- procedure FILE_OPEN (Status: out FILE_OPEN_STATUS; file F: TEXT;
  -- External_Name: in STRING;
  -- Open_Kind: in FILE_OPEN_KIND := READ_MODE);
  -- procedure FILE_CLOSE (file F: TEXT);
  -- procedure READ (file F: TEXT; VALUE: out STRING);
  -- procedure WRITE (file F: TEXT; VALUE: in STRING);
  -- function ENDFILE (file F: TEXT) return BOOLEAN;
  type SIDE is (RIGHT, LEFT); -- For justifying output data within fields.
  -- The predefined operators for this type are as follows:
  -- function "=" (anonymous, anonymous: SIDE) return BOOLEAN;
  -- function "/=" (anonymous, anonymous: SIDE) return BOOLEAN;
  -- function "<" (anonymous, anonymous: SIDE) return BOOLEAN;
  -- function "<=" (anonymous, anonymous: SIDE) return BOOLEAN;
  -- function ">" (anonymous, anonymous: SIDE) return BOOLEAN;
  -- function ">=" (anonymous, anonymous: SIDE) return BOOLEAN;
  subtype WIDTH is NATURAL; -- For specifying widths of output fields.

  -- Standard text files:
  file INPUT: TEXT open READ_MODE is "STD_INPUT";
  file OUTPUT: TEXT open WRITE_MODE is "STD_OUTPUT";
  -- Input routines for standard types:
  procedure READLINE (file F: TEXT; L: inout LINE);
  procedure READ (L: inout LINE; VALUE: out BIT; GOOD: out BOOLEAN);
  procedure READ (L: inout LINE; VALUE: out BIT);
  procedure READ (L: inout LINE; VALUE: out BIT_VECTOR; GOOD: out BOOLEAN);
  procedure READ (L: inout LINE; VALUE: out BIT_VECTOR);
```

```
procedure READ (L: inout LINE; VALUE: out BOOLEAN; GOOD: out BOOLEAN);
procedure READ (L: inout LINE; VALUE: out BOOLEAN);
procedure READ (L: inout LINE; VALUE: out CHARACTER; GOOD: out BOOLEAN);
procedure READ (L: inout LINE; VALUE: out CHARACTER);
procedure READ (L: inout LINE; VALUE: out INTEGER; GOOD: out BOOLEAN);
procedure READ (L: inout LINE; VALUE: out INTEGER);
procedure READ (L: inout LINE; VALUE: out REAL; GOOD: out BOOLEAN);
procedure READ (L: inout LINE; VALUE: out REAL);
procedure READ (L: inout LINE; VALUE: out STRING; GOOD: out BOOLEAN);
procedure READ (L: inout LINE; VALUE: out STRING);
procedure READ (L: inout LINE; VALUE: out TIME; GOOD: out BOOLEAN);
procedure READ (L: inout LINE; VALUE: out TIME);
-- Output routines for standard types:
procedure WRITELINE (file F: TEXT; L: inout LINE);
procedure WRITE (L: inout LINE; VALUE: in BIT;
                 JUSTIFIED: in SIDE:= RIGHT; FIELD: in WIDTH := 0);
procedure WRITE (L: inout LINE; VALUE: in BIT_VECTOR;
                 JUSTIFIED: in SIDE:= RIGHT; FIELD: in WIDTH := 0);
procedure WRITE (L: inout LINE; VALUE: in BOOLEAN;
                 JUSTIFIED: in SIDE:= RIGHT; FIELD: in WIDTH := 0);
procedure WRITE (L: inout LINE; VALUE: in CHARACTER;
                 JUSTIFIED: in SIDE:= RIGHT; FIELD: in WIDTH := 0);
procedure WRITE (L: inout LINE; VALUE: in INTEGER;
                 JUSTIFIED: in SIDE:= RIGHT; FIELD: in WIDTH := 0);
procedure WRITE (L: inout LINE; VALUE: in REAL;
                 JUSTIFIED: in SIDE:= RIGHT; FIELD: in WIDTH := 0;
                 DIGITS: in NATURAL:= 0);
procedure WRITE (L: inout LINE; VALUE: in STRING;
                 JUSTIFIED: in SIDE:= RIGHT; FIELD: in WIDTH := 0);
procedure WRITE (L: inout LINE; VALUE: in TIME;
                 JUSTIFIED: in SIDE:= RIGHT; FIELD: in WIDTH := 0;
                 UNIT: in TIME:= ns);
-- File position predicate:
-- function ENDFILE (file F: TEXT) return BOOLEAN;
end TEXTIO;
```

D

# Projects

For each of these projects, choose an appropriate FPGA or CPLD as a target device and carry out the following steps:

1. Work out an overall design strategy for the system and draw block diagrams. Divide the system into modules if appropriate. Develop an algorithm, SM charts, or state graphs as appropriate for each module. Unless otherwise specified, your design should be a synchronous system with appropriate circuits added to synchronize the inputs with the clock.
2. Write synthesizable VHDL code for each module, simulate it, and debug it. To avoid timing problems in the hardware, use signals instead of variables and make sure the code synthesizes without latches. Use test benches when appropriate to verify correct operation of each module.
3. Integrate the VHDL code for the modules, simulate, and test the overall system.
4. Make any needed changes and synthesize the VHDL code for the target device. Simulate the system after synthesis.
5. Generate a bit file for the target device and download it. Verify that the hardware works correctly.

## P1. Push-Button Door Lock

Design a push-button door lock that uses a standard telephone keypad as input. Use the keypad scanner designed in Chapter 4 as a module. The length of the combination is 4 to 7 digits. To unlock the door, enter the combination followed by the # key. As long as # is held down, the door will remain unlocked and can be opened. When # is released, the door is relocked. To change the combination, first enter the correct combination followed by the * key. The lock is then in the "store" mode. The "store" indicator light comes on and remains on until the combination has been successfully changed. Next enter the new combination (4 to 7 digits) followed by #. Then enter the new combination a second time followed by #. If the second time does not match the first time, the new combination must be entered two times again. Store the combination in an array of eight 4-bit registers or in a small RAM. Store the 4-bit key codes followed by the code for the # key. Also provide a reset button that is not part of the keypad. When the reset button is pushed, the system enters the "store" state and a new combination may be entered. Use a separate counter for counting the inputs as they come in. A 4-bit code, a key-down signal (*Kd*), and a valid data signal (*V*) are available from the keypad module.

### P2. Synchronous Serial Peripheral Interface

Design an SPI (synchronous serial peripheral interface) module suitable for use with a microcontroller. The SPI allows synchronous serial communication with peripheral devices or with other microcontrollers. The SPI contains four registers— *SPCR* (SPI control), *SPSR* (SPI status), *SPDR* (SPI data), and *SPSHR* (SPI shift register). The following diagram shows how two SPIs can be connected for serial communications. One SPI operates as a master and one as a slave. The master provides the clock for synchronizing transmit and receive operations. When a byte of data is loaded into the master *SPSHR*, it initiates serial transmission and supplies a serial clock (*SCK*). Data is exchanged between the master and slave shift registers in eight clocks. As soon as transmission is complete, data from each *SPSHR* is transferred to the corresponding *SPDR*, and the SPI flag (*SPIF*) in the *SPSR* is set.

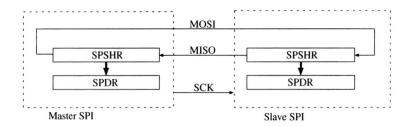

Master SPI                        Slave SPI

The function of the pins depends on whether the device is in master or slave mode:

*MOSI*—output for master, input for slave
*MISO*—input for master, output for slave
*SCK*—output for master, input for slave

The *SPDR* and *SPSHR* are mapped to the same address. Reading from this address reads the *SPDR*, but writing loads the *SPSHR*. *SPSR* bit 7 is the SPI flag (*SPIF*). *SPSR* may also contain error flags, but we will omit them from this design. The following sequence will clear *SPIF*:

Read *SPSR* when *SPIF* is set.
Read or write to the *SPDR* address.

The *SPCR* register contains the following bits:

*SPIE*—enable SPI interrupt
*SPE*—enable the SPI
*MSTR*—set to '1' for master mode, '0' for slave mode

*SPR1* and *SPR0*—set *SCLK* rate as follows:

$$SPR1\&SPR0 = 00 \qquad SCK \text{ rate} = \text{Sysclk rate/2}$$
$$SPR1\&SPR0 = 01 \qquad SCK \text{ rate} = \text{Sysclk rate/4}$$
$$SPR1\&SPR0 = 10 \qquad SCK \text{ rate} = \text{Sysclk rate/16}$$
$$SPR1\&SPR0 = 11 \qquad SCK \text{ rate} = \text{Sysclk rate/32}$$

## P3. Bowling Score Keeper

The digital system shown below will be used to keep score for a bowling game. The score-keeping system will score the game according to the following (regular) rules of bowling: A game of bowling is divided into ten frames. During each frame, the player gets two tries to knock down all of the bowling pins. At the beginning of a frame, ten pins are set up. If the bowler knocks all ten pins down on his or her first throw, then the frame is scored as a *strike*. If some (or all) of the pins remain standing after the first throw, the bowler gets a second try. If the bowler knocks down all of the pins on the second try, the frame is scored as a *spare*. Otherwise, the frame is scored as the total number of pins knocked down during that frame.

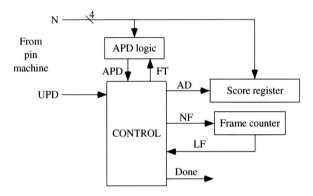

The total score for a game is the sum of the number of pins knocked down plus bonuses for scoring strikes and spares. A strike is worth 10 points (for knocking down all ten pins) plus the number of pins knocked down on the next two *throws* (not frames). A spare is worth 10 points (for knocking down ten pins) plus the number of pins knocked down on the next throw. If the bowler gets a spare on the tenth frame, then he or she gets one more throw. The number of pins knocked down from this extra throw are added to the current score to get the final score. If the bowler gets a strike on the last frame, then he or she gets two more throws, and the number of pins knocked down are added to the score. If the bowler gets a strike in frame 9 and 10, then he or she also gets two more throws, but the score from the first bonus throw is added into the total *twice* (once for the strike in frame 9, once for the strike in frame 10), and the second bonus throw is added in once. The maximum score for a perfect game (all strikes) is 300. An example of bowling game scoring follows:

| Frame | First Throw | Second Throw | Result | Score |
|-------|-------------|--------------|--------|-------|
| 1 | 3 | 4 | 7 | 7 |
| 2 | 5 | 5 | spare | 7 + 10 = 17 |
| 3 | 7 | 1 | 8 | 17 + 7 (bonus for spare in 2) + 8 = 32 |
| . . . | . . . | . . . | . . . | 87 |
| 9 | 10 | — | strike | 87 + 10 = 97 |
| 10 | 10 | — | strike | 97 + 10 (for this throw) + 10 (bonus for strike in 9) |
| — | 6 | 3 | — | 117 + 6 (bonus for strike in 9)<br>+ 6 (bonus for strike in 10)<br>+ 3 (bonus for strike in 10) = 132 |

The score-keeping system has the form shown in the preceding table. The control network has three inputs: *APD* (All Pins Down), *LF* (Last Frame), and *UPD* (update). *APD* is 1 if the bowler has knocked all ten pins down (in either one or two throws). *LF* is 1 if the frame counter is in state 9 (frame 10). *UPD* is a signal to the network that causes it to update the score. *UPD* is 1 for exactly one clock cycle after every throw the bowler makes. There are many clock cycles between updates.

The control network has four outputs: *AD, NF, FT,* and *Done. N* represents the number of pins knocked down on the current throw. If *AD* is 1, *N* will be added to the score register on the rising edge of the next clock. If *NF* is 1, the frame counter will increment on the rising edge of the next clock. *FT* is 1 when the first throw in a frame is made. *Done* should be set to 1 when all ten frames and bonus throws, if applicable, are complete.

Use a 10-bit score register and keep the score in BCD form rather than in binary. That is, a score of 197 would be represented as 01 1001 0111. The lower two decimal digits of the register should be displayed using two 7-segment LED indicators, and the upper 2 bits can be connected to two single LEDs. When *ADD* = 1 and the register is clocked, *N* should be added to the register. *N* is a 4-bit binary number in the range 0 through 10. Use a 4-bit BCD counter module for the middle BCD digit. Note that in the lower 4 bits, you will add a binary number to a BCD digit to give a BCD digit and a carry.

## P4. Simple Microcomputer

Design a simple microcomputer for 8-bit signed binary numbers. Use a keypad for data entry and a 256 × 8 static RAM memory. The microcomputer should have the following 8-bit registers: *A* (accumulator), *B* (multiplier), *MDR* (memory data register), *PC* (program counter), and *MAR* (memory address register). The *IR* (instruction register) may be 5 to 8 bits, depending on how the instructions are encoded. The *B* register is connected to the *A* register so that *A* and *B* can be shifted together during the multiply. Only one 8-bit adder and one complementer is allowed. The microcomputer should have a 256-word-by-8-bit memory for storing instructions and data. It should have two modes: (a) memory load and (b) execute program. Use a DIP switch to select the mode.

Memory load mode operates as follows: Select mode = 0 and reset the system. Then press two keys on the keypad followed by pushing a button to load each word in memory. The first word is loaded at address 0, the second word at address 1, and so on. Data should be loaded immediately following the program. Execution mode operates as follows: Select mode = 1 and press reset. Execution begins with the instruction at address 0.

Each instruction will be one or two words long. The first word will be the opcode, and the second word (if any) will be an 8-bit memory address or immediate operand. One bit in the opcode should distinguish between memory address or immediate operand mode. Represent negative numbers in 2's complement. Implement the following instructions:

| | |
|---|---|
| LDA <memadd> | load $A$ from the specified memory address |
| LDA <imm> | load $A$ with immediate data |
| STA <memadd> | store $A$ at the specified memory address |
| ADD <memadd> | add data from memory address to $A$, set carry flag if carry, set $V$ if 2's complement overflow |
| ADD <imm> | add immediate data to $A$, set carry flag if carry, set $V$ if overflow |
| SUB <memadd> | subtract data from memory address from $A$, set carry flag if borrow, set $V$ if 2's complement overflow |
| SUB <imm> | subtract immediate data from $A$, set carry flag if borrow, set $V$ if overflow |
| MUL <memadd> | multiply data from memory address by $B$, result in $A$ & $B$ |
| MUL <imm> | multiply immediate data by $B$ |
| SWAP | swap $A$ and $B$ |
| PAUSE | pause until a button is pressed and released (*Note*: A register should always be displayed on LEDs.) |
| JZ <target addr> | jump to target address if $A = 0$ |
| JC <target addr> | jump to target address if carry flag ($CF$) is set |
| JV <target addr> | jump to target address if overflow flag ($V$) is set |

The control module should be implemented as a linked state machine, with a separate state machine for the multiplier control. Try to keep the number of states small. (A good solution should have about ten states for the main control.) The multiplier control should use a separate counter to count the number of shifts. Assume that the clock speed is slow enough so that memory can be accessed in one clock period.

## P5. Stack-Based Calculator

Design a stack-based calculator for 8-bit signed binary numbers. Input data to the calculator can come from a keypad or from DIP switches with a separate push-button to enter the data. The calculator should have the following operations:

| | |
|---|---|
| enter | push the 8-bit input data onto the stack |
| 0 – clear | clear the top of the stack, reset the stack counter, reset overflow, and so on. |
| 1 – add | replace the top two data entries on the stack with their sum |

2 – sub    replace the top two data entries on the stack with their difference (stack top—next entry)

3 – mul    replace the top two data entries on the stack with their product (8 bits $\times$ 8 bits to give 8-bit product)

4 – div    replace the top two data entries on the stack with their quotient (stack top / next entry) (8 bits divided by 8 bits to give 8-bit quotient)

5 – xchg    exchange the top two data entries on the stack

6 – neg    replace the top of the stack with its 2's complement

Negative numbers should be represented in 2's complement. Provide an overflow indicator for 2's complement overflow. This indicator should also be set if the product requires more than 8 bits including sign or if divide by 0 is attempted.

Implement a stack module that has four 8-bit words. The stack should have the following operations: push, pop, and exchange the top two words on the stack. The top of the stack should always be displayed on eight LEDs. Include an indicator for stack overflow (attempt to push a fifth word) and stack underflow (attempt to pop an empty stack or to exchange the top of stack with an empty location).

Design the control unit for the calculator using linked state machines. Draw a main SM chart with separate SM charts for the multiplier and divider control. When you design the arithmetic unit, try to avoid adding unnecessary registers. You should be able to implement the arithmetic unit with three registers (8 or 9 bits each), an adder, two complementers, and so on.

### P6. Floating-Point Arithmetic Unit

Design a floating-point arithmetic unit. Each floating-point number should have a 4-bit fraction and a 4-bit exponent, with negative numbers represented in 2's complement. (This is the notation used in the examples in Chapter 7.) The unit should accept the following floating-point instructions:

001    FPL—load floating-point accumulator (fraction and exponent)

010    FPA—add floating-point operand to accumulator

011    FBS—subtract floating-point operand from accumulator

100    FPM—multiply accumulator by floating-point operand

101    FPD—(optional) divide floating-point accumulator by floating-point operand

The result of each operation (4-bit fraction and 4-bit exponent) should be in the floating-point accumulator. All output should be properly normalized. The accumulator should always be displayed as hex digits on 7 segment LEDs. Use an LED to indicate an overflow.

The input to the floating-point unit will come from a $4 \times 4$ hexadecimal keypad, using a scanner similar to the one designed in Chapter 4. Each instruction will be represented by three hex digits from the keypad—the opcode, the fraction, and the exponent. For example, FPA $1.011 \times 2^{-3}$ is coded as 2 B D = 0010 1011 1101. Assume that all inputs are properly normalized or zero. Your design should include the following modules: fraction unit, exponent unit, control module, and 4-bit binary to seven-segment display conversion logic.

## P7. Tic-Tac-Toe Game

Design a machine to play the defensive game of tic-tac-toe using an FPGA. Input will be a $3 \times 3$ keypad, a reset button, and a switch SW1. If SW1 is off, the machine should always win if possible, or draw (nobody wins) if winning is not possible. If SW1 is on, part of the machine's logic should be bypassed so that the player can win occasionally. Output will be a $3 \times 3$ array of LEDs with a red and a green LED in each square. Use two LEDs to indicate player wins or machine wins. If the game is a draw, light both LEDs. Since the machine is playing a defensive game, the human player will always move first. Each time the player moves, the machine should wait two seconds before making its move. Your VHDL code should represent a synchronous digital system that makes efficient use of available hardware resources.

Here is one strategy for playing the game: (player = X, machine = O)

1. Player moves first.
2. Machine makes an appropriate initial move. If player starts in center, machine plays corner; otherwise, machine plays center.
3. After each subsequent move by the player, the machine checks the following in sequence:
   (a) Two O's in a row: machine plays in the third square and wins.
   (b) Two X's in a row: machine plays in the third square to block player.
   (c) If it is the machine's second move, a special move may be required: If player's first two moves are opposite corners, the machine's second move must be side. If player's first move is center, the machine's *second* move should be corner if rule (b) does not apply.
   (d) Two intersecting rows each contain only one X: Machine plays in the square at the intersection of the two rows (this blocks the player from forcing a win).
   (e) If there is no better move, play anywhere.

The preceding rules obviously apply only when the appropriate squares are empty.


## P8. CORDIC Computing Unit

CORDIC (coordinate rotation digital computer) is a computing technique that uses two-dimensional planar rotation to compute trigonometric functions. This algorithm has a wide variety of applications, ranging from your calculator to global positioning systems. The algorithm is perfect for digital systems since computation is merely a set of repeated adds and shifts. For details of this algorithm, review the paper[1]. "A Survey of CORDIC Algorithms for FPGA-Based Computers," located at http://www.andraka.com/files/crdcsrvy.pdf.

Implement the CORDIC algorithm using an FPGA. Your implementation must correctly produce the sine or cosine of an input angle ranging from $-179$ to $+180$ degrees, inclusive. You will only be required to satisfy 8-bit precision. Input will be received in decimal format via a keypad. Three decimal digits will be input (most

[1]R. Andraka, "A Survey of CORDIC Algorithms for FPGA-Based Computers," in *Proceedings of the 1998 ACM/SIGDA Sixth International Symposium on Field Programmable Gate Arrays*, pp.191–200, February 22–24, 1998.

significant digit first) followed by a sign. The angle should be initially represented in BCD and then converted to binary (negative angles represented 2's complement). Designate two special keys for sine and cosine. Output will be displayed on a set of four 7-segment LEDs.

The following pseudocode demonstrates the basics of the CORDIC algorithm. Read the document referenced above and then iterate through this process by hand to help you understand this algorithm.

```
for i = 0 to n          // n-bit precision
    dx = x/(2^i)        // x is 16-bit register representing
                           fractional values. It should be
                        // initialized to .607 (1001_1011_0111_
                           0001). After the algorithm
                        // completes, x holds cos(a). dx is also
                           16 bits.
    dy = y/(2^i)        // y is a 16-bit register representing
                           fractional values. It should
                        // be initialized to 0 (0000_0000_0000_
                           0000). After the
                        // algorithm completes, y holds sin(a). dy
                           is also 16 bits.
    da = arctan(2^-i)   // pre-calculated values in a lookup table
                        // these values should be represented as
                           follows: upper
                        // 8 bits whole number part, lower 8 bits
                           fractional part
                        // a is the input angle represented with
                           at least 10 bits.
                        // All input angles are whole numbers.
    if (a >= 0) then
            x = x - dy; a = a - da; y = y + dx;
    else
            x = x + dy; a = a + da; y = y - dx;
    end if
end loop
```

When you work through this algorithm, notice that it does not produce the negative and positive values associated with sine and cosine. Create separate logic to determine the sign. The algorithm shown above only works for $-90$ to $+90$ input angles. You can simplify your design if you do all calculations in the first quadrant (e.g., sin(105) is the same as sin(75)).

## P9. Calculator for Average and Standard Deviation

Design a special-purpose calculator to calculate the average and standard deviation of a set of test scores. Input will be from a decimal keypad and output will be an LCD display. Each test score will be an integer in the range 0 to 100. The number of scores will be in the range 1 to 31.

**Entry sequence:** For each score, enter one, two, or three digits followed by E (enter). After all scores have been entered, press A to calculate the average and then press D to compute the standard deviation. The average and standard deviation should be displayed with one digit after the decimal point.

The formula for the standard deviation is

$$\text{s.d} = \sqrt{\frac{\sum\limits_{i=1}^{N}(x_i - A)^2}{N}} = \sqrt{\frac{\sum\limits_{i=1}^{N}x_i^2}{N} - A^2}$$

where $A$ is the average. Use the latter form because it is not necessary to store the $N$ scores.

Your design should have three main modules: input, computation, and display. The computation module computes the average and standard deviation of the input data. All computation should be done with binary integers. The input data will be scaled up by a factor of 10 and converted to binary by the input module. The outputs will be converted to decimal and scaled down by a factor of 10 by the display module.

The input module should include a keypad scanner similar to the one designed in Chapter 4. Every time a key is pressed, the scanner will debounce and decode the key. It will then output a 4-bit binary code for the key that was pressed, along with a valid signal (V). This input module will process the digits from the keypad scanner and convert the input number to binary. This module should perform the following tasks:

1. If the input is a digit in the range 0 through 9, store it in a register. Ignore invalid inputs.
2. After one, two, or three digits have been entered followed by E, check to see that the number is within range ($\leq 100$). If not, turn on an error signal.
3. If the input number is in range, append BCD 0, which in effective multiplies by 10. Example: If the entry sequence is 7, 9, E, the BCD register should contain 0000 0111 1001 0000 (790).
4. Convert the BCD to binary and signal the computation unit when conversion is complete.
5. When the A or D key is pressed, generate a signal for the computation module.

The computation module should have one register to accumulate the sum of the inputs and another to accumulate the sum of the squares of the inputs. The data input should be a binary integer with the decimal range 0 to 1000 (score $\times$ 10). Assume three input control signals: V1 (valid data), A (compute and output the average), S (compute and output the standard deviation). Ignore S unless computation of the average has been completed. The computation module should include a square root circuit which will find the square root of a 18-bit binary integer to give a 9-bit integer result. Refer to Reference 35 for a binary square root algorithm. When testing the computation module, be sure to include the worst cases: largest average with 31 inputs (s.d. should be 0), largest standard deviation with 30 inputs (average should be 500).

The display module should drive a two-line LCD display. This module serves two functions: First it displays each number as it is being input, and second it displays the average and standard deviation. During input, each valid decimal digit should be shifted into the display. When E is pushed, the input number will remain displayed until another key is pushed. After the average has been computed, the display module should convert it to BCD and output it to the first line of the LCD display. After the standard deviation has been computed, it should be converted to BCD and displayed on the second line.

### P10. Four-Function Decimal Calculator

Design a four-function hand-held calculator for decimal numbers and implement it using an FPGA. The input will be a keypad and the output will be an LCD display. When you implement your design on the FPGA, optimize for area since speed is unimportant for a hand calculator. General operation of the calculator should be similar to a standard four-function calculator.

The main calculator input keypad has 16 keys to be labeled as follows:

$$7 \quad 8 \quad 9 \quad \div$$
$$4 \quad 5 \quad 6 \quad *$$
$$1 \quad 2 \quad 3 \quad -$$
$$0 \quad . \quad = \quad +$$

Use one additional key for the clear function. The input and output will be a maximum of eight decimal digits and a decimal point with an optional minus sign. Assume that at any time, any key may be pressed. Either take appropriate action or ignore the key press. If more than eight digits are entered, extra digits are ignored.

If the answer requires more than eight digits, some digits to the right of the decimal point are truncated.

Example: 123.45678 + 12345.678 = 12469.134

If more than eight digits are required to the left of the decimal point, display the letter E to indicate an error. For numbers less than 1, display a 0 before the decimal point.

Your calculator should have three modules. The input module scans, debounces, and decodes the keypad. The main module accepts digits and commands from the input module and processes them. The display module displays the input numbers and results on an LCD display.

The main module should have two 8-digit BCD registers, A and B. Register A should have an associated counter that counts the number of digits (ctrA), another counter that counts the number of digits to the right of the decimal point (rctA), and a sign flip-flop (signA). Register B should have similar associated hardware. As each decimal digit is entered, its BCD code should be shifted into A. The result of each computation should be placed in A. The display module should always display the contents of A, along with the associated decimal point and sign. When the first digit of a new number is entered into A, the previous contents of A should be transferred to B. Although input and output is sign and magnitude BCD, internal computations should be done using 2's complement binary arithmetic. A typical sequence of calculations to add A and B is

1. Adjust A and B to align the decimal points.
2. Convert A and B to binary (Abin and Bbin).
3. Add Abin and Bbin
4. Convert the result to BCD, store in A.
5. If an overflow occurs, correct it if possible, else set the E (error) flag.

The display module should output signals to the LCD to properly display the contents of the A register. After initializing and clearing the LCD, it should display "E" if the error flip-flop is set. Otherwise it should output a minus sign if signA = '1', followed by up to eight digits with the decimal point in the correct place. Leading zeros should be replaced by blanks.

# References

References 14, 19, 21, 27, 28, 39, 41, 46, and 48 are general references on digital logic and digital system design. References 2, 3, 4, 15, 16, 20, 24, 30, 40, 42, 44, 47, 49, and 50 provide information on PLDs, FPGAs, and CPLDs. References 10, 21, 31, 38, 43, 45, and 52 provide a basic introduction to VHDL. References 5, 6, 8, 9, 17, 18, 22, 23, 33, 34 and 51 cover more advanced VHDL topics. References 1, 7, 11, 29, 32, and 36 relate to hardware testing and design for testability. The MIPS ISA and architectures of several MIPS processors are described in references 13, 25, 26 and 37. Reference 37 provides an excellent introduction to various computer organization topics, the understanding of which will help in learning the material presented in Chapter 9.

1. Abromovici, M., Breuer, M., and Friedman, F. *Digital Systems Testing and Testable Design*. Indianapolis, Ind. Wiley–IEEE Press, 1994.
2. Actel Corporation, Actel Technical Documentation, www.actel.com/techdocs/
3. Altera Corporation, Altera Literature, www.altera.com/literature/lit-index.html
4. Atmel Corporation, Atmel Products, www.atmel.com/products
5. Armstrong, James, and Gary, G. *Structured Logic Design with VHDL*. Upper Saddle River, N.J.: Prentice Hall, 1993.
6. Ashenden, Peter J. *The Designer's Guide to VHDL*, 2nd ed. San Francisco, Calif.: Morgan Kaufmann, an imprint of Elsevier, 2002.
7. Bardell, P. H., and McAnney, W. H. "Self-Testing of Logic Modules," *Proceedings of the International Test Conference*, Philadelphia, PA November, 1982, pp. 200–204.
8. Berge, F., and Maginot, R. J. *VHDL Designer's Reference*. Boston, Mass.: Kluwer Academic Publishers, 1992.
9. Bhasker, J. *A Guide to VHDL Syntax*. Upper Saddle River, N.J.: Prentice Hall, 1995.
10. Bhasker, J. *A VHDL Primer*, 3rd ed. Upper Saddle River, N.J.: Prentice Hall, 1999.
11. Bleeker, H., van den Eijnden, P., and de Jong, Frans. *Boundary Scan Test— A Practical Approach*. Boston, Mass.: Kluwer Academic Publishers, 1993.
12. Brayton, Robert K. et al. *Logic Minimization Algorithms for VLSI Synthesis*. Boston, Mass.: Kluwer Academic Publishers, 1984.
13. Britton, Robert. *MIPS Assembly Language Programming*. Upper Saddle River, N.J.: Prentice Hall, 2003.
14. Brown, Stephen and Vranesic, Zvonko. *Fundamentals of Digital Logic with VHDL Design*, 2nd ed. New York: McGraw-Hill, 2005.

15. Brown, Stephen D., Francis, Robert J., Rose, Jonathan, and Vranesic, Zvonko G. *Field-Programmable Gate Arrays*. Boston, Mass.: Kluwer Academic Publishers, 1992.

16. Chan, P., and Mourad, S. *Digital Design Using Field Programmable Gate Arrays*. Upper Saddle River, N.J.: Prentice Hall, 1994.

17. Chang, K. C. *Digital Design and Modeling with VHDL and Synthesis*. Los Alamitos, Calif.: IEEE Computer Society Press, 1997.

18. Cohen, Ben. *VHDL—Coding Styles and Methodologies*. Boston, Mass.: Kluwer Academic Publishers, 1995.

19. Comer, David J. *Digital Logic and State Machine Design*, 3rd ed. New York: Oxford University Press, 1995.

20. Cypress Semiconductor Programmable Logic Documentation, www.cypress.com

21. Dewey, Allen. *Analysis and Design of Digital Systems with VHDL*. Toronto, Ontario, Canada: Thomson Engineering, 1997.

22. *IEEE Standard Multivalue Logic System for VHDL Model Interoperability (Std_logic_1164)*. New York: The Institute of Electrical and Electronics Engineers, 1993.

23. *IEEE Standard VHDL Language Reference Manual*. New York: The Institute of Electrical and Electronics Engineers, 1993.

24. Jenkins, Jesse H. *Designing with FPGAs and CPLDs*. Upper Saddle River, N.J.: Prentice Hall, 1994.

25. Kane, Gerry, and Heinrich, Joseph. *MIPS RISC Architecture*. Upper Saddle River, N.J.: Prentice Hall, 1991.

26. Kane, Gerry. *MIPS RISC Architecture*. Upper Saddle River, N.J.: Prentice Hall, 1989.

27. Katz, Randy H. *Contemporary Logic Design*, 2nd ed. Upper Saddle River, N.J.: Prentice Hall, 2004.

28. Kohavi, Z., *Switching and Finite Automata Theory*. New York: McGraw-Hill, 1979.

29. Larsson, Erik *Introduction to Advanced System-on-Chip Test Design and Optimization*. Springer, 2005.

30. Lattice Semiconductors, www.latticesemi.com

31. Mazor, Stanley, and Langstraat, Patricia. *A Guide to VHDL*, 2nd ed. Boston, Mass.: Kluwer Academic Publishers, 1993.

32. McCluskey, E. J. *Logic Design Principles with Emphasis on Testable Semicustom Circuits*. Upper Saddle River, N.J.: Prentice Hall, 1986.

33. Navabi, Zainalabedin. *VHDL—Analysis and Modeling of Digital Systems*, 2nd ed. New York: McGraw-Hill, 1997.

34. Ott, Douglas E., and Wilderotter, Thomas J. *A Designer's Guide to VHDL Synthesis*. Boston, Mass.: Kluwer Academic Publishers, 1994.

35. Parhami, Behrooz. *Computer Arithmetic: Algorithms and Hardware Design*. New York: Oxford University Press, 2000.

36. Parker, Kenneth P. *The Boundary Scan Handbook*, 3rd ed. New York: Springer, 2003.

37. Patterson, David A., and Hennessey, John L. *Computer Organization and Design: The Hardware Software Interface*, 3rd ed. San Francisco, Calif.: Morgan Kaufmann, an imprint of Elsevier, 2005.

38. Perry, Douglas. *VHDL: Programming by Example*, 4th ed. New York: McGraw-Hill, 2002.

39. Prosser, Franklin P., and Winkel, David E. *The Art of Digital Design: An Introduction to Top-Down Design*, 2nd ed. Englewood Cliffs, N.J.: Prentice Hall, 1987.

40. QuickLogic Corporation, Products and Services, www.quicklogic.com

41. Roth, Charles H. *Fundamentals of Logic Design*, 5th ed. Toronto, Ontario, Canada: Thomson Engineering, 2004. (Includes Direct VHDL simulation software.)

42. Rucinski, Andrzej, and Hludik, Frank. *Introduction to FPGA-Based Microsystem Design*. Texas Instruments, 1993.

43. Rushton, Andrew, *VHDL for Logic Synthesis*, 2nd ed. New York: John Wiley & Sons Ltd., 1998.

44. Salcic, C., and Smailagic, A. *Digital Systems Design and Prototyping Using Field Programmable Logic*, 2nd ed. Boston, Mass.: Kluwer Academic Publishers, 2000. (Includes VHDL software for Altera products.)

45. Skahill, Kenneth, and Cypress Semiconductor. *VHDL for Programmable Logic*. Reading, Mass.: Addison-Wesley, 1996. (Includes VHDL software for Cypress products.)

46. Smith, M. J. S. *Application-Specific Integrated Circuits*. Reading, Mass.: Addison Wesley, 1997.

47. Tallyn, Kent. "Reprogrammable Missile: How an FPGA Adds Flexibility to the Navy's TomaHawk," *Military and Aerospace Electronics*, April 1990. Article reprinted in *The Programmable Gate Array Data Book*, San Jose, Cal.: Xilinx, 1992.

48. Wakerly, John F. *Digital Design Principles and Practices*, 4th ed. Upper Saddle River, N.J.: Prentice Hall, 2006.

49. Xilinx, Inc. *Xilinx Documentation and Literature*, www.xilinx.com/support/library.htm

50. XILINX, Inc. *The Programmable Logic Data Book*, 1996. www.xilinx.com

51. Yalamanchili, S. *Introductory VHDL: From Simulation to Synthesis*. Upper Saddle River, N.J.: Prentice Hall, 2001. (Includes XILINX Student Edition Foundation Series Software.)

52. Yalamanchili, S. *VHDL: A Starter's Guide*, 2nd ed. Upper Saddle River, N.J.: Prentice Hall, 2005.

# Index

0-hazard, 13–14
1-hazard, 13–14
22V10 (22CEV10), 153–155
2's complement
    floating point format using, 361–362
    fractions, 219–221
    multiplier, 223
4-valued logic system, 400
9-valued logic system, 405

Actel, 181–182, 325–326
Active-high signal, 409
Active-low signal, 409
Add-and shift multiplier, design of, 210–216
Adders, 2, 63–66, 158–160, 192–200
    BCD, 192–194
    carry look-ahead, 195–200
    full, 2
    Full Adder, four-bit VHDL module, 63–66
    parallel, CPLD implementation of with accumulator, 158–160
    ripple-carry, 194–195
    32-bit, 194–200
    VHDL design of, 192–200
Algorithmic State Machine (ASM) charts, see State Machine (SM) charts
Alias declaration, 193, 548
Altera, 140, 157, 161, 325
And function for std-logic, 406–407
AND gates, 1–2
ANSI/IEEE Standard 1149.1, 479–480, 483
Antifuse FPGAs, 165, 168
Application-specific integrated circuit (ASIC), 54, 138
Architecture declaration, 61–63, 547
Area-Time (AT) product, 348
Arithmetic components, synthesis of, 345–347

Arithmetic instructions, MIPS ISA, 432–434
Arithmetic logic unit (ALU), 190, 442, 444
Array attributes, 396–397, 398–399, 546–547
    predefined in VHDL, 546–547
    use of, 398–399
    vector addition, use of in, 398–399
Arrays, 114–117, 158, 216–217
    interconnect (IA), 158
    look-up table (LUT) method, 115–117
    matrices, 115
    multiplier, VHDL design of, 216–219
    unconstrained, 115
    VHDL and, 114–117
Array declaration, 114–115, 548
Array multiplier, 216–218
ASIC, see Application-specific integrated circuit (ASIC)
ASM Chart, see SM chart
Assert statement, 119–122, 238, 550, 552
Associative law, 4, 6
Asynchronous design, 40–41
ATPG, see Automatic test pattern generator
Attributes, 395–399, 546–547
    array, 396–397, 398–399, 546–547
    predefined in VHDL, 546–547
    signal, 395–396, 397–398, 546
    use of, 397–399
    VHDL, 395–399, 546–547
Automatic test pattern generators (ATPGs), 468

Baud rate, 527, 536
BCD, see Binary-coded-decimal (BCD)
BCD to binary conversion, 252
Bed-of-nails test fixture, 479

Behavioral description, 53, 55, 101–107, 110
    CAD design entry, 53
    modeling a sequential machine, 103–107
    time-to-market criterion, 110
    VHDL, 55, 101–107
Behavioral modeling in VHDL, 55, 102–106
Biased notation, IEEE 754 floating-point formats, 363
Big-endian memory, 452
BILBO, 494–500
Binary-coded-decimal (BCD), 19–25, 191–192, 192–194
    adder, VHDL design of, 192–194
    Mealy machine conversion to excess-3 code, 19–25
    seven-segment display decoder, VHDL design of, 191–192
Binary dividers, VHDL design of, 239–249
Binary multipliers, 265–267, 277–279
    derivation of SM chart, 265–267
    implementation of SM chart, 277–279
BIST, see Built-in self-test (BIST)
Bit-vector, 60
BlockRAM, 327
Boolean algebra, 3–6
    DeMorgan's law, 3
    laws and theorems of, 4
    logic design and, 3–6
    simplification using, 5–6
Booth's algorithm, 254
Boundary scan, 479–490
    ANSI/IEEE Standard 1149.1 instructions, 479–480, 483
    BYPASS, 483
    EXTEST, 483, 484
    IC connection steps, 485–486
    INTEST, 483, 484

**571**

Boundary scan (*Continued*)
Joint Test Action Group (JTAG), 479
PC boards, testing, 479–490
register (BSR), 479–480
RUNBIST, 483
SAMPLE/PRELOAD, 483, 484
test-access port (TAP), 479–482
VHDL code for, 487–490
Bowling Score Keeper, 559
Buffer mode, VHDL modules, 62, 66–67
Buffers, tristate logic, 41, 401–402
Built-in logic block observer (BILBO) technique, 495–500
Built-in self-test (BIST), 468, 490–500
built-in logic block observer (BILBO) technique, 495–500
checkerboard patterns, 491
linear-feedback shift registers (LFSRs), 492–493
march test, 491
multiple-input signature register (MISR), 491–492, 493–494
pseudo-random pattern generator (PRPG), 492
self-testing using an MISR and parallel SRSG (STUMPS), 494–495
shift register sequence generator (SRSG), 494
signature bits, 491
taps, 492
test bench for, 499–500
test-per-clock sheme, 495
test-per-scan scheme, 494–495
use of, 468, 490–491
VHDL code for BILBO registers, 497–499
Busses, tri-state logic, 41–42
BYPASS, boundary scan instruction, 483

CAD, *see* Computer-aided design (CAD)
Calculator
for average and standard deviation, 564
four-function decimal, 566
stack-based, 561
Carry chains, FPGAs, 321–323
Carry look-ahead adder, 195–199
Cascade chains, FPGAs, 323
Case statement, 90, 265, 270, 340–344, 551
SM charts and, 265, 270
synthesis of a, 340–344

Central processing unit (CPU), VHDL code for, 451–457
Channel routing, 176
Characteristic equation, 15
Checkerboard patterns, BIST, 413, 491
CISC, *see* Complex Instruction Set Computing
Clock gating, 37–41
Clock skew, 37, 39, 175
CMOS, 148, 153, 179
Code, 77–81
analyzer, 77
compilation of, 77–81
elaboration, 78
simulation, 53, 77–81
synthesis of, 81, 84–87
VHDL, 77–81
Code converters, 19–25, 26–28
binary-coded-decimal (BCD) to excess-3, 19–25
Mealy machine design of, 19–25
Moore machine design of, 26–28
nonreturn-to-zero (NRZ) to Manchester, 26–28
Combinational circuits, 57–60
concurrent statements, 57–60
VHDL description of, 57–60
Combinational logic, 1–3, 12–14, 468–472
bridging faults, 472
dynamic hazards, 13
full adders, 2
gates, 1–2
hazards in combinational circuits, 12–14
logic design and, 1–3
maxterm expansion, 3
minterm expansion, 2
path sensitization, 471
propagation delays, 12–13
static hazards, 13
stuck-at-faults, 468–472
sum of products (SOP), 2
testing, 468–472
truth tables, 2
Command file examples, 65, 105, 110, 275, 461
Commutative law, 4, 6
Complex digital systems, VHDL design of, 507–544
Complex Instruction Set Computing (CISC), 429, 431
Complex programmable logic devices (CPLDs), 54, 139, 156–160
CAD technology and post-synthesis simulation, 54
erasable (EPLDs), 156
implementation of parallel adder with accumulator, 158–160

interconnect array (IA), 158
types of and capacities, 157
Xilinx CoolRunner, example of, 157–158
Component declaration, 64–66, 549
Component instantiation, 66, 550
Components, VHDL modules, 63–64, 66
Computer-aided design (CAD), 51–54
behavioral description, 53
design entry, 52
design flow in, 52
design requirements, 52
design specification, 52
formulation of design, 52
hardware description languages (HDLs), 52–53
mapping, 54
netlist, 53
placing, 54
post-synthesis simulation, 53
routing, 54
schematic capture, 52
simulation, 53
structural description, 53
synthesis, 53
technology of, 53–54
Concurrent statements, 57–60, 87–90, 549–551
combinational circuits and, 57–60
multiplexer models using, 87–90
VHDL language for, 549–551
VHDL models and, 57–60, 87–90
Conditional assignment statement 87–88, 549
Conditions, 32–35, 261
SM charts, 261
timing, 32–35
Consensus theorem, 4–6
Constant declarations, 114, 548
Constant parameter, VHDL, 393–394
Content addressable memories (CAMs), 181–182
Control circuits, design of state graphs for, 204–205
Control signal (CS), 37–41
Control signal gating, 38
Control store, 284
Control transfer instructions, MIPS ISA, 435–438
Controller, 37, 206, 234–235
Conversion functions, 93, 554
CoolRunner, 157–159
CORDIC, 563
Counters, modeling using VHDL processes, 95–101

CPLDs, *see* Complex programmable logic devices (CPLDs)
Critical path, synthesis and, 217, 347

D flip-flops, 15
Data flow modeling in VHDL, 55, 102, 107
Data memory unit, MIPS subset data path design, 445
Data path, 37, 190–191, 205, 442–447
aritmetic logic unit (ALU), 442, 444
data memory unit, 445
decode unit instruction, 444
defined, 190–191
destination register, 444
execution unit instruction, 444–447
fetch unit instruction, 443–444
MIPS subset, design of, 442–447
overall microprocessor design of, 445–447
program counter (PC), 443–444
register file, 444
scoreboard, design of, 205
source registers, 444
synchronous design and, 37
Data types, VHDL, 82–84
Dataflow description, 55, 101–102, 107–108
modeling a sequential machine, 107–108
VHDL, 55, 101–102, 107–108
Debouncing, design and, 208–210, 233
Decision box, SM charts, 261
Declarations in VHDL, 61–64, 547–549
Decode unit instruction, MIPS subset data path design, 444
Dedicated arithmetic units, FPGAs, 179–180, 332–333
Dedicated memory, FPGAs, 179, 326–332
block RAM, 327
distributed, 328
LUT-based, 327–328
TriMatrix, 327
VHDL models, 328–332
Delay (D) flip-flops, 14–15
Delays, *see* Timing
Delta (Δ) delay, 58, 78–79
DeMorgan's law, 3
Denormalized numbers, IEEE 754 floating-point standard, 366–367
Design for testability (DFT), 468–506

Design translation, FPGAs, 339–353
mapping, 348–349
optimizations of area, power and delay, 347–348
placement, 348, 349–353
routing, 348, 349–353
synthesis, 339–348
Design, 1–50, 156, 183–185, 190–259, 310–360, 507–544. *See also* Computer-aided design (CAD); Field programmable gate arrays (FPGAs)
add-and shift multiplier, example of, 210–216
array multiplier, example of, 216–219
BCD adder, example of, 192–194
BCD to seven-segment display decoder, example of, 191–192
binary dividers, examples of, 239–249
complex digital systems, examples of, 507–544
controller, 37, 206, 234–235
data path, 37, 190–191, 205
debouncing, 208–210, 233
decoders, 233–234
dividers, signed and unsigned, 239–249
FPGAs, flow for, 183–185
implementing using field programmable gate arrays (FPGAs), 310–360
keypad scanner, example of, 231–238
logic, 1–50
PLDs, flow for, 156
RAM memory model, 519–526
scoreboard and controller, example of, 205–208
signed integer/fraction multiplier, example of, 219–231
single pulser, 209–210
small digital systems, examples of, 190–259
state graphs for control circuits, 204–205
synchronization, 208–210
synchronous, 36–41
test benches for, 227–229, 237–238, 246–248, 516–518
32-bit adder, example of, 194–200
traffic light controller, example of, 201–203
universal asynchronous receiver (UART), example of, 526–539
using NAND and NOR gates, 10–12

VHDL models, 190–259, 507–544
wristwatch, example of, 507–518
Destination register, MIPS subset data path design, 444
Dice game, 267–275, 279–283, 292–297
derivation of SM chart, 267–275
implementation of SM chart, 279–283
microprogramming the controller, 292–297
single-address microcode for, 294–297
two-address microcode for, 292–294
Digital signal processing blocks, FPGAs, 180
Distinguishing sequence, 475
Distributed memory, 328
Distributed memory, FPGAs, 328
Distributive law, 4, 6
Dividers, 239–249
signed, design of, 242–249
unsigned, design of, 239–242
Division, 239–248
Don't cares, 7–8
Door lock, 557
Double precision format, IEEE 754, 365–366
Dynamic hazards, 13

Edge triggered, 15, 37–40
Elaboration, 77
Elsif statements, 71
Embedded processors, FPGAs, 180–182
Encoded state assignment, 19
Energy-Delay (ED) product, 348
Entity declaration, 61–63, 547
Entrance path, SM charts, 261
Enumeration type declaration, 82, 548
EPROM/EEPROM programming technology, FPGAs, 167–168
Equations, *see* Dataflow descriptions
Equivalent gate count, 337–338
Equivalent states, 28–30
defined, 28
implication table method, 29–30
sequential circuits and, 28–30
state equivalence theorem, 28–29
Erasable CPLDs (EPLDs), 156
Essential prime implicant, 8
Excitation table, 23

Execution, 444–447, 447–448
  flow of, 447–448
  MIPS subset implementation, 444–447, 447–448
  unit instruction, 444–447
Exit path, SM charts, 261
Exit statement, 118, 552
Exponents, 363–364, 366–374, 370–371
  adder, 370–371
  IEEE 754 floating-point formats use of, 363–364
  special cases of IEEE 754 standard for, 366–367
EXTEST, boundary scan instruction, 483, 484

Falling edge, 37
Feedback, SM block with, 263
Fetch unit instruction, MIPS subset data path design, 443–444
Field programmable gate arrays (FPGAs), 54, 138, 139–140, 160–185, 310–360
  Actel Fusion VersaTile, 325–326
  Altera, 325
  applications of, 182–183
  CAD technology and post-synthesis simulation, 54
  carry chains, 321–323
  cascade chains, 323
  dedicated memory, 326–332
  dedicated multipliers, 332–333
  dedicated specialized components, 179–182
  design flow for, 183–185
  design translation, 339–348
  designing with, 310–360
  equivalent gate count, 337–338
  gates, maximum versus usable, 337–338
  hierachical architectures, 164
  I/O blocks, programmable, 177–179
  implementing functions in, 310–316, 316–321
  interconnects, programmable, 173–177
  introduction to, 138, 139–140, 160–161
  logic block architectures, programmable, 169–173
  logic blocks, examples of, 324–326
  mapping, 348–349
  matrix-based (symmetrical array) architectures, 163, 164
  one-hot state assignment, 336–337

organization of, 161–165
placement, 348, 349–353
programmability cost, 333–335
Programmable Electronics Performance Company (PREP) benchmarks, 338
programming technologies, 165–169
routing, 348, 349–353
row-based architectures, 163–164
sea-of gates architecture, 164–165
Shannon's decomposition, 316–321
slice, 319–321, 324
synthesis, 339–348
types of and capacities, 161
Xilinx, 324–325
File declaration, 417, 549
Files, VHDL, 417–421
Flash memories, 142
Flip-flops, 14–16, 19, 22–24, 25–26, 31, 33–34, 69–73
  characteristic equation, 15
  delay (D), 14–15
  excitation table, 23
  hold times, 31, 33–34
  J-K, 15, 72–73
  Mealy machine state assignment of, 19, 22–24
  modeling using VHDL, 69–73
  Moore machine state assignment of, 25–26
  set-reset (S-R), 16
  setup time, 31, 33–34
  state assignment 19, 22–24, 25–26
  toggle (T), 14, 15
Floating-point arthmetic, 361–388
  addition, 377–383
  division, 383–384
  IEEE 754 formats, 363–367
  multiplication, 367–377
  numbers, representation of, 361–367
  subtraction, 383
  2's complement, 361–362
Floating-point Arithmetic Unit, 562–563
For loops, 118, 390–391, 552
FPGAs, see Field programmable gate arrays (FPGAs)
Fraction multiplier, floating-point multiplication, 371
Fractional part, IEEE 754 floating-point formats, 363
Full Adder, four-bit module, 63–66
Function declaration, 390, 548

Function implementation, 310–316, 316–321
  FPGAs, 31–316, 316–321
  look-up tables (LUTs), 311–316
  Shannon's decomposition, 316–321
Functions, 389–393, 402–405, 547, 550
  call, 390, 551
  predefined, 547
  signal resolution, 402–405
  VHDL, 389–393, 402–405, 547

GAL, see Generic Array Logic
Gate arrays, see Mask programmable gate arrays (MPGAs)
Gated control signal, 38
Gated D latch, 16
Gates, 1–2, 10–12, 171–173, 337–338
  bubbles at, 10
  combinational logic and, 1–2
  conversion of, 11–12
  equivalent gate count, 337–338
  FPGA capacity, 337–338
  logic blocks based on, FPGA use of, 171–173
  maximum versus usable, 337–338
  NAND, designing with, 10–12
  NOR, designing with, 10–12
Generic array logic (GALs), 139, 153–156
Generate statements, 415–417, 550
Generics, VHDL, 413–414
Glitches, 35–36, 37–38
  control signals (CS), 37–38
  defined, 35
  sequential circuits, 35–36
Glue logic, FPGAs, 183
Greedy algorithms, 350
Guard and round bits, IEEE 754 floating-point standard, 367

Handel-C, 55
Hardware accelerators/coprocessors, FPGAs, 183
Hardware description languages (HDL), 52–53, 54–57
  computer-aided design (CAD) and, 52–53
  Handel-C, 55
  learning, 56
  System C, 55–56
  System Verilog, 55
  Verilog, 55
  VHDL, 55, 56
Hardwiring, SM charts, 284

Hazard, 12–13
HDL, *see* Hardware description
   languages (HDL)
Hierachical architectures,
   FPGAs, 164
Hold times, 31, 33–34

I-format, MIPS instruction,
   438–441
I/O blocks, programmable in
   FPGAs, 177–179
I/O standards, 178–179
Identifiers, VHDL, 59
IEEE 1164 standard, 405–408,
   408–410
   9-valued logic system,
     405–408
   SRAM model using, 408–410
IEEE 754 floating-point
   formats, 363–367
   biased notation, 363
   denormalized numbers,
     366–367
   double precision, 365–366
   exponent, 363
   fractional part, 363
   infinity, 367
   not a number (NaN), 367
   overflow, 363, 369
   rounding, 367
   sign-magnitude system, 363
   single precision, 363–365
   underflow, 363, 369
   zero, 366
IEEE standard libraries, 91–94,
   407–408, 553
IEEE standard libraries, 91–94,
   553–554
   NUMERIC_BIT, 553
   NUMERIC_STD, 554
If statements, 70–71, 265, 270,
   344–345, 416–417, 551
   conditional generate
     statement using, 416–417
   SM charts and, 265, 270
   synthesis of, 344–345
   VHDL language for, 551
Implication table method of
   state equivalence, 29–30
Inertial delays, 75–77
Infinity, IEEE 754 floating-point
   standard, 367
Inout mode, VHDL modules,
   62, 66
Input-output block, 177–179
Instruction encoding, MIPS,
   438–441
Instruction Set Architecture
   (ISA), 429, 432–438
Interconnect array (IA), 158
Interconnects, 173–177
   clock skew, 175
   direct, 173–174
   general purpose, 173

global lines, 174–175
nonsegmeted channel routing
   architecture, 176
programmable in FPGAs,
   173–177
row-based FPGAs, in,
   175–177
Interface-signal declaration, 61,
   547
INTEST, boundary scan
   instruction, 483, 484
ISA, *see* Instruction Set
   Architecture, 429
Iterative circuit, converting
   sequential circuits to,
   473–474
Iterative improvement
   algorithms, 350

J-format, MIPS instruction,
   438–441
J-K flip-flops, 15, 72–73
Joint Test Action Group
   (JTAG), 479
JTAG Standard, *see* ANSI/IEEE
   Standard 1149.1

K-map, *see* Karnaugh maps
Karnaugh maps, 7–10
   don't cares, 7–8
   map-entered variables,
     simplification using, 9–10
   minimum sum of
     products, 7–9
   prime implicants, 7–8
Keypad scanner, design of,
   231–238

Large scale integration (LSI), 51
Latch creation, unintentional in
   synthesis, 342–344
Latches, 16
Lattice Semiconductor, 140, 156,
   157, 161
LE, *see* Logic Element
Leading edge, *see* rising edge
LFSR, *see* Linear Feedback Shift
   Register
Library declaration, 549
Libraries, 90–94, 553–554
   IEEE standard, 553–554
   VHDL, 90–94
Linear-feedback shift registers
   (LFSRs), 492–493
Link path, SM charts, 261
Linked state machines, 297–299
Little-endian memory, 452
Load/store achitecture, RISC,
   430–431
Logic blocks, examples of in
   FPGAs, 324–326
Logic design, 1–50
   Boolean algebra, 3–6
   combinational, 1–3, 12–14

equivalent states, 28–30
flip-flops, 14–16
hazards in combinational
   circuits, 12–14
Karnaugh maps, 7–10
latches, 14–16
Mealy sequential circuits,
   17–25
Moore sequential circuits,
   25–28
NAND gates, 10–12
NOR gates, 10–12
review of fundamentals of,
   1–50
sequential circuit timing,
   30–41
state tables, reduction of,
   28–30
tristate, 41–42
Logic Element, 325
Logical instructions, MIPS ISA,
   434
Long lines, 174
Look-up tables (LUTs),
   115–117, 142, 169–171,
   287–297, 311–316,
   327–328
   array matrices, VHDL,
     115–117
   distributed memory and,
     327–328
   FPGA memory, LUT-based),
     327–328
   FPGAs, implementing func-
     tions in, 311–316
   method (ROM method),
     115–117, 142
   programmable logic blocks,
     (LUT-based) for FPGAs,
     169–171
Loops, 117–119, 390–391, 552
   for statements, 118, 390–391,
     552
   infinite, 117–118
   while statements, 119, 552
LSI, 51
LUTs, *see* Look-up tables
   (LUTs)

Macrocells, 154–155, 156–158
   CLPD function blocks,
     156–158
   GAL output logic, 154–155
Main control unit, floating-point
   multiplication, 371–377
Manchester code, 27
Map-entered variables, 9–10,
   280–281
Mapping designs, 54, 348–349
   CAD, 54
   FPGAs, 348–349
   standard cell approach, 349
March test, BIST, 491
Mask programmable gate arrays
   (MPGAs), 138

Matrices, 115–117
Matrix-based (symmetrical array) architectures, FPGAs, 163, 164
Maxterm expansion, 3
Mealy sequential circuits, 17–25, 103–110
  code converter, BCD to excess-3, 19–25
  design of, 17–25
  excitation table, 23
  general model of, 17
  sequence detector, 17–19
  state assignment, 19, 22
  state graph, 17–18
  transition table, 19
  VHDL modeling of, 103–110
Medium-speed systems, FPGAs, 183
Memory, 138, 141–145, 284, 326–332, 434–435, 445, 450–451, 452, 518–526
  access instructions, MIPS ISA, 434–435
  big-endian, 452
  control store, 284
  data unit, data path design of, 445
  dedicated, FPGAs, 326–332
  distributed, FPGAs, 328
  little-endian, 452
  microprogramming, 284
  RAM models, 519–526
  read-only (ROM), 138, 141–145
  RISC microprocessor design, 445, 450–451
  testing, 491
  timing models, VHDL design of, 518–526
  VHDL model for, 408–413, 450–451
Microcode, 286–289, 292–297
  dice controller, implementation of, 292–297
  single-qualifier, single-address, 289–290, 294–297
  two-address, 286–289, 292–294
Microcomputer, Simple 560-561
Microinstruction, 285, 296–297
Microprocessors, see MIPS Processors, Reduced Instruction Set Computing (RISC)
Microprogramming, 283–297
  control store, 284
  memory, 284
  microcode, 286–289, 292–297
  microinstruction, 285, 296–297
  sequencing, 284
  single-qualifier, single-address microcode, 289–290, 294–297

SM qualifiers, 287–289, 289–292
  state machine (SM) charts and, 283–297
  two-address microcode, 286–289, 292–294
Minterm expansion, 2
MIPS Processor, 430, 432–448, 453–463
  arithmetic instructions, 432–434
  complete processor moel, 457–458
  control transfer instructions, 435–438
  data path design for subsets, 442–447
  I-format, 438–441
  instruction encoding, 438–441
  Instruction Set Architecture (ISA), 432–438
  introduction to, 430
  J-format, 438–441
  logical instructions, 434
  memory access instructions, 434–435
  nop (no operation) instructions, 430
  opcode (operations), 429, 439
  R 14000, 431
  R 2000, 430
  R-format, 438–441
  RISC processors and, 430, 432–448
  signals for model of processor, 453
  subset implementation, 441–448
  test bench for processor model, 459–461
  testing processor model, 458–463
  three-address format, 432
  unconditional jump instructions, 437
  VHDL code for subset implementation, 454–457
MISR, see Multiple Input Signature register
Mode, VHDL modules, 62
Modules, 61–67
  architecture declaration, 61–63
  components, 63–64, 66
  entity declaration, 61–63
  Full Adder, 63–66
  VHDL, 61–67
Moore sequential circuits, 25–28
  code converter, NRZ to Manchester, 26–28
  sequence detector, 25–26
  state assignment, 25–26
  transition table, 26

MPGA, see Mask programmable gate arrays (MPGAs)
MSI, 51
Multiple-input signature register (MISR), 491–492, 493–494
Multiplexers (MUX), 87–90, 171–173
  case statement, using, 90
  concurrent statements, using, 87–90
  logic blocks based on, FPGA use of, 171–173
  process statements, using, 90
  VHDL models for, 87–90
Multiplicand, 210, 219–222
Multipliers, 210–216, 216–219, 219–231, 265–267, 277–279, 332–333
  add-and-shift, VHDL design of, 210–216
  array, VHDL design of, 216–219
  binary, 265–267, 277–279
  dedicated, FPGAs, 332–333
  signed integer/fraction, VHDL design of, 219–231
Multivalued logic, 400–405, 405–408, 410–413
  bidirectional tristate bus, 410
  data register, 410
  4-valued system, 400–408
  IEEE 1164 standard, using, 405–408
  9-valued system, 405–408
  read/write system, 410–413
  signal resolution functions, 402–405
  SRAM models, 408–410, 410–413
MUX, see Multiplexers (MUX)

Named association, VHDL, 414–415
NaN, see Not a number
NAND gates, 10–12
NATURAL subtype, 117
Negative logic, 1
Netlist, synthesis output, 53, 339
NMOS, 148
Nonreturn-to-zero (NRZ) code to Manchester, 26–28
Non-segmented tracks, 176
Nop (no operation) instructions, MIPS, 30
NOR gates, 10–12
Normalized floating point, 362–367
Not a number (NaN), IEEE 754 floating-point standard, 367
NOT gates, 1–2
NRZ code, see Non return to zero code

Numeric_bit package, 91–94, 407–408, 553
Numeric_std package, 91–94, 407–408, 553

One-hot state assignment, 19, 336–337
Opcode (operations), MIPS, 429, 439
Operators, VHDL, 82–84, 546
OR gates, 1–2
Output box, SM charts, 261
Overflow, IEEE 754 exponents, 363, 369
Overloaded operators, creating in VHDL, 399–400

PAL, *see* Programmable array logic (PAL)
Package declaration, 549
Parallel load, 96
Parameters, VHDL, 393–394
Parity, 115, 390, 526
Path sensitization, 471
PC boards, *see* Boundary scan
Placing designs, 54, 348, 349–353. *See also* Routing
    CAD, 54
    FPGAs, 348, 349–353
PLAs, *see* Programmable logic arrays (PLAs)
PLDs, *see* Programmable logic devices (PLDs)
PMOS, 148
Port map statements, 219, 229, 414–415
POS, *see* product of sum
POSITIVE subtype, 117
Post-synthesis simulation, 53
PREP Benchmarks, 338
Prime implicants, 7–8
Priority encoder, 143
Procedure declaration, 393, 548
Procedures, VHDL, 393–394, 551
    call, 393, 551, 552
    parameters and, 393–394
    VHDL use of, 393–394
Process statements, 67–68, 90, 550
    multiplexer modeling using, 90
    sequential statements and, 67–68
    VHDL language for, 550
Product of sums, 8
Program counter (PC), MIPS subset data path design, 443–444
Programmability cost, FPGAs, 333–335
Programmable array logic (PAL), 53, 138–139, 151–153
    CAD technology and post-synthesis simulation, 53
    implementation of, 151–153

overview of as PLDs, 138–139
Programmable Electronics Performance Company (PREP) benchmarks, 338
Programmable logic arrays (PLAs), 53, 138, 146–150
    CAD technology and post-synthesis simulation, 53
    implementation of, 146–150
    overview of as PLDs, 138
Programmable logic devices (PLDs), 137–189
    application-specific integrated circuit (ASIC), 54, 138
    classification of, 137–138
    comparison of, 140
    complex (CPLDs), 54, 139, 156–160
    design flow for, 156
    factory, 137–138, 160
    field programmable gate arrays (FPGAs), 137, 138, 139–140, 160–185
    generic array logic (GALs), 139, 153–156
    introduction to, 137–189
    mask programmable gate arrays (MPGAs), 138
    mask programmable gate arrays (MPGAs), 138
    programmable array logic (PAL), 138–139, 151–153
    programmable logic arrays (PLAs), 138, 146–150
    read-only memory (ROM), 138, 141–145
    simple (SPLDs), 53, 139, 140–156
Programming technologies, 165–169
    antifuse, 168
    comparison of, 168–169
    EPROM/EEPROM, 167–168
    FPGAs, 165–169
    SRAM, 165–167
Propagation delays, 12–13, 31
    combinational logic and, 12–13
    defined, 31
    dynamic hazards, 13
    hold times, 31
    sequential circuit timing and, 31
    setup time, 31
    static hazards, 13
PRPG, *see* Pseudo Random Pattern Generator
Pseudo-random pattern generator (PRPG), 492

Qualifiers, SM charts and micro-programming, 287–289, 289–292

Race, 40
R-format, MIPS instruction, 438–441
Random-access memory (RAM), 408–409, 519–526. *See also* Static RAM (SRAM)
    memory timing models, VHDL design of, 519–526
    use of, 408–409
Rapid prototyping, FPGAs, 182
Read-only memory (ROM), 115–117, 138, 141–145. *See also* Look-up tables (LUTs)
    address, 141
    flash memories, 142
    look-up tables (LUTs), 142
    method of implementation, 115–117, 142
    programmable logic device, use as a, 138, 141–145
    types of, 142
    word, 141
Reconfigurable circuits and systems, FPGAs, 183
Reduced Instruction Set Computing (RISC), 429–467
    central processing unit (CPU), VHDL code for, 451–457
    complete MIPS processor model, 457–458
    design features, 430–431
    execution, 444–447, 447–448
    Instruction Set Architecture (ISA), 429, 432–438
    load/store architecture, 430–431
    memory, 434–435, 445, 450–451
    microprocessor, 429–467
    MIPS Technologies, 430, 432–448
    philosophy, 429–432
    processor model signals, 453
    register file, 444, 449–450
    register-register architectures, 430
    single-instruction computer, 432
    subset implementation, 441–448, 454–4578
    testing MIPS processor model, 458–463
    VHDL code for RISC subset implementation, 454–457
    VHDL models, 449–463
Register file, 444, 449–450
    MIPS subset data path design, 444
    RISC microprocessor design, 444, 449–450
    VHDL model for, 449–450

Register-register architectures, RISC, 430
Register transfer language (RTL) models, 102–13
Registers, modeling using VHDL processes, 95–101
Reject, 76
Report statement, 119–122, 552
Reserved words, VHDL, 59
Resolution function, 401, 402–405
Return-to-zero (RZ) code, 26–27
RISC, *see* Reduced Instruction Set Computing (RISC)
Rising edge, 15, 39–40
ROM method, 142–145
ROM, *see* Read-only memory (ROM)
Round bits, 367
Rounding, IEEE 754 floating-point standard, 367
Routing designs, 54, 348, 349–353
CAD, 54
FPGAs, 348, 349–353
greedy algorithms, 350
iterative improvement algorithms, 350
simulated annealing, 350–351
Row-based architectures, FPGAs, 163–164
RUNBIST, boundary scan instruction, 483
RZ code, *see* Return to zero code, 26

s-a-0, *see* Stuck at 0
s-a-1, *see* Stuck at 1
SAMPLE/PRELOAD, boundary scan instruction, 483, 484
Scan data input (SDI), 477
Scan data output (SDO), 477
Scan path (design) testing, 468, 476–479
Schematic capture, 52
Scoreboard, design of, 205–208
Sea-of gates architecture, FPGAs, 164–165
Sea of tiles, 165, 172, 181–182
Segmented tracks, 176
Selected signal assignment, 89, 550
Self-testing using an MISR and parallel SRSG (STUMPS), 494–495
Sensitivity list, 67
Sequence detector, 17–19, 25–26
Mealy machine design of, 17–19
Moore machine design of, 25–26
Sequencing memory, 284
Sequential circuits, 17–25, 25–28, 28–30, 30–41, 473–476
clock gating, 37–41
clock skew, 37, 39

control signals (CS), 37–41
distinguishing sequence, 475
equivalent states and, 28–30
glitches in, 35–36
iterative circuit, converting to, 473–474
maximum clock frequency of operation, 31–32
Mealy, 17–25
Moore, 25–28
propagation delays, 31
strongly connected state graph, 474–475
stuck-at-faults, 475–476
synchronous design, 36–41
testing, 473–476
timing conditions, 32–35
timing in, 30–41
Sequential statements, 67–69, 70–71, 551–552
if and elsif statements, 70–71
process statements, 67–68
VHDL language for, 551–552
VHDL processes, 67–69
Sensitivity list, 67
Set-reset (S-R) flip-flops, 16
Setup time, 31, 33–34
Severity statements, 119–120
Shannon's expansion theorem, 316
Shannon's decomposition, FPGAs, 316–321
Shift register sequence generator (SRSG), 494
Sign-magnitude system, IEEE 754 floating-point formats, 363
Signal assignment statements, 58, 549–551
Signal attributes, 395–396, 397–398, 546
creating signals, 396, 546
predefined in VHDL, 546
returning values, 395, 546
use of, 397–398
Signal declarations, 111–114, 548
Signal parameter, VHDL, 393–394
Signal resolution, VHDL, 400–408
Signals, MIPS processor models, 453
Signature bits, BIST, 491
Signed integer/fraction multiplier, VHDL design of, 219–231
Signed type, 91–93
Simple programmable logic devices (SPLDs), 53, 138–139, 140–156
CAD technology and post-synthesis simulation, 53
generic array logic (GALs), 139, 153–156
implementation of, 140–156

programmable array logic (PAL), 53, 138–139, 151–153
programmable logic arrays (PLAs), 53, 138, 146–150
Simplification, 5–6, 9–10
Boolean algebra, using, 5–6
Karnaugh map-entered variables, using, 9–10
Simulated annealing, FPGA design routing, 350–351
Simulation, 53, 77–81
delta ($\Delta$) delay, 78–79
design conceptualization, 53
discrete event, 78
event, 78
initializing phase, 78
multiple processes, 79–81
post-synthesis, 53
scheduling a transaction, 78
VHDL code, 78–81
Single-instruction computer, 432
Single-precision format, IEEE 754, 363–365
Slew rate, FPGAs, 177
Slice, FPGAs, 319–321, 324
SM charts, *see* State Machine (SM) charts
Small digital systems, VHDL design of, 190–259
Small scale integration (SSI), 51
SOP, *see* sum of product
Source registers, MIPS subset data path design, 444
Spartan, 140, 161, 319–321
SPLD, *see* Simple Programmable Logic Device
SPLDs, *see* Simple programmable logic devices (SPLDs)
S-R flip-flops, 16
SRAM, *see* Static RAM (SRAM)
SRAM FPGAs, 165–168
SSI, 51
Standard Logic, *see* Std_logic
State assignment, 19, 22–24, 25–26, 82, 336–337
encoded, 19
enumeration type, 82
flip-flop values and, 19, 22–24, 25–26
FPGAs, 336–337
Mealy machine design, 19, 22–24
Moore machine design, 25–26
one-hot, 19, 336–337
transition table, 19, 26
State box, SM charts, 261
State graphs, 17–18, 204–205, 264, 474–475
control circuits, use of for, 204–205
conversion of, to SM charts, 264

distinguishing sequence, 475
Mealy sequential circuit
  design and, 17–18
strongly connected state
  graph, 474–475
testing sequential circuits,
  474–475
State Machine (SM) charts,
  260–309
  binary multipliers, 265–267,
    277–279
  blocks, 261–264
  case statements in, 265, 270
  decision box, 261
  derivation of, 265–275
  dice game, 267–275,
    279–283
  feedback, SM block with, 263
  hardwiring, 284
  implementation of, 277–279,
    279–283
  introduction to, 260–265
  link path, 261
  linked, 297–299
  microprogramming, 283–297
  output box, 261
  parallel blocks, 263–264
  qualifiers, 287–292
  realization of, 275–279
  ROM method of implemen-
    tation, 287–279
  serial blocks, 263–264
  state box, 261
  state graph, conversion to,
    264
  timing charts, 264–265
State tables, 18, 26, 28–30
  equivalent states and, 28–30
  Mealy machine design using,
    18
  Moore machine design using,
    26
  reduction of, 28–30
Static Hazard, 13
Static RAM (SRAM), 165–167,
  408–410, 410–413
  IEEE 1164 standard, using,
    408–410
  models, 408–410, 410–413
  multivalued logic and,
    408–410, 410–413
  programming technology,
    FPGAs, 165–167
  Read/write system, using,
    410–413
Std_logic, 93–94, 407–408, 553
Sticky bits, IEEE 754 floating-
  point standard, 367
Strongly connected, 474
Structural description, 53, 55,
  101–102, 108–110
  CAD design entry, 53
  modeling a sequential
    machine, 108–110
  VHDL, 55, 101–102, 108–110

Stuck-at-0 fault, 469–470
Stuck-at-1 fault, 469–470
Stuck-at-faults, 468–472,
  475–476
  combinational circuits,
    468–472
  sequential circuits, 475–476
STUMPS architecture, 494
Subset implementation, 441–448,
  454–4578
  data path, design of,
    442–447
  flow of execution, 447–448
  MIPS, 441–448
  VHDL code for, 454–457
Subtype, 117
Subtype declaration, 117, 548
Sum of products, 7
Sum of products (SOP), 2, 7–9
  combinational logic and, 2
  Karnaugh maps and, 7–9
  minimum, 7–9
Synchronization, design and,
  208–210
Synchronous clear, 96
Synchronous design, 36–41. *See
  also* Design
  architecture, 37
  clock enable, 38
  clock gating, 37–41
  clock skew, 37, 39
  control signals (CS), 37–41
  controller, 37
  data path, 37
  rising-edge devices, 39–40
Synchronous Serial Peripheral
  Interface, 558
Synthesis, 53, 81, 84–87,
  339–348
  Area-Time (AT) product,
    348
  arithmetic components, of,
    345–347
  CAD conversion, 53
  case statement, of a, 340–344
  critical path, 347
  defined, 53
  design translation, 339–348
  Energy-Delay (ED) product,
    348
  examples of, 84–87, 341-347
  FPGAs, 339–348
  if statement, of a, 344–3454
  latch creation, unintentional,
    342–344
  netlist, 53, 339
  VHDL code, 81, 84–87
System C, 55–56
System Verilog, 55

T flip-flops, 15
TAP, 479–490
Taps, defined, 492
Test-access port (TAP),
  479–482

Test benches, 120–122, 227–229,
  237–238, 246–248, 459–461,
  499–500, 516–518
  assert statements in, 120–121
  BILBO system, 499–500
  binary dividers, 246–248
  keypad scanner, 237–238
  MIPS processor model,
    459–461
  port map statement, 229
  report statements in, 120–121
  signed integer/fraction multi-
    plier, 227–229
  use of, 120–122, 227
  wristwatch design
    module, 516–518
Test-per-clock scheme, 495–500
Test-per-scan scheme, 494–495
Testing, 458–463, 468–506. *See
  also* Test bench
  automatic test pattern
    generators (ATPGs), 468
  boundary scan, 479–490
  bridging faults, 472
  built-in self-test (BIST), 468,
    490–500
  combinational logic, 468–472
  coverage of test vectors, 472
  design for testability (DFT),
    468–506
  hardware testing, 468–506
  MIPS processor model,
    458–463
  path sensitization, 471
  scan path (design), 468,
    476–479
  sequential logic, 473–476
  stuck-at-faults, 468–472,
    475–476
TEXTIO package, VHDL,
  417–421, 555–556
Three-address format, MIPS
  ISA, 432
Tic-Tac-Toe Game, 563
Time-to-market criterion, 110
Timing, 30–41, 75–77, 78–79,
  264–265, 347–348, 518–526
  Area-Time (AT) product,
    348
  charts for SM charts, 264–265
  clock skew, 37
  conditions, 32–35
  delta (Δ) delay, 78–79
  design translation, optimiza-
    tion of in FPGAs, 347–348
  Energy-Delay (ED) product,
    348
  glitches in sequential circuits,
    35–36
  hold time, 31, 33–34
  inertial delays, 75–77
  maximum clock frequency of
    operation, 31–32
  memory timing models,
    VHDL, 518–526

Timing (*Continued*)
    propagation delays, 31
    RAM memory models,
        519–526
    sequential circuits and, 30–41
    setup time, 31, 32–34
    synchronous design, 36–41
    transport delays, 75–77
    VHDL design and, 518–526
Toggle (T) flip-flops, 14, 15
Traffic light controller, design of,
    201–203
Trailing edge, *see* falling edge
Transactions in VHDL, 395
Transition tables, 19, 26
Transparent D-latch, 16
Transport delays, 75–77
Tristate logic, 41–42
    buffers, 41
    busses, 41–42
    in VHDL, 401–402
Truncate, IEEE 754
    floating-point standard, 367
Type, 62, 545–546
    predefined VHDL, 545–546
    VHDL modules, 62

UART, *see* Universal
    Asynchronous Receiver
    Transmitter
Ultra large scale integration
    (ULSI), 51
Unbiased rounding, IEEE 754
    floating-point standard, 367
Unconditional jump instructions,
    MIPS ISA, 437
Underflow, IEEE 754 exponents,
    363, 369
Universal asynchronous receiver
    (UART), 526–539
    baud rate generator for, 534,
        536–537
    receiver for, 530–535
    transmitter for, 526–530
    VHDL code for, 537–539
    VHDL design of, 526–539
Unsigned type, 91–93
Usable gates, 337
Use statement, 91, 549

Variables, 111–113
Variable assignment statement,
    111, 551
Variable declarations, 111–114,
    548
Variable parameter, VHDL,
    393–394
Verilog, 55

VersaTile blocks, 325–326
Very high speed integrated
    circuit (VHSIC), 52
Very large scale integration
    (VLSI), 51
VHDL, 51–136, 389–428,
    545–552. *See also* Design
    and function, std-logic,
        406–407
    arrays, 114–117, 396–397,
        398–399
    assert statement, 119–122,
        552
    attributes, 395–399, 546–547
    behavioral description, 53,
        55, 101–107, 110
    combinational circuits and,
        57–60
    compilation of code, 77–81
    computer-aided design
        (CAD), 51–54
    concurrent statements,
        57–60, 549–551
    constant declarations, 114,
        548
    counters, modeling using,
        95–101
    data types, 82–84
    dataflow description, 55,
        101–102, 107–108
    declarations, 114, 547–549
    files, 417–421
    flip-flops, modeling using,
        69–73
    functions, 389–393, 547
    generate statements,
        415–417, 550
    generics, 413–414
    hardware description
        languages (HDL), 52–53,
        54–57
    identifiers, 59
    IEEE 1164 standard, using,
        405–408, 408–410
    if statements, 70–71, 416–417,
        551
    inertial delays, 75–77
    introduction to, 51–136
    language, 545–552
    large scale integration (LSI),
        51
    libraries, 90–94
    loops, 117–119, 390–391, 552
    modules, 61–67
    multiplexers, models for,
        87–90
    multivalued logic, 400–408
    named association, 414–415

operators, 82–84, 546
    overloaded operators,
        creating, 399–400
    parameters, 393–394
    port map statements, 219,
        414–415
    procedures, 393–394
    registers, modeling using,
        95–101
    report statement, 119–122,
        552
    reserved words, 59
    sequential statements, 67–69,
        551–552
    signal attributes, 395–396,
        397–398
    signal declarations, 111–114,
        548
    signal resolution, 400–408
    simulation, 53, 77–81
    small scale integration (SSI),
        51
    static RAM (SRAM) mod-
        els, 408–410, 410–413
    structural description, 53, 55,
        101–102, 108–110
    synthesis, 53, 81, 84–87
    test benches, 120–122
    TEXTIO package, 417–421
    transport delays, 75–77
    types, 545–546
    ultra large scale integration
        (ULSI), 51
    variable declarations,
        111–114, 548
    very high speed integrated
        circuit (VHSIC), 52
    very large scale integration
        (VLSI), 51
    wait statements, 73–75, 551
Virtex, 140, 161, 324, 327–329

Wait statements, 73–75, 551
While loops, 119, 552
Wristwatch, 507–518
    implementation of, 508–516
    specifications for, 507
    test bench, 516–518
    VHDL design of, 507–518

X01Z logic, 400
Xilinx, 140, 157–158, 161,
    324–325
XOR gates, 1–2

Zero, IEEE 754 floating-point
    standard, 366